W9-BNI-305

# Our Changing Planet
## The View From Space

For over forty years now, satellites have been orbiting above the Earth, quietly monitoring the state of our planet. Unseen by most of us, they are providing information on the many changes taking place on Earth, from natural processes such as land movements, volcanic eruptions, and the ebb and flow of the seasons, to human-caused changes such as the growth of cities, deforestation, the spread of pollutants in the atmosphere and oceans, and the depletion of the ozone layer over the poles.

Led by four editors with support from a production team at NASA Goddard Space Flight Center, many of the world's top remote sensing scientists showcase spectacular and beautiful satellite imagery and provide informed essays on the science behind the images and the implications of what is shown. The images and text highlight numerous examples, on local, regional, and global scales, of both natural changes and the many effects that human activity is having on the Earth.

*Our Changing Planet: The View from Space* is a stunningly attractive and informative book for anyone interested in environmental issues and the beauty of our home planet. It portrays the astounding range and scientific utility of information that can be derived from satellites and demonstrates the great benefit that satellites provide for monitoring both natural and human-caused changes occurring throughout the world. It will provide inspiration for students, teachers, environmentalists, the general public, and scientists alike.

Pre-publication praise for *Our Changing Planet: The View from Space:*

'I've had the good fortune to visit some of the most beautiful parts of our home planet, and witness first hand the effect human activity is having on our world. This book allows you to explore some of these effects through an awe-inspiring collection of images of the Earth from space. This wonderful volume provides a masterful association of imagery and explanation. You will have difficulty closing it once opened. I give it my highest recommendation.'

*Sir Ranulph Fiennes, internationally renowned explorer and adventurer*

'Our understanding of the world radically changed when we first saw it from space. It became at once precious yet vulnerable, romantic yet realistic, and above all unified as the physical and living environments came indissolubly together. The pictures in this remarkable book could almost tell the story by themselves; but they are supported by a well written collection of essays which together explain the underlying science and how the system actually works. As tiny ephemeral creatures on the surface, we can also see the alarming effects our activities are having, whether on the land, in the seas or in the atmosphere, so that this epoch in the Earth's long history can be appropriately named the Anthropocene. We still have to reckon with the consequences. This is an excellent guide to greater understanding of the fundamental issues of our time.'

*Sir Crispin Tickell, Director of the Policy Foresight Programme at the James Martin Institute for Science and Civilization at Oxford University, former British Ambassador to the United Nations, former President of the Royal Geographical Society*

'Great, readable book with spectacular space views of how our home planet is changing – the land, atmosphere, oceans and ice.'

*James Hansen, Director of the NASA Goddard Institute for Space Studies*

'We have been advised for years to act locally and think globally and nothing helps us think more about the globe than the wonder of viewing the whole and all its changeable parts from space. This book tells us what is happening to our world in graphic detail and it is truly an eye opener.'

*'William Ruckelshaus, First Administrator of the U.S. Environmental Protection Agency'*

'This collection of stunning images of our planet from space serves as a focus for a discussion of geologic, atmospheric and oceanic processes. The emphasis is on change, and here the images of shrinking glaciers are particularly impressive. Much of scientific progress has been associated with increased magnification of ever smaller subjects, down to the molecular and atomic dimensions. This volume illustrates rather the opposite: of how much can be learned from integrated views of large areas, up to the planetary dimension.'

*Walter Munk, Professor Emeritus Scripps Institution of Oceanography*

# Our Changing Planet

## The View From Space


EDITED BY

MICHAEL D. KING
*NASA Goddard Space Flight Center*

CLAIRE L. PARKINSON
*NASA Goddard Space Flight Center*

KIM C. PARTINGTON
*Polar Imaging Ltd*

ROBIN G. WILLIAMS
*La Jolla, California*

CAMBRIDGE UNIVERSITY PRESS
Cambridge, New York, Melbourne, Madrid, Cape Town, Singapore, São Paulo

Cambridge University Press
The Edinburgh Building, Cambridge CB2 8RU, UK

Published in the United States of America by Cambridge University Press, New York

www.cambridge.org
Information on this title: www.cambridge.org/9780521828703



First published 2007

Printed in Singapore

*A catalog record for this publication is available from the British Library*

ISBN 978-0-521-82870-3 hardback



Cover image: View of Earth's horizon as the Sun sets over the Pacific Ocean.
This image was taken by a crewmember onboard the
International Space Station (ISS), July 21, 2003. Anvil tops of thunderclouds are visible.
Image courtesy of the Image Science and Analysis Laboratory,
NASA Johnson Space Center, photograph number ISS007-10808

# Contents

# Foreword

Astronaut Piers Sellers on a spacewalk during the STS-112 mission, October 2002. (Image courtesy the Image Science & Analysis Laboratory, NASA Johnson Space Center.)

There are six and a half billion people on Earth right now, and this number will likely grow to about nine and a half billion by 2050. All of us inhabitants grow up with our own perception of the world around us, how it works and our place in it. Gandhi said that when a person dies, a whole Universe dies with him, and somehow we all know this to be true— every one of us has a unique view of the environment outside of ourselves; its form, its meaning, and its purpose. And this view changes for each one of us as we grow, as it has changed for mankind as a whole as humans have evolved and understood more about the Universe around them.

For small children, the world is a small and exciting place consisting of a few rooms, a garden, and perhaps a street or fields, all inhabited by gigantic people. This little Universe is simply there to be explored and enjoyed. As we grow up, our horizons expand, through direct experience and from learning, to encompass the whole Earth and, dimly, the Universe beyond that. But still, places, scales and people that we have seen or met for ourselves seem

*Our Changing Planet*, ed. King, Parkinson, Partington and Williams
Published by Cambridge University Press 

more real and tangible, which is why Gandhi was demonstrably correct: the Universe in Einstein's head was surely different than the Universes inside the heads of a rock star, a desert nomad, or an astronaut. In general, more knowledge and experience tend to nudge us towards common alignments in our thinking, so that, as Science progresses, we tend to share more similar, and hopefully more realistic, perceptions of the world. This can only be for the good, as generally shared views based on science usually lead to better discussions and better decisions. This book makes a solid contribution towards this common alignment, as it brings to us realistic views of our world which modify the sometimes abstract and vague pictures we have in our heads. For example, a school kid once asked me if the countries of the world all had different colors, like on the atlas in the classroom—this book should help him out.

The book has a compilation of stunning images, but there are dozens of beautiful coffee table books out there full of Earth images. I should know; I have a stack of them at home. These images appeal to the human and the aesthetic in each of us, as the world is a staggeringly beautiful place at almost every scale we choose to observe. But this book is more than just a succession of attractive images; it penetrates deeply into the science of what is being observed: not just the what, but also the how, the why, the where and when. For example, the images of global temperature (pages 54 and 55), when analyzed, combined and replotted as trends against time, clearly show the stratospheric cooling and tropospheric warming that are the signatures of global warming. So, beauty and aesthetics—certainly; knowledge, insight, and understanding—emphatically.

Satellites have revolutionized the way we see our home planet. The first interplanetary spacecraft sent back fuzzy but intriguing images that changed the way we saw these far-off places forever: Venus turned out to be a high-pressure, boiling hell—no princesses or dinosaurs to be seen anywhere. And Mars turned out to be airless, arid, and very cold—no canals, no princesses and, disappointingly, no signs of current life either. The point is that whatever ideas any of us had about these places prior to the arrival of the early spacecraft were irrevocably changed—generally in the direction of objective reality—by the first bits of data received back. It should have been no surprise then that Earth-viewing satellites would change the way we see our planet and likewise correct and converge the many and varied ideas people had about its form and functioning. But it was a surprise—in fact there was a continuous succession of surprises—to both the public and scientists alike. Most people expected that satellites would mainly confirm what we have already learned from 500 years of scientific investigation, but the impact of the satellite viewpoint went far beyond that. The orbital view combines an incredible enlargement of scale—the term "God's-eye view" is not far off the mark—while still providing detail that relates back to our everyday experience. For example, we have all seen pictures of hurricanes from space (page 36). From orbit, we can see them in their entirety; huge, cloudy cartwheels sliding across the oceans. Prior to the satellite age, people knew very well what hurricanes were, and how they formed, behaved and their size and shape, but nonetheless, the first actual pictures mesmerized scientists and laymen. People could take in everything about the scale and structure of these immense phenomena, but at the same time could see the constituent clouds that provide yardsticks back to the scale of our everyday lives. (Another school kid I met, eight years old, was looking at a picture of a hurricane and pointed at a little cloud at the edge of one spiral arm. She immediately understood everything about the size of the hurricane from that one cloud and even pointed out that she would be a little dot down here, under this cloud, near that island. She also, without prompting, understood that she was seeing the clouds from above, able in her mind to turn her ground-based idea of a cloud upside down and see it from the other side, with the Earth below it.) So images from satellites enlarge our view, literally, of the greater world. All of us, from kids to scientists, can thereby gain a deeper understanding and appreciation of the scale and beauty of Earth, how it works, how it is changing, and what its future state might be.

Satellites and their instruments are extraordinary, almost miraculous, devices, among the most intricate and complex things that man has devised, incredibly expensive and unbelievably powerful and useful. Consider that a satellite instrument is a combination of precision optics, propulsion and navigation units, and delicate electronics but must nevertheless withstand the rigors of launch—vibration and acceleration to the unthinkable velocity of 8 kilometers per second—and then survive a lifetime of coasting through an airless void with extremes of heat and cold, flashing into day and night every 50 minutes or so. It is surprising that they work at all, let alone that they can be lofted into orbit in the first place. But once there, the satellite provides an extraordinary perch from which to view the world. In a matter of a few hours, it can map the whole planet, using the same well-calibrated instrument, and send the stream of pointillist data back to our laboratories almost as soon as they are collected. We can drive the point home with a concrete example: biologists and scientists working on the global carbon budget want to know how much green material there is on the Earth at any given time, which translates to knowing roughly how many green leaves there are. Not only that, they want to know how this pattern of greenness changes over time; with the seasons, with different weather conditions, and as a result of human activities—deforestation, logging, fire, cultivation. Now we can envisage sending out a million biologists every day to do leaf density surveys, bringing back their results every night, and transmitting their reports to NASA's Goddard Space Flight Center where an army of researchers would process and reduce the data to a time series of vegetation density maps. Once the problems of ensuring that the data collectors all use the exact same measurement technique, are equally conscientious, and can maintain their enthusiasm over several decades are ironed out, it may be possible to finally produce a time series of maps. (Funding would be a challenge, organization a nightmare, and quality control a fantasy, but we will skip over these management problems for the purposes of this thought experiment.) Instead, all of this is avoided by having the data collected by a couple of satellites which provide us with wall-to-wall global estimates of vegetation density at intervals of a few days (pages 134–135), using a consistent, vandal-proof pair of instruments and a single measurement methodology. These data are being quietly collected somewhere above you right now and are being sent to laboratories to be studied and analyzed at leisure. Thus it is the synoptic quality of satellite instruments that is so unique and so powerful, and which make them such cheap tools when compared to other methods of global observation, like our million diligent biologists.

The book starts with images of the atmosphere. The static pictures tell us much about the structure of the atmosphere but time-series data reveal the dynamics of the atmosphere as well as information about its interactions with other parts of the Earth system. The transport of heat from the equator to the poles is done by the large-scale circulation, some by the stately and steady flow of ocean currents and some by the violent and episodic hurricanes. Both are key to maintaining the current climate; both are thought to be sensitive to global warming. The large-scale transport of dust particles in the atmosphere—almost completely ignored before being revealed by satellite observations—can exert a direct influence on the climate by absorbing sunlight and warming the air, or by reflecting sunlight and thereby contributing to reduced warming. Satellites have shown us how much of the atmospheric dust burden is directly due to cultivation. The concentration of carbon monoxide (CO), a by-product of incomplete combustion, is now routinely measured from orbit—we can see the traces of industrial output and field burning all over the world.

Images of ship tracks and aircraft contrails again bring us to the junction of a new larger scale with the familiar scales of our own experience. We can see a recognizable picture of the United States (page 73), busily crisscrossed by aircraft contrails. Each tiny contrail leads back to an invisible aircraft containing a couple of hundred souls flitting from somewhere to somewhere else. Were you flying on a commercial aircraft on January 29, 2004? Look at page 73—you might be able to see yourself. (Once, while spacewalking outside the International Space Station, I looked down to see a brilliant sunlit Atlantic Ocean dotted with tiny clouds hovering above the water, each one slant-tethered to its own perfect, slightly displaced cloud shadow. Then I saw a minute jet contrail slowly and bravely pushing its way across the huge ocean towards me. I could not see the aircraft at the apex of that long, white streak, but I wondered what the people on board were thinking and doing. After another few seconds, the scene with its clouds and contrail slipped below the horizon behind me, into my past and into their future.)

The ozone hole has been cited, convincingly, as the first global environmental crisis that was recognized, discussed, and dealt with by the global community. Satellites played a key role in understanding and monitoring the problem, but perhaps more importantly, in convincing people everywhere that a problem existed and needed to be addressed. Most of the public and politicians did not understand the chemistry and dynamics of the Antarctic ozone hole, but everyone could grasp the size and the growth of the problem just by looking at the images (see page 78). The sequence of images allowed non-scientists to leapfrog over a bewildering tonnage of scientific discussion and analysis to see, to really see, what was happening. And no truths were lost in the process.

When I watch people looking at images of the land surface for the first time, I almost always see the same thing. First, they turn the pages quickly, but after a while they slow down, as their minds synchronize with the things they are seeing, aligning the images with almost-forgotten geography lessons, or with familiar places seen from a different vantage point. ("Hey—that's where I used to live.") A favorite is the Nile valley (page 92)—the river snakes its way through the desert, flanked by green banks of cultivated land, ultimately fanning out into an emerald delta as it reaches the sea. Desert, mountains, lakes, greenery, seas are shown as they really are, not conceptualized abstractly like on a map. The image of the great deserts of North Africa and the Middle East (page 130) shows us striking patterns and structures—great swirls of sand washing up against mountain ranges and rock plains.

Satellites have allowed us to study the oceans in a completely new way. Satellite-based radar altimeters detect the bumps and dips in the ocean's surface caused by currents, tides, and flows over subsurface structures. These data, combined with satellite-derived sea surface temperature and wind data, allow us to map the great currents. What have we learned? Again, we have achieved a short cut to our understanding of how the oceans work. El Niño can be explained with words and graphs, but a couple of images (pages 160–161) do a better job in getting across the size, scope, and pattern of this dynamic phenomenon. Likewise, a graph of sea level rise (page 163), distilled from billions of bits of data into a single line, speaks volumes: it's tough to argue with hard data that even a fourth-grader can understand. "Dad, it's going up, isn't it." (Note the lack of a question mark.)

If our million biologists would have to be a hardy and motivated bunch to complete their weekly leaf surveys, the army of glaciologists required to map the cryosphere would need to be of very stern stuff indeed. Instead, satellites have allowed us to assay the global ice volume without unnecessary heroics and to determine how it changes seasonally and over the decades. It is now clear that many of the great Greenland glaciers have accelerated towards the sea, almost in unison, lubricated by increases in meltwater flowing between the rocks and the glacier bottoms. Likewise, the Arctic Ocean ice cap is unevenly but decidedly shrinking, with ramifications throughout the Arctic climate and ecosystem.

Satellites show us directly the evidence of Man's hand at work on the planet. When we look at a multiyear trace of $CO_2$ concentration, we see a clear annual cycle—the dips in the trace correspond to the Northern Hemisphere spring and summer, when continental vegetation pulls $CO_2$ down from the atmosphere for photosynthesis (plant growth), and the peaks represent the release of $CO_2$ back to the atmosphere by dying and decomposing vegetation in the fall and winter. This cycle is closely matched by the time series of images of vegetation greenness over the land. Each spring, we can see the "green wave" creep northwards over North America and Eurasia, reach a high tide mark in the late summer, and then slowly flow back towards the equator during the fall. The combination of atmospheric $CO_2$ data, with their precise inferences about the amounts of $CO_2$ being exchanged, with the spatial patterns provided by the satellite data, has given us a powerful new way to understand the global biosphere. This much we know: the growing season in the Northern Hemisphere has stretched out by 12 to 18 days over the period 1982–1999 (page 283). In addition, we now think that the land biosphere has acted as a net sink for $CO_2$ over the same period, burying more carbon than it has returned to the atmosphere.

Finally, satellites can tell us a lot about the inhabitants of this planet. Pages 292–293 show a synthetic night time view of the Earth. The city lights outline the continents and dot the interiors, and to a great extent tell us who is consuming how much energy and where. North America, Europe, and Japan are all brightly lit—the southern half of the densely populated United Kingdom is almost a continuous wash of light, much to the annoyance of amateur astronomers there. Most of Africa is dark—just a few widely spaced cities can be seen. The patterns of industrialization in India and China show up unambiguously. All the people of Earth, snapshotted onto a pair of pages.

Sometimes it's good to stand back and take in the "big picture." This book will allow you to see the world as visualized by billions of dollars worth of satellite hardware, and analyzed by several hundred dedicated scientists over several years of painstaking work. You can read it in the time it takes you to fly across the country, followed by your own tiny contrail which will probably be observed through a satellite lens or a spacewalker's visor several hundred kilometers above you.

Piers J. Sellers
Over the Atlantic between Scotland and Greenland
Continental Flight CO5 (London Gatwick to Houston)

Artist's rendition of the Upper Atmosphere Research Satellite (UARS) superimposed on a sunrise view of the Atlantic Ocean off the coast of Jacksonville, Florida on February 23, 2003. NASA's UARS satellite is one of 48 different international Earth-observing satellites used throughout this book. It was deployed from Space Shuttle Discovery on September 14, 1991 and raised to an orbital altitude of 600 km, where it provided observations of the Earth's atmospheric chemical composition until the end of 2005, circling the globe for over 5000 days. (Photograph from the International Space Station, NASA image number ISS006-E-32544.)

# Preface

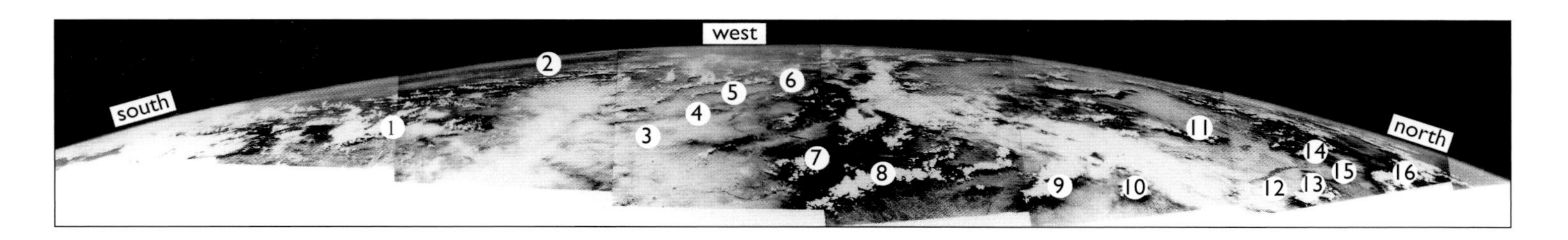


Panorama of the Earth taken from a V-2 rocket fired from White Sands, New Mexico, on July 26, 1948. The area shown is approximately 2 million km$^2$, photographed from an altitude of 100 km. The V-2 rockets were the world's first ballistic missiles. Parts of almost 100 rockets were captured from Germany at the end of World War II and transported to the United States, where they were instrumental in initiating the post-war 'space race.' (Image courtesy Johns Hopkins University Applied Physics Laboratory.)

The numbers on the photograph correspond to the following locations: 1-Mexico, 2-Gulf of California, 3-Lordsburg, New Mexico, 4-Peloncillo Mountains, 5-Gila River, 6-San Carlos Reservoir, 7-Mogollon Mountains, 8-Black Range, 9-San Mateo Mountains, 10-Magadalena Mountains, 11-Mt. Taylor, 12-Albuquerque, New Mexico, 13-Sandia Mountains, 14-Valle Grande Mountains, 15-Rio Grande, 16-Sangre de Cristo Range.

Just over 60 years ago, in 1946, an unmanned V-2 rocket was launched into space from White Sands, New Mexico, and recorded the first picture of Earth from space, demonstrating the feasibility of observing our planet from beyond. By the early 1960s, Earth-orbiting satellites were providing black and white photographs of clouds and other major phenomena in the Earth's atmosphere. The famous photographs of the Earth taken by Apollo astronauts on their way to the moon provided iconic images of the 1960s and early 1970s in which our planet appears small and isolated, helping to spur the environmental movement of the late 20th century, epitomized with the creation of Earth Day on April 22, 1970. Since then, the public has experienced an increasingly sophisticated and ever-present exposure to images of the Earth from space. The capabilities of satellites and their Earth-imaging instruments have developed enormously, to the point where their use is integrated fully into many areas of life, from news reports to weather forecasting, and from scientific studies and environmental assessments to ship routing and oil and gas exploration.

It is in the monitoring of the dynamics and long-term changes of our planet that some of the most fascinating insights are now coming to light and, some 60 years after the first grainy photographs of the Earth from space, we take a look in this book at how satellites are starting to reveal significant changes occurring on and around our planet. Earth-orbiting satellites have become an essential tool for monitoring the Earth and its changing environment, from natural hazards, such as hurricanes and tropical cyclones, tornadoes, floods, tsunamis, fires, dust storms, and volcanic eruptions, to long-term changes such as the retreat of glaciers and Arctic sea ice and the reduction of tropical rain forests. Satellites, with their sophisticated instrumentation, also provide evidence of, and insight into, more subtle changes, for example in the Earth's protective ozone layer, urban and rural air quality, cloud cover and properties, land surface characteristics arising from land cover and land use change, sea level variation, land and sea surface temperature, ocean biology, atmospheric greenhouse gases and their concentration, rainfall, sea surface elevation to within 2–3 centimeters accuracy, and lightning. Some of these phenomena have been monitored from space for over 25 years, with ever-increasing sophistication, whereas the monitoring of others has only become possible in the last 5–10 years.

With the wide variety of satellites encircling the Earth, provided by many countries and government agencies, and their various orbits and instrumentation, satellites provide an

*Our Changing Planet*, ed. King, Parkinson, Partington and Williams
Published by Cambridge University Press © Cambridge University Press 2007

The classic photograph of the Earth taken on December 7, 1972 by the Apollo 17 crew while traveling toward the moon, often referred to as 'the blue marble.' This photograph extends from the Mediterranean Sea to Antarctica and was the first time the Apollo trajectory made it possible to photograph the south polar ice cap. Almost the entire coastline of Africa is visible, with the Arabian Peninsula in view at the northeastern edge of Africa. (Image courtesy the Image Science & Analysis Laboratory, NASA Johnson Space Center, photograph number AS17-148-22727.)

invaluable means of monitoring our planet with a consistent measurement capability across national and other political boundaries. They play an essential role in alerting us to potentially detrimental or even catastrophic changes taking place on our planet that require actions from our politicians and other policymakers. They also provide important information for verifying compliance with international treaties and other agreements. Furthermore, satellite observations regularly reveal features of the planet that take scientists by surprise and remind us that we remain a long way from fully understanding the behavior of the complex web of physical, chemical, and biological processes that take place on our home planet.

This book attempts to present, in a readily understandable manner, a compilation of phenomena and changes observed through satellites, thereby providing both an indication of the impressive capabilities of satellites orbiting the Earth today and an indication of how our planet is changing. The core of the book is divided into the following five main sections: The Dynamic Atmosphere, The Vital Land, The Restless Ocean, The Frozen Caps, and Evidence of Our Tenure. These sections contain, in turn, a collection of articles contributed by prominent scientists worldwide, well illustrated with satellite images and data products, selected photographs, time series of changes in the environment (where appropriate), and historical photographs or drawings relevant to the subject at hand. The five main sections are followed by four appendices that provide, in turn, a description of

Satellite image of the same scene as the Apollo 17 photograph to the left. Satellites of today enable digital images of the entire globe to be produced every day. This image was created from digital data acquired on December 7, 2006, thus having the same solar illumination and seasonal meteorology as the Apollo 17 photograph, but with clouds and haze that occurred in 2006. (Data from the MODIS instruments on the Terra and Aqua satellites.)

"Satellites and Satellite Orbits," including how the "magic" of remote sensing and global coverage is achieved, a Glossary, a List of Acronyms, and a List of Contributors. In total, the book's illustrations employ data from 48 satellites and many more sensors, contributed by agencies and other organizations in the United States (U.S.), Japan, and Europe. The U.S. organizations include the National Aeronautics and Space Administration (NASA), the National Oceanic and Atmospheric Administration (NOAA), the U.S. military, and several commercial companies.

The book covers many aspects of environmental remote sensing from space and chronicles the state of our planet circa 2007. We hope that it will prove useful to scientists, policymakers, students of the environment, and anyone who shares a concern and interest in the Earth and the changes occurring on its surface and in its atmosphere.

Michael D. King
Claire L. Parkinson
Kim C. Partington
Robin G. Williams

# Acknowledgments

This publication required the considerable time and effort of several people who were absolutely crucial to the book's completion.

Winnie Humberson contributed her design and production talents during the initial stages of the book, with additional design support from Sterling Spangler. Winnie established the overall look of the publication.

Sally Bensusen joined the team as production manager and designer during the second phase of the book. She was responsible for layout and design and for making sure that the book made it to publication on time. Extensive design and layout assistance from Debbi McLean was essential in making this possible.

The images and maps in this book are the product of an expert team of visualizers. The majority of these were created by Jesse Allen, Marit Jentoft-Nilsen, Reto Stöckli, Rob Simmon, and Mark Malanoski, unless otherwise noted. Debbi McLean and Sally Bensusen created several of the illustrations.

This team, from production manager to visualizer, was not only technically skilled but also a delight to work with and was a major reason for the high quality of the figures and the overall appearance of the book.

We would like to thank the University of Puerto Rico—Mayaguez for granting time for one of the editors (Robin Williams) to work on this project.

Finally, Matt Lloyd, publisher at Cambridge University Press, was instrumental in helping our team through the process of publication. We are grateful for his clear guidance and unwavering patience.

A severe winter storm pummeled the northeastern United States and southeastern Canada, February 14–15, 2007. To the east off the coast are cloud streets sculpted by winds that drove icy air from the north into a layer of moist air above the open waters of the Atlantic Ocean. (Data from the MODIS instrument on the Terra satellite; image courtesy MODIS Rapid Response Team.)

Anvil clouds over the South Pacific Ocean, December 2, 2000. (Photograph from the International Space Station, NASA image number ISS006-E-5820.)

This photograph, acquired February 1984 by an astronaut aboard the Space Shuttle, shows a series of mature thunderstorms located near the Parana River in southern Brazil. (NASA Earth from Space image number STS41B-41-2347.)

# The Dynamic Atmosphere: Introduction

Michael D. King

The Earth's atmosphere is a thin veil that extends up to 1000 km from the Earth's surface into space, but the bulk of the atmosphere (99.9999%) by mass resides below 100 km. It is a layer of gases surrounding the planet that consists of roughly 78% nitrogen and 21% oxygen, with important trace amounts of other gases such as ozone, carbon dioxide, methane, nitrogen oxides, and water vapor. Its temperature varies significantly with altitude, and it is generally divided into 5 distinct layers, from the troposphere, derived from the Greek work 'tropos' meaning to turn or mix, where man lives and the bulk of weather occurs, to the stratosphere, extending from an altitude of about 7–17 km to about 50 km and containing the layer of important stratospheric ozone that protects man from harmful ultraviolet radiation from the sun, to three higher layers, known respectively as the mesosphere, thermosphere, and exosphere.

Many aspects of the Earth's atmosphere are undergoing significant alterations as a direct consequence of human activity. Some very familiar alterations to atmospheric composition, in particular the increase in carbon dioxide concentration since the industrial revolution and

Atmospheric temperature as a function of height and latitude for January 2004, showing the lower layers of the Earth's atmosphere and the decrease in temperature with height in the troposphere and the temperature increase in the stratosphere. (Data from the AIRS and AMSU instruments on the Aqua satellite.)

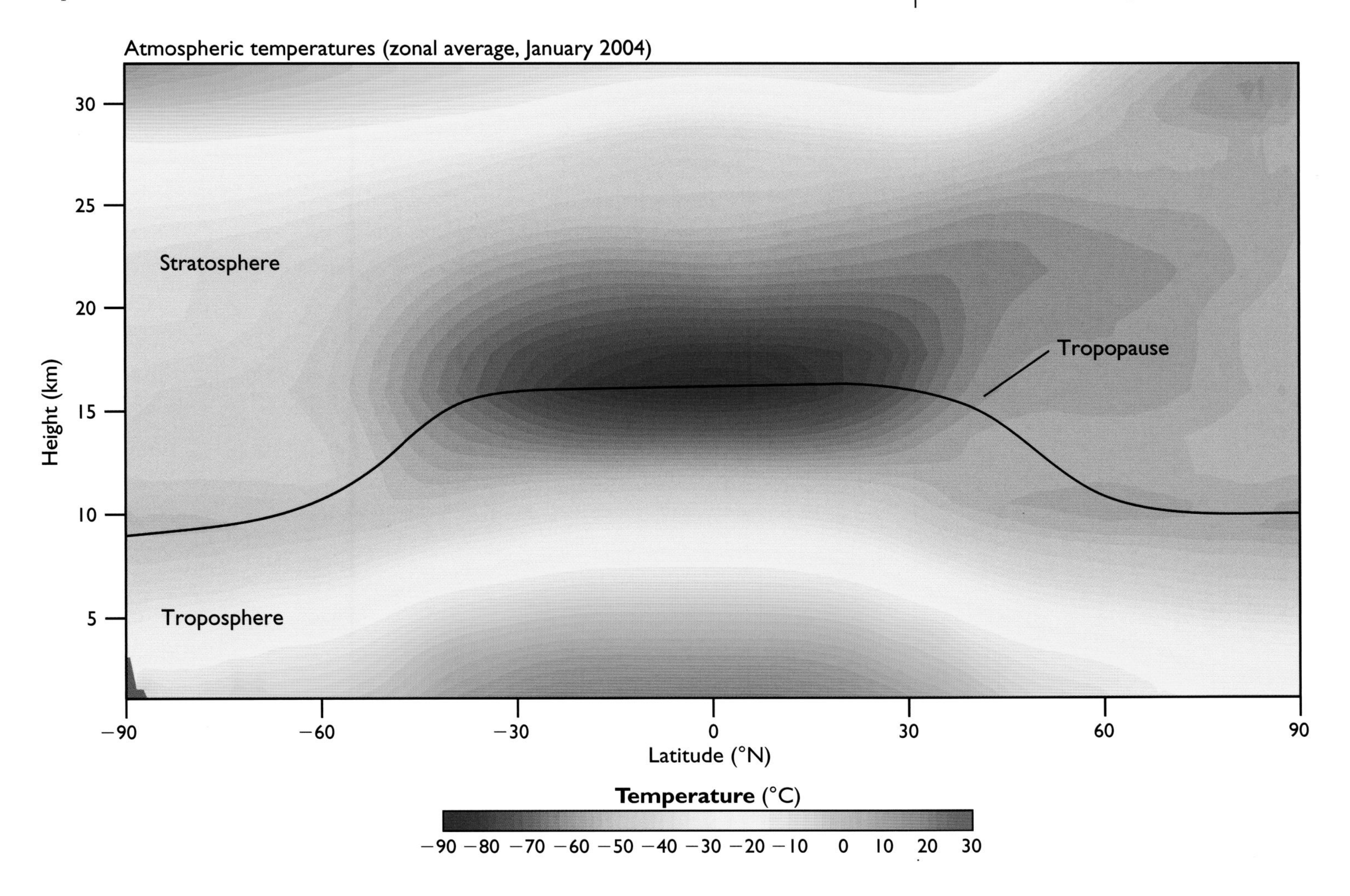


*Our Changing Planet*, ed. King, Parkinson, Partington and Williams
Published by Cambridge University Press 

the decrease in stratospheric ozone over Antarctica in the Austral spring (the so-called ozone hole), were first observed and long-term trends established from ground-based observations. Satellite and aircraft data, however, have also significantly increased our understanding. For instance, satellites have played a key role in establishing the size and transformation of the ozone hole in the Antarctic, as well as ozone changes in the Arctic and mid-latitudes, and the connection between human produced chlorofluorocarbons (CFCs) and ozone loss was first established from NASA aircraft observations in the late 1980s. Other atmospheric characteristics, though perhaps changing with time, have not been observed globally prior to the advent of the satellite era or were not possible to observe prior to the development of sophisticated instrumentation available for the first time in just the past 5–10 years or so. Regarding other aspects of the Earth's atmosphere, though observed for perhaps 20 years from space, further technology improvements permitted increasing quality and quantification of atmospheric characteristics as we approached and entered the 21st century.

Humans as well as Nature have had impacts on the environment through the long-range transport of dust, smoke, and other aerosol particles in the Earth's atmosphere, and many of these impacts are observable from satellite observations. Several categories of dust impacts on the ocean are illustrated under "The Restless Ocean" in chapters on the ocean biosphere and red tides, both of which are directly impacted by the iron content in dust. In this section, 15 chapters illustrate the many ways that satellite data are being used to reveal characteristics of the Earth's atmosphere, including clouds, air pollution, severe storms, and atmospheric composition.

Illustrating the application of satellite imagery to the study of Earth's dominant atmospheric phenomenon, as viewed from space, the chapter on "Clouds: Are the Shutters of the Universe Changing?" describes satellite observations of global cloud cover as well as high clouds from the late 1990s to the first several years of the 21st century. This chapter shows that the total cloud cover of the Earth is around 75%, and has undergone no appreciable change over this period, though high clouds do show a noticeable increase in occurrence. There is no effect of El Niño or La Niña events on total cloud cover worldwide, though the distribution of clouds changes spatially between these two states of the climate system, especially in the tropical Pacific.

The chapter entitled "Cloud Optical and Microphysical Properties" provides an overview of liquid water and ice clouds, illustrating both the spatial distribution of these clouds in the Earth's atmosphere and the size of the liquid water drops and ice crystals found within these two types of clouds. This chapter is followed by a chapter on "Clouds and the Earth's Radiation Budget," describing the balance between the incoming solar radiation from the sun that is absorbed by the Earth-atmosphere system and the outgoing longwave radiation emitted by the Earth. Though close to balance over the globe when averaged over the year, there were periods of excessive loss of energy following the eruption of the Mt. Pinatubo volcano in 1991. The "Water Vapor" chapter highlights the global distribution of water vapor in the Earth's atmosphere in January, April, July, and October and indicates both the fluid nature of water vapor in the atmosphere and how it varies over the world's oceans from 1987 to 2005. This chapter demonstrates that water vapor over the world's oceans has responded to an overall warming of the Earth-atmosphere system that has led to an increase of water vapor with time, with additional fluctuations in response to global climatic events such as El Niños and La Niñas.

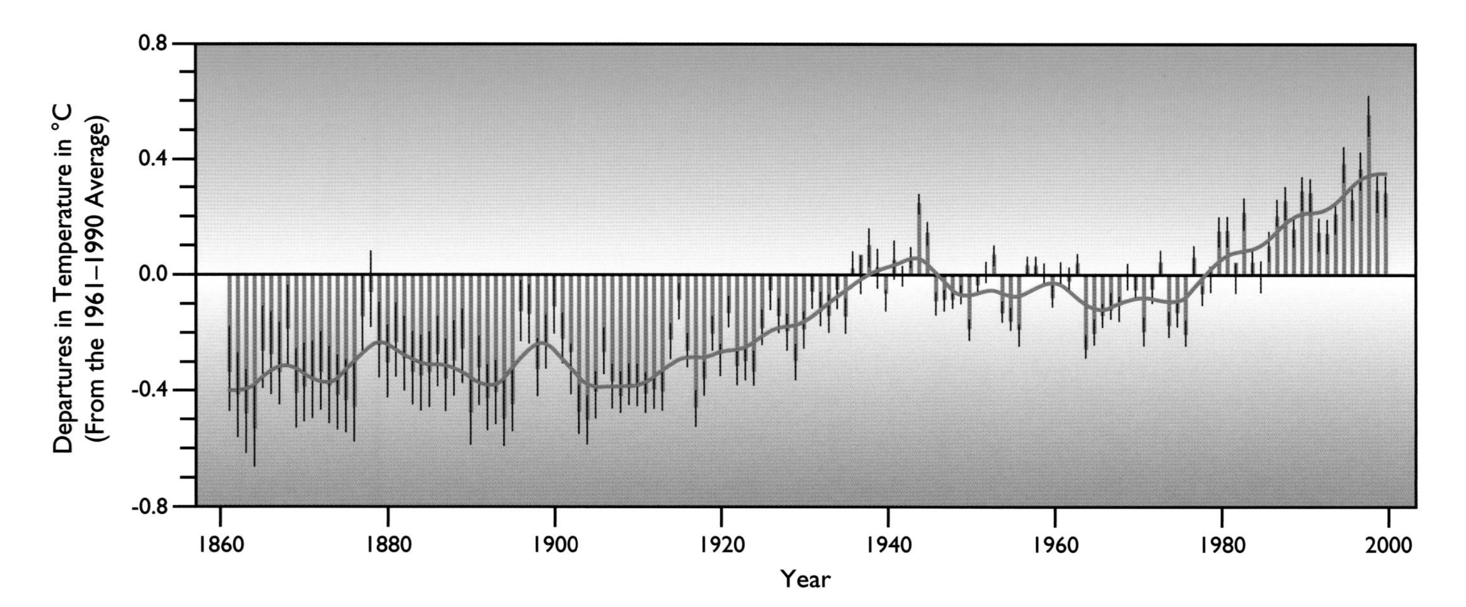


Combined annual land-surface air and sea surface temperature anomalies (°C) from 1861–2000, relative to 1961 to 1990. The Earth's surface temperature is shown year by year (red bars) and approximately decade by decade (solid red line). There are uncertainties in the annual data (thin black whisker bars represent the 95% confidence range) due to data gaps, random instrumental errors and uncertainties, uncertainties in bias corrections in the ocean surface temperature data and also in adjustments for urbanization over the land. Over both the last 140 years and 100 years, the best estimate is that the global average surface temperature has increased by 0.6 ± 0.2°C. (Adapted from *Climate Change 2001: The Scientific Basis*, Intergovernmental Panel on Climate Change, Cambridge University Press.)

Several chapters center on precipitation, which is essential to human sustenance. This group of chapters begins with an overview of "A World of Rain," which shows the global distribution of precipitation from 28 years of satellite observations. During this period there have been several El Niños and La Niñas, and the differences in precipitation patterns between these events are dramatically illustrated, with areas of excessive precipitation and excessive droughts during El Niño years clearly identified. This chapter also illustrates the seasonal and spatial distribution of precipitation during the Asian monsoon in the Indian Ocean and western tropical Pacific, the planet's greatest seasonal shift in rainfall. The chapter "Hurricanes: Connections with Climate Change" discusses Nature's extreme rainfall process of the tropics and subtropics, known in various regions of the world as hurricane, typhoon, or cyclone. The capability of space-based radar on the Tropical Rainfall Measuring Mission (TRMM) to monitor the three-dimensional distribution of rainfall within hurricanes, has only been available since 1997 and is used in the production of some of the rain rate analysis illustrated in this chapter. Since many North Atlantic hurricanes are 'born' off the west coast of Africa and transported across the Atlantic, this chapter is illustrated with both dust storms off the Cape Verde Islands during hurricane formation and subsequent transport, and tracking, from the TRMM satellite, as Hurricane Isabel approaches the mainland in the United States in 2003. Due to the large tropical ocean extent of hurricane and typhoon formation and development, satellites are an essential component in formulating warnings and improving landfall predictions of numerical weather prediction centers worldwide.

Another natural phenomenon that contributes to injuries and loss of life worldwide is lightning, which is summarized in a "Lightning" chapter. Lightning has been viewed as a curious phenomenon of nature for millennia; and Benjamin Franklin in the 1700s invented the lightning rod and designed an experiment to use a kite to see if lightning was an electrical discharge. Detecting lightning from space, and hence globally, did not occur until 1995, and subsequent analysis, presented in this chapter, shows the global and seasonal distribution of lightning worldwide. The space-based observations have revealed that about 90% of all lightning occurs over land, with the ocean lightning occurring primarily along warm ocean currents with convective activity, such as the Gulf Stream of the North Atlantic and the Agullas Current off South Africa.

The tracking of atmospheric temperature from space is highlighted in the chapter entitled "Warming and Cooling of the Atmosphere." This chapter demonstrates how the lower stratosphere has been cooling and the troposphere has been warming in the past 25 years. These trends were interrupted by two large volcanic eruption events that injected sulfate particles (sulfuric acid) into the stratosphere, causing a veil around the planet that reflected sunlight back to space, cooling the lower atmosphere (the troposphere) while warming the lower stratosphere. This chapter also shows the spatial distribution of tropospheric warming and stratospheric cooling in °C/decade.

This section continues with 2 chapters highlighting natural and manmade 'pollution,' one entitled "Dust in the Wind" that illustrates the long-range transport of dust from the Taklimakan Desert in China as it crosses the Pacific and arrives in the United States. Though dust storms in China (Taklimakan and Gobi deserts in particular) are 'natural,' they have been increasing in the last two decades in large part due to land use practices and the desertification of Inner Mongolia. Dust storms in north Africa and the middle East are also widespread and seasonal, and these are illustrated and discussed in this chapter. In addition to dust from natural and transformed land practices, the chapter on "Atmospheric Pollution: A Global Problem" turns attention to the global distribution of carbon monoxide (CO), a trace gas that results from incomplete combustion, either from fires or fossil fuel burning by industry, domestic heating, and motor vehicles. It is also a precursor gas to the formation of tropospheric ozone that is an oxidant and bad for plants and human lungs. This chapter uses space-based observations, not available prior to 2000, to map the global distribution of CO, which shows the expected high concentration in the Northern Hemisphere due to industrial activity but also the very sizeable CO maximum that occurs in southern Africa and in Brazil due to biomass burning in the Austral spring (September–November).

This section then turns to 2 chapters on additional human impacts on the Earth's atmosphere. The first, "Ship Tracks," concerns ships at sea emitting sulfate particles from their smoke stacks that serve as cloud condensation nuclei. The resulting ship-modified clouds are brighter and, especially, consist of larger numbers of smaller cloud drops that reflect more solar radiation back to space. Using satellite observations with the right optical sensitivity, it is possible to detect the presence of ships at sea beneath clouds due to the residual effect they leave in the cloud's reflectance as seen from satellites. Another phenomenon readily observable from space and due entirely to man's modern activities are "Airplane Contrails," in which 'condensation trails' are produced in the exhaust of commercial aircraft at altitudes in the upper troposphere where aircraft operate.

The section concludes with 3 chapters highlighting atmospheric chemistry and the effects that mankind has on atmospheric composition. "Weekly Cycle of Nitrogen Dioxide Pollution from Space" shows the amazing characteristic that nitrogen dioxide ($NO_2$), a short-lived, manmade chemical of the lower atmosphere, leaves a daily signal that reflects human activity. All major cities of the world, industrial coal-fired power plants, and regions of high lightning activity, are clearly evident from space-based observations. Furthermore, in the eastern United States and Europe (especially the Poe Valley of Italy), the $NO_2$ signal is much reduced on Sunday versus mid-week, whereas in the Islamic countries of the Middle East the $NO_2$ level is lowest on Friday and in Jewish Jerusalem it is lowest on Saturday.

A subsequent chapter on the "The Ozone Hole" highlights this well-documented and satellite-observed phenomenon of significant stratospheric ozone loss over Antarctica every

Austral spring (especially late September and early October). The spatial and interannual variation of the ozone hole is illustrated, coupled with a time series showing the size of the ozone hole in comparison to the area of Antarctica and North America.

This section concludes with a chapter on "The Chlorine Threat to Earth's Ozone Shield" that centers on the deleterious effect that chlorine monoxide (ClO), a biproduct of manmade chlorofluorocarbons, has on stratospheric ozone both in the Arctic and the Antarctic. This gas is released only during daylight and in the presence of polar stratospheric clouds that form at high altitudes in polar regions due to the very cold temperatures and atmospheric composition. As a direct consequence of the Montréal Protocol and its various amendments and adjustments, production of CFCs was stopped worldwide in the 1990s and the ClO and hence ozone concentrations are expected to return to 1980 levels by 2070 or so, thereby reducing skin cancer cases, as shown in this chapter.

The chapters in this section illustrate a representative, though not exhaustive, sample of the many characteristics of the Earth's atmosphere that are readily observable from space. Other phenomena not illustrated in this section include (i) stratospheric sulfur dioxide ($SO_2$) arising from volcanic eruptions, (ii) tropospheric $SO_2$ arising from coal-fired power plants and copper smelters, (iii) the spatial distribution of manmade and natural aerosol particles, and (iv) the vertical distribution of many atmospheric constituents. As this section demonstrates, satellite imagery can show more readily than any other means the large-scale state of the atmosphere, its many dynamical processes, and the evolution of atmospheric conditions worldwide.

Measurements of the total ozone content of the atmosphere at Halley, Antarctica, have been made from ground-based instruments since 1956 and satellites since 1980. These observations of the minimum total ozone during September and October show an acute drop in total atmospheric ozone in the early- and mid-1980s, commonly referred to as the ozone hole. (Ground-based measurements from the Dobson ozone spectrophotometer and satellite measurements from the TOMS instrument on the Nimbus 7, Meteor 3, and Earth Probe satellites and the OMI instrument on Aura.)

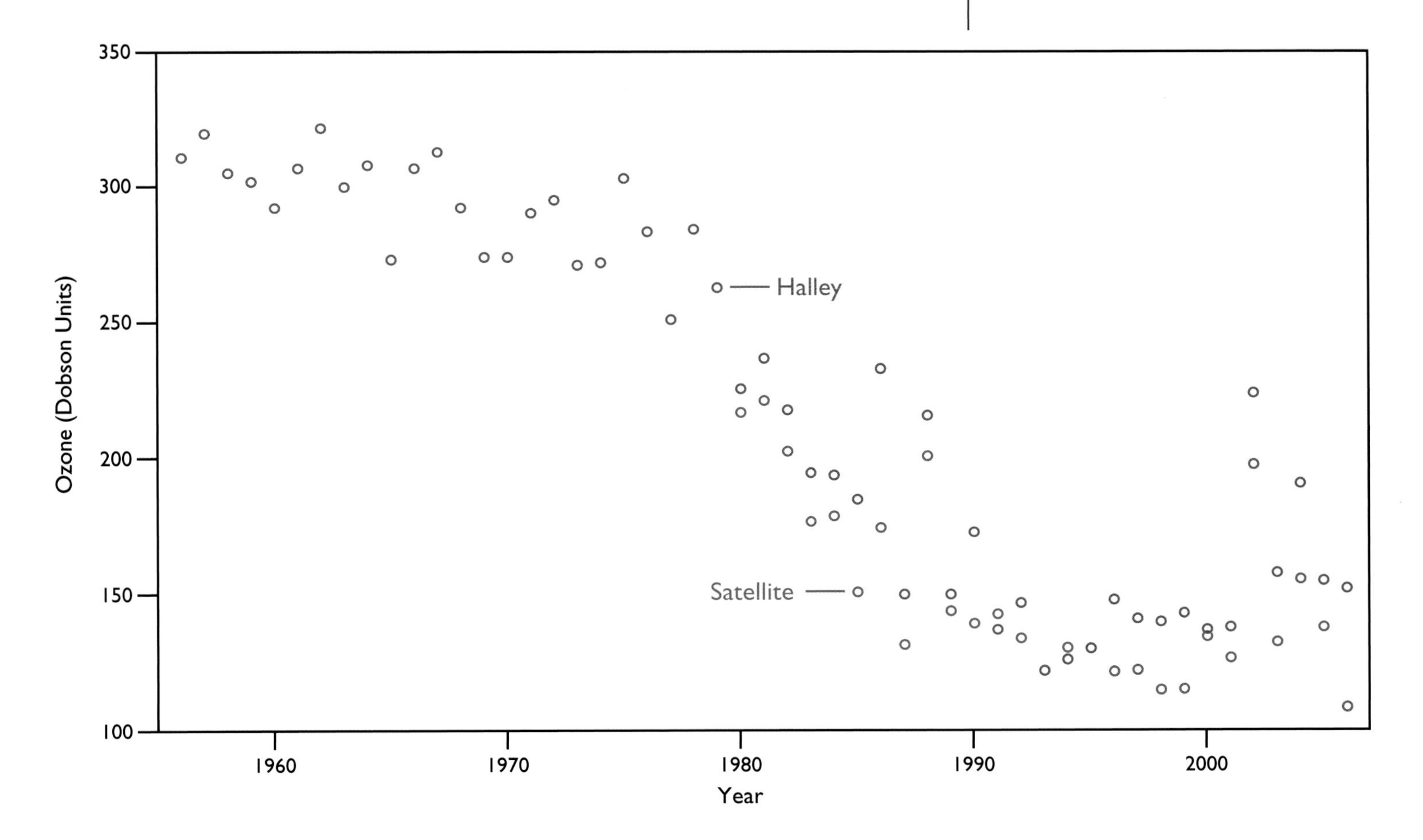

Donald P. Wylie
W. Paul Menzel

# Clouds: Are the Shutters of the Earth Changing?

Photograph by Nasir Khan.

The father of satellite meteorology—Verner Suomi (1915–1995). (Photograph from University of Wisconsin archives.)

Clouds are a strong modulator of the amount of solar heating and thermal cooling of the Earth system. This is why Vern Suomi called them "the shutters of the Earth." We are currently in a warming trend that is often attributed to the increase in carbon dioxide ($CO_2$) in the atmosphere. But clouds cover approximately three-fourths of the Earth and a small increase in cloud cover could offset the warming from increased $CO_2$ in the atmosphere.

Global cloud observations using satellites were first advocated by Vern Suomi, often called the "father of satellite meteorology." It has been found that clouds vary greatly over the Earth, just as our weather does, and their effect on warming and cooling has to be inferred from the combined effect of all clouds in all places. Thick liquid water clouds are cooling the Earth through their reflection of sunlight. Thin ice clouds, called cirrus, allow sunlight to enter the Earth system but trap thermal radiation attempting to leave. Cirrus clouds are warming the Earth.

The International Satellite Cloud Climatology Project (ISCCP) has collected the largest global cloud data set using visible and infrared measurements from the international suite of weather satellites. As a supplement, multi-spectral infrared measurements from the National Oceanic and Atmospheric Administration polar orbiting High Resolution Infrared Radiation Sounders (HIRS) have been used for enhanced cirrus detection. Using regions of the infrared spectrum with differing sensitivity to atmospheric carbon dioxide, the HIRS measurements probe the atmosphere to different depths and reveal thin ice clouds high in the atmosphere.

*Our Changing Planet*, ed. King, Parkinson, Partington and Williams
Published by Cambridge University Press 

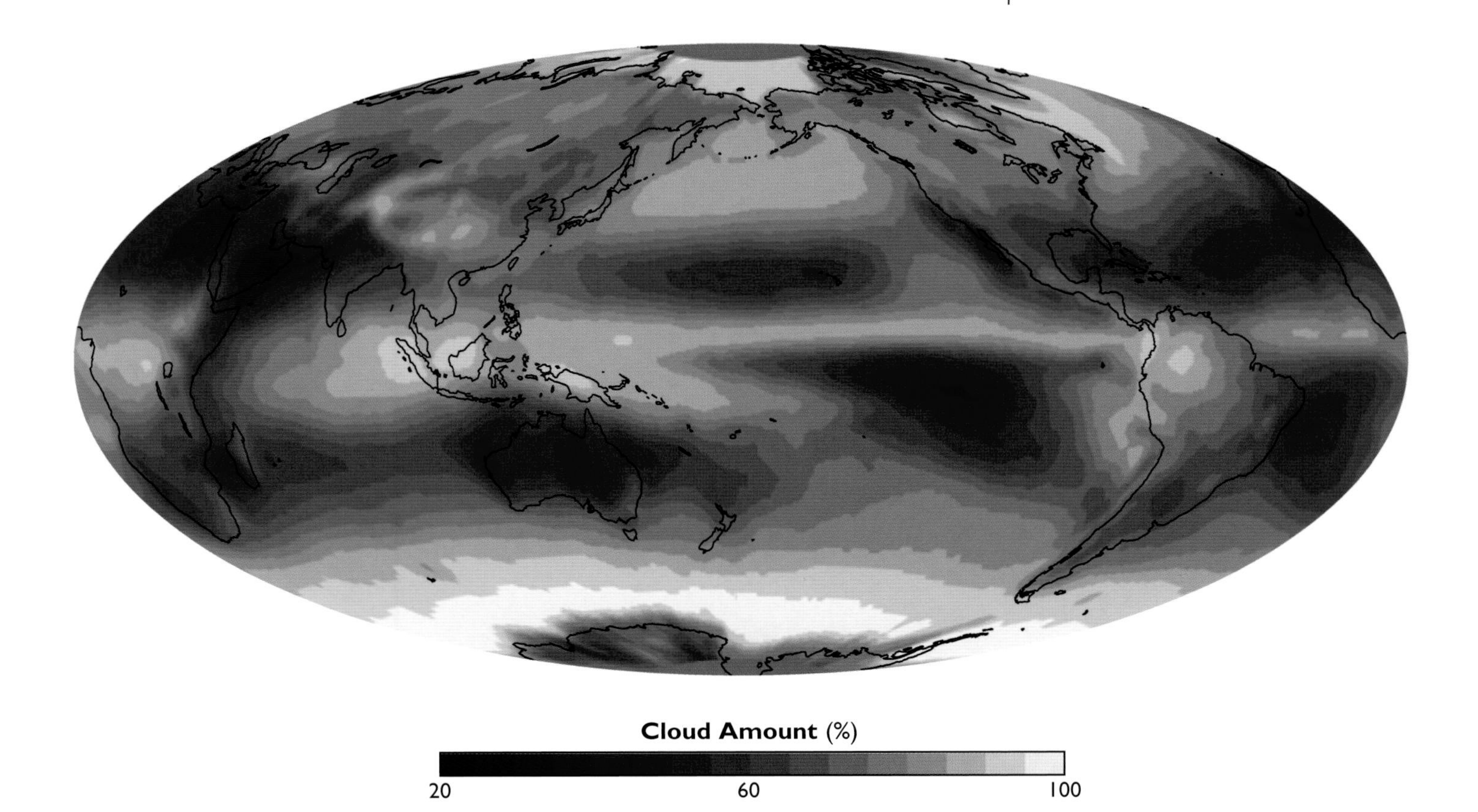


Annual mean cloud amount from 1979 through 2002. (Data from the HIRS instruments on the NOAA 5 to NOAA 14 satellites.)

Since 1979, HIRS measurements have found clouds most frequently in two locations (i) the Intertropical Convergence Zone (ITCZ) in the deep tropics where trade winds converge, and (ii) the middle to high latitude storm belts where low pressure systems and their fronts occur. In between are latitudes with fewer clouds and rain, called subtropical deserts over land and subtropical high pressure systems over oceans.

The decadal average cloud cover has not changed appreciably from the 1980s to the 1990s. Small increases occurred in the tropics, mainly in the Arabian Sea, Indochina, and the Indonesian Islands. Small decreases occurred in the subtropics, especially the eastern Sahara and in the central Pacific Ocean from Hawaii westward. The decreasing trend in Antarctica is uncertain because cloud detection itself is very difficult in the cold temperatures of Antarctica.

High cloud cover has changed some in the Northern Hemisphere winter season between the 1980s and the 1990s. Increases of 10% for clouds above 6 km altitude have occurred in the western Pacific, Indonesia, and over northern Australia. Other fairly large increases occurred in western North America, Europe, the Caribbean, western South America, and the Southern Ocean north of Antarctica. Decreases in high clouds occurred mainly in the tropical South Pacific, Atlantic and Indian Oceans south of the ITCZ.

While jet aircraft have been suspected of increasing cirrus cloud cover from their contrails, these data do not reveal such a trend. Increases of high clouds seem to occur in areas of high air traffic, such as central and western North America and Europe, as well as areas of rare air traffic, such as the Southern Ocean around Antarctica. It appears that high cloud cover changes are mostly caused by larger weather systems.

Globally averaged frequency of cloud detection (excluding the poles where cloud detection is less certain) has stayed relatively constant at 75%; there are seasonal fluctuations but

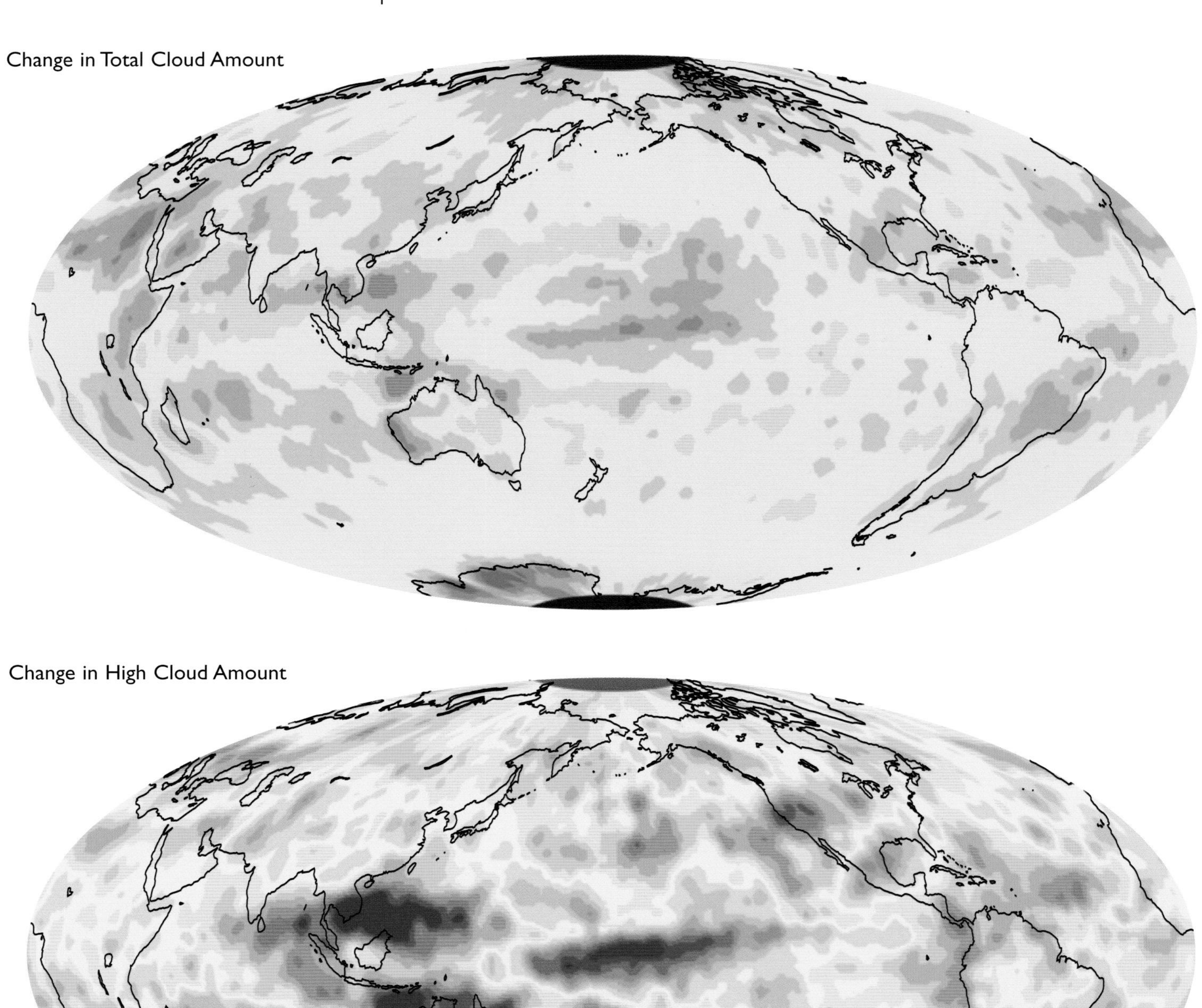


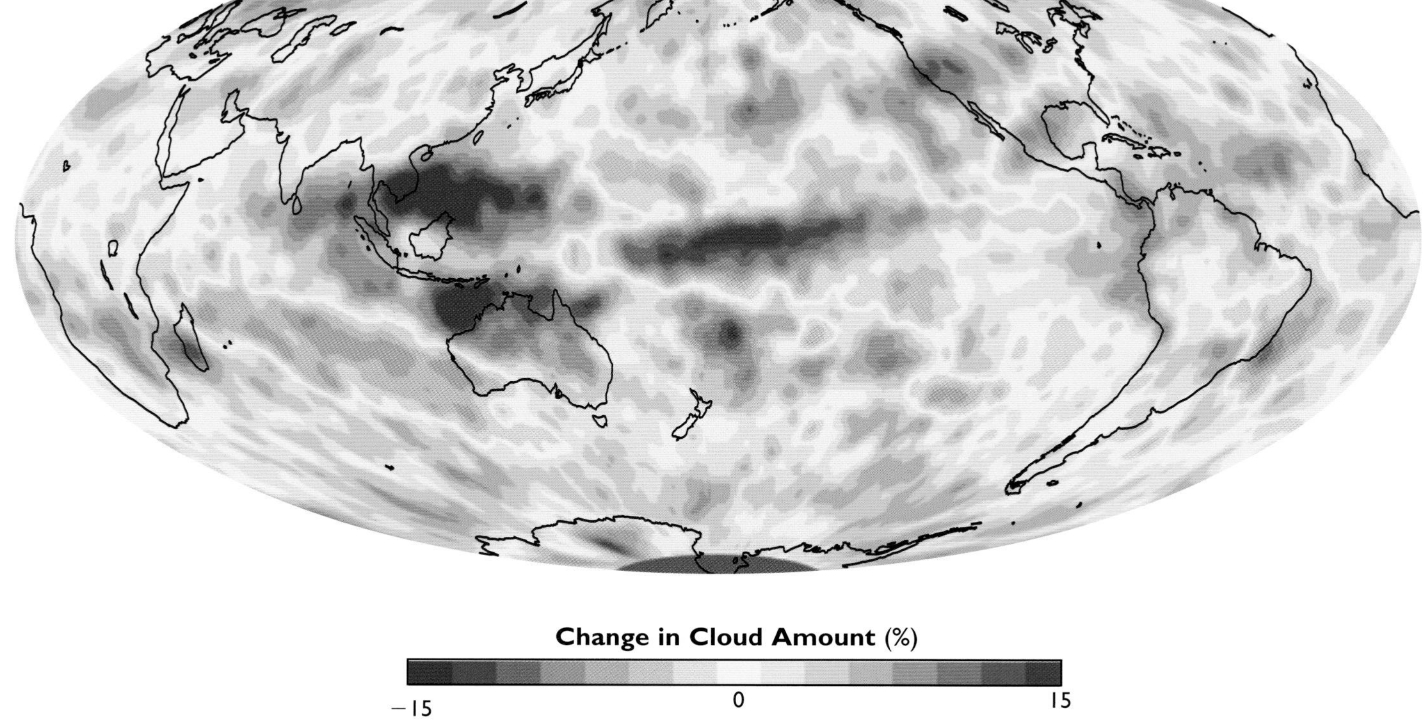


Change in total cloud amount (top) and change in high cloud amount (bottom) during Northern Hemisphere winters (December, January, and February) from the 1980s to the 1990s. (Data from the HIRS instruments on the NOAA 5 to NOAA 14 satellites.)

no general trends. High clouds in the upper troposphere (above 6 km) are found in roughly one third of the HIRS measurements; a small increasing trend of ~2% per decade is evident.

The most significant feature of these data may be that the globally averaged cloud cover has shown little change in spite of dramatic volcanic and El Niño events. During the four El Niño events between 1980 and 2001, winter clouds moved from the western Pacific to the

El Niño Years

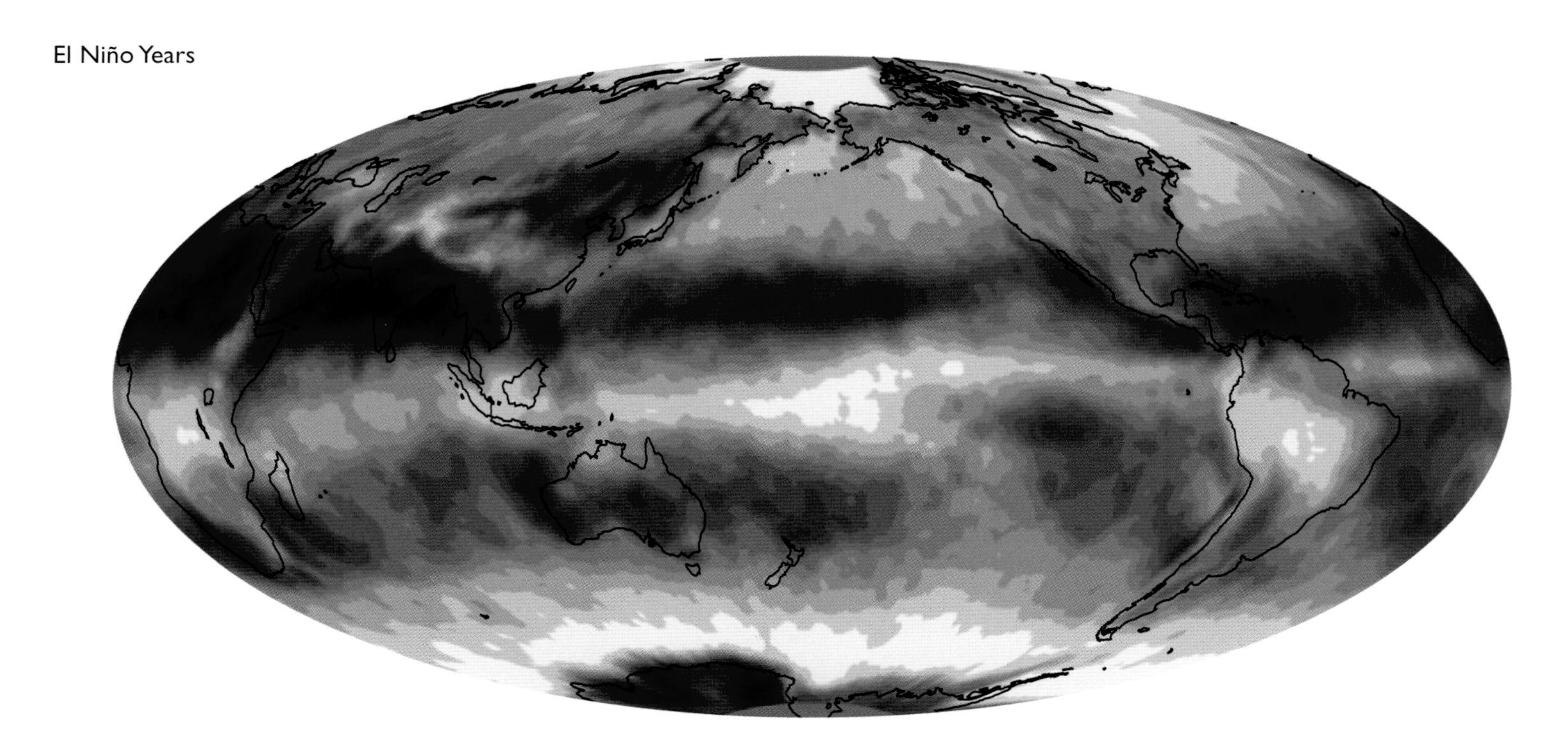

All Years Except El Niño Years

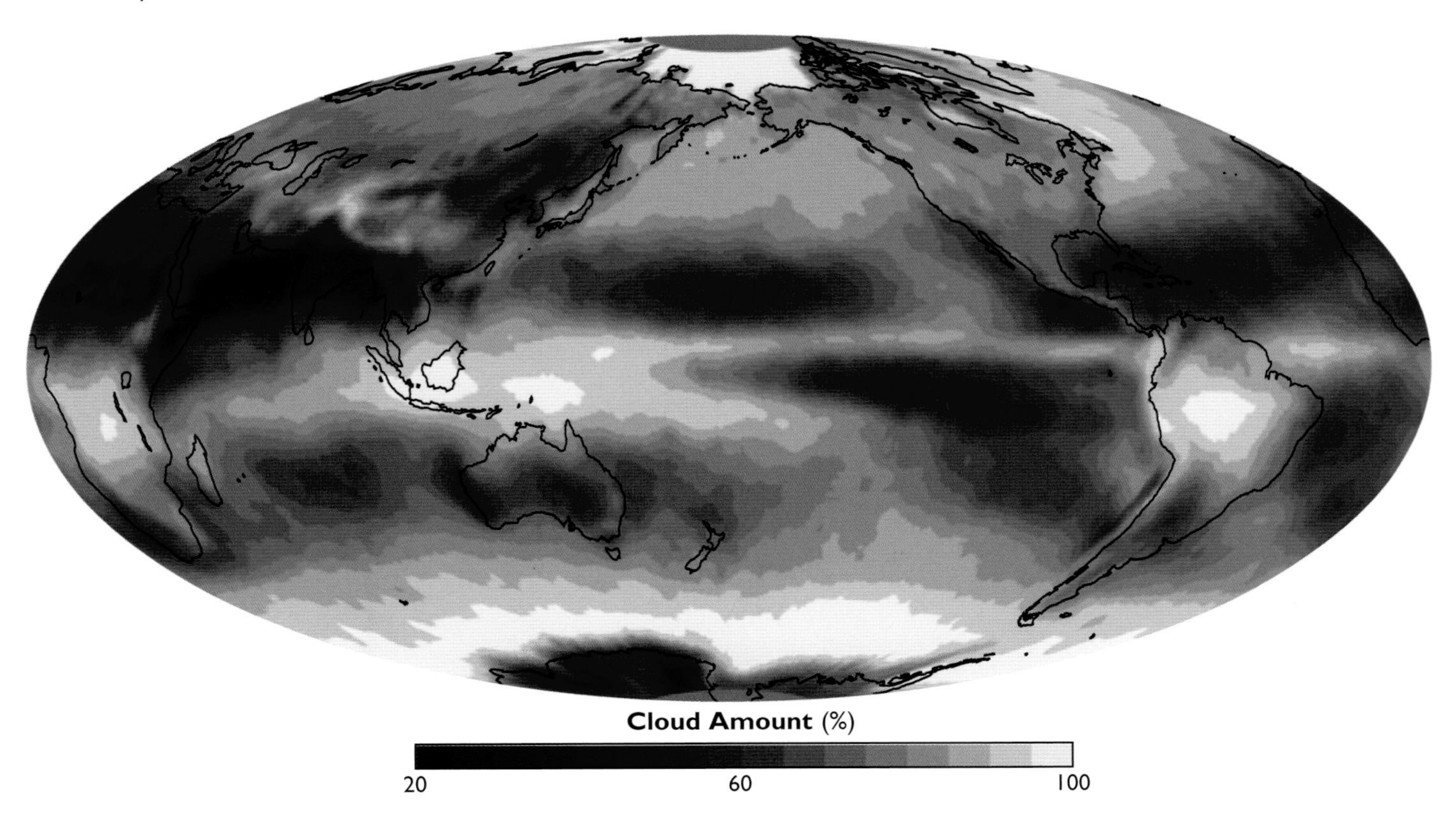


central Pacific Ocean, but their global average in the tropics did not change. El Chichón and Mt. Pinatubo spewed volcanic ash into the stratosphere that took 1–2 years to fall out, but cloud cover was not affected significantly.

These cloud data show overall constancy, with a small increase in high clouds in the 1990s as compared to the 1980s. The high thin clouds capture some of the Earth's infrared radiation

Cloud amount during El Niño years (top) compared with all other years (bottom) during Northern Hemisphere winters (December, January, and February) from the 1980s to the 1990s. (Data from the HIRS instruments on the NOAA 5 to NOAA 14 satellites.)

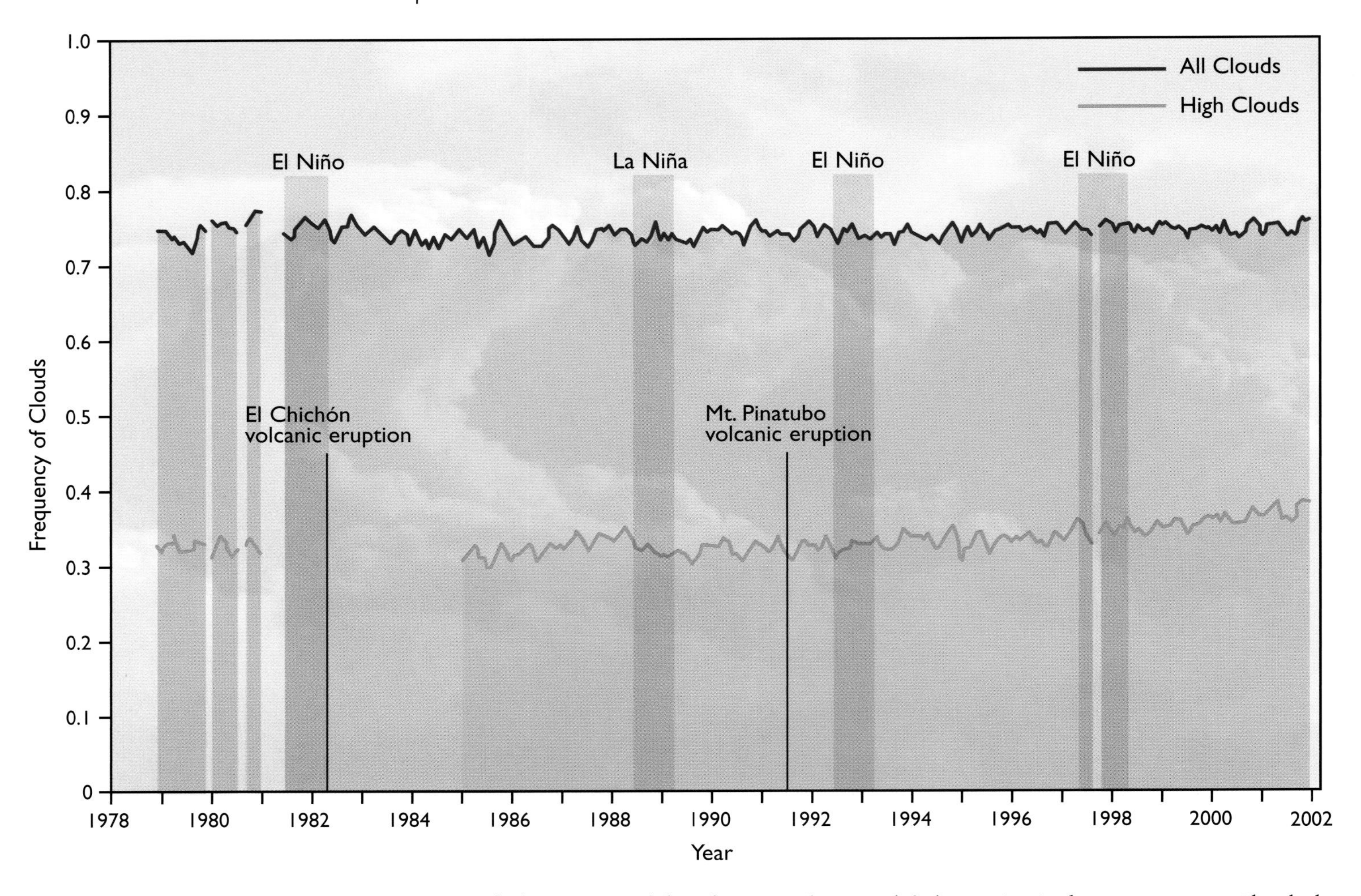


Although there is no trend in the frequency of occurrence of all clouds over this time period, there is a noticeable increase in the occurrence of high clouds. Even strong El Niño events have little impact on the frequency of occurrence of clouds worldwide. (Data from the HIRS instruments on the NOAA 5 to NOAA 14 satellites.)

similarly to $CO_2$ and thus they contribute to global warming in the same manner. Clouds do not appear to be off-setting global warming by increasing their reflection of incoming solar radiation; they are possibly enhancing it with modest increases of thermally trapping high thin ice clouds.

# Cloud Optical and Microphysical Properties

Michael D. King

Deep convective cumulonimbus capillatus incus cloud composed of upper layer ice crystals in the anvil overlying lower-level cumulus congestus clouds composed of liquid water droplets. (Photograph taken from the NASA ER-2 aircraft by pilot Jim Barrilleaux over Florida on July 19, 2002.)

Clouds occur in the Earth's atmosphere in both liquid water and ice phases. John Aitken, a Scottish physicist who did research on atmospheric dust and the formation of dew, cyclones, and evaporation, first reported in 1880 that "when water vapor condenses in the atmosphere, it always does so on some solid nucleus; that the dust particles in the air form the nuclei on which it condenses; and if there was no dust in the air there would be no fogs, no clouds, no mists, and probably no rain." Atmospheric dust is today referred to as cloud condensation nuclei, and they are ever present in the Earth's atmosphere as a result of breaking waves, dust storms, atmospheric chemical transformation in urban and industrial areas, and smoke from natural and manmade fires. But what kind of clouds does the Earth have today? This can

*Our Changing Planet*, ed. King, Parkinson, Partington and Williams
Published by Cambridge University Press 

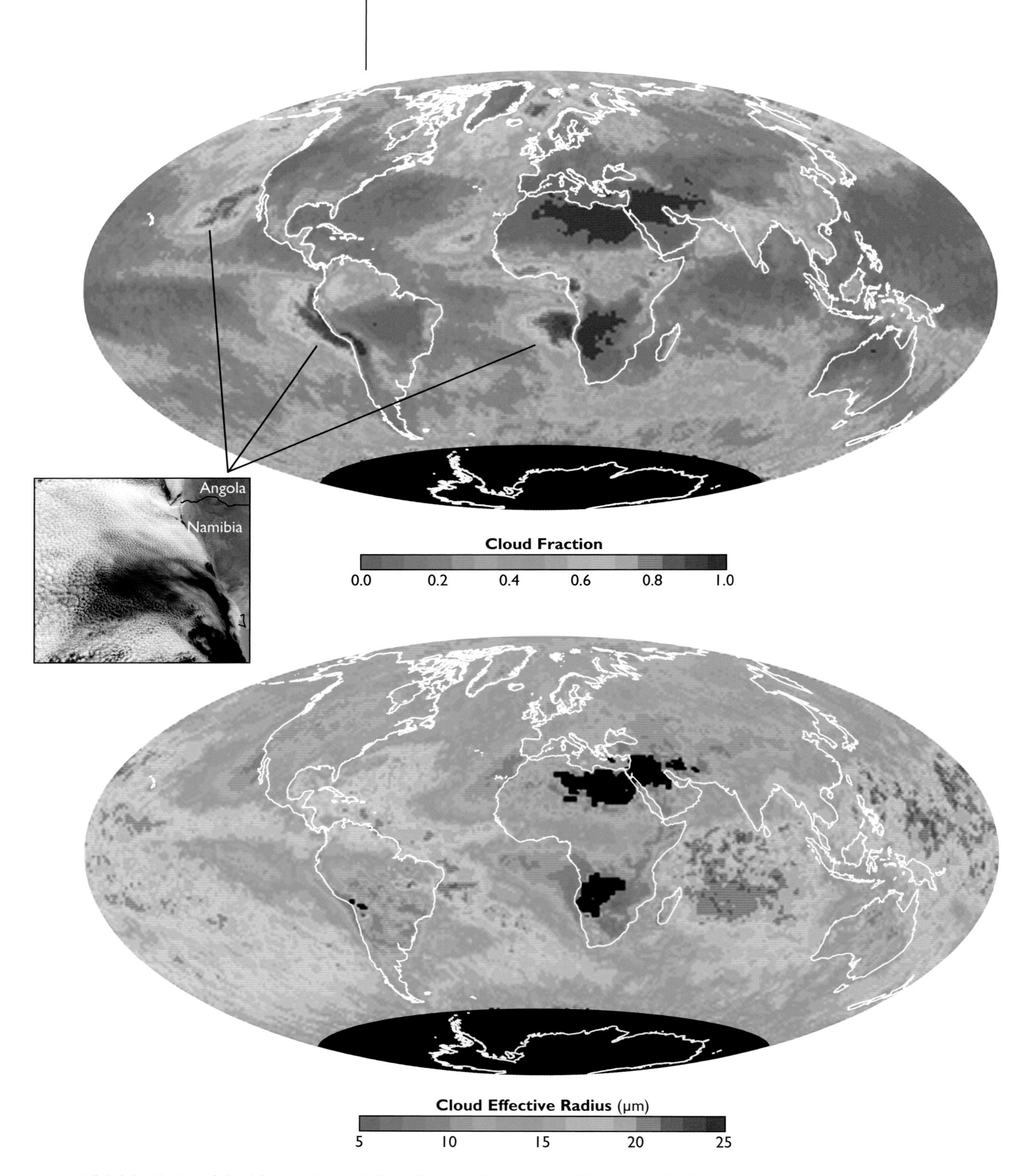


Global distribution of cloud fraction (top) and cloud effective radius (bottom) of liquid water clouds for July 2006. (Data from the MODIS instrument on the Terra satellite.) Inset: Marine stratocumulus clouds off Angola and Namibia on July 15, 2006. (Data from the MODIS instrument on the Terra satellite.)

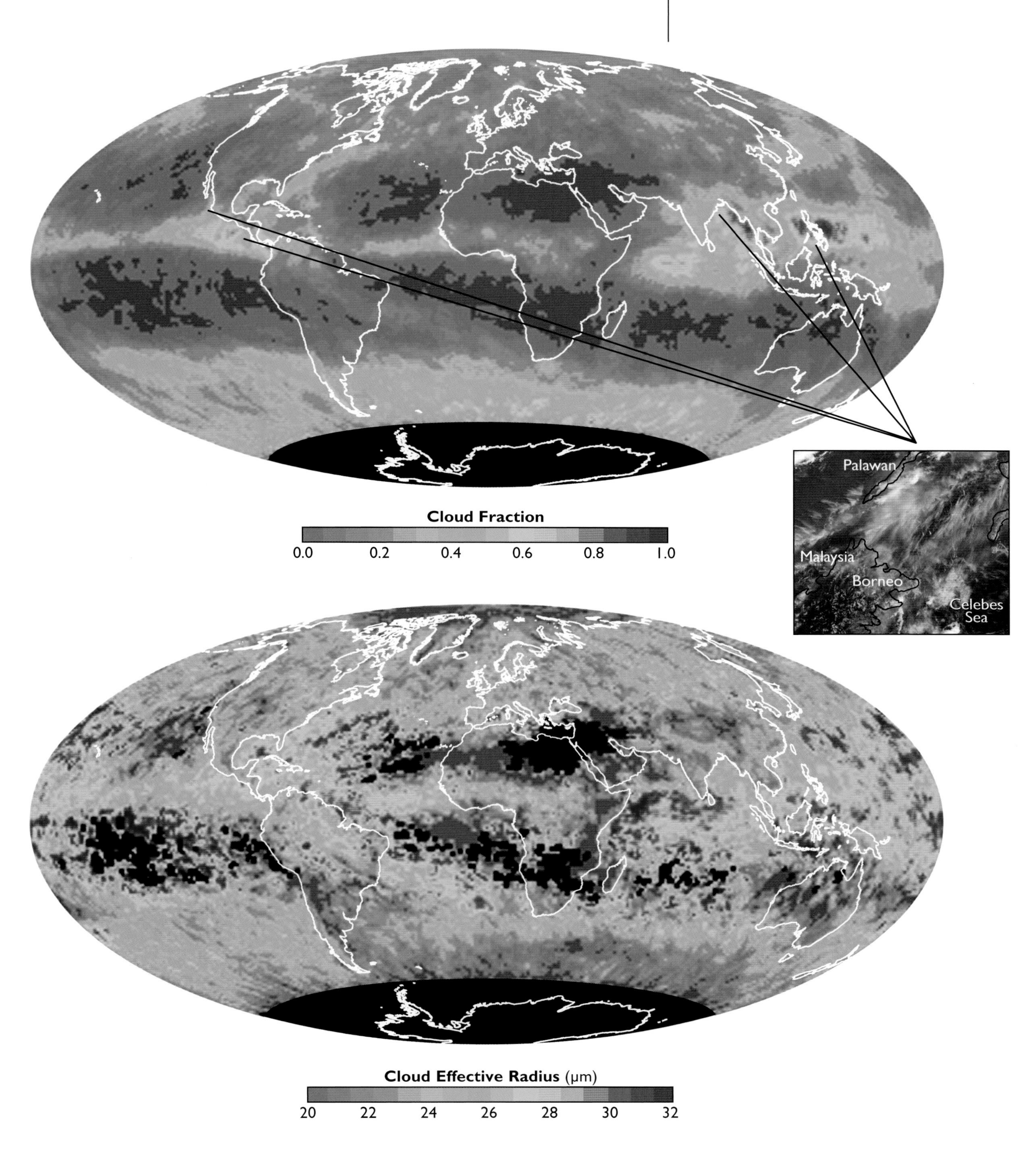


Global distribution of cloud fraction (top) and cloud effective radius (bottom) of ice clouds for July 2006. (Data from the MODIS instrument on the Terra satellite.) Inset: Deep convective clouds and thin ice clouds off the Philippines and Indonesia on July 14, 2006. (Data from the MODIS instrument on the Terra satellite.)

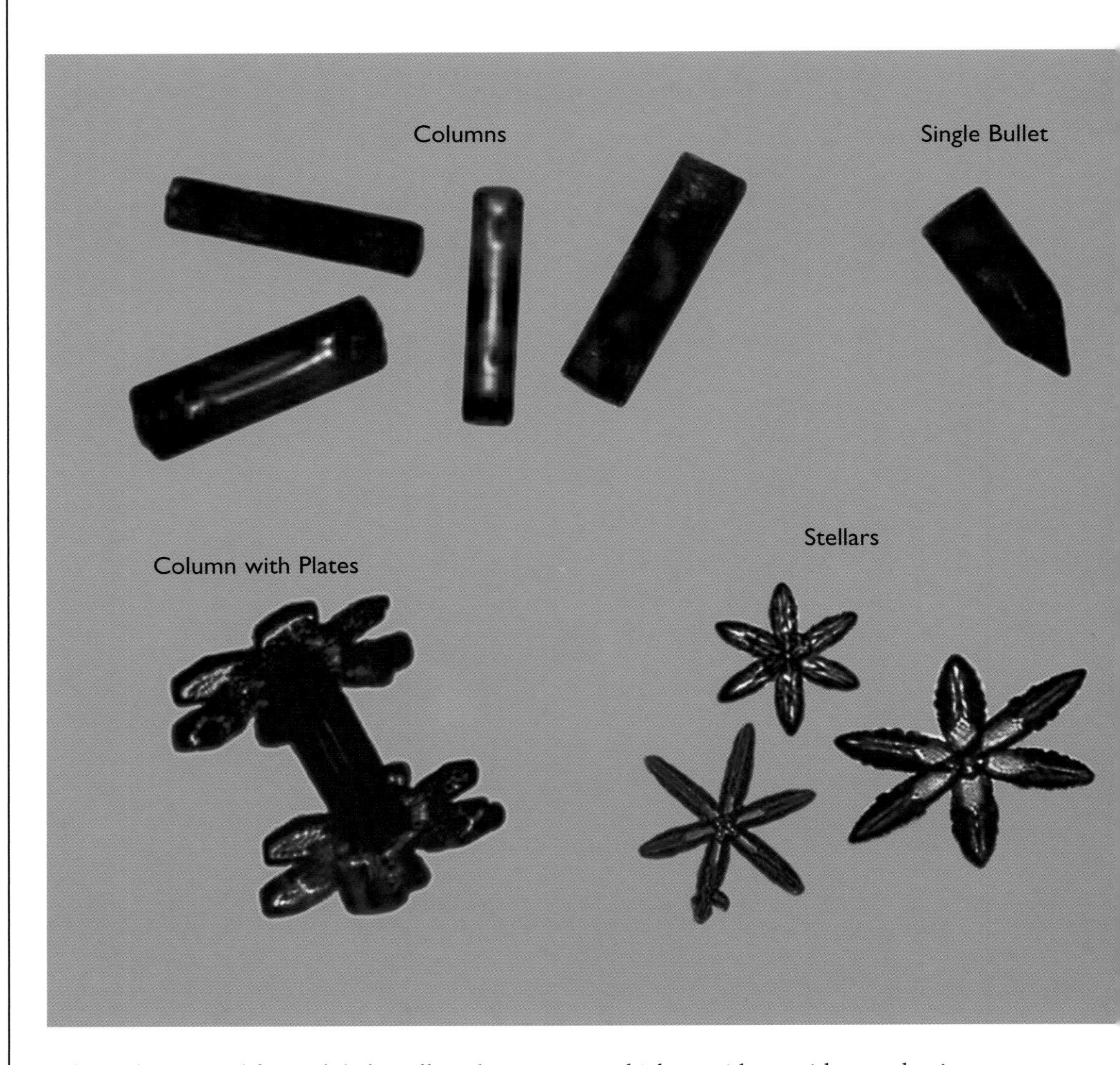


only be determined from global satellite observations, which provide us with a mechanism to determine the presence, spatial distribution, and frequency of occurrence of both liquid water and ice clouds.

Low-level clouds that occur at temperatures above 0°C, such as marine stratocumulus, altocumulus, and cumulus humilus clouds, are composed of liquid water droplets. Areas where these types of clouds dominate include oceanic regions off the west coasts of the United States, Peru, and Namibia, where they occur 60–90% of the time, depending on time of year. High-level clouds are often quite thin, such as cirrostratus and cirrus uncinus (mare's tale clouds), but can also occur as deep convective cumulonimbus clouds. These clouds are composed exclusively of ice particles, especially in the upper layers of the clouds. Thin ice clouds are readily observed in the Earth's atmosphere when they contribute to atmospheric optical phenomena such as sundogs or haloes, a clear indication of the presence of hexagonal ice crystals. Areas where these kinds of ice clouds dominate include the intertropical convergence zone and deep convective clouds in the tropics, such as over the Congo basin and the western tropical Pacific. These clouds are composed of complex ice crystal shapes such as columns, plates, bullet rosettes, dendrites, and aggregates like graupel or hail, and form at temperatures less than 0°C. Liquid water clouds can, and often do, form at temperatures below freezing, and are then referred to as supercooled clouds. At temperatures less that −39°C, however, only ice crystals can exist in the Earth's atmosphere.

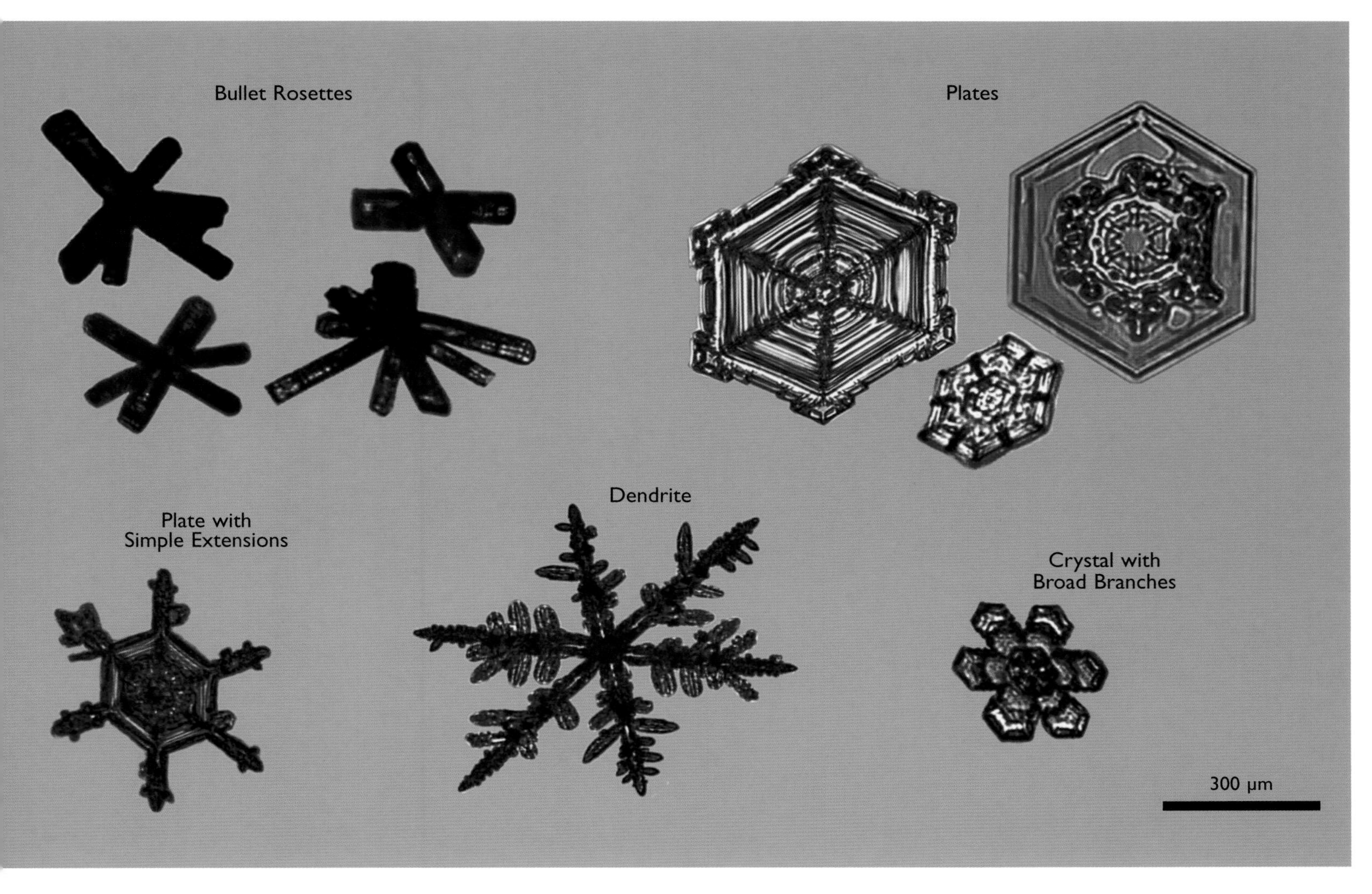


Digital images of ice crystals in clouds. A broad range of temperatures and humidities in clouds lead to the growth of different ice crystal shapes, which scatter light in widely different patterns. (Data from the CPI instrument on the SPEC Learjet, provided by Paul Lawson.)

Using a complex process to determine the presence of clouds and their thermodynamic phase, one finds that clouds occur over all latitudes and over a wide range of altitude, but it is possible from space to determine the presence, height, and optical thickness (opacity) of clouds, as well as their phase and cloud drop (or ice crystal) size. What size droplets exist in water clouds around the world? It turns out that the effective radius (characteristic size) of cloud drops ranges largely between 7 and 10 μm for low-level marine stratocumulus clouds to perhaps 20 μm for cumulus humilus and cumulus congestus clouds over the tropical oceans. It is also readily observed from space that liquid water cloud droplets are larger for clouds over the ocean than over the land, due primarily to the fact that cloud drops form on particles in the Earth's atmosphere, as first noted by John Aitken, and there are more particles over land than over the ocean. With more particles to distribute water vapor around, clouds over the land tend to have more cloud drops but they are smaller in size. Ice clouds also show a preference for larger crystals over ocean than over land, but ice particles in clouds are typically between 20 and 35 μm in effective radius, somewhat larger than water drops, but not in large numbers at precipitation size particles. Most clouds, whether they are liquid water or ice clouds, do not precipitate.

The rugged terrain of the eastern Pacific island of Guadalupe reaches a maximum elevation of 1.3 km and disturbs the flow of air around the island, made visible in this June 11, 2000 image by the marine stratocumulus clouds that are below the altitude of the island peak. Turbulent atmospheric flow patterns known as von Karman vortex streets form in the wake of an obstacle. Guadalupe is a volcanic Mexican island located 260 km west of Baja California. (Data from the MISR instrument on the Terra satellite.)

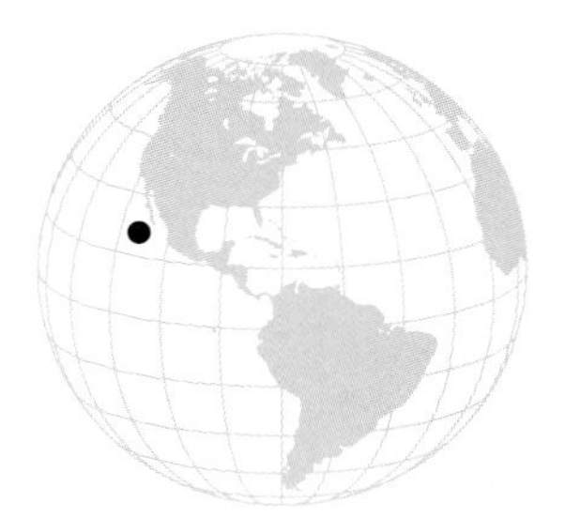

# Clouds and the Earth's Radiation Budget

Norman G. Loeb
Takmeng Wong

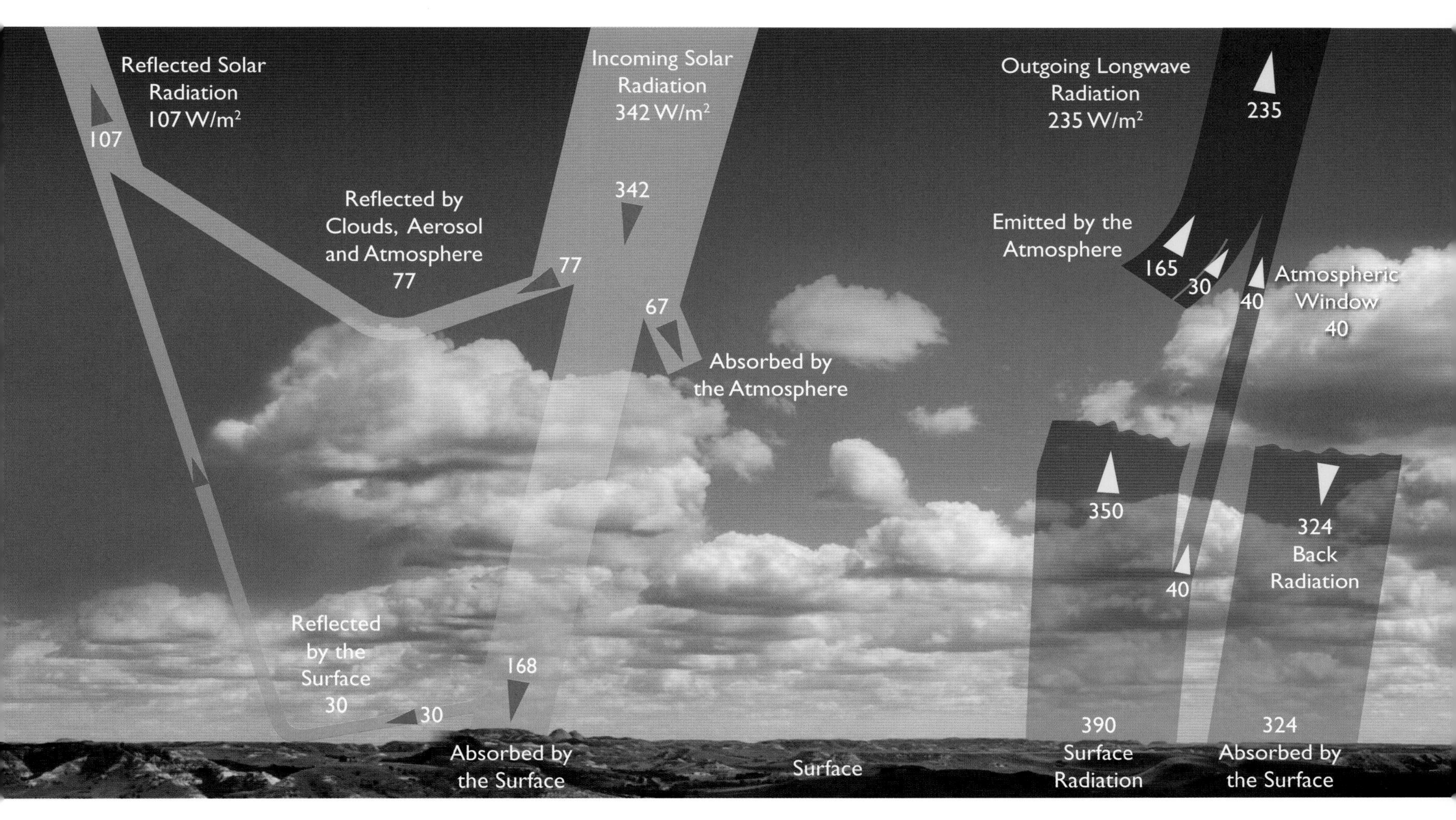


The Earth's annual global mean energy budget at the top-of-atmosphere, surface and within the atmosphere. (Illustration adapted from J. T. Kiehl and K. E. Trenberth, 1997, *Bulletin of the American Meteorological Society*, vol. 78, pp. 197–208.)

Energy from the Sun arrives at the Earth as electromagnetic waves of light that can travel up to 300,000 km/s in a vacuum. The electromagnetic waves carry radiant energy, or radiation, that can either be reflected, absorbed, or transmitted upon contact with an object. The radiant energy transported by an electromagnetic wave depends upon its wavelength, which is the distance along a wave from one crest to another. Most of the radiant energy from the Sun reaches the Earth at visible wavelengths (radiation we can see). The Earth also emits radiant energy, but because the Earth is much cooler than the Sun, most of its radiation is emitted at wavelengths in the infrared region, which cannot be seen by humans. Because they occur in different parts of the electromagnetic spectrum, reflected solar and terrestrial infrared radiation are often referred to as shortwave (SW) and longwave (LW) radiation, respectively.

The exchange of radiant energy between the Sun, Earth, and space is fundamental to climate. The Earth's average temperature remains relatively constant from year to year because the Earth-atmosphere system reflects and emits as much radiant energy to space as it absorbs

*Our Changing Planet*, ed. King, Parkinson, Partington and Williams
Published by Cambridge University Press 

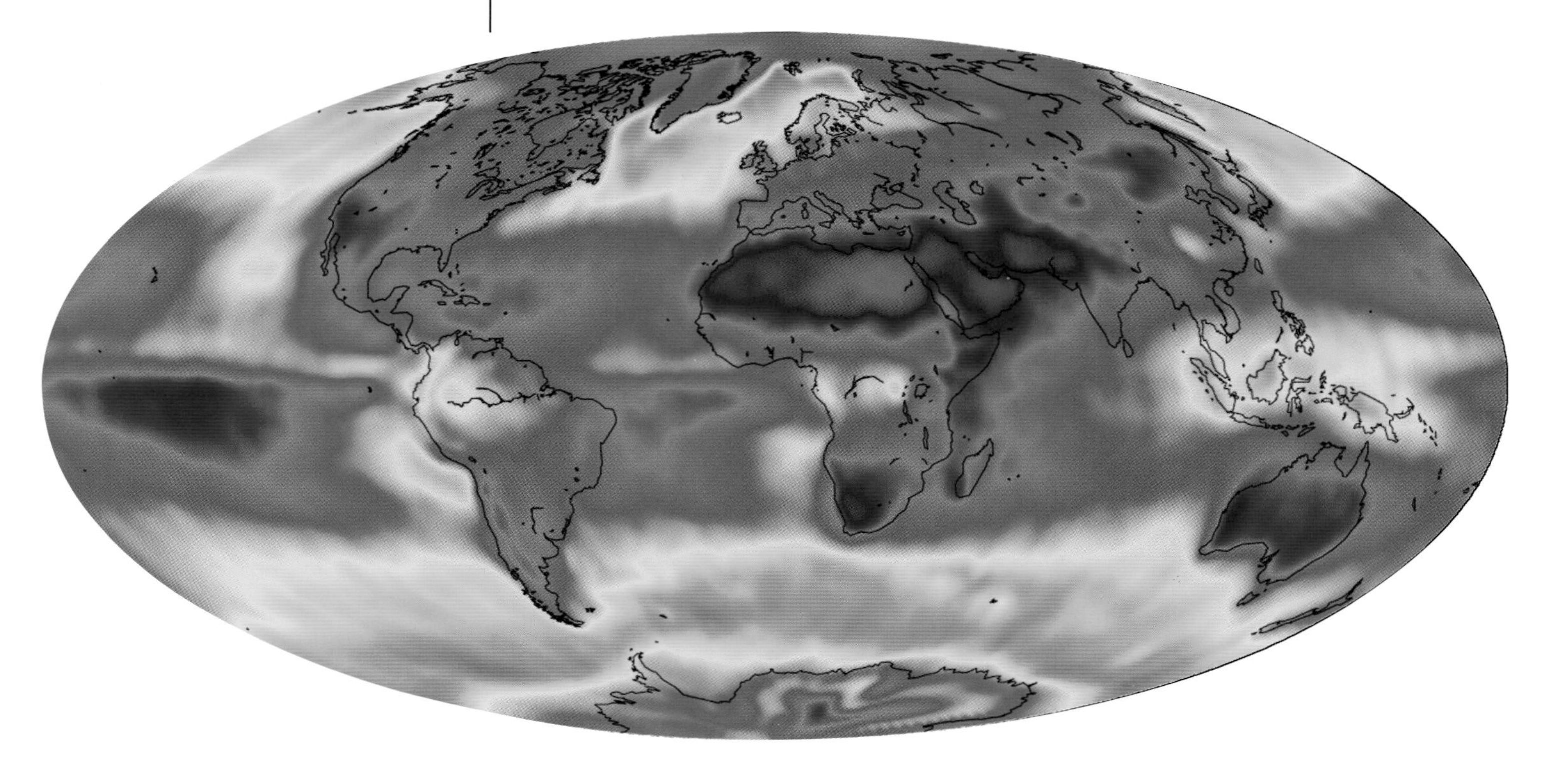

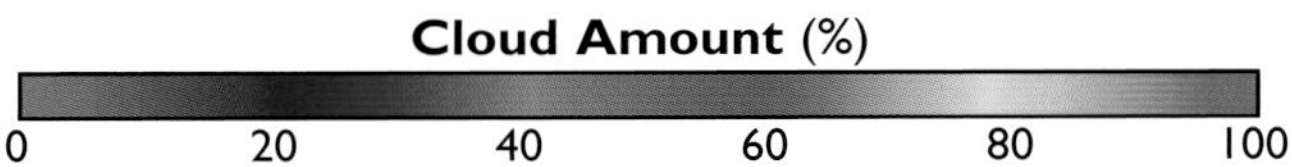


Annual mean cloud amount from March 2000 through February 2001. (Data from the MODIS instrument on the Terra satellite.)

from the Sun. Without this radiative balance, the Earth's temperature would change. If we assume that 342 W/m$^2$ of solar energy reaches the top of the atmosphere, approximately 107 W/m$^2$ (or 31%) is reflected to space by clouds, aerosols, air molecules, and the Earth's surface. The atmosphere absorbs 67 W/m$^2$ (20%) of the Sun's energy, and the remaining 168 W/m$^2$ (49%) is absorbed at the surface. The latter is used to evaporate water and heat the lower atmosphere. To maintain a radiative balance at the top of the atmosphere, the Earth-atmosphere system loses 235 W/m$^2$ of infrared radiant energy to space. The bulk of this (195 W/m$^2$) comes from emission by air molecules and clouds, and the remaining 40 W/m$^2$ is emitted by the surface. That only 40 W/m$^2$ of infrared radiation from the surface escapes to space is remarkable given that a staggering 390 W/m$^2$ is emitted at the surface. The reason so little surface infrared radiation gets through the atmosphere is because atmospheric gases (mainly water vapor and carbon dioxide) and clouds absorb and then re-emit most of the radiation back down to the surface. This process, known as the atmospheric greenhouse effect, is the reason why the average observed temperature at the surface of the Earth is 15°C instead of −18°C, the temperature at the Earth's surface in the absence of a greenhouse effect.

Predicting global climate change due to natural or manmade perturbations requires an understanding of how the various components of the climate system influence the Earth's radiation balance. Clouds have an important modulating role at both visible and terrestrial infrared wavelengths. They reduce the amount of solar radiation absorbed by the climate system by increasing the Earth's albedo (the fraction of incident solar radiation reflected by the Earth), and decrease the loss of terrestrial infrared radiation to space through the greenhouse effect. One way of quantifying the influence of clouds on the Earth's radiation balance is to compare the reflected solar and emitted infrared radiation under all-sky conditions with that which would be observed if the Earth were cloud-free. This difference, called net

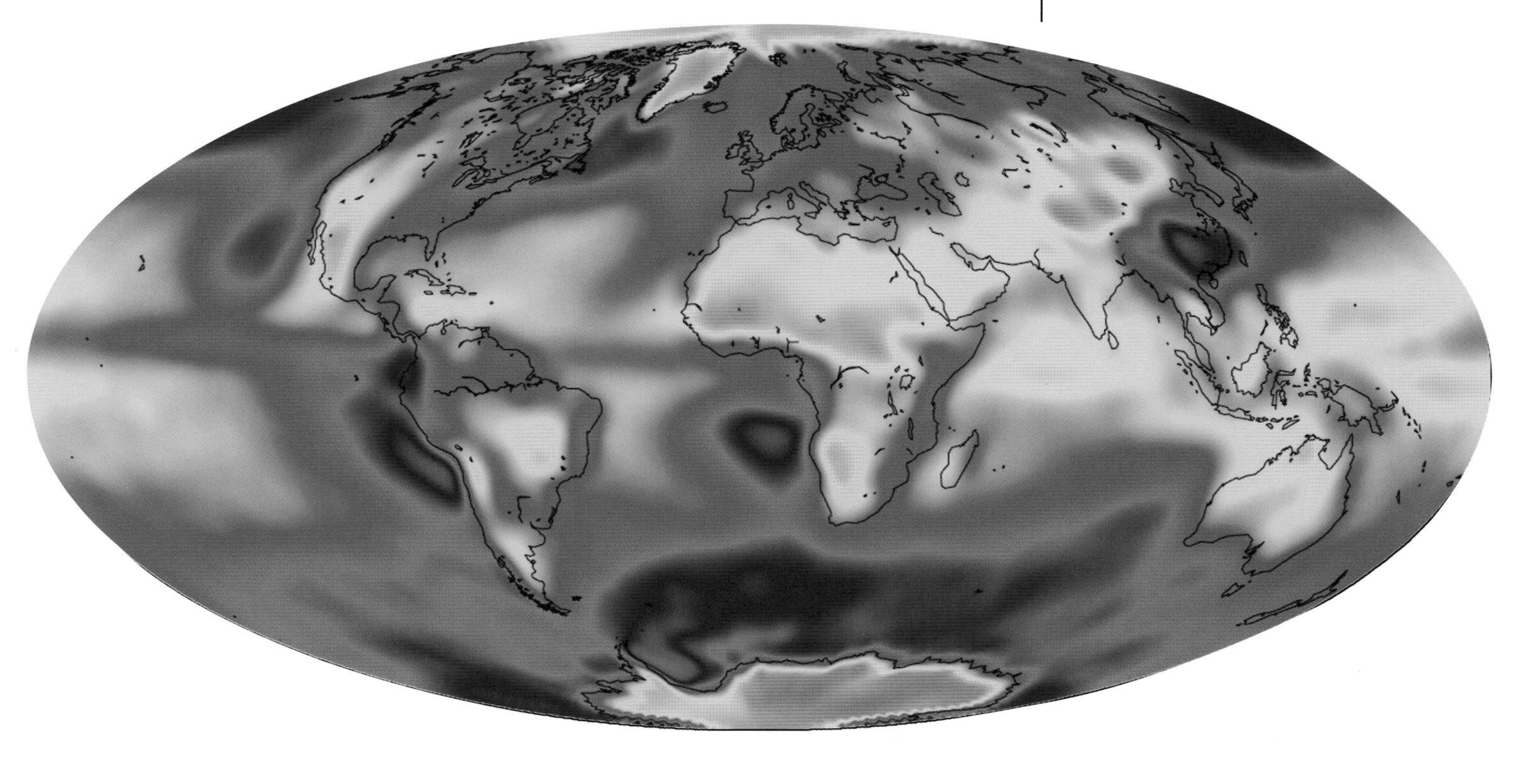

**Net Cloud Radiative Forcing** (W/m²)

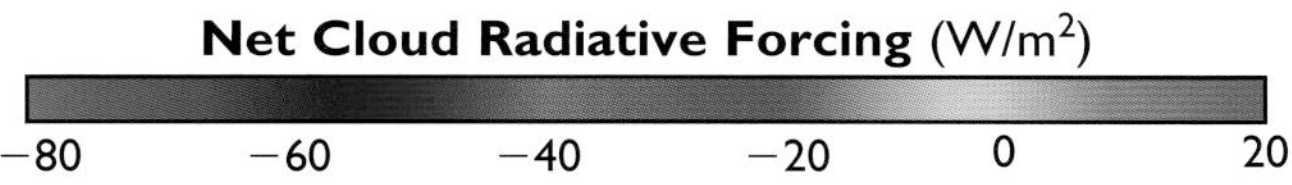


Annual mean net cloud radiative forcing for March 2000 through February 2001. (Data from the CERES instrument on the Terra satellite.)

cloud radiative forcing, varies substantially on a regional basis, depending on the physical properties of clouds such as cloud amount, cloud water content, cloud height, and cloud droplet sizes.

In regions where the net cloud radiative forcing is negative, clouds have a cooling effect on climate since they are more effective at blocking solar radiation from reaching the surface through reflection than they are at 'trapping' infrared radiation by decreasing the loss of terrestrial infrared radiation to space. Low stratiform clouds over the southern oceans between 30°S and 60°S and off the west coasts of the Americas and Africa have the strongest cooling effect. In tropical regions where thunderstorm clouds are frequently observed, such as in the western tropical Pacific Ocean and in the Intertropical Convergence Zone (convergence of low level airflow near the equator, often accompanied by a band of thunderstorms near the equator), the net cloud forcing is close to zero. While these clouds are highly reflective since they are thick, their tops are also very cold compared to the surface. As a result, the cooling and warming effects of these clouds nearly cancel and the net cloud radiative forcing is close to zero.

As the Earth undergoes changes in its climate, the amount of cloud as well as the physical properties of clouds may also change in ways we don't yet fully understand. Will a warmer climate lead to more or less cloud? Will the global distributions of cloud thickness and cloud height change? Satellite measurements indicate that clouds have an overall net cooling effect on the Earth's climate. If cloud properties change in response to global warming, will this cause further warming or cooling as the Earth tries to maintain a radiation balance? Such questions are currently being addressed by scientists through observations and computer models of the Earth's climate system. Because of the importance of clouds and their influence on climate, this area of research is a high priority for the climate research community.

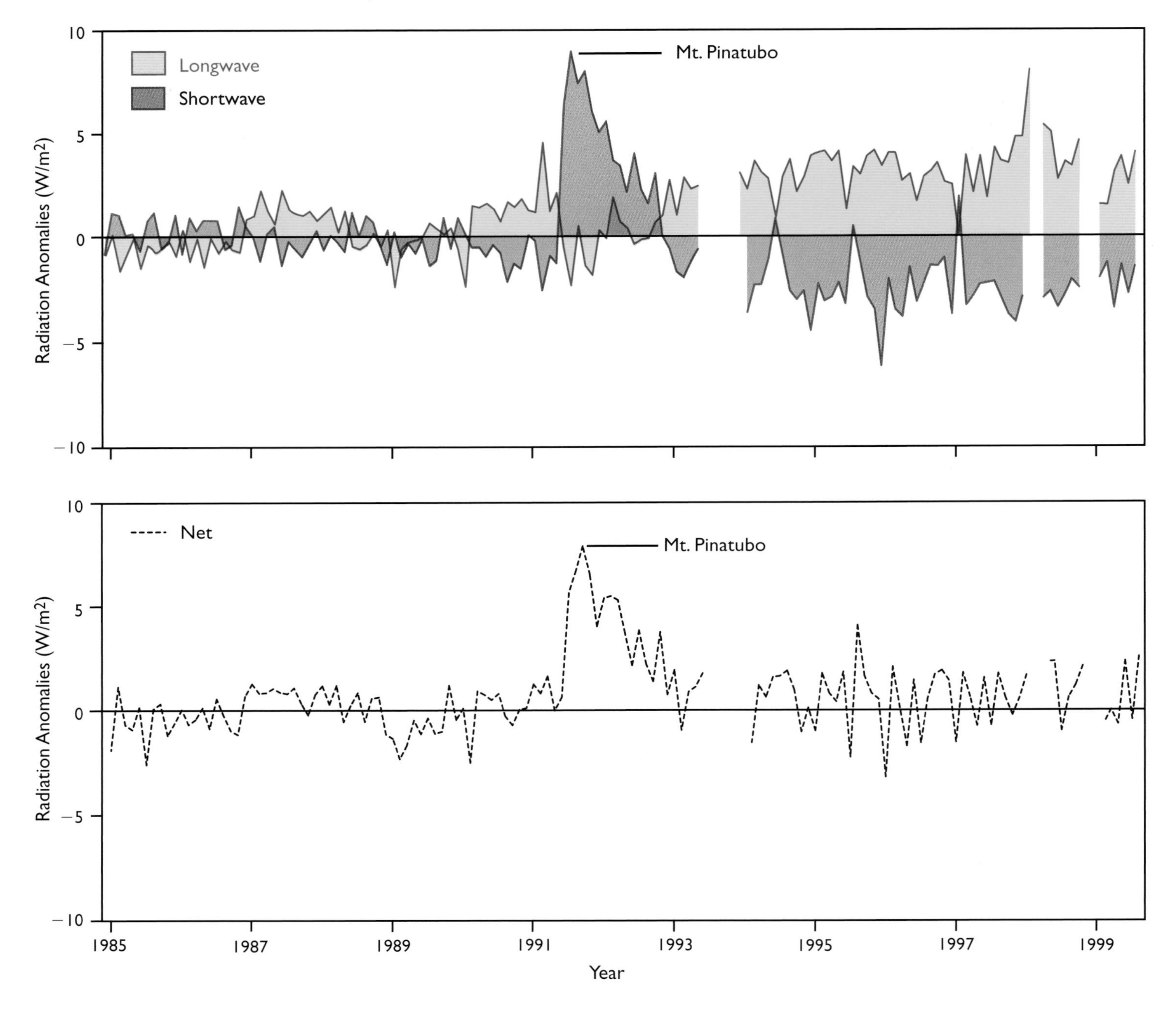


Comparison of the observed longwave (red), shortwave (blue) and net (black) radiation at the top of the atmosphere for the tropics (20°N–20°S) as deviation from the mean for 1985–1990. (Data from the ERBE instruments on the ERBS satellite and the CERES instrument on the TRMM satellite.)

Measurements from the past 20 years are revealing unexpected changes in the Earth's radiation budget believed to be due to changes in cloud properties. A 15-year time series of differences in mean longwave, shortwave and net radiation for the tropics between 20°N and 20°S from the ERBS and TRMM satellites reveal changes over time in the energy emitted, reflected and absorbed by the Earth-atmosphere system. The spike in the observations in 1991 is due to the Mt. Pinatubo volcanic eruption. Beginning in the early-to-mid 1990s, observed tropical mean longwave radiation shows a steady increase with time while shortwave radiation decreases, and net radiation increases. It is believed that this large decadal variability is caused by changes in the annual average and seasonal cloudiness in the tropics. Recent global measurements on the Terra satellite are providing new and exciting data that demonstrate the link between changes in cloud amount and changes in the Earth's radiation budget.

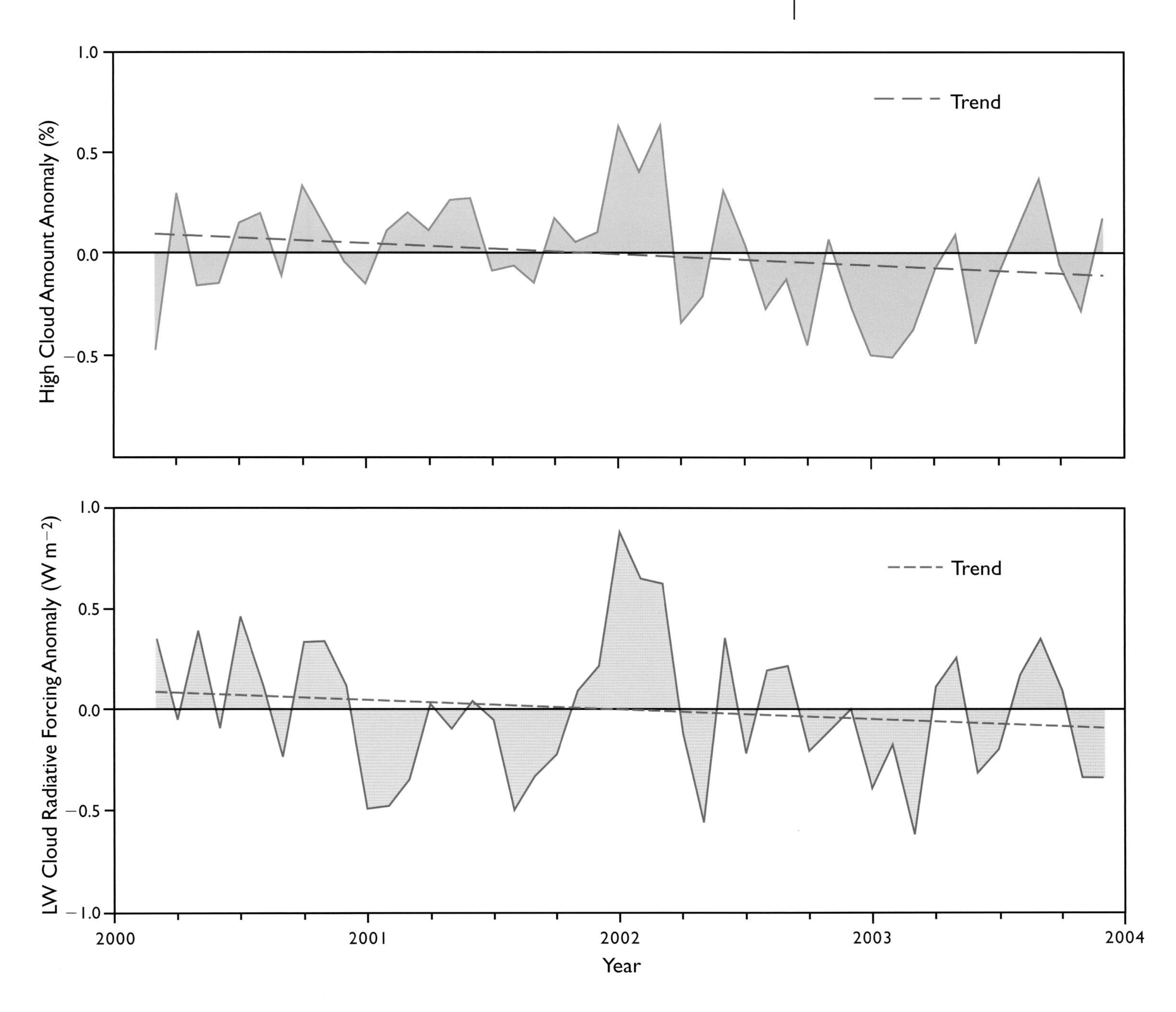


The first four years of Terra data reveal a decrease in longwave cloud radiative forcing that is accompanied by a corresponding decrease in high-level cloud amount. These data show a slight decrease in high-level cloud amount and a corresponding decrease in longwave cloud radiative forcing.

Global anomalies in high-level cloud amount (top) and longwave cloud radiative forcing (bottom) from 2000 to 2004. During this period, high-level cloud amount has decreased by 0.21% and longwave cloud radiative forcing has decreased by 0.20 $W/m^2$. (Data from the MODIS and CERES instruments on the Terra satellite.)

Brian J. Soden

# Water Vapor

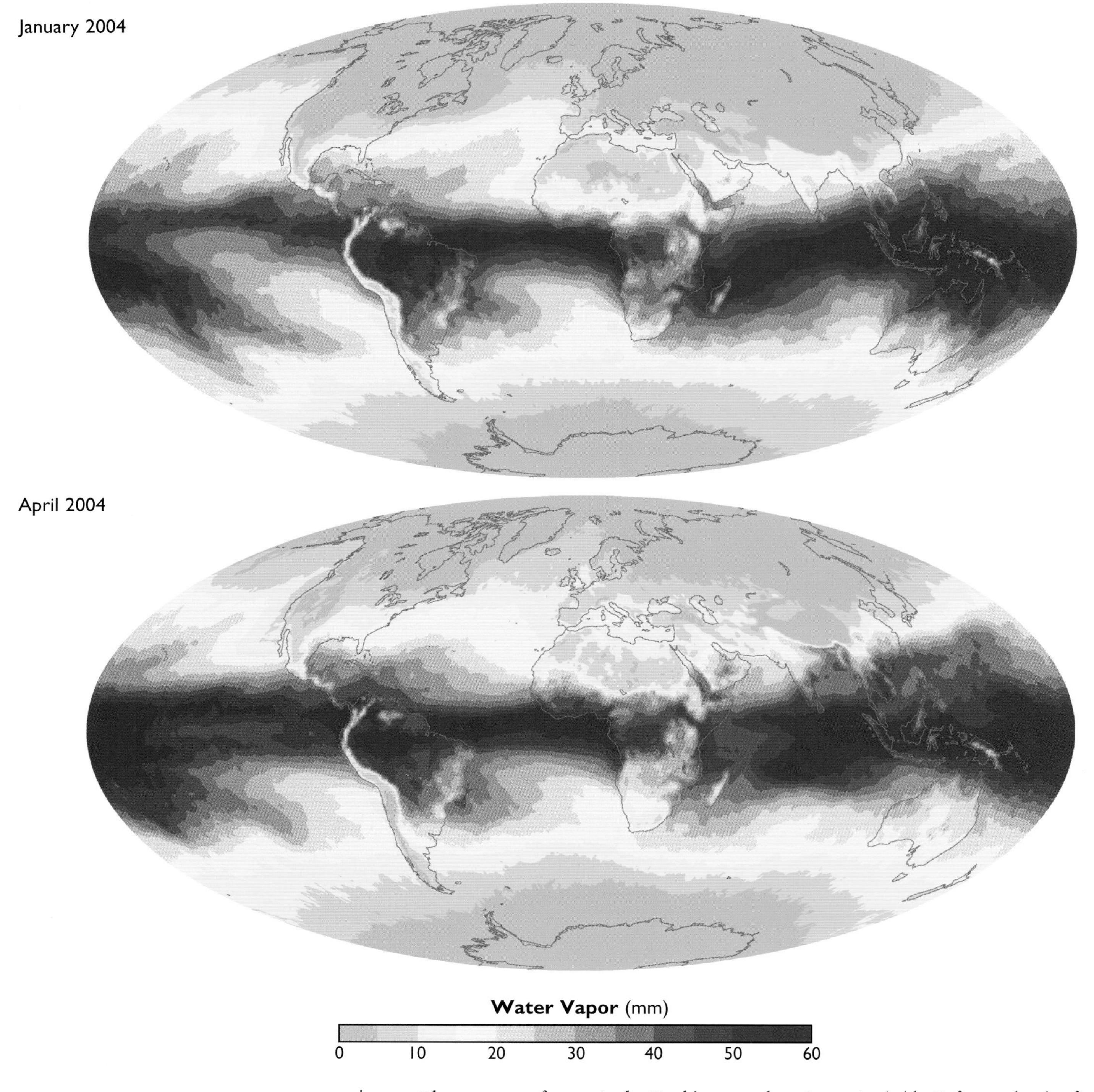


The presence of water in the Earth's atmosphere is unmistakable. It forms clouds of various colors, shapes, and sizes; it falls from the sky as rain or snow; and it produces breathtaking optical phenomena such as rainbows and halos. Yet it is the water that lies unseen in the form of gaseous water vapor that exerts the most profound influence on our planet's weather and climate.

*Our Changing Planet*, ed. King, Parkinson, Partington and Williams
Published by Cambridge University Press 

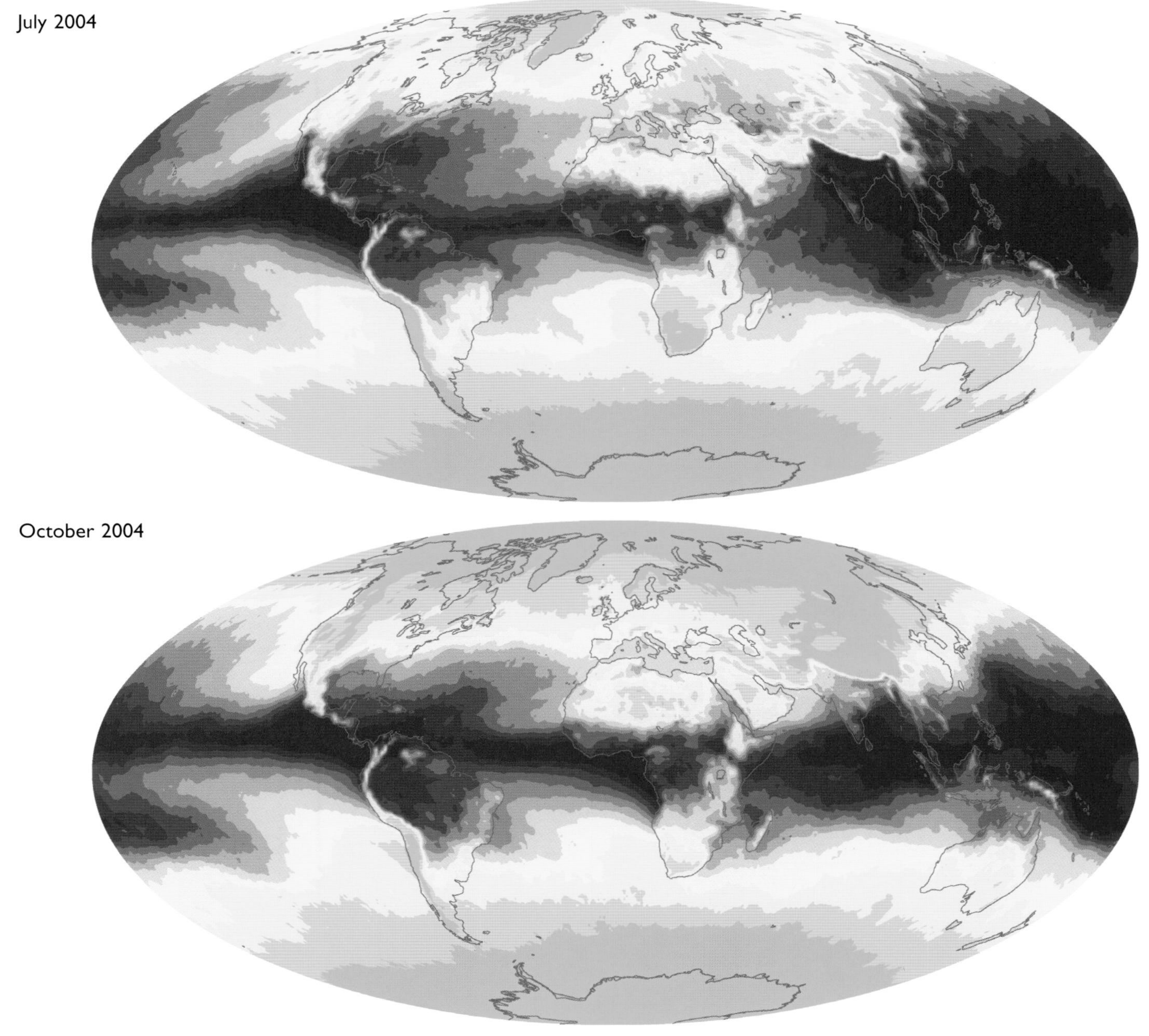


The seasonally changing global distribution of atmospheric water vapor. Total column water vapor is shown separately for January, April, July, and October 2004, which shows a noticeable 'river' of moisture around the tropics. (Data from the AIRS and AMSU instruments on the Aqua satellite.)

All substances, including water, can exist in three phases: solid, liquid, and gas. However, water is special because it's the only substance that exists in all three phases under normal conditions on Earth. When present in sufficient quantity, gaseous water vapor condenses to form liquid cloud droplets. If the temperatures are cold enough, the cloud droplets will freeze

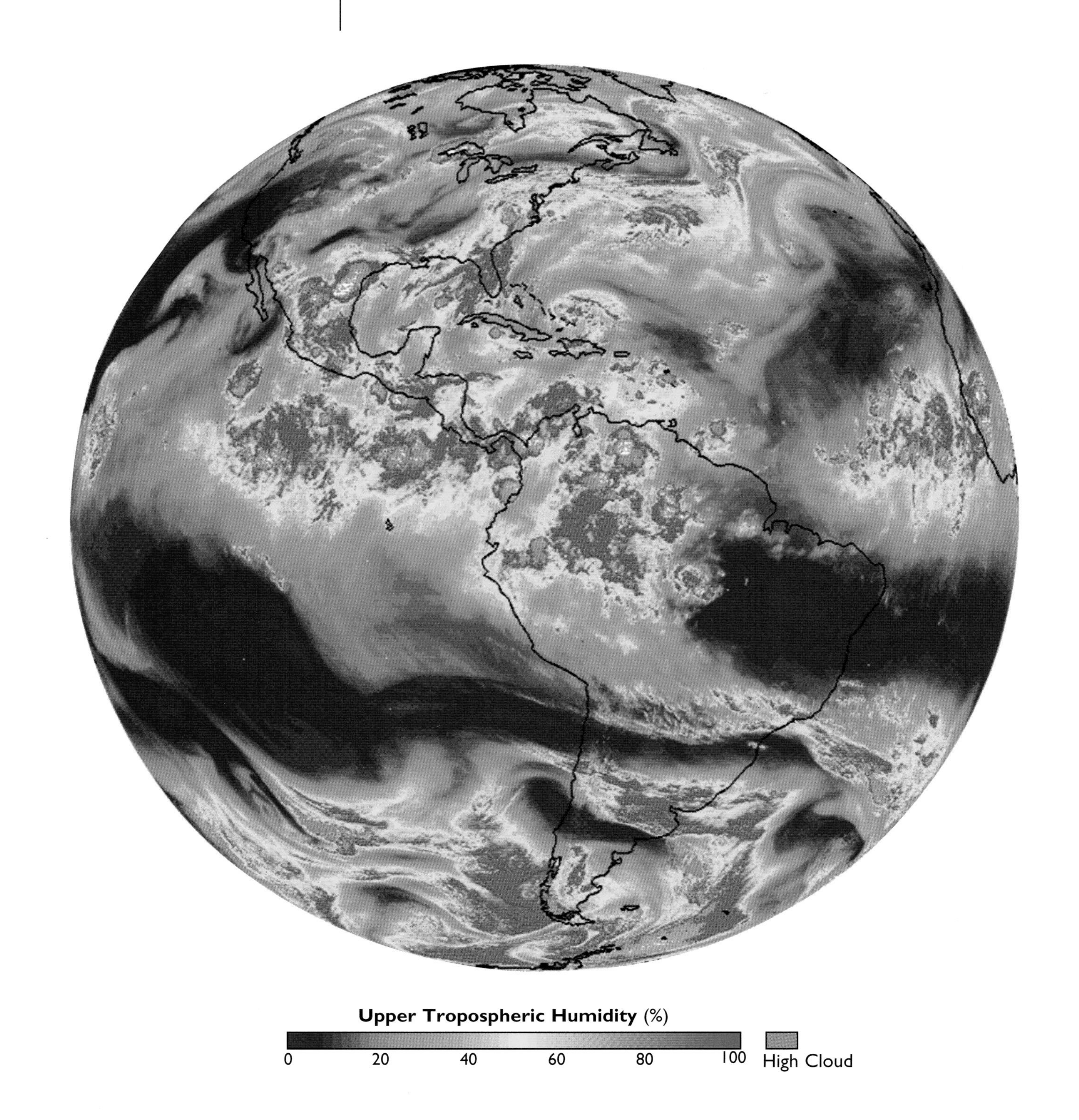


Upper atmosphere moisture. Upper tropospheric relative humidity and high cloud amount on June 26, 1999. (Data from the GOES Imager instrument on the GOES-8 satellite.)

into solid ice particles that may eventually fall to the surface, arriving as either snow or rain, again depending upon the temperature.

The conversion of water vapor into a liquid or solid involves the union of countless water vapor molecules and the formation of bonds to hold these molecules together. In the process of forming these bonds, a tremendous amount of energy is released. Consider an average hurricane that condenses 20,000 million tons of water a day. The energy released from converting this water vapor into rain is equivalent to that contained in one-half million atomic bombs. In this way, water vapor provides the fuel that drives the planet's weather.

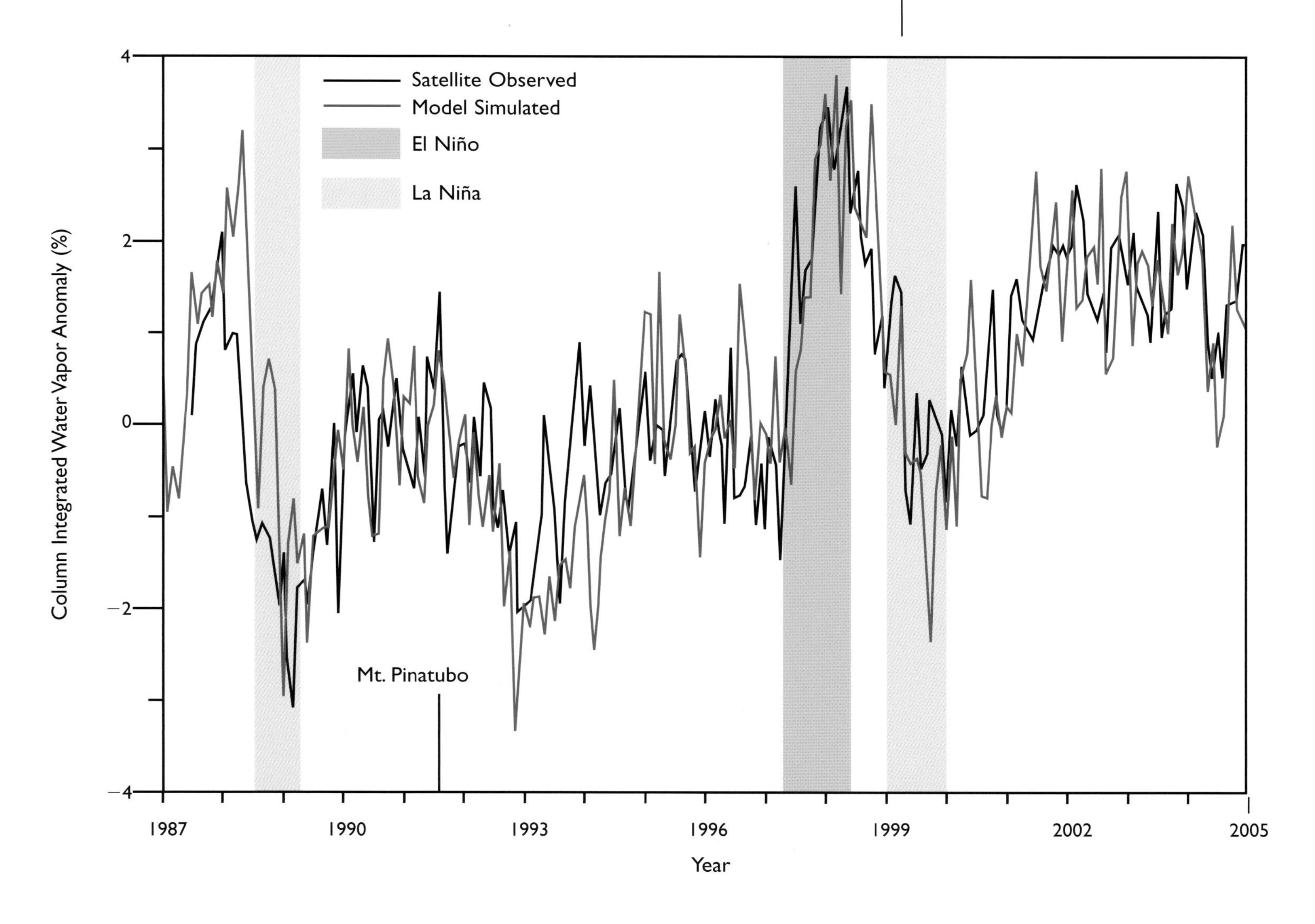


Time series of the globally averaged (ocean-only) concentration of atmospheric water vapor expressed as a percentage deviation from its mean seasonal value. Results are shown for both satellite observations (black) and from climate model simulations (red). A positive trend is apparent from the 1980s to present, owing to the increase in ocean surface temperatures over this period. The large year-to-year deviations reflect the occurrence of anomalously warm El Niño events (1997–1998) and cold La Niña events (1988–1989, 1999–2000), and global cooling following the 1991 eruption of Mt. Pinatubo (1992–1993). (Adapted from B.J. Soden *et al.*, 2005, *Science*, vol. 310, pp. 841–844.)

Given its ubiquitous nature, one might be surprised at how rare water vapor molecules are. On average, fewer than one out of every one hundred air molecules are water vapor. However the concentration of water vapor is highly variable in both space and time. Consider satellite measurements of the distribution of water vapor and clouds. Note the presence of deep convective clouds (white), detraining cirrus anvils (gray), the convective moistening of adjacent regions of high relative humidity (red), and the gradual reduction in relative humidity as air is carried towards the subtropics, ultimately resulting in water vapor concentrations less than 10% of their saturated value (blue).

Yet the small concentrations of water vapor belie its importance. Water vapor is the dominant greenhouse gas, trapping more of the planet's heat than any other atmospheric constituent. Its concentration also depends strongly upon temperature. As the climate warms from the burning of fossil fuels, the concentrations of water vapor are expected to increase. This moistening of the atmosphere, in turn, absorbs more heat and further raises the temperature. Through this process, water vapor not only keeps our planet warm today but also hastens the rate at which our planet will warm in the future.

# A World of Rain

Robert F. Adler
George J. Huffman
Scott Curtis

1979–2006

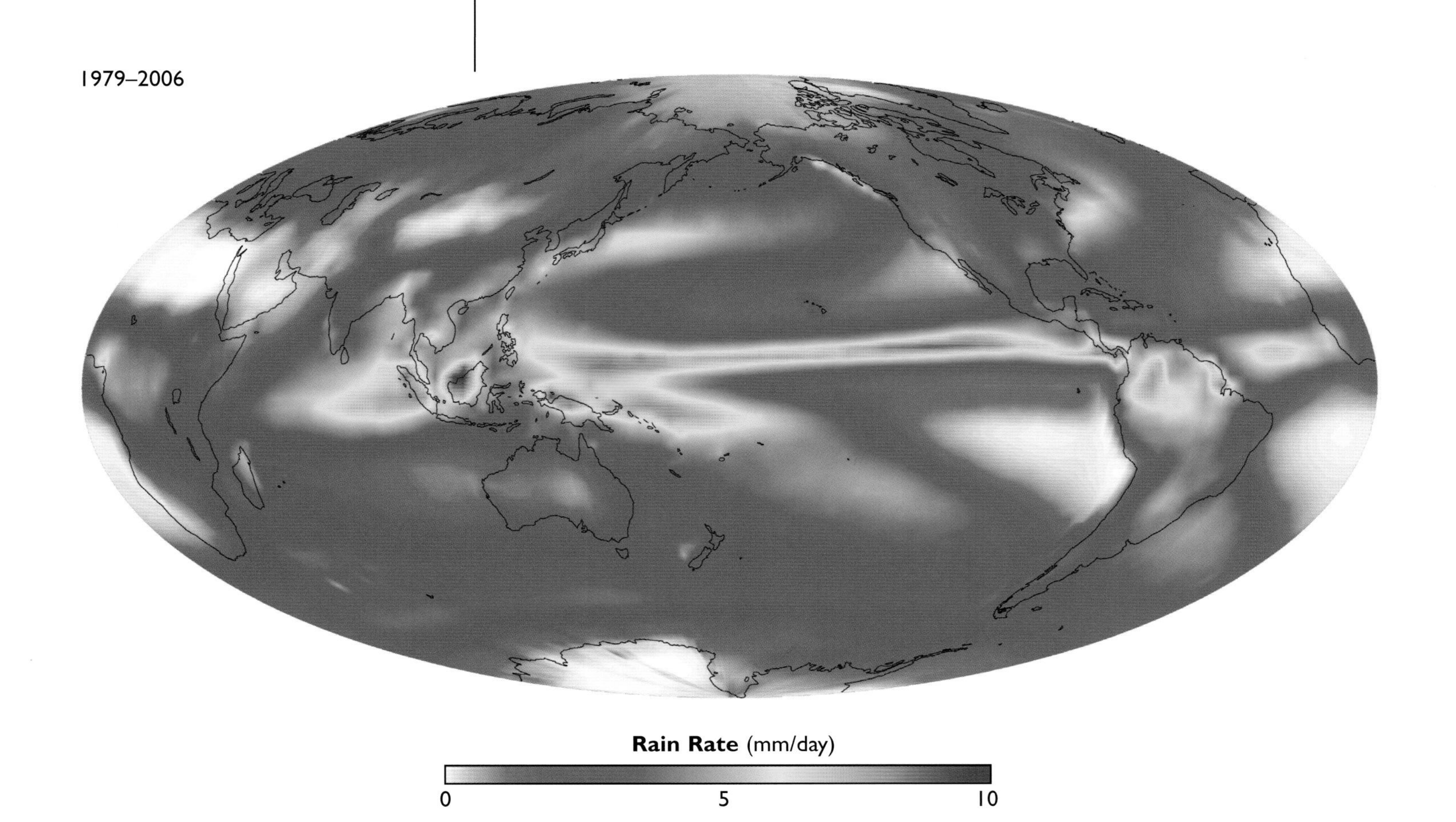


Average precipitation over the globe between 1979 and 2006 showing the high rainfall in the tropics, the dry sub-tropics, and the precipitation patterns associated with mid-latitude storm tracks. (Data from the Global Precipitation Climatology Project.)

From ancient times onward people have been concerned about the amount and variation of rainfall, because these factors determine the local rain climatology (or average conditions) and the dangerous extremes of climate (floods and drought). The distribution of forest, desert, and arable land (and even people) around the globe is determined to a large extent by the relative abundance or absence of precipitation. During the last century or so scientists have developed increasingly sophisticated methods of measuring precipitation over some land areas, including rain gauges and ground-based radars. They realized that precipitation (rain and snow) was just one component of a continuous cycling of water from ocean and land surfaces into the atmosphere (evaporation), the movement and transformation in the atmosphere of that gaseous water vapor into liquid water droplets and solid ice particles that make up clouds (condensation), and the completion of the cycle through rainfall and snowfall (precipitation) from the atmosphere back to the surface. Also, in order to understand the place-to-place and time variations of precipitation critical for human life and prosperity, they needed to understand the global (both land and ocean) distribution of precipitation. To obtain coverage over oceans and remote land areas and complete that global picture of rainfall, scientists turned to the use of satellites.

*Our Changing Planet*, ed. King, Parkinson, Partington and Williams
Published by Cambridge University Press 

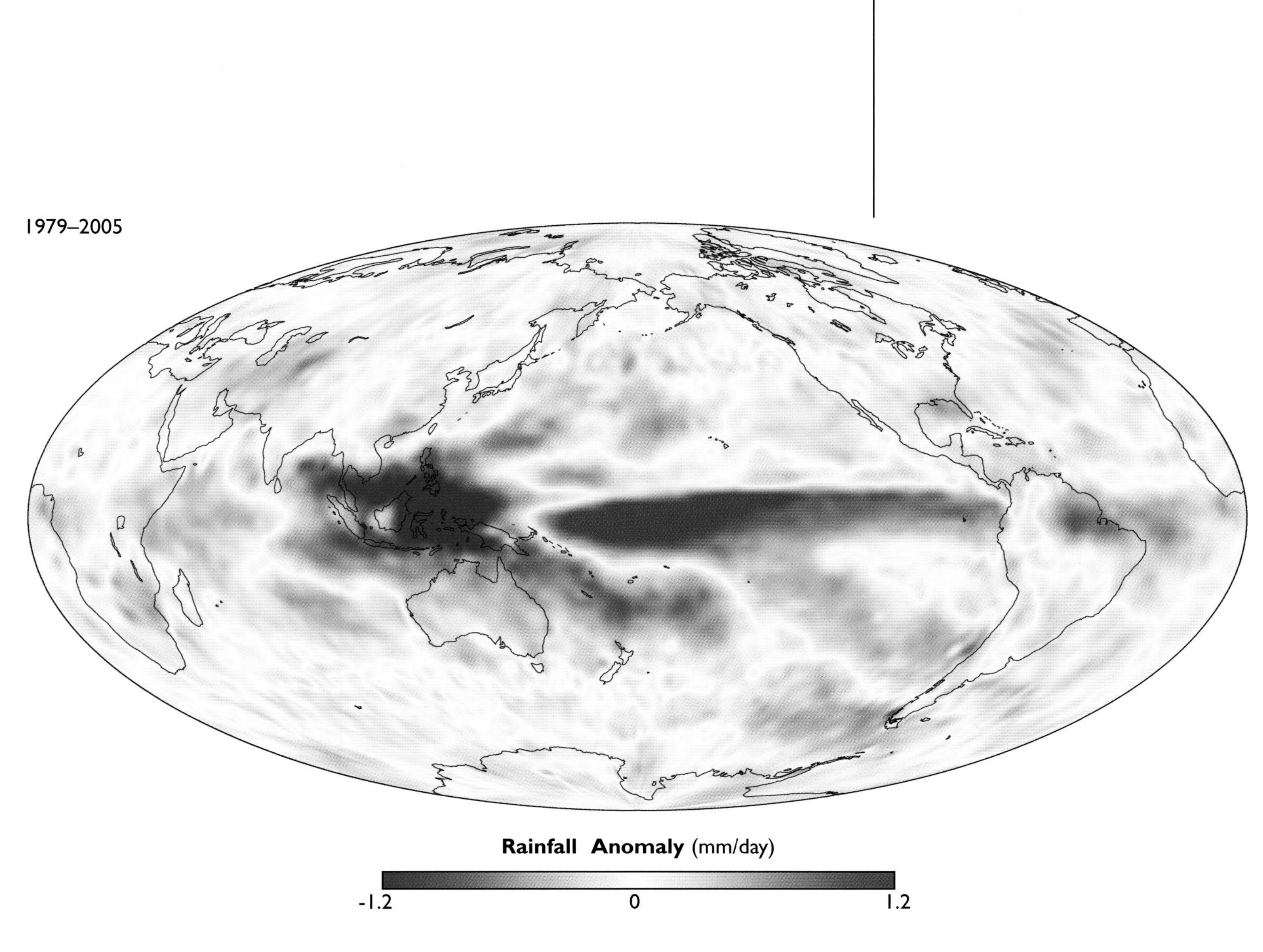


The differences between the average precipitation in El Niño years and the average precipitation in La Niña years between 1979 and 2005 demonstrates the global impact of these events, which are focused in the tropical Pacific Ocean. Red shading indicates much more precipitation during El Niño years, while blue shading indicates higher precipitation during La Niña years. A number of features can be seen to extend from the tropics to middle latitudes and even to polar regions, indicating the long-distance connections of climate variations. (Data from the Global Precipitation Climatology Project.)

During the past 25 years information from a number of satellites has been compiled to give a better understanding of how the precious commodity of rain is distributed across our planet. If a year's worth of global precipitation (rain and melted snow) were spread evenly over the globe, the average depth of water would be waist deep, about 1 m. However, precipitation is not evenly distributed across our planet, but has striking variations from place to place over land (hence the existence of forests and deserts) and also over oceans. Tropical jungles such as the Amazon basin in South America and the Congo basin in central Africa are dependent on heavy annual rain, as deep as a room (3 m). Deserts, such as the Sahara in northern Africa, have very little rainfall (ankle deep or less over a year). Over oceans there are also regions of very heavy average rainfall, but also 'ocean deserts,' areas devoid of rainfall despite the tremendous reservoir of water just below. Across the ocean a very narrow, east-to-west belt of heavy rain exists just north of the Equator and connects with the jungle rainfall in Indonesia, South America, and Africa. Just slightly farther away from the Equator in both directions are areas of much less rainfall, where the major deserts of the world are located. These rainfall features circling the globe are associated with rising air motion (heavy rainfall) near the Equator and descending air motion (light rainfall) slightly away from the

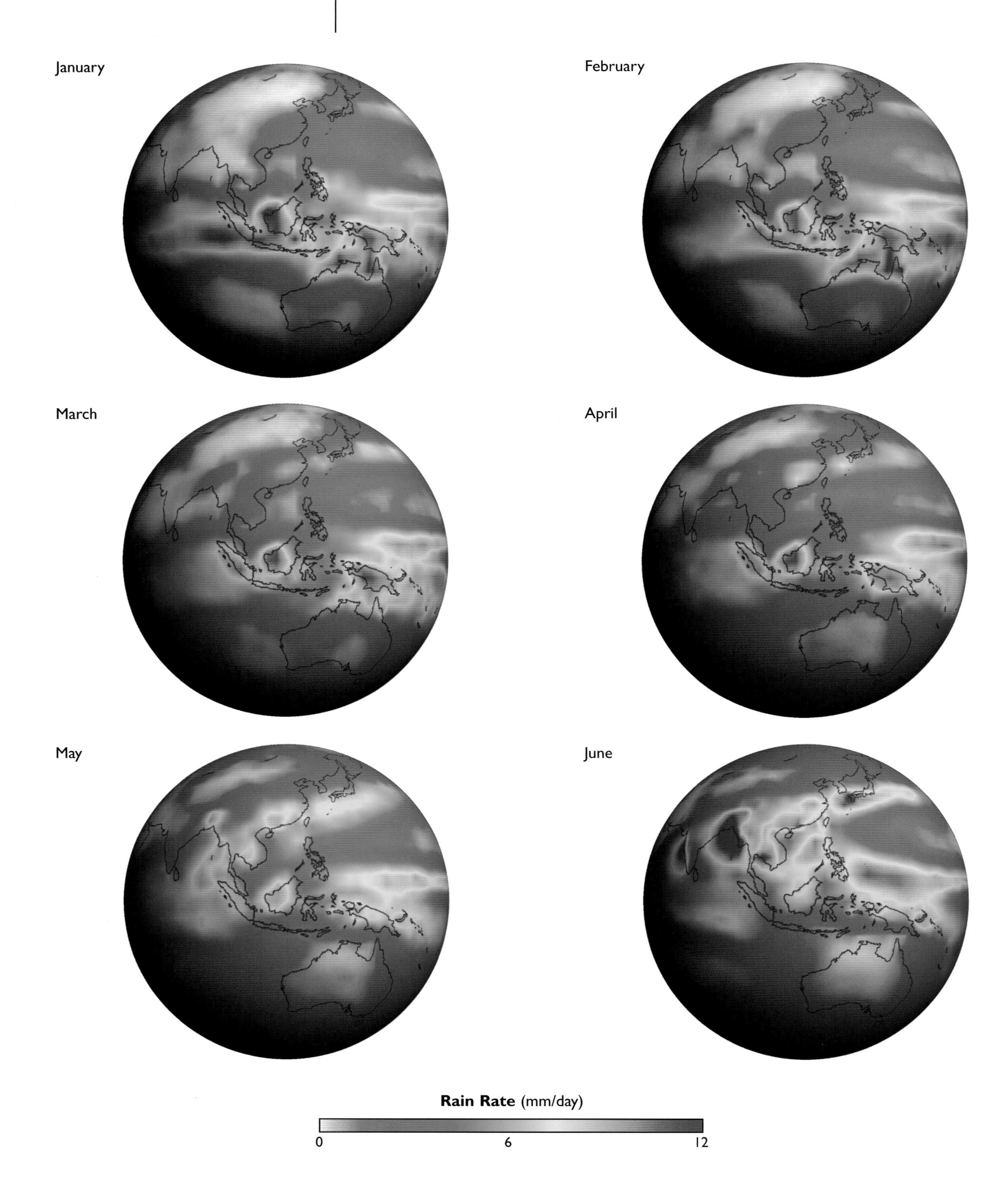
January
February
March
April
May
June
Rain Rate (mm/day)
0
6
12

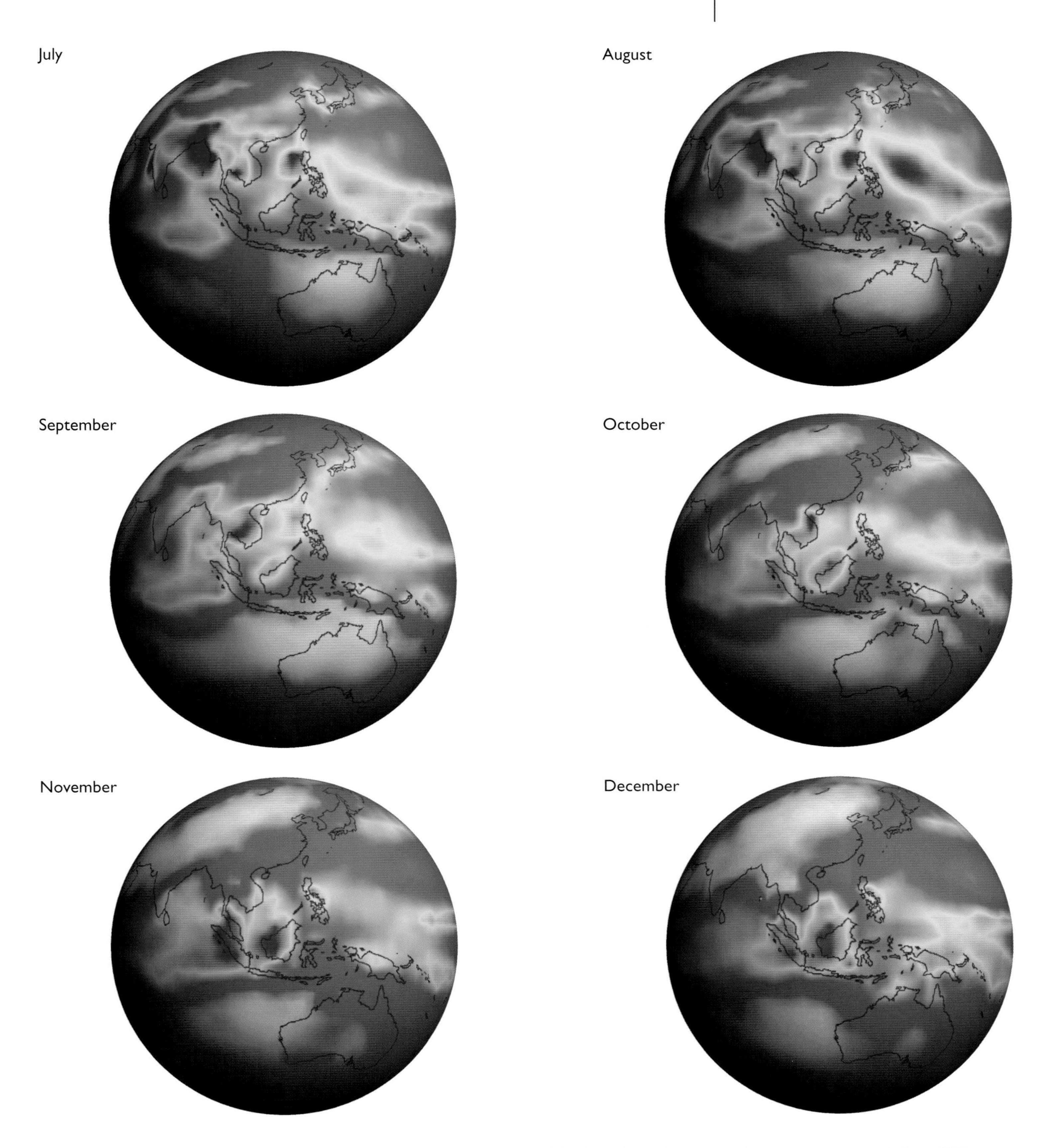


The Asian Monsoon is the planet's greatest seasonal shift in rainfall. Each monthly globe shows data averaged over a 26-year period, from 1979 to 2005. During winter in the Northern Hemisphere (January–March), the Sun is south of the Equator along with heavy rain, even reaching into northern Australia, while India and Indochina are dry. When the seasons reverse, South Asia is one of the wettest places on Earth and it is dry south of the Equator. (Data from the Global Precipitation Climatology Project.)

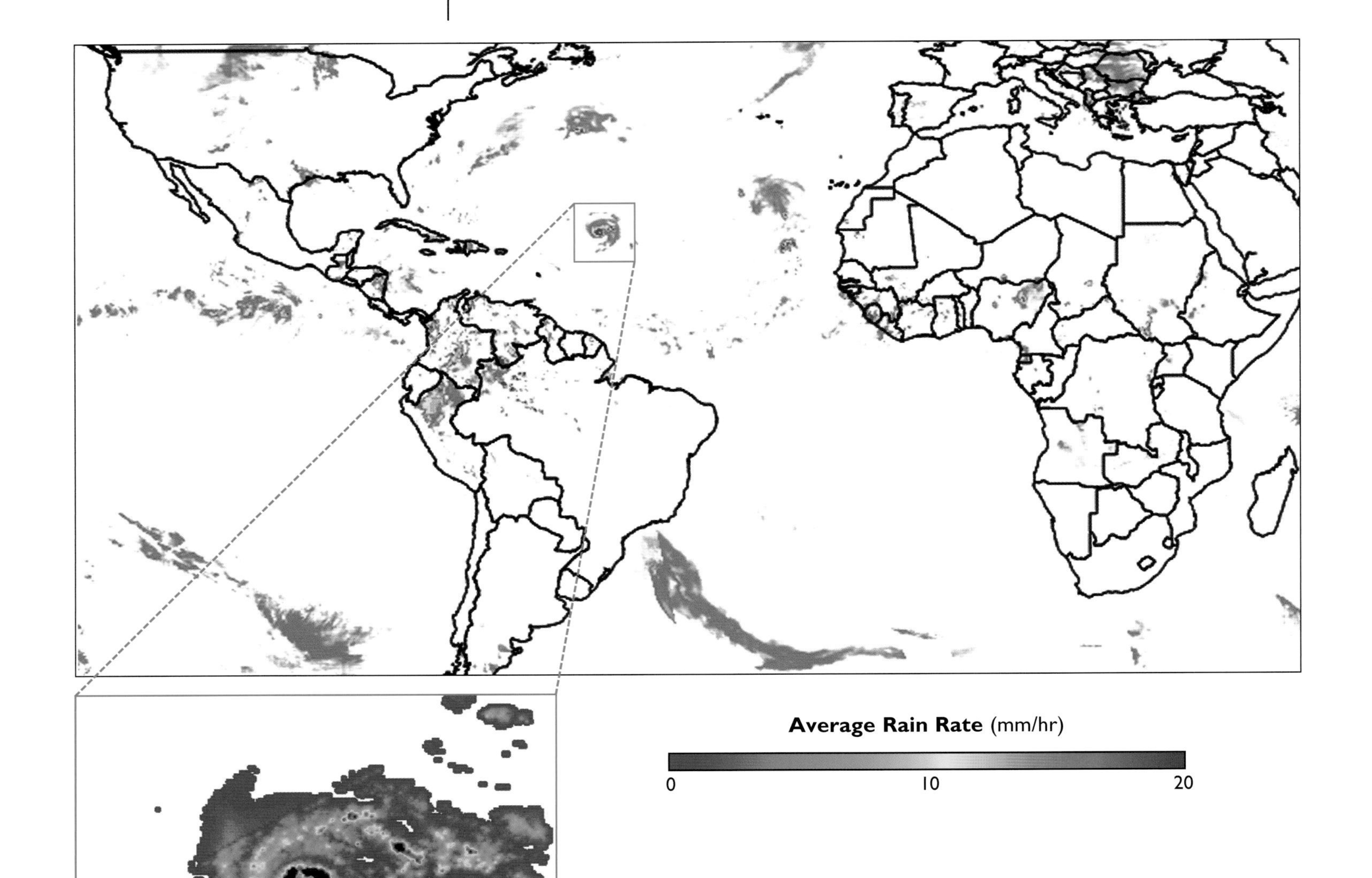

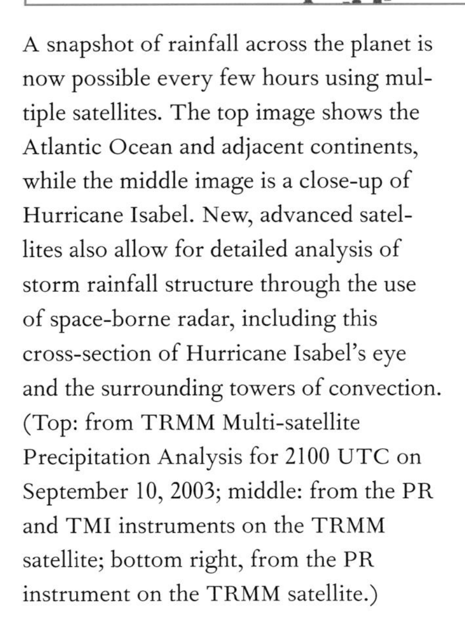



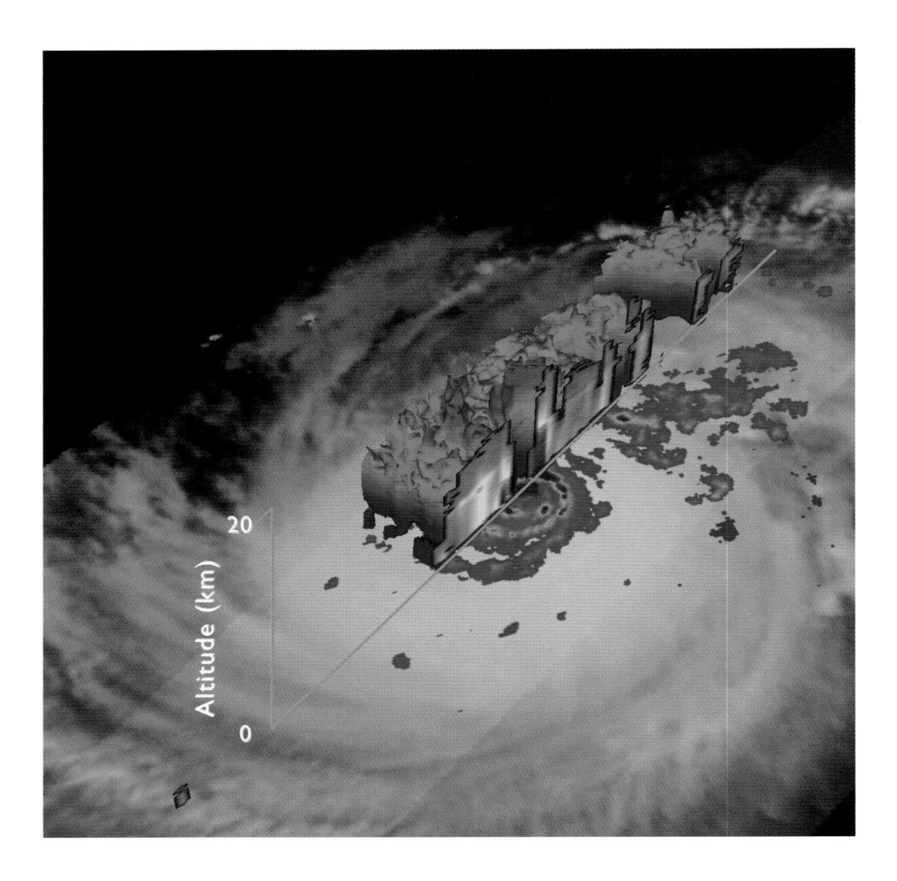



A snapshot of rainfall across the planet is now possible every few hours using multiple satellites. The top image shows the Atlantic Ocean and adjacent continents, while the middle image is a close-up of Hurricane Isabel. New, advanced satellites also allow for detailed analysis of storm rainfall structure through the use of space-borne radar, including this cross-section of Hurricane Isabel's eye and the surrounding towers of convection. (Top: from TRMM Multi-satellite Precipitation Analysis for 2100 UTC on September 10, 2003; middle: from the PR and TMI instruments on the TRMM satellite; bottom right, from the PR instrument on the TRMM satellite.)

Equator in the sub-tropics. At middle latitudes west-to-east moving storms leave a trail of rainfall. In the Northern Hemisphere, these rainfall swaths are broken up by dry areas over the continents with their associated mountains. In the Southern Hemisphere there is a more continuous ring of high-latitude rainfall around the globe, due to the lack of continents at these latitudes.

The satellite rainfall information has also been used to understand how precipitation varies from month to month and year to year at particular locations and how these variations are related to flood and drought. Rainfall varies in many locations during the year because of the changing position of the Sun as we move through the seasons. Especially over land, the rain follows the Sun because the increased solar heating of the surface evaporates more water and destabilizes the atmosphere, and therefore helps produce convective rain and thunderstorms. When these seasonal variations in rain are strong they are often referred to as Monsoons, as in India and the rest of South Asia. Another factor in rainfall variation year-to-year across the globe is the notorious El Niño phenomenon and its mirror-image counterpart La Niña. These complex ocean-atmosphere phenomena produce anomalously high rainfall in the central Pacific Ocean associated with warmer than normal ocean temperatures (El Niño), while the opposite occurs for La Niña. However, what happens in the Pacific Ocean along the Equator during an El Niño has repercussions across our planet. A pattern of above- and below-normal rain is created around the Equator. Narrow, poleward extensions of those features form great arcs across the world, propagating El Niño/La Niña impacts far from the source of the disturbance in the central Pacific Ocean.

New satellites and techniques are now allowing scientists to probe rainfall systems even further. Long used as ground-based instruments to measure rain, radars are now occasionally flown on research satellites. These precipitation radars are being used to make more accurate measurements of surface rain and to give a three-dimensional view of weather systems. This type of accurate information is also being used as a standard to calibrate rain information from other satellites. The combination of information from all the satellites permits nearly global analysis of precipitation every 3 hours, which is proving valuable for tracking weather systems such as hurricanes, detecting flood zones around the world, and monitoring the availability of surface moisture for crops.

Jeffrey B. Halverson

# Hurricanes: Connections with Climate Change

Complex rain structure of Hurricane Isabel on September 15, 2003. Outermost swath (white colors) indicate cloud structure while inner swath shows rainfall intensity. (Data from the VIRS, TMI, and PR instruments on the TRMM satellite.)

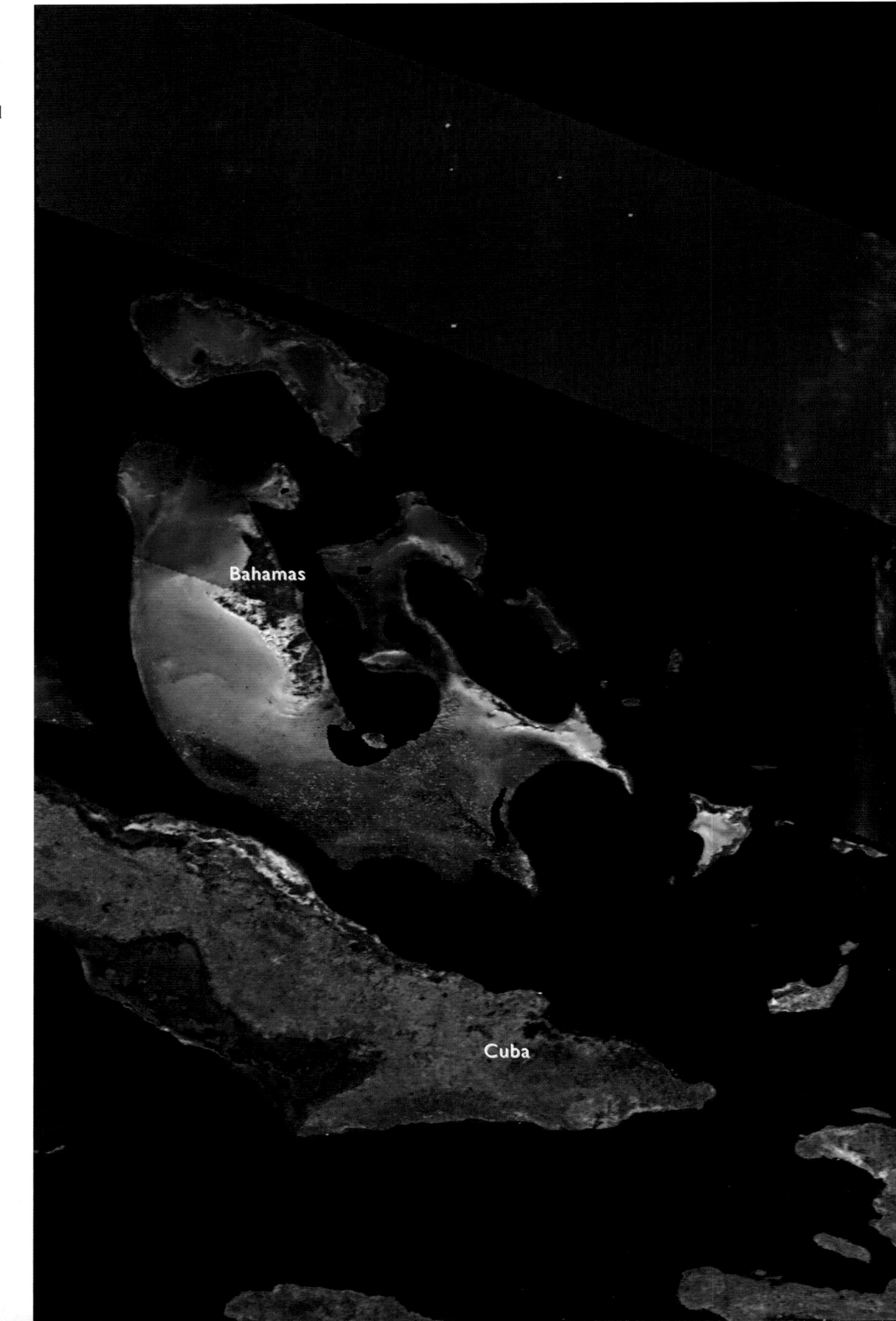


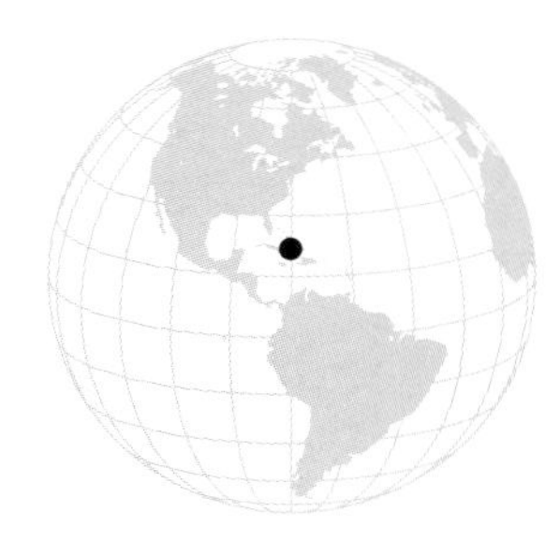

*Our Changing Planet*, ed. King, Parkinson, Partington and Williams
Published by Cambridge University Press 

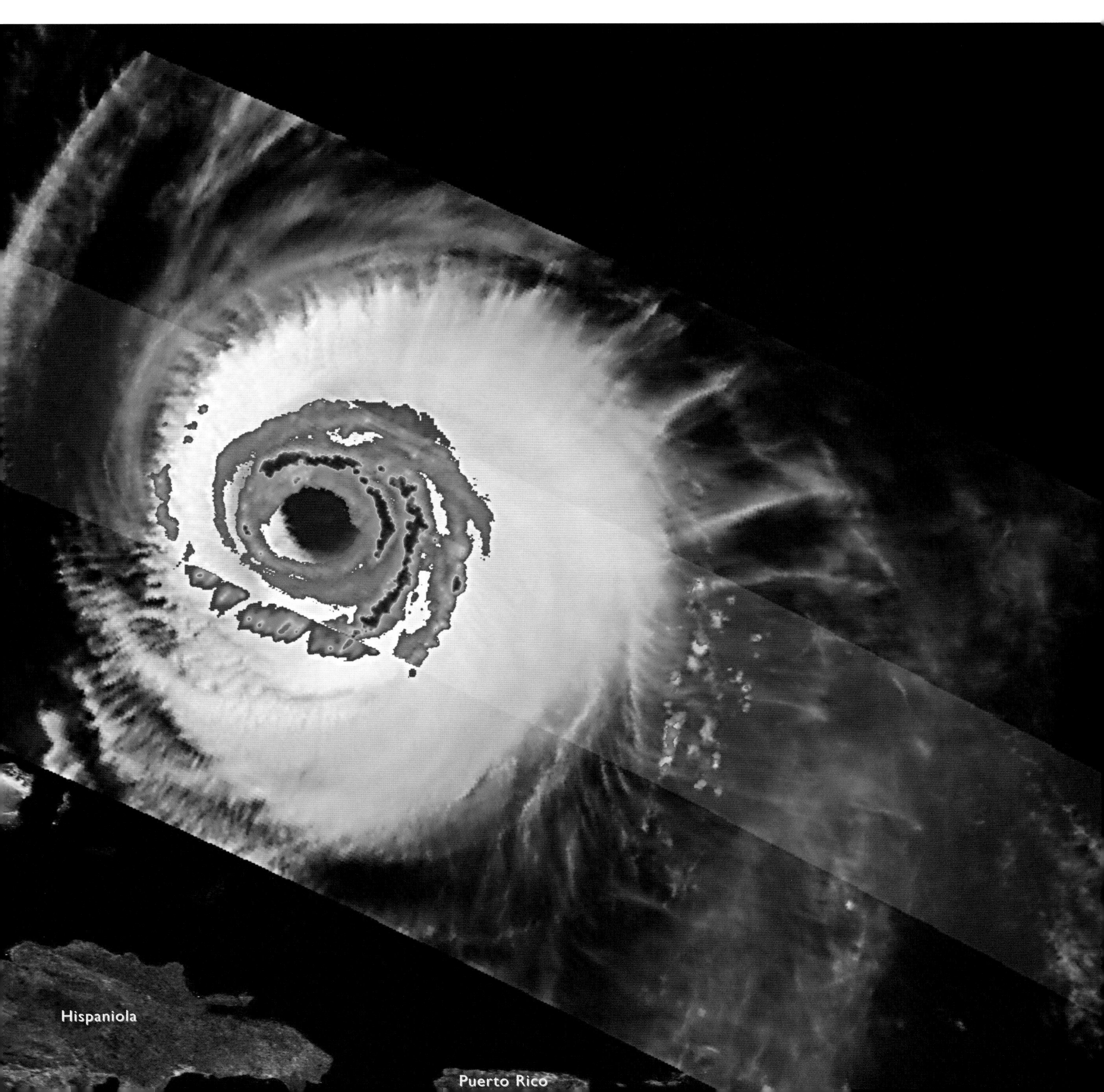
Rain Rate (mm/hr)
0
10
20
30
40
50
Hispaniola
Puerto Rico

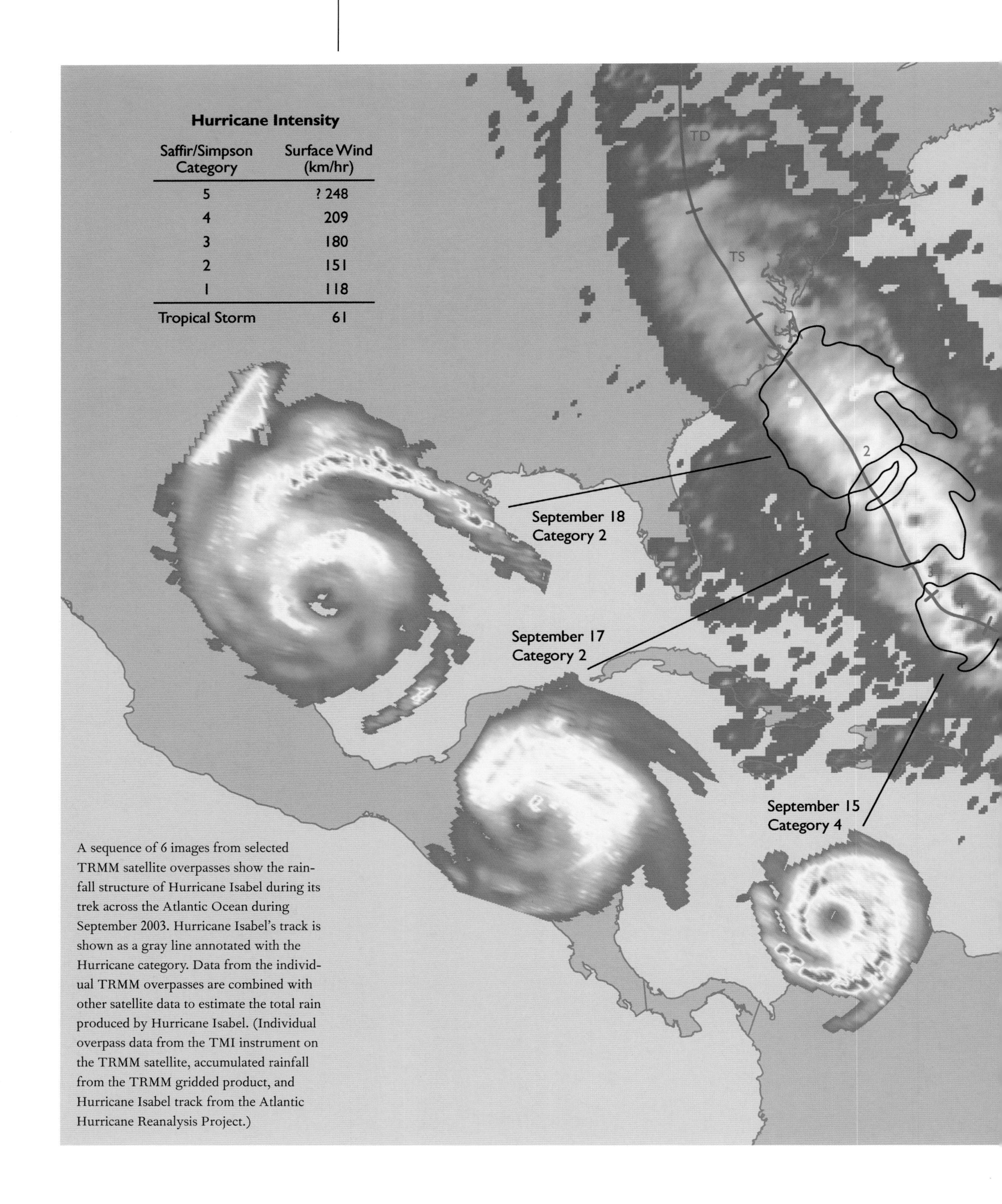

**Hurricane Intensity**

| Saffir/Simpson Category | Surface Wind (km/hr) |
|---|---|
| 5 | ? 248 |
| 4 | 209 |
| 3 | 180 |
| 2 | 151 |
| 1 | 118 |
| Tropical Storm | 61 |



A sequence of 6 images from selected TRMM satellite overpasses show the rainfall structure of Hurricane Isabel during its trek across the Atlantic Ocean during September 2003. Hurricane Isabel's track is shown as a gray line annotated with the Hurricane category. Data from the individual TRMM overpasses are combined with other satellite data to estimate the total rain produced by Hurricane Isabel. (Individual overpass data from the TMI instrument on the TRMM satellite, accumulated rainfall from the TRMM gridded product, and Hurricane Isabel track from the Atlantic Hurricane Reanalysis Project.)

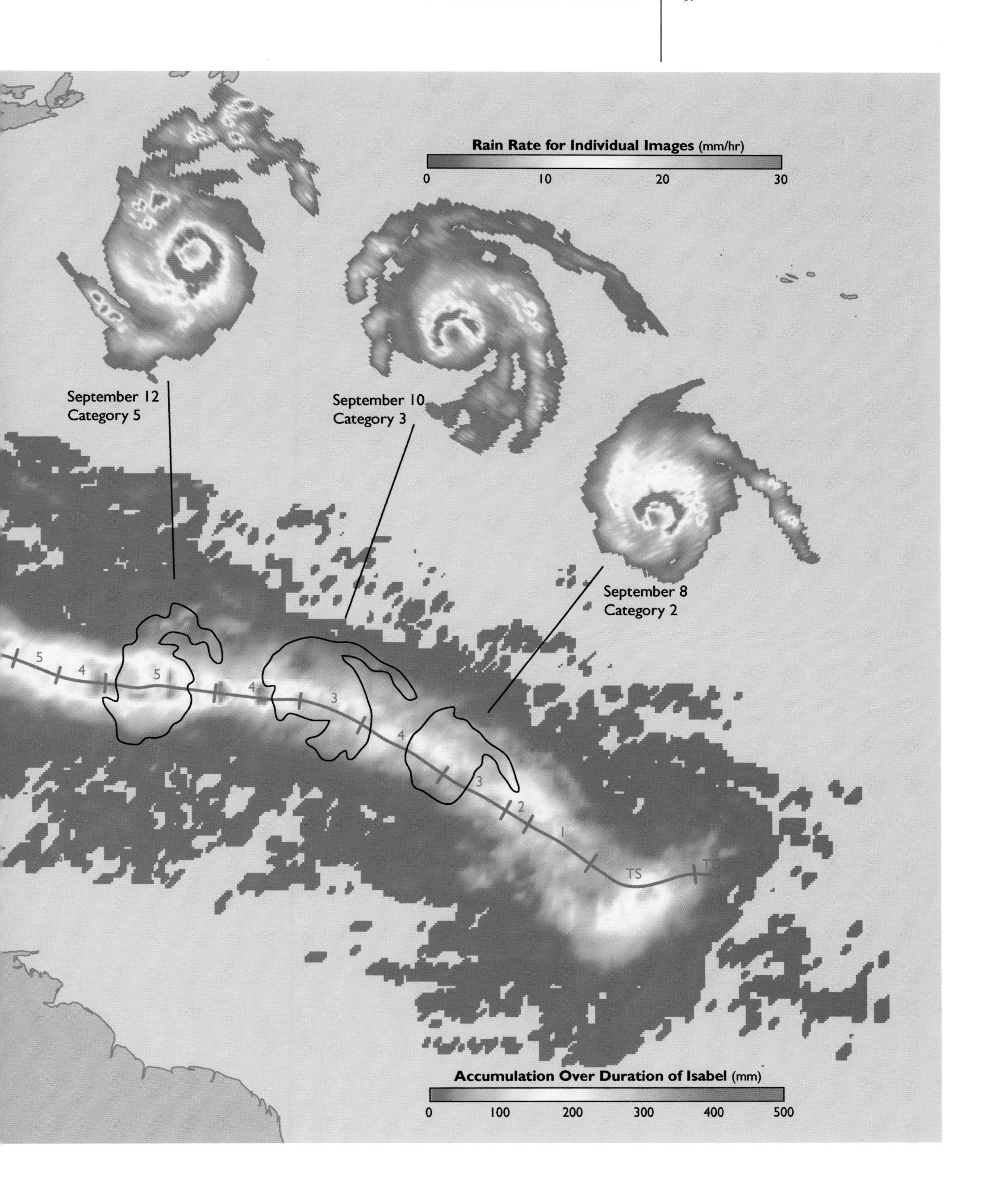
Rain Rate for Individual Images (mm/hr)
0
10
20
30
September 12
Category 5
September 10
Category 3
September 8
Category 2
5
4
5
4
3
4
3
2
1
TS
Accumulation Over Duration of Isabel (mm)
0
100
200
300
400
500

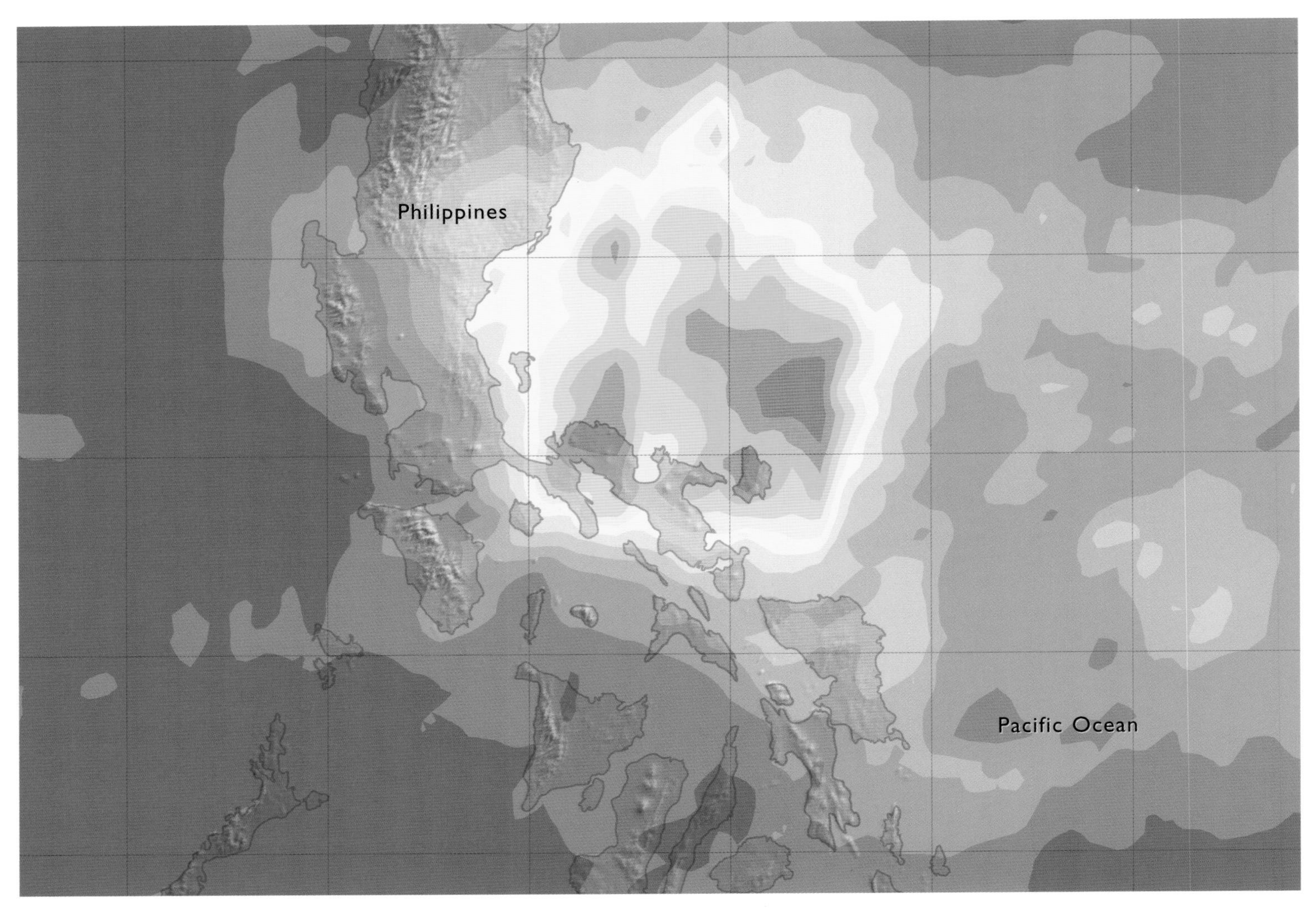


A multi-satellite precipitation analysis of heavy rains falling over the Philippines in a 2-week period (November 16 to December 3, 2004) during which a series of 4 tropical cyclones passed over the island nation. (Data from TRMM Multi-Satellite Precipitation Analysis.)

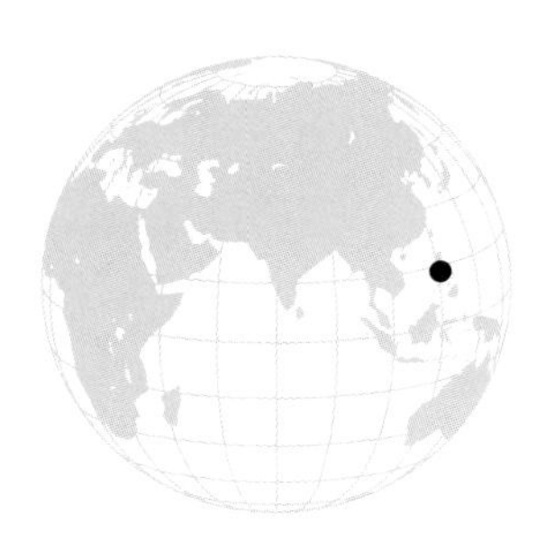

Hurricanes, otherwise known as tropical cyclones and typhoons, are the most extreme manifestation of tropical storm clouds. Hurricanes consist of thunderstorms organized in spiral or circular bands, surrounding a central low-pressure vortex of intense wind. Considering that 1000–2000 ordinary thunderstorm cells are active at any instant around the globe, hurricanes are quite rare, as only 80 to 90 of them develop worldwide in any given year. While hurricanes are quite infamous for their powerful, gusty winds, storm surge, and inland flooding, they play a beneficial role in many parts of the world by supplying much-needed rainfall—up to 30–40% of the annual rainfall in some portions of the lower latitudes.

In the Atlantic Ocean, an average of about 10 named storms is expected to develop each year. However, of all the tropical ocean basins that give rise to hurricanes, the Atlantic is considered marginal for hurricane formation. The ocean undergoes substantial variation in the number of storms experienced annually. Part of the year-to-year variation is linked to the El Niño Southern Oscillation (ENSO), a vast coupling between tropical oceans and the atmosphere. In El Niño conditions, atmospheric pressure, winds, ocean temperature, and rainfall undergo marked changes on an irregular cycle that spans 2–5 years. Major shifts in the tropical atmospheric circulation cause winds to blow more strongly across the near-equatorial

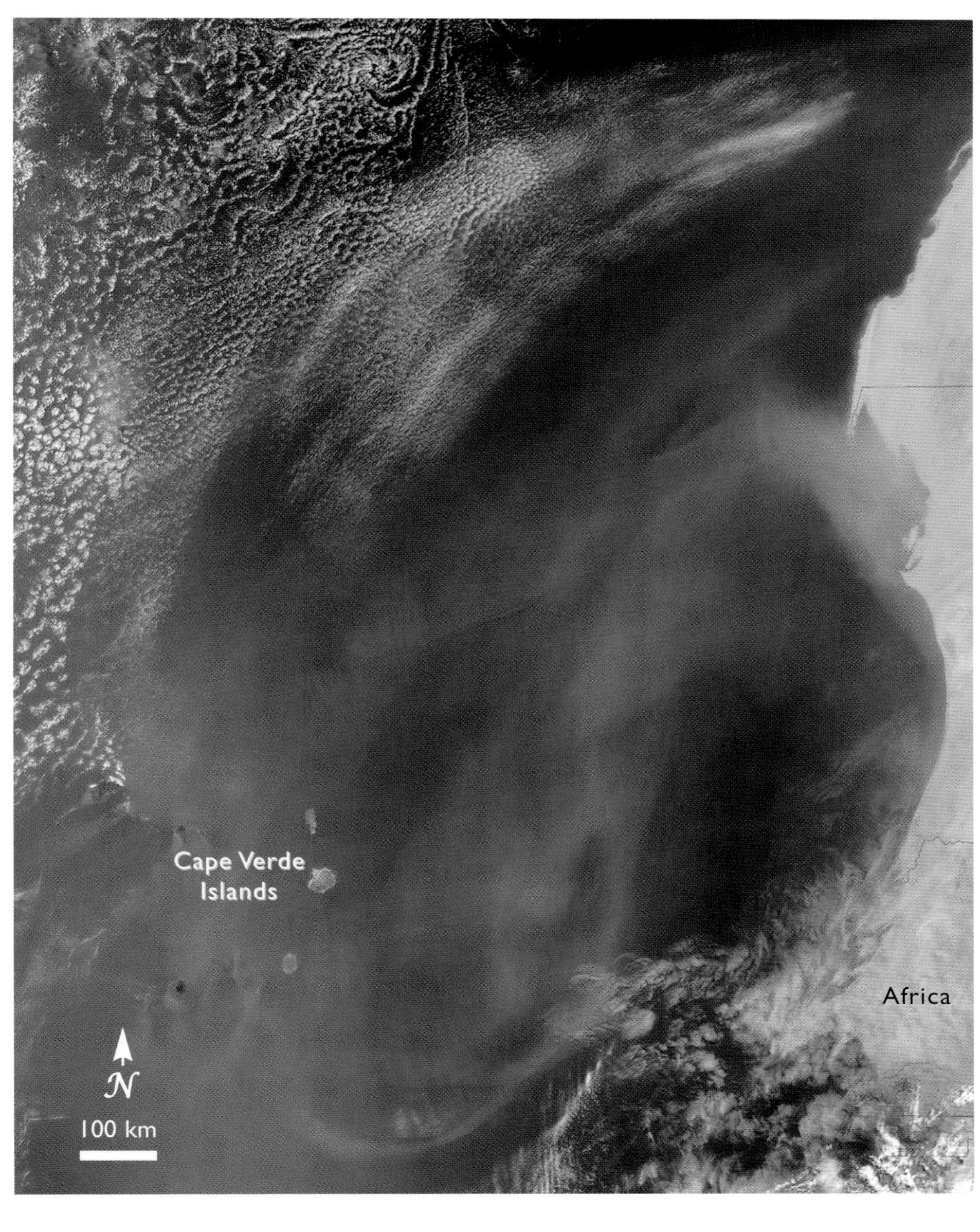


Giant plume of Saharan dust on May 19, 2005, which eventually traveled 1,600 km into the Atlantic Ocean. (Data from the MODIS instrument on the Aqua satellite.)

Atlantic Ocean. These strong currents disrupt the early, fragile cores of the hurricane vortex, and so limit the number and intensity of Atlantic hurricanes.

Recent research has uncovered a unique aspect of the Atlantic basin that influences hurricane development—enormous clouds of dust lofted into the atmosphere off the African continent. Large plumes of Saharan dust occasionally interact with hurricane 'seedlings' or waves in the easterly trades that stream off Africa. When these disturbances are embedded within dust clouds, or dust tendrils are pulled into the vortex, hurricane formation is suppressed. Storms may be weakened by some combination of processes, including unusually dry air associated with the dust, the strong winds aloft that carry the dust, or the influence of the dust particles themselves on microscopic water droplets. It remains to be shown by the emerging discipline of 'tempestology' whether prolonged periods of African drought—which would favor an abundance of dust crossing the Atlantic—are associated with reductions in hurricanes impinging on the United States.

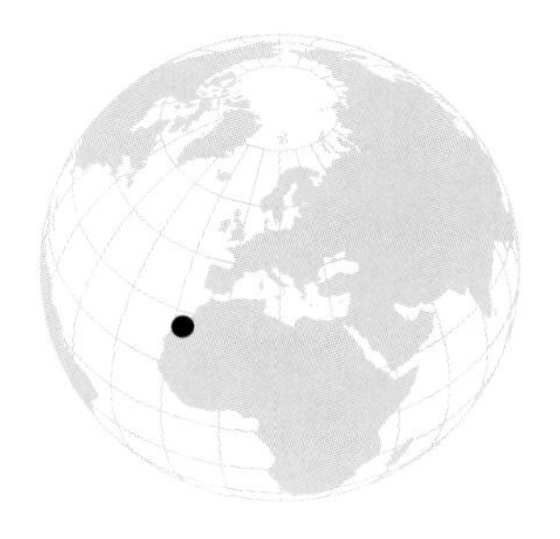

Against this backdrop of variable factors, there is an underlying rhythm governing Atlantic hurricane intensity and frequency that occurs over decades. When one examines the

On September 25, 1998, for the first time since 1893, there were 4 hurricanes in the Atlantic. This brightness temperature image from the 6.7 μm band of the GOES-8 satellite shows the water vapor content of the Earth's middle atmosphere. The color scale ranges from red, indicating cold temperatures caused by high clouds or moist air, to blue, indicating warm temperatures or dry air. (Data from the GOES Imager on the GOES-8 satellite.)

historical record of Atlantic hurricane activity, there are distinct cycles of heightened occurrence that last 20–30 years. These alternate with more quiescent periods of similar duration. For instance, the 1940s through 1960s saw a period of intense hurricane activity, with many damaging storms impacting the United States. Conditions slowed during the 1970s and 1980s. Starting in 1995, with a near-record 19 named storms, it appears that the Atlantic has entered a resurgence of storm frequency. The cycles of storm frequency appear to be related to slowly varying patterns of ocean surface temperature, as hurricanes are fueled by warm ocean water and the moisture evaporated from it. The Atlantic cycles are also phased with periods of alternating drought and abundant rainfall across the African Sahel. Years of rainfall that are below average imply that weaker seedlings or precursor disturbances are emerging from Africa, while years of abundant rain signal more vigorous waves. The more powerful seedlings are likely to survive the turbulent growth process across the Atlantic and thus evolve into mature hurricanes.

The recent acceleration in Atlantic hurricane activity points to an ominous trend. The upswing is coupled with an exponential increase in beachfront population, real estate and urbanization along the eastern seaboard of the United States. Not only are more storms predicted to develop and potentially strike the United States in coming years, but rapidly growing segments of society have placed themselves in harm's way.

There are possibly larger links between hurricanes and climate. Because of the Earth's spherical shape, there is an excess of solar energy received in the deep tropics, and a deficit at the poles. The fluid systems of the planet (atmosphere and oceans) redistribute the energy imbalance by exporting heat out of the tropics. The oceans accomplish roughly half of the poleward transport, partly in the form of a giant overturning termed the thermohaline circulation. Warm, tropical water heated by the Sun flows northward, where it cools and sinks at the poles, returning south as a deep water current. Tantalizing new research suggests that Atlantic hurricanes may in part maintain the poleward flow of warm water. The thermohaline circulation helps regulate the Earth's climate, and the historical record suggests that the current operates in an unsteady manner. Sudden changes in the 'speed' of the thermohaline circulation may hasten abrupt shifts in climate. It is hypothesized, for instance, that an increase in Atlantic tropical cyclones will accelerate the thermohaline circulation, effectively cooling the tropics and warming the higher latitudes.

Perhaps the final question to ask is this: "What effect will global warming have on hurricanes?" There is ample evidence to suggest that the atmosphere has slowly warmed by a small amount over the past 30 or so years. Several recent scientific papers of note have begun to identify possible connections between hurricane activity and the global increase in oceanic temperature. If the oceans warm, one would predict that hurricanes will increase in intensity, frequency and coverage. This follows from the fact that warm surface water provides the basic flux of energy into the storm. However, warmer oceans are only part of the story. The tropical atmosphere will respond in different ways to greenhouse-enhanced warming—impacting the strength of tropical winds, the seedlings that give rise to new hurricanes, the thermodynamic propensity of the atmosphere to sustain thunderstorms, rates of evaporation off the oceans, even perhaps El Niño. Several of these factors might mitigate any tendency of the oceans to increase levels of hurricane activity. Regardless of what the prediction models currently state, changes in Atlantic (and global) hurricane activity arising from natural vagaries in Earth's complex climate system are substantial. These variations will likely outweigh much slower and subtle trends that may arise from global warming.

Steven J. Goodman
Dennis E. Buechler
Eugene W. McCaul, Jr.

# Lightning

Lightning flashes over Guntersville Lake, Alabama. (Photograph by Bob Blankenship.)

Mankind has long regarded lightning with both fear and awe. Early cultures attributed its occurrence to supernatural origins. The Greeks both marveled at and feared lightning bolts hurled by Zeus. For the Vikings, lightning was the result of Thor's hammer striking an anvil while he rode his chariot across the clouds. Early statues of Buddha show him carrying a thunderbolt with arrows at each end. Native Americans believed that lightning was caused by the flashing feathers of a mystical bird whose flapping wings produced the sound of thunder.

We now know that lightning is a visually spectacular electrical discharge that occurs within thunderstorms or between thunderstorms and the Earth's surface. Monitoring the lightning flash rate of thunderstorms aids forecasters in predicting and detecting severe storms, since lightning rates are often observed to increase prior to surface severe weather events, such as damaging winds, tornadoes, and large hail. Thunderstorms are

*Our Changing Planet*, ed. King, Parkinson, Partington and Williams
Published by Cambridge University Press 

Lightning from a storm system extending from Argentina to southern Brazil on the evening of April 23, 2003. (Photograph from the International Space Station, NASA Image Exchange, image number ISS006-E-48196.)

important contributors to the Earth's global circulation and also influence regional atmospheric circulations.

Satellites are well placed to monitor global lightning activity. While lightning discharges are spectacular when viewed from Earth's surface, they are equally impressive viewed from orbiting spacecraft. Indeed, astronauts often comment that one of the most noticeable features on Earth is the lightning activity.

In 1995, a satellite carrying a NASA-developed lightning instrument called the Optical Transient Detector (OTD) was placed into low Earth orbit 750 km above the Earth. The OTD was the first to provide a detailed storm-scale view of lightning activity during both day and night on a near-global basis during its 5-year mission life. Launched in 1997 and still collecting data in 2007, an improved lightning detector, the Lightning Imaging Sensor (LIS), was included on the NASA Tropical Rainfall Measuring Mission (TRMM) satellite observatory along with a suite of instruments, including the first spaceborne weather radar. From its lower orbital altitude of 350 km, later boosted to 400 km in 2001 to extend the mission lifetime, the LIS provided even higher resolution images of lightning than its predecessor. Data from these instruments have been used to create detailed global maps of lightning activity as well as to determine that the annually averaged global flash rate on Earth is 44 flashes per second.

Perhaps the most striking feature of the global lightning maps is that lightning occurs primarily over land areas, with about 90% over land and only 10% over the world's oceans. This is due to the differing characteristics of thunderstorms and their environment that occur over the two surfaces, with thunderstorms over land being more conducive to lightning development.

Coastal areas (for example, the southeastern United States coast bordering the Gulf of Mexico and Florida, Panama, Cuba, Bangladesh, the Indonesian Archipelago, and Italy) have high lightning activity owing to the presence of ample moisture and sea breezes that help lift the moist air to initiate thunderstorm development. The east-west band of mountains extending from Europe across Asia limits the poleward flow of tropical moisture, resulting

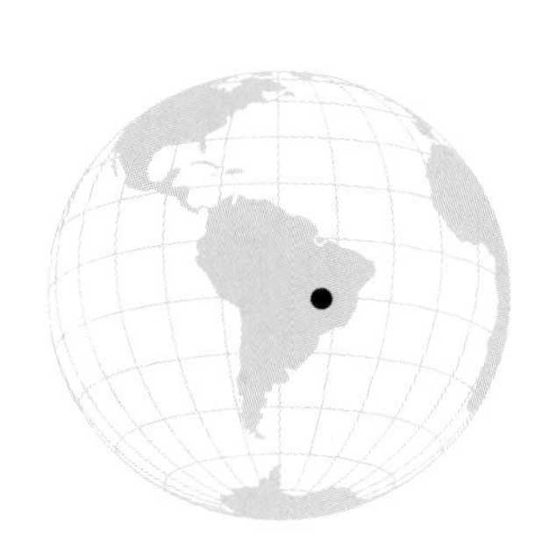

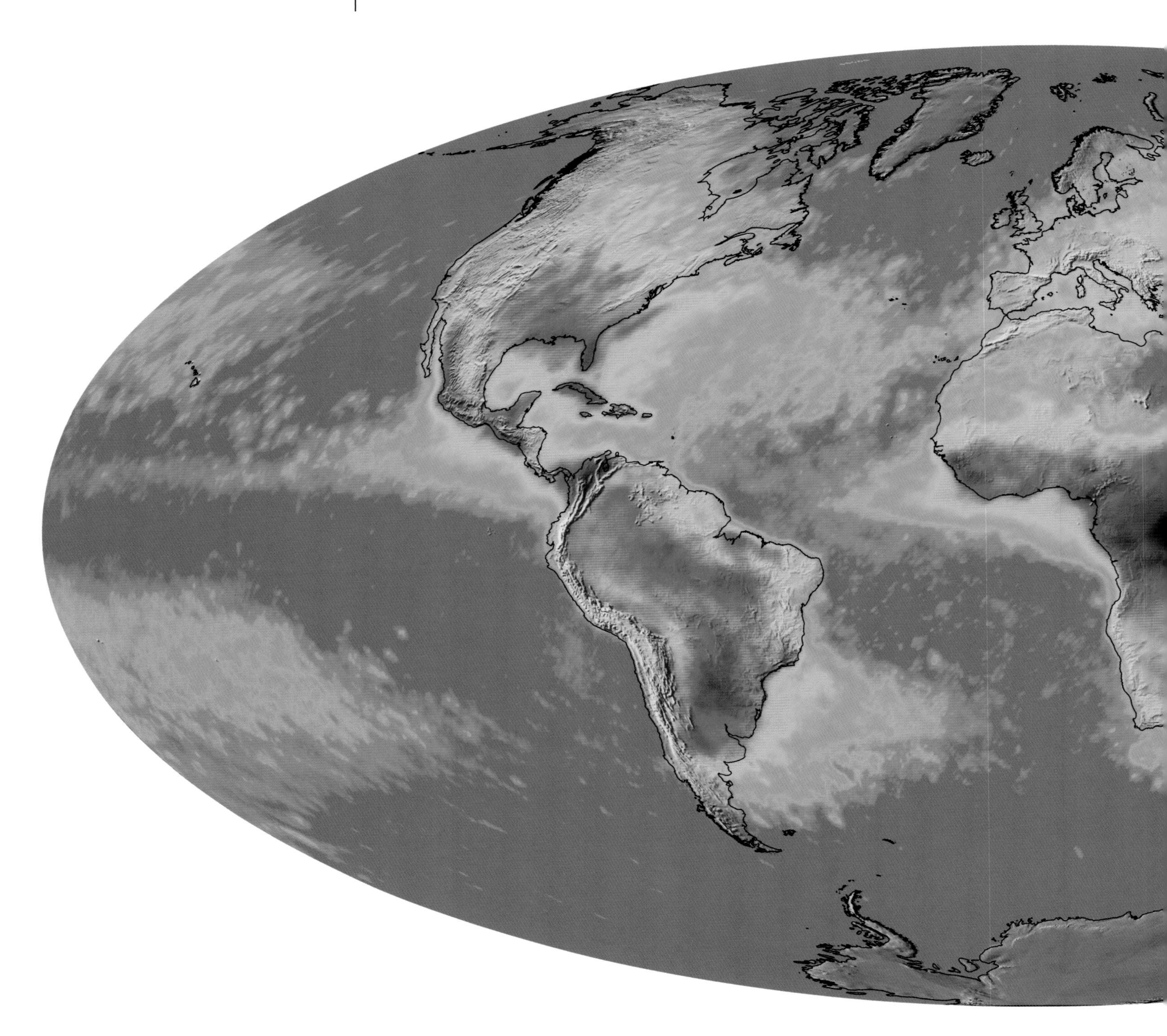

Mean annual global lightning flash rate (flashes per km$^2$ per year) derived from a combined 8 years from April 1995 to February 2003. (Data from the OTD instrument on the OrbView-1 satellite and the LIS instrument on the TRMM satellite.)

in a sharp drop in lightning activity north of the mountains. This is especially noticeable in Northern Pakistan, where the Himalayas block the poleward flow of moist tropical air. Other areas of enhanced lightning activity include regions frequented by migrating weather systems (e.g., mid-latitude cyclones) or widespread convergence (e.g., the Intertropical Convergence Zone).

Seasonally, lightning activity is more prevalent in the summer months (June, July, August for the Northern Hemisphere, December, January, February for the Southern Hemisphere)

than during the winter. Lightning over the oceans tends to occur primarily in the winter months, such as over the Mediterranean Sea and Gulf of Mexico during December, January, February, where thunderstorms develop as cold winter air flowing over the warmer ocean surface produces instability. Lightning activity occurs in all seasons in tropical regions such as central Africa.

The equatorial Congo Basin is the 'hot spot' of the planet, with a maximum annual lightning flash rate of 158 flashes per $km^2$ per year (latitude 2.75°S, longitude 27.75°E) in

December, January, February

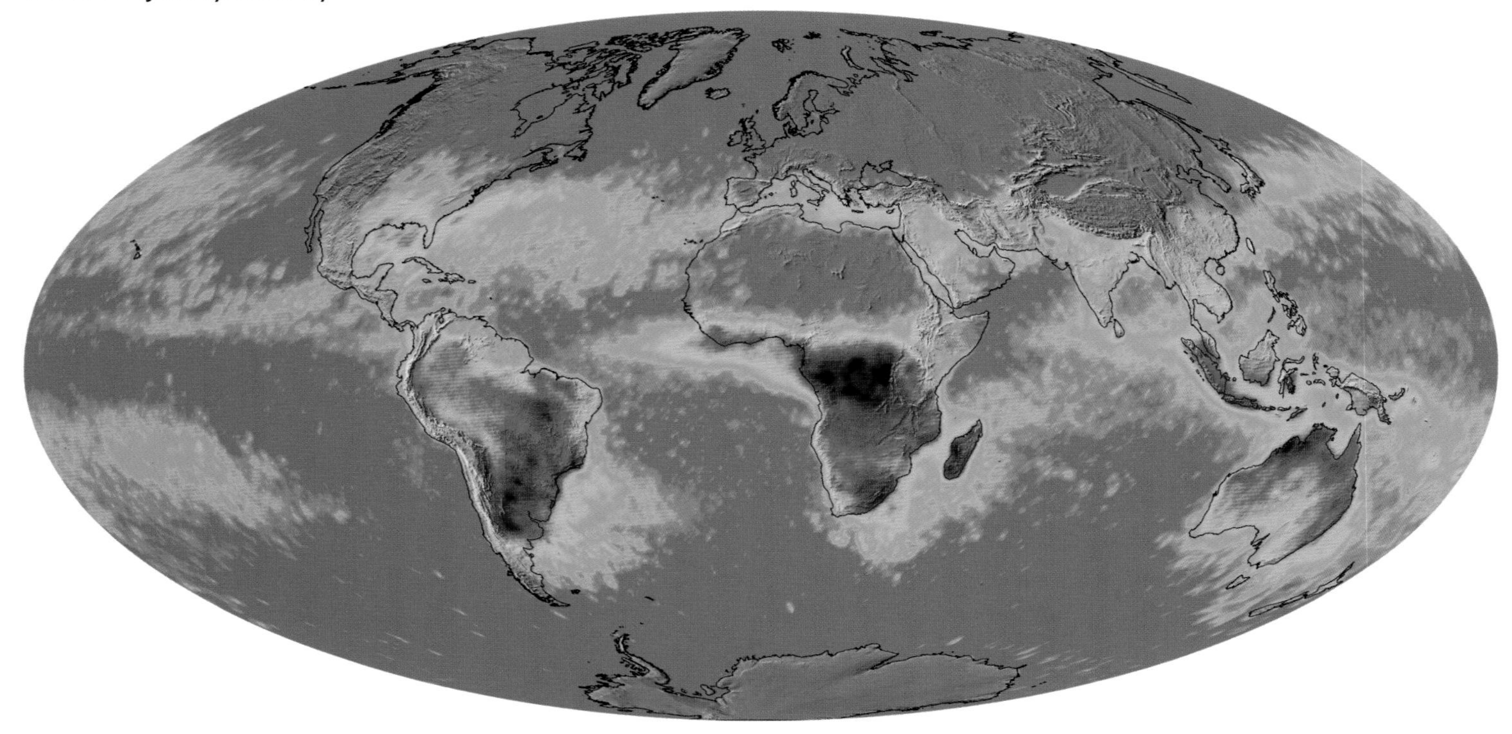

March, April, May

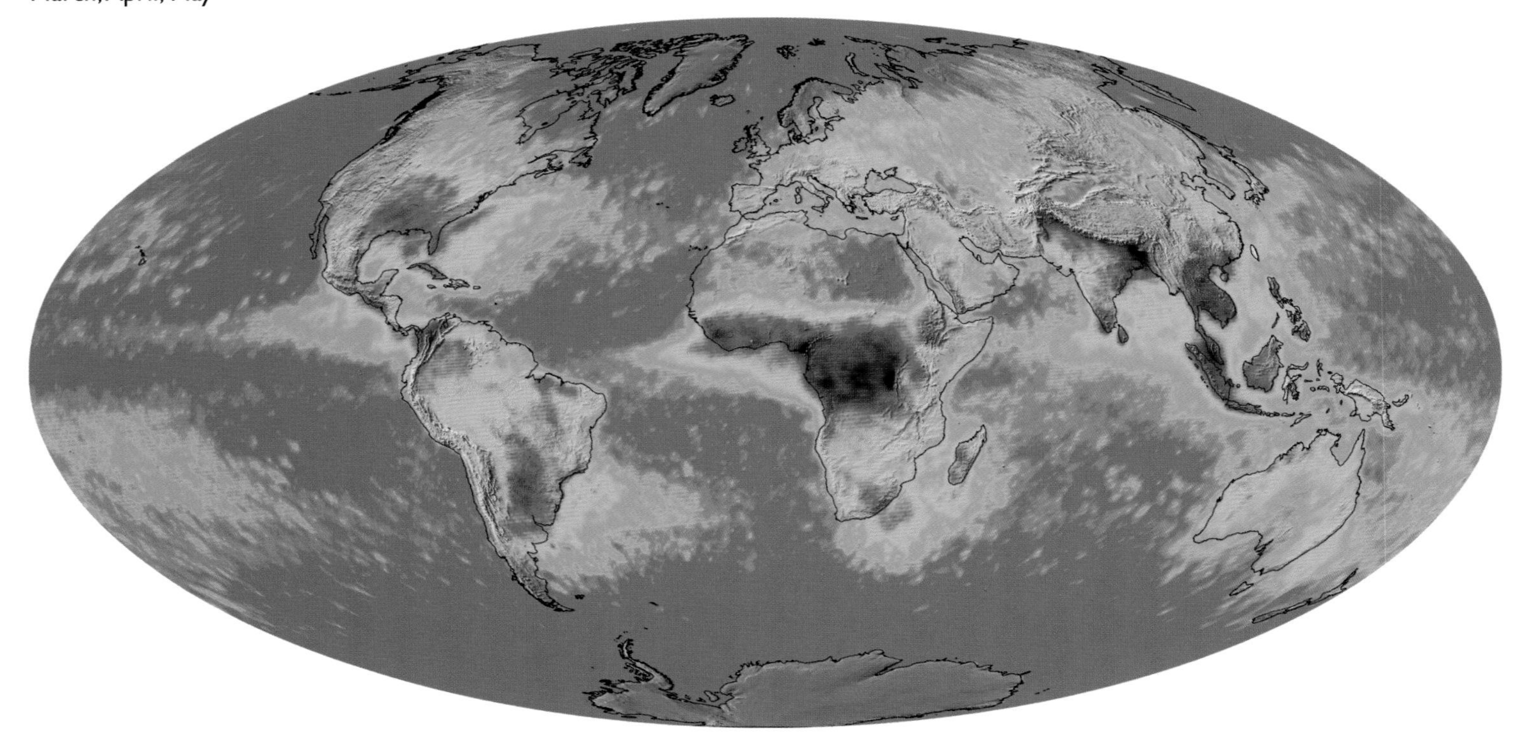

**Annualized Lightning Flash Rate** (flashes per km$^2$/yr)

0.1 | 1 | 10 | 80

June, July, August

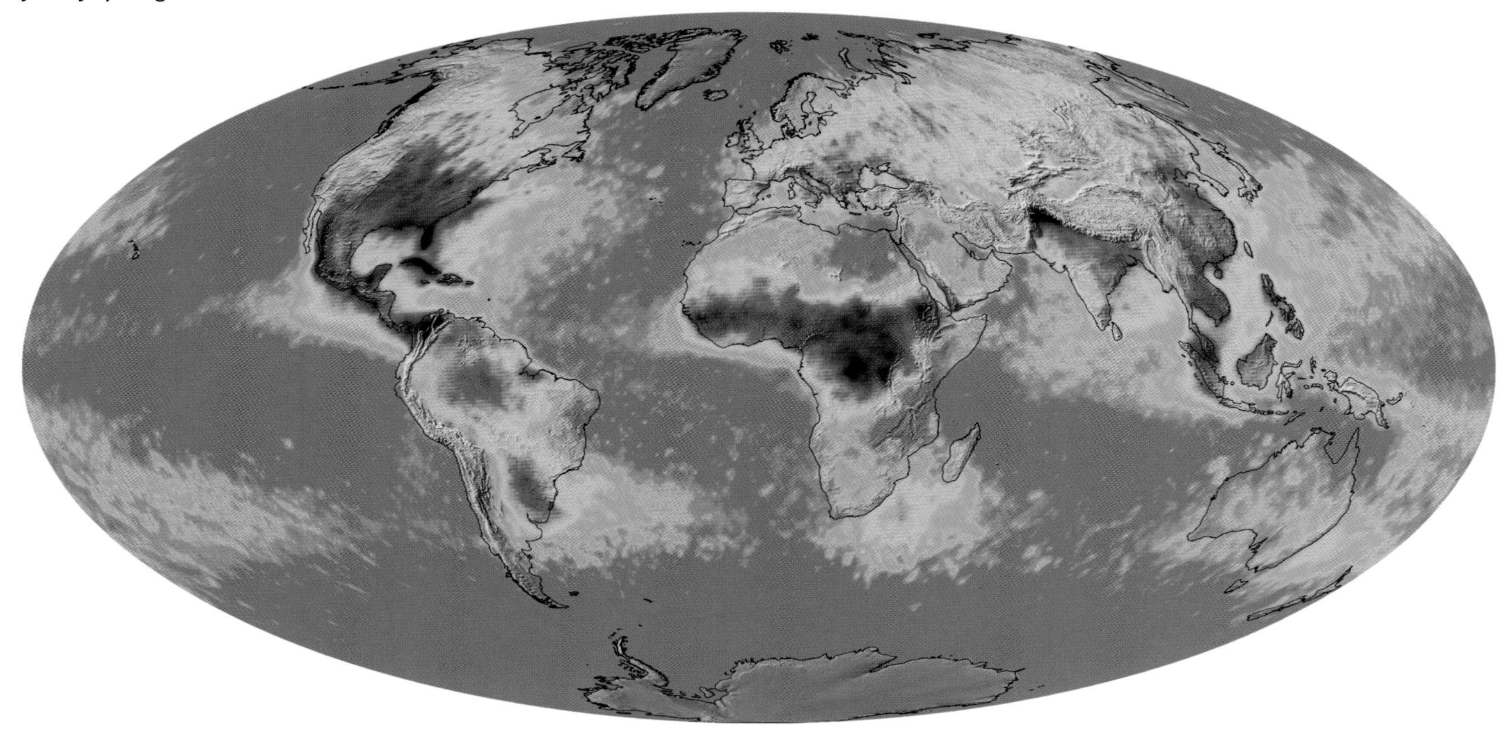

September, October, November

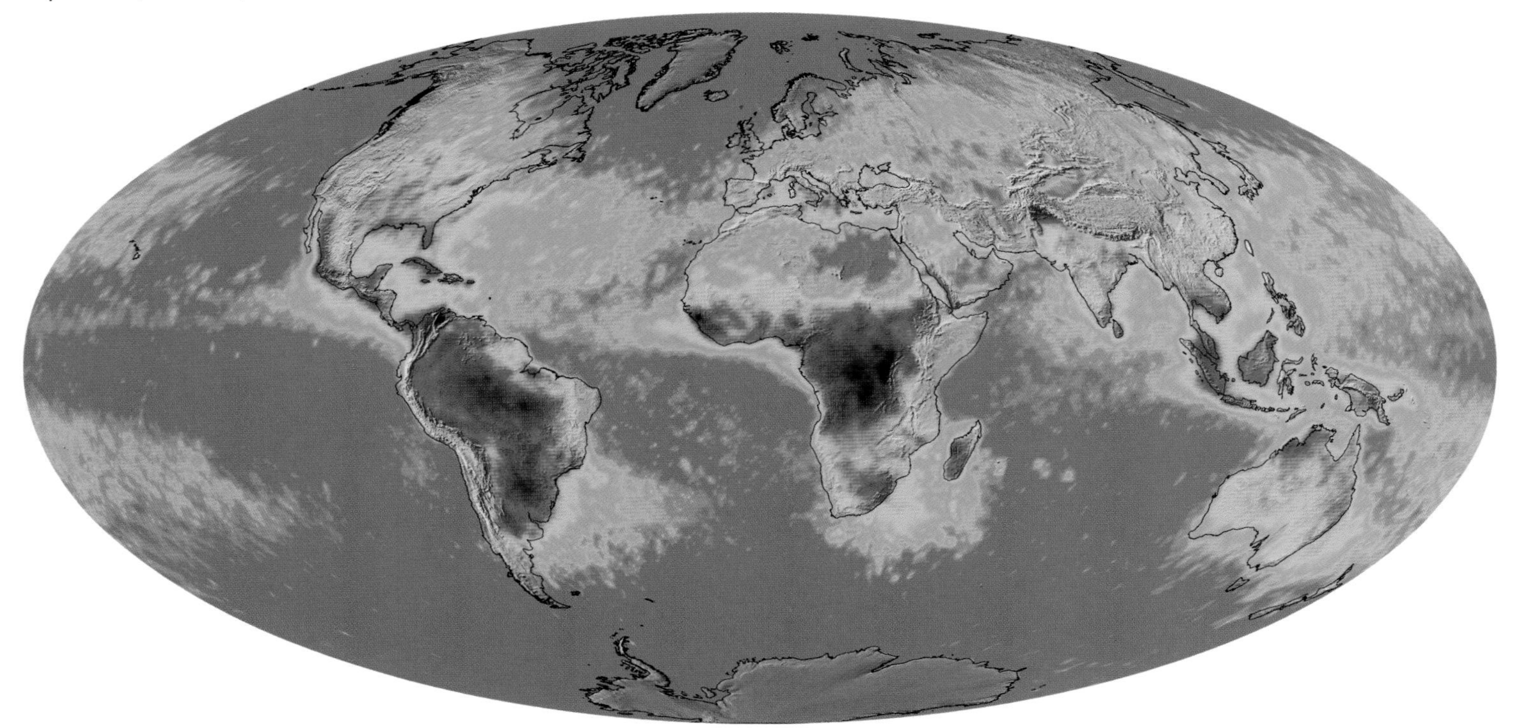

mountainous terrain near Kifuka in the Democratic Republic of the Congo. Lightning occurring in all seasons contributes to this peak in Central Africa. The weather station at Kamembe, Rwanda, approximately 125 km east of Kifuka, reports thunder on 221 days per year, on average. In North America, thunderstorms and lightning are most prevalent over Florida (59 flashes per $km^2$ per year), with most of this activity occurring during the summer and in the afternoon. Similarly, the annual peak in Northern Australia (53 flashes

Equivalent annualized lightning flash rate for each season (flashes per $km^2$ per year). Note that these rates are scaled to a full year for comparison with the annual map. Divide values by 4 to arrive at the actual seasonal rate. (Data from the LIS instrument on the TRMM satellite, and the OTD instrument on the OrbView-1 satellite.)

per km$^2$ per year) is due primarily to summertime thunderstorms. The peak annual flash rate in South America occurs in Colombia with a value of 110 flashes per km$^2$ per year. The peak annual lightning flash rate in Asia occurs over Northern Pakistan (87 flashes per km$^2$ per year), most of which occurs during the spring, summer and autumn months. In Europe, an annual maximum of 28 flashes per km$^2$ per year occurs over Northern Italy.

Lightning is indeed prevalent and dangerous. Lightning that strikes the Earth's surface can be harmful to people and property, sometimes causing injury or death. Lightning can also start fires that have destroyed homes or spread over thousands of acres of forest or prairie. Sometimes, lightning will strike an aircraft in flight, occasionally resulting in an aircraft crash. Approximately 80 deaths each year in the United States alone are attributed to lightning strikes, which is comparable to the number killed by tornadoes. Worldwide, the annual lightning death toll may be more than 24,000. Because lightning generally kills only about 10–30% of its victims, the global annual casualty rate from lightning strikes has been estimated to be as large as 240,000.

Lightning casualty rates depend on both the population density and lightning flash density, as well as on the degree of urbanization, the quality of housing and other structures in which people seek shelter during storms, their education and level of weather awareness, and on the amount of outdoor activity they pursue. As the world population increases, especially in lightning-prone areas, it is likely that lightning-related casualty rates will also increase, unless suitable educational and other preventive measures are taken.

More than a century ago, when the United States was still predominantly rural in character, lightning casualty rates were approximately 6 per million people. By the end of the 20th century, this rate gradually fell to less than 0.5 per million people. Most of this decrease has been attributed to improved building construction including conducting plumbing systems and improved electrical grounding, increased urbanization, and a decrease in the number of people engaged in outdoor agricultural activity during storm season, improved weather awareness and education about lightning danger, and improved availability of expert medical care. However, one other factor, outdoor recreation, has shown an increase in recent decades, and lightning casualties associated with outdoor recreation have also increased. It is likely that the lightning casualty rate in less-developed nations, although poorly documented, still lies in the range of 6 per million people. While lightning kills only a minority of its victims, nearly 90% suffer from permanent disabilities and serious long-term health problems. Some of these problems do not begin to manifest themselves until months or years after the lightning strike. Most of these problems are neurological in nature.

Lightning clearly has the ability to impact, both directly and indirectly, the quality of human life. In addition to the safety and human health impacts related above, accurate observations of global lightning activity may also allow scientists to better monitor events such as El Niños and global temperature change. Thus, there are many reasons why it is important to understand the amount and trends of lightning activity, especially on a global basis. The current satellite instruments are providing measurements that scientists can use today and in the future to monitor longer-term trends in global lightning activity and associated weather.

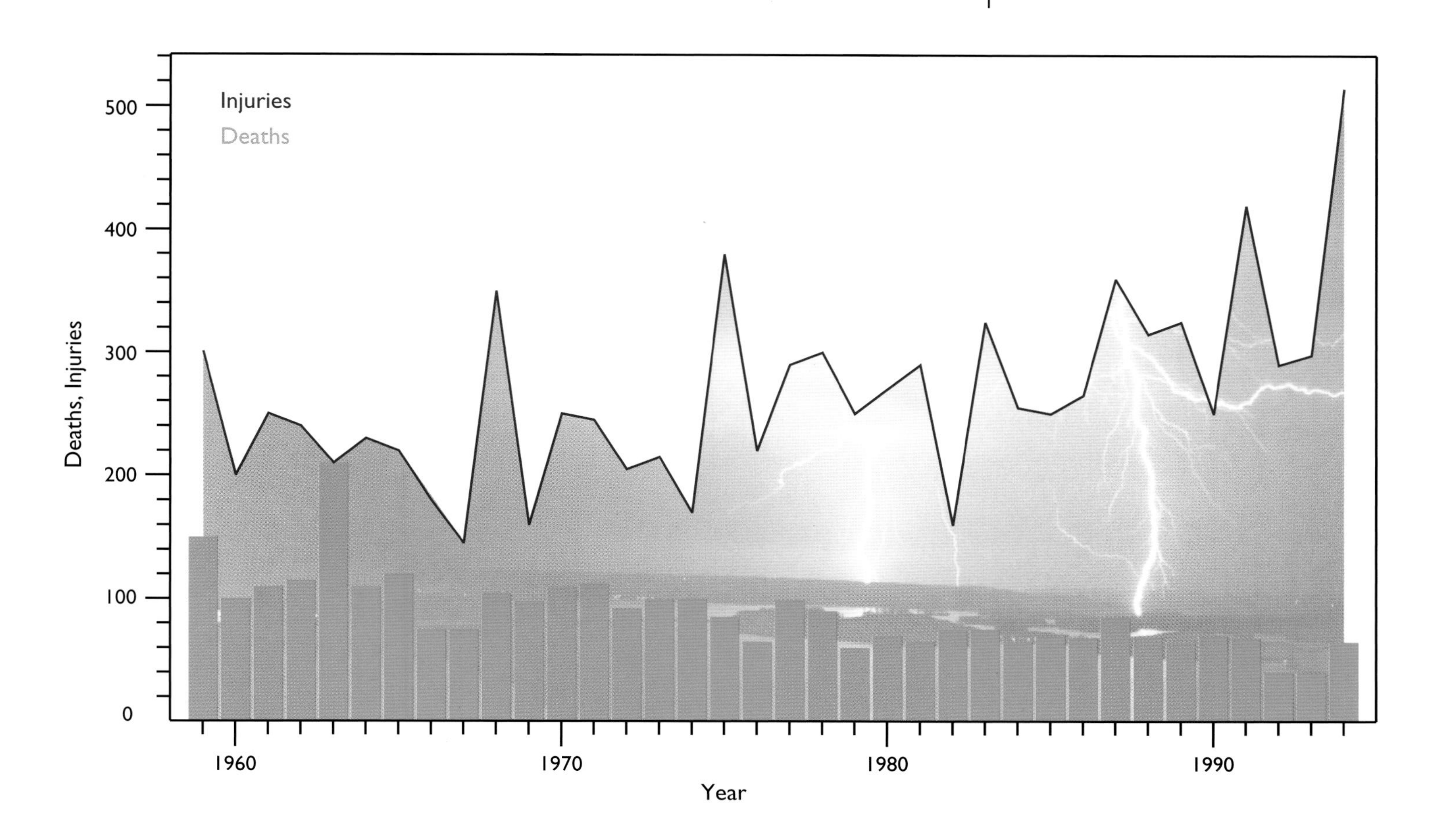


Lightning casualties in the United States from 1959 to 1994. (Adapted from E.B. Curran, R.L. Holle, and R.E. Lopez, 2000, *Journal of Climate*, vol. 13, pp. 3448–3453.)

**The following lightning safety guidelines can minimize your chances of being struck by lightning:**

Be aware of the weather forecast before heading out. Be alert, lightning can strike over 10 miles from a thunderstorm.

Use the 30/30 rule when outside:
- When a lightning flash occurs, count the time until you hear the thunder. If the time is 30 seconds or less, seek safe shelter immediately.

- Stay in the safe shelter until 30 minutes after the last rumble of thunder.

- A safe shelter is inside a fully enclosed, properly grounded, substantial building, or a vehicle with a solid metal roof and rolled-up windows. In buildings, avoid windows, plumbing, and plugged-in electrical devices.

- When outdoors, avoid higher elevations, open areas, tall isolated objects, water-related activities, and open vehicles. Avoid unprotected open structures like picnic pavilions, rain shelters, and bus stops. DO NOT GO UNDER A TREE TO KEEP DRY DURING A THUNDERSTORM! Trees can conduct dangerous electrical currents from lightning to ground, so during thunderstorms, stay twice as far away from a tree as it is tall.

- There will often be a few seconds of warning before a lightning strike hits: hair standing up, tingling skin, light metal objects vibrating, visible corona discharge, and/or hearing a crackling or 'kee-kee' sound. If you are in a group, spread out so there are several body lengths between each person. Be the lowest point around, but do not lie flat on the ground. Instead, use the lightning crouch—put your feet together, squat down, tuck your head, and cover your ears.

Call emergency services if lightning hits someone. Death from lightning usually occurs from cardiac arrest or stopped breathing at the time of the strike. The recommended first aid is CPR or mouth-to-mouth resuscitation.

Lightning strike during a winter thunderstorm over Oceanside, California on February 25, 2005. (Photograph by Steven Vanderburg, NOAA.)

# Warming and Cooling of the Atmosphere

QIANG FU
CELESTE M. JOHANSON

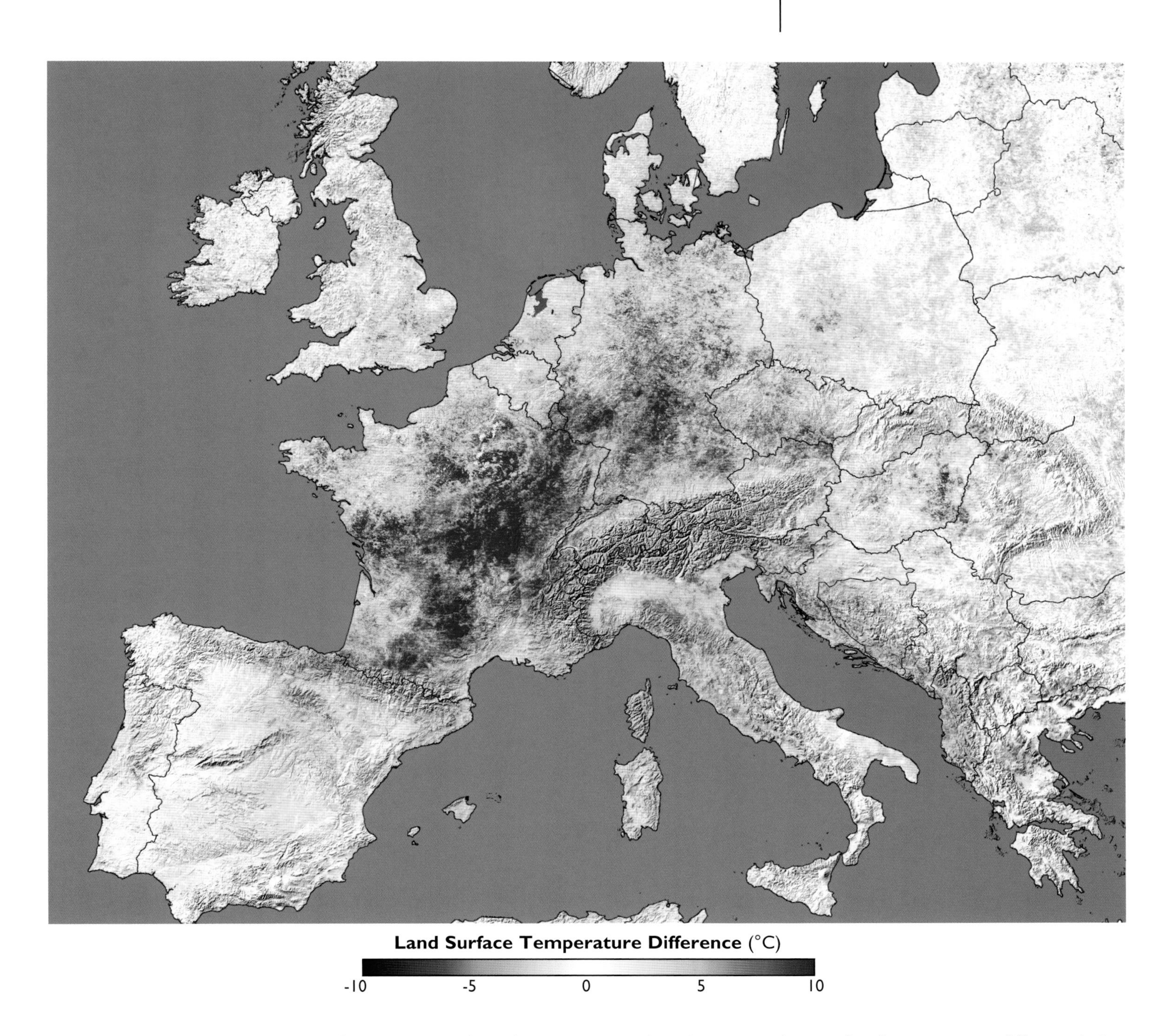


Land surface temperature difference during the heat wave in western Europe for July 20 through August 20, 2003 compared to data from the average of 2000, 2001, 2002 and 2004 for the same period. (Data from the MODIS instrument on the Terra satellite.)

Thermometer observations over the past century show that temperatures have been rising at the Earth's surface. This warming can be largely explained by the increase of greenhouse gases in the atmosphere, principally carbon dioxide. The increase of greenhouse gases was also blamed for some extreme weather events such as the 2003 heatwave in Europe. Scientists found that human influence has at least doubled the risk of the occurrence

*Our Changing Planet*, ed. King, Parkinson, Partington and Williams
Published by Cambridge University Press 

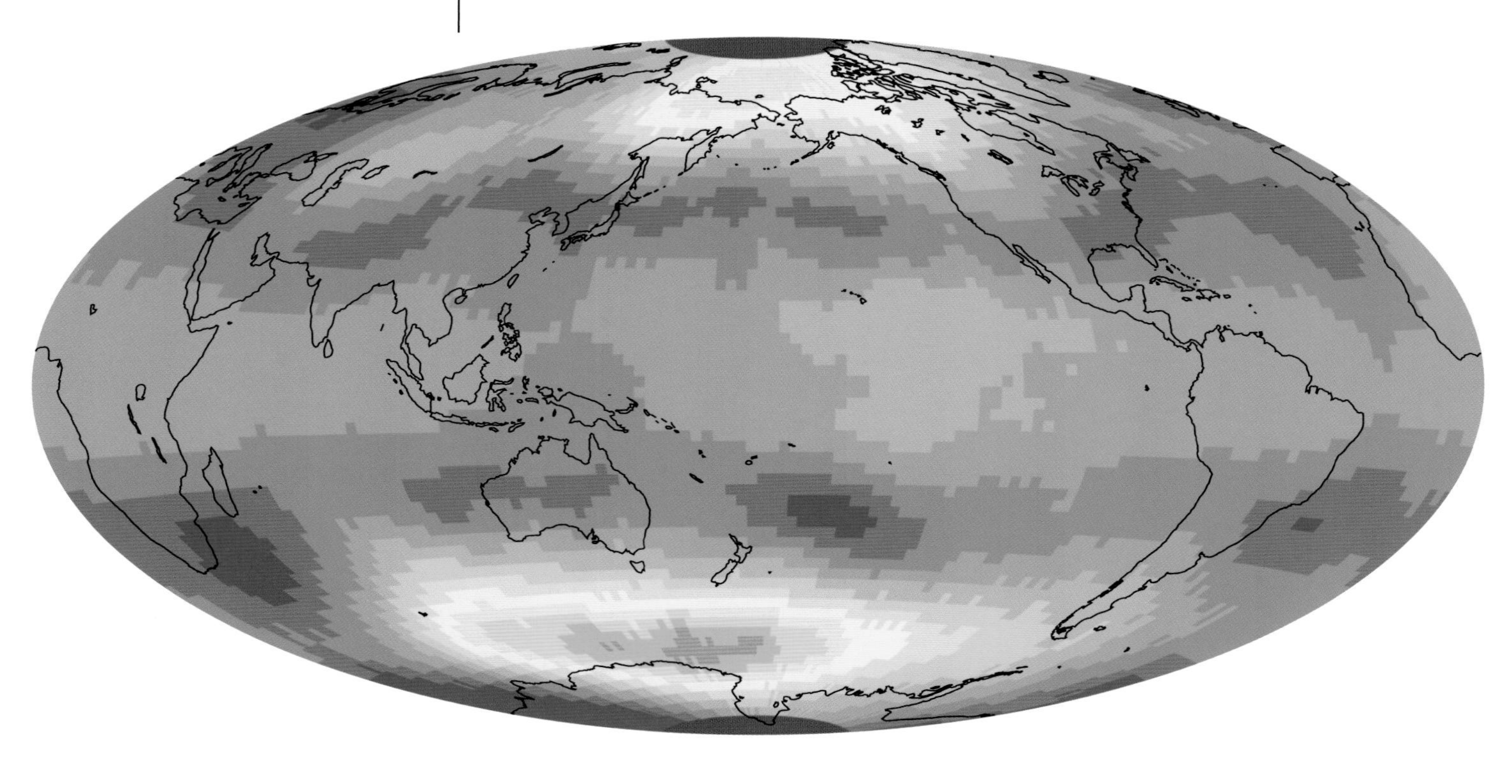

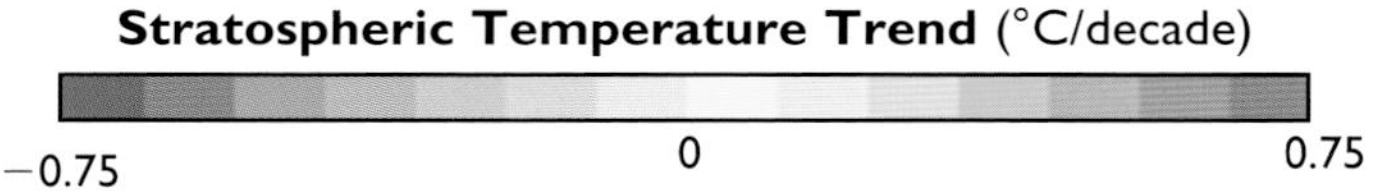


Global distribution of temperature trends in the lower stratosphere (top) from January 1979 to December 2004. The plot (below) shows the global mean lower stratospheric temperature changes from 1979 to 2004. The straight line is the linear trend of the data, which shows a cooling of -0.36°C per decade in the lower stratosphere. Two large peaks are due to stratospheric warmings following volcanic eruptions of El Chichón (1982) and Mt. Pinatubo (1991). (Data from the MSU instrument on the Tiros-N and NOAA 6 to NOAA 14 satellites.)

of such a heat wave. But how do human activities influence temperatures in the atmosphere?

Satellite observations provide a global record of temperature for several atmospheric layers since 1979. The data clearly show that the stratosphere, a section of our atmosphere between about 12 and 50 km above the Earth's surface, has been cooling in recent decades. Scientists explain this cooling as a result of stratospheric ozone depletion and the insulating effect of increasing greenhouse gases in the atmosphere. Two periods when the stratosphere warmed (1982–1983 and 1991–1992) are a direct result of the injection of aerosol particles into the lower stratosphere by the El Chichón and Mt. Pinatubo volcanic eruptions. In contrast, scientists expect that the Earth's troposphere, the lower layer of atmosphere where most weather occurs, should be warming at least as much as the surface due to the greenhouse effect. But they are still debating whether or not satellite measurements show this warming.

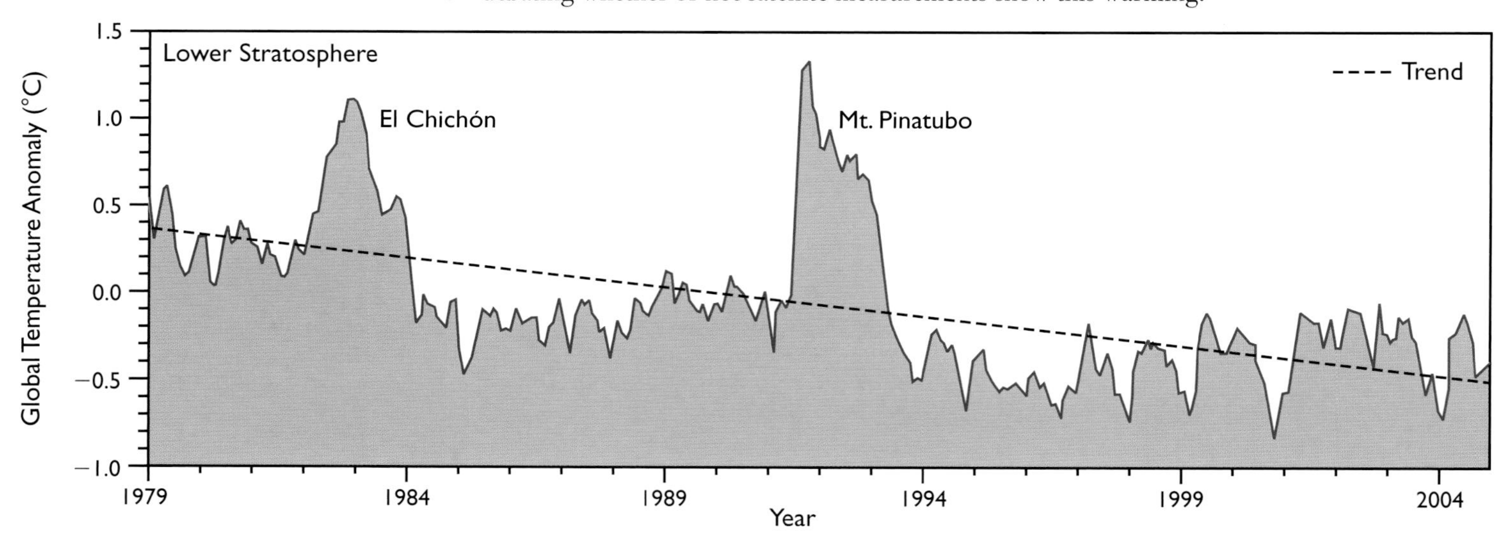

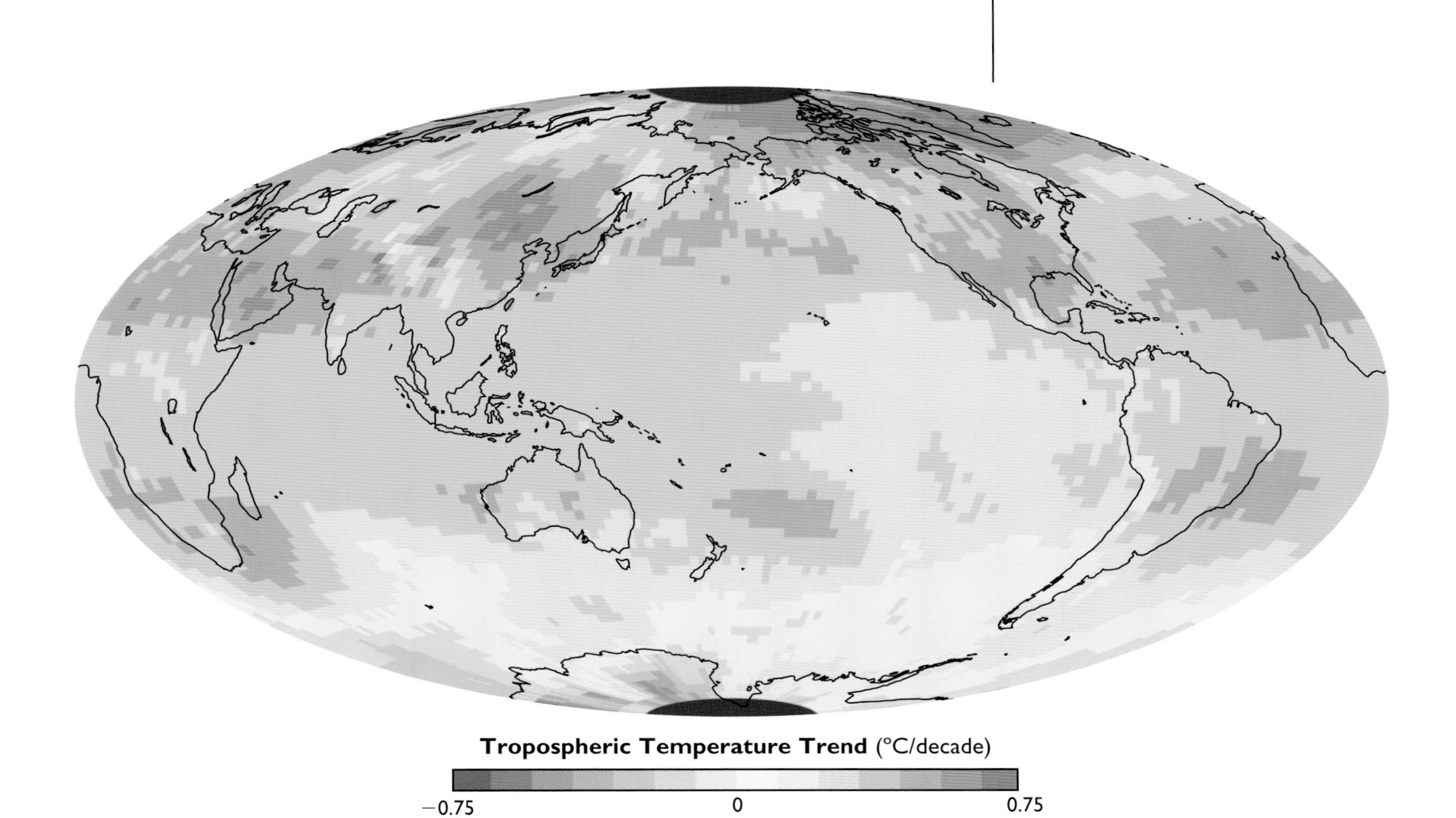


Atmospheric temperatures are derived from satellite measurements of microwave radiation emitted by oxygen. For a certain wavelength range, the signal comes mostly from the troposphere and therefore mostly reflects the temperature of the troposphere. However, a small part of this signal comes from the stratosphere. Since the stratosphere has been cooling rapidly, this tends to mask whatever warming trend may exist in the troposphere. Recently, scientists have devised an effective way to correct for the stratospheric effect. When this correction is made, the satellite data suggest that the troposphere has been warming a bit faster than the Earth's surface, which is just what computer models predict. The validation of these models with observations is important for establishing their credibility to forecast future climate change.

Global distribution of temperature trends in the troposphere (top) from January 1979 to December 2004. The plot (below) shows the global mean tropospheric temperature changes from 1979 to 2004. The straight line is the linear trend of the data, which shows a warming of +0.19°C per decade in the troposphere. The interannual variabilities in tropospheric temperatures are largely related to El Niño Southern Oscillation events. (Data from the MSU instrument on the Tiros-N and NOAA 6 to NOAA 14 satellites.)

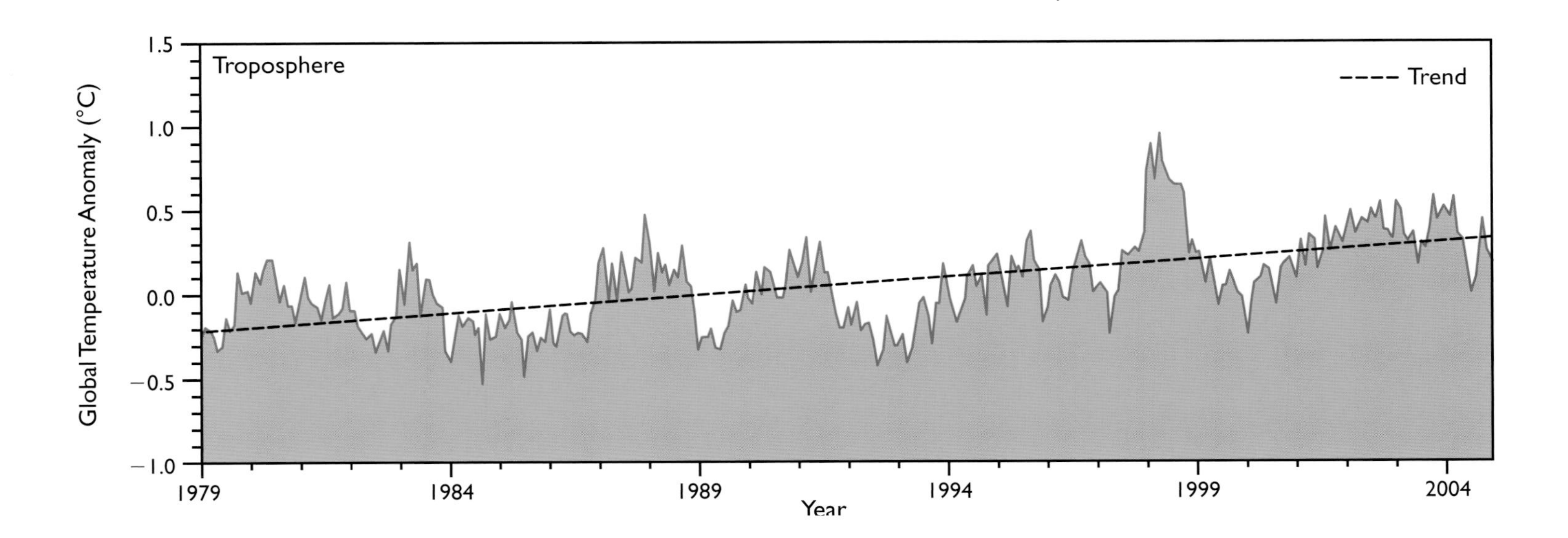

N. Christina Hsu
Si-Chee Tsay
Michael D. King
David J. Diner

# Dust in the Wind

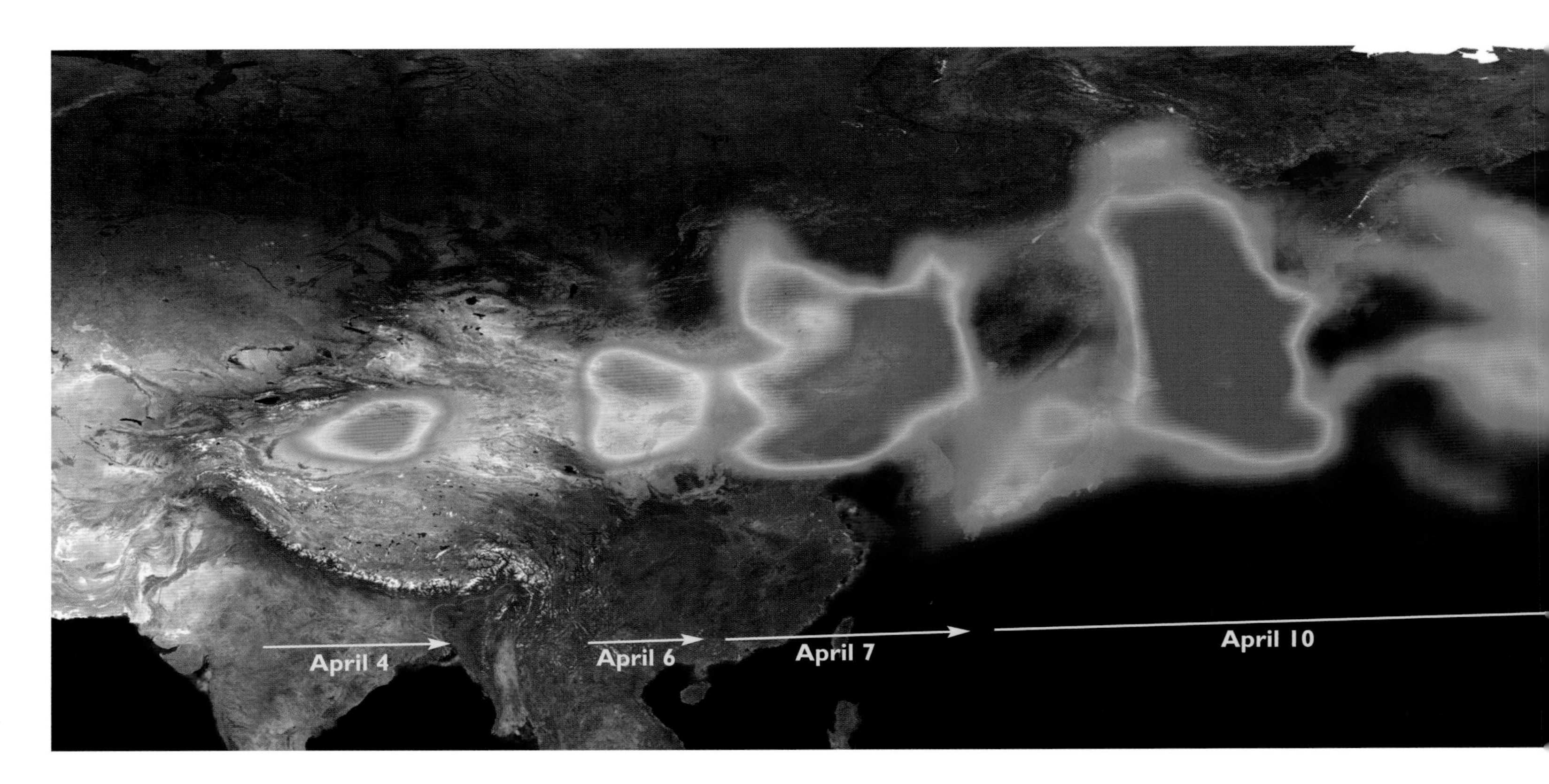


A composite of aerosol index measurements taken from April 4–16, 2001 dramatically illustrates the transport of dust from a huge storm in China that moves across the Pacific Ocean and North America to the Atlantic Ocean. The arrows indicate the size and location of the dust cloud observed on the date given below. Red areas indicate high aerosol index values and correspond to the densest part of the dust cloud. Yellows and greens are moderately high values. (Data from the TOMS instrument on the Earth Probe satellite.)

When strong winds blow through arid regions, immense amounts of dust often get lifted from the surface and injected into the atmosphere. These dust storms, like many other types of natural hazards, often impact human activities in dramatic ways when passing over inhabited areas, causing breathing problems, delaying flights, pushing grit through windows and doors, forcing people to stay indoors, and generally creating havoc.

The ability to monitor and study dust storms increased significantly in the 1970s when satellite instruments started to provide unprecedented views on their size, scale, and movement. The unique vantage point provided by satellites helped lead to discoveries of how these storms affect the environment, and us, in less obvious but even more important ways. For example, the ability to detect and track the movement of dust from satellites indicated a link between the outbreak of meningitis in the Sahel and active dust storm periods in the Sahara. Also, by showing that mineral dust provides a source of iron to fertilize phytoplankton growth by being routinely transported from the Saharan desert into the Atlantic Ocean, this capability illustrated the vital biogeochemical role dust storms play in the ecosystem. This same dust-laden air also supplies crucial nutrients for the soil of rain forests in South America.

Satellite measurements of dust characteristics also provided much needed information used to determine how dust particles affect the climate by redistributing solar energy within the Earth's atmosphere and by changing the thermal contrast between land and ocean. When

*Our Changing Planet*, ed. King, Parkinson, Partington and Williams
Published by Cambridge University Press 

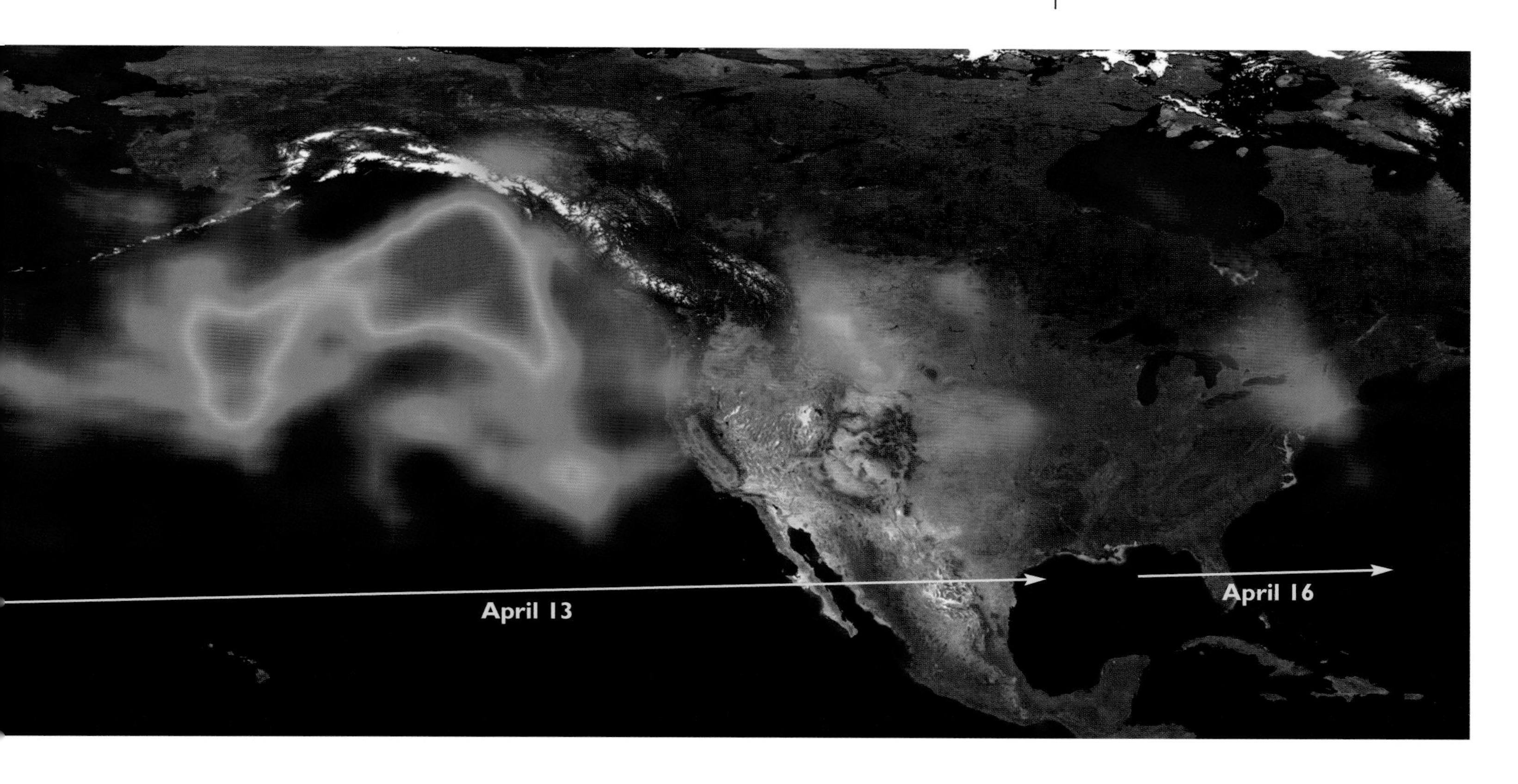


interacting with sunlight, these tiny particles not only absorb solar radiation, but also reflect it back to space. The result is a net warming effect in the dust-laden atmosphere and a net cooling effect on the Earth's surface. Exactly how different types of dust modify Earth's radiation budget depends on how the absorption of sunlight by particular types of dust changes with wavelength, in other words the dust's color. Asian dust is yellowish, Saharan dust is

Street scenes over the air base in Al Asad located near the western desert of Iraq during the passage of a dust storm on April 26, 2005. The passage of such storms can change both the atmospheric visibility and the actual color of the sky in a matter of minutes. (Photograph courtesy of Gunnery Sgt. Shannon Arledge.)

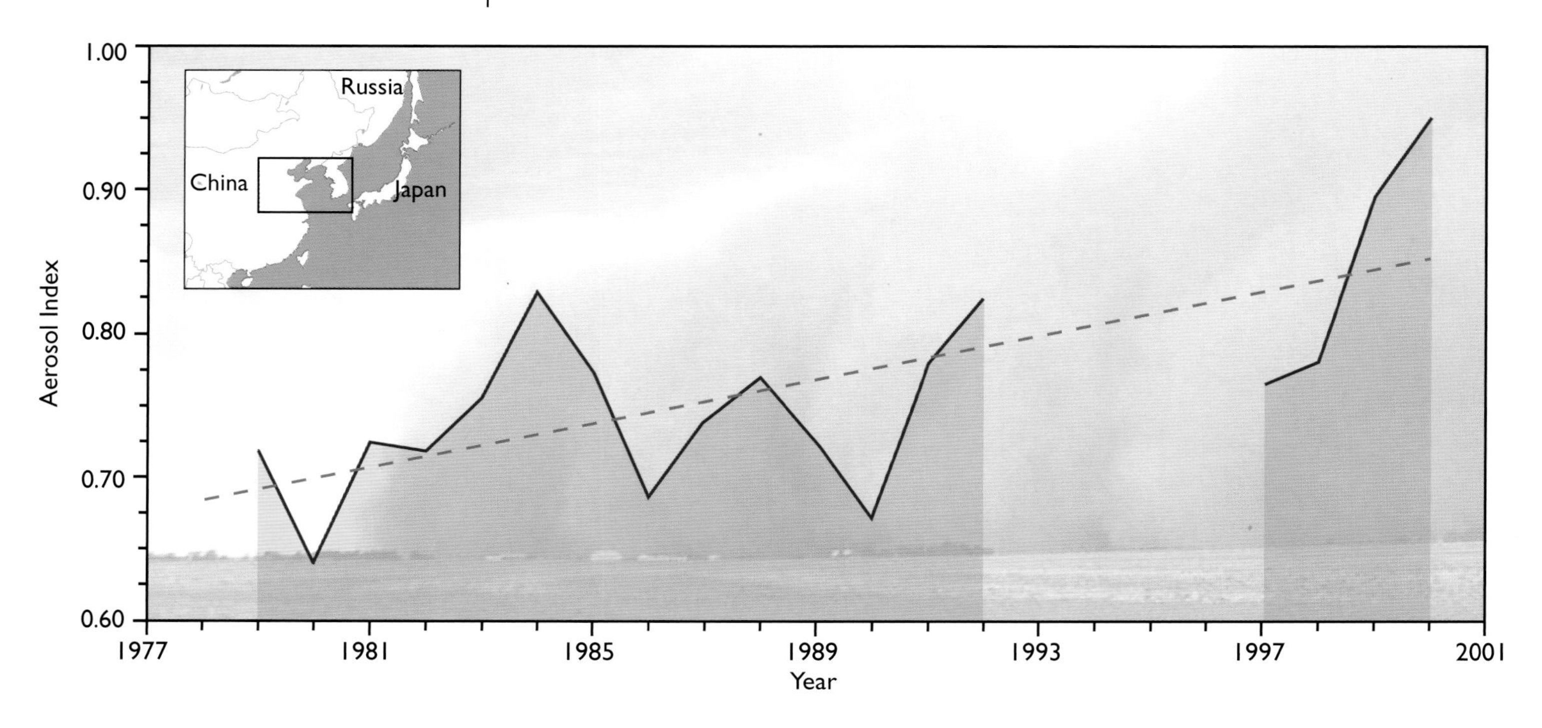


Time series of the 1977–2001 springtime (March, April, and May) averaged aerosol index over East Asia (32°N–40°N, 112°E–130°E). (Data from the TOMS instruments on the Nimbus 7 and Earth Probe satellites.)

generally more brownish, while some dust, such as that found in the Bodele Depression in Chad, even looks white.

One of the more dramatic results provided by satellite measurements was the discovery that fine airborne dust particles can travel remarkably long distances, impacting all of the environments along their transport pathway. In the spring, dust from storms occurring over Asia are caught by the prevailing winds of the jet stream and observed to travel from China across the Pacific to the west coast of North America, sometimes even crossing the United States and Canada to reach the Atlantic Ocean. These dust clouds have a significant impact on the springtime air quality over North America.

Although dust storms are natural events, surface disturbances associated with human activities may also contribute to the trend in the amount of mineral dust transported in the atmosphere. During the last 3 decades, the ability to monitor the size and distribution of dust storms from space has been critical in shedding light on how man has affected the development and magnitude of these storms. Uncontrolled land use such as overplowing, overgrazing, and overdevelopment, coupled with periods of drought, often leads to a massive deterioration of land cover, intensifying the frequency of occurrence of large-scale dust storms. In one example, satellite observations suggest an increase in both the intensity and frequency of dust storms over East Asia that parallels the manmade development, and consequent change in the qualities of the land surface, occurring in northwestern China during the same time period.

Since the beginning of the new millennium, measurements from instruments on NASA's Earth Observing System (EOS), and new techniques developed to analyze them, have added many capabilities needed to help improve the mapping and prediction of dust storms. Information obtained from the MODIS and MISR sensors now allow scientists to study the optical and microphysical properties of mineral dust from sources to sinks on a much finer scale than previously available from satellite.

In particular, measurements made by different instruments passing overhead at different times have been instrumental in studying the creation and evolution of dust plumes over

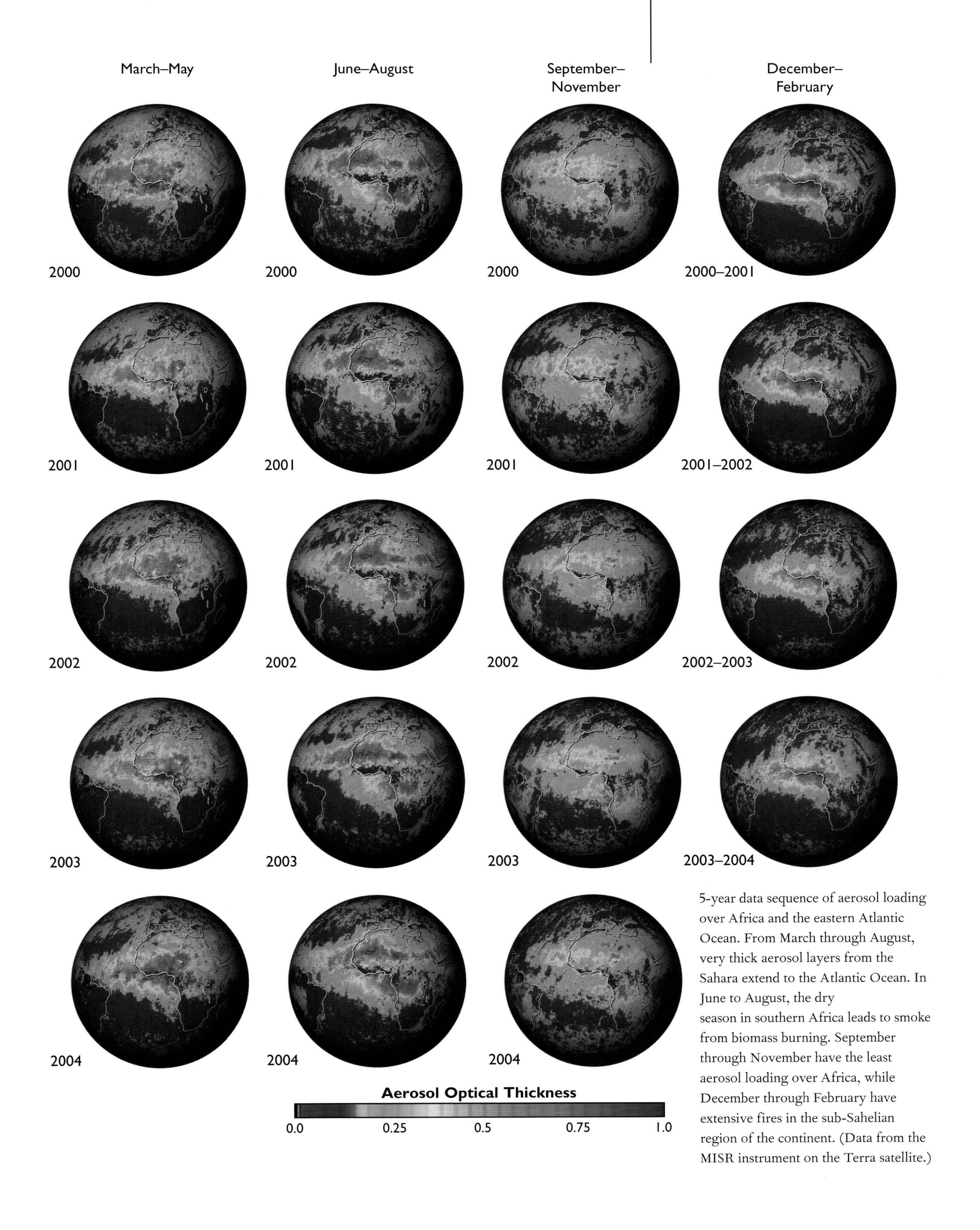


5-year data sequence of aerosol loading over Africa and the eastern Atlantic Ocean. From March through August, very thick aerosol layers from the Sahara extend to the Atlantic Ocean. In June to August, the dry season in southern Africa leads to smoke from biomass burning. September through November have the least aerosol loading over Africa, while December through February have extensive fires in the sub-Sahelian region of the continent. (Data from the MISR instrument on the Terra satellite.)

time. A comprehensive understanding of the properties of these tiny particles as well as their temporal and spatial distribution is imperative to understanding how the Earth's atmosphere maintains its current state of equilibrium and how anthropogenic activities could potentially destroy that balance.

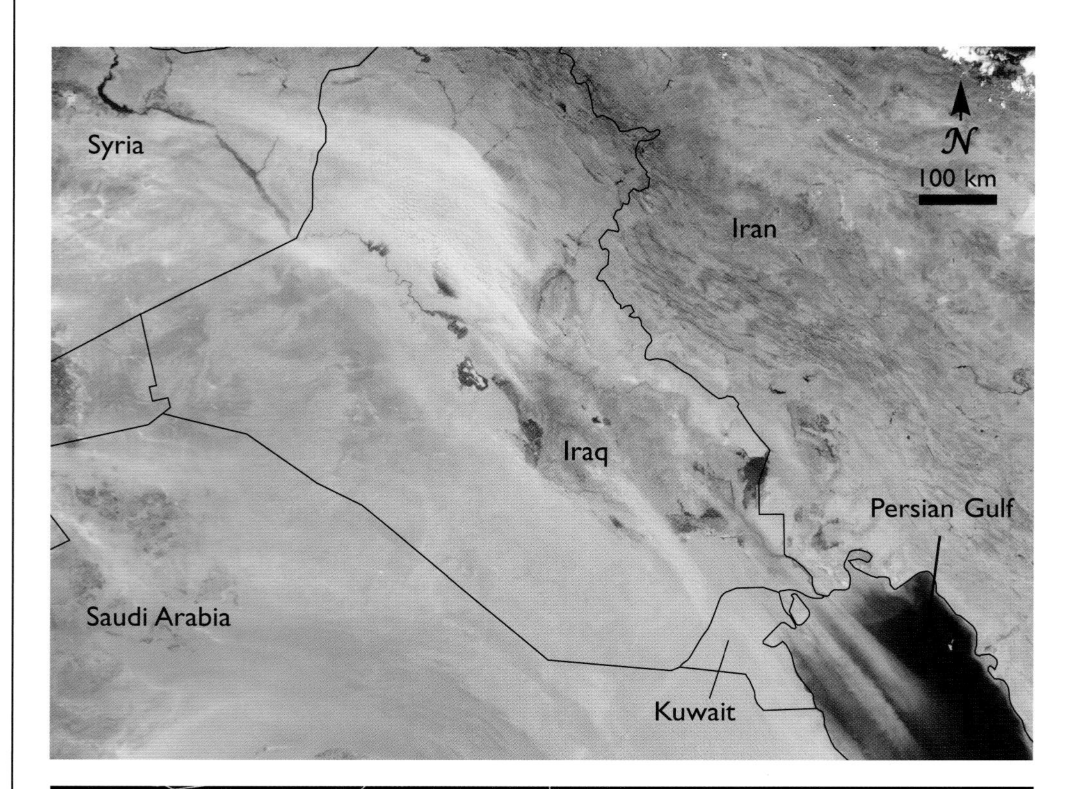


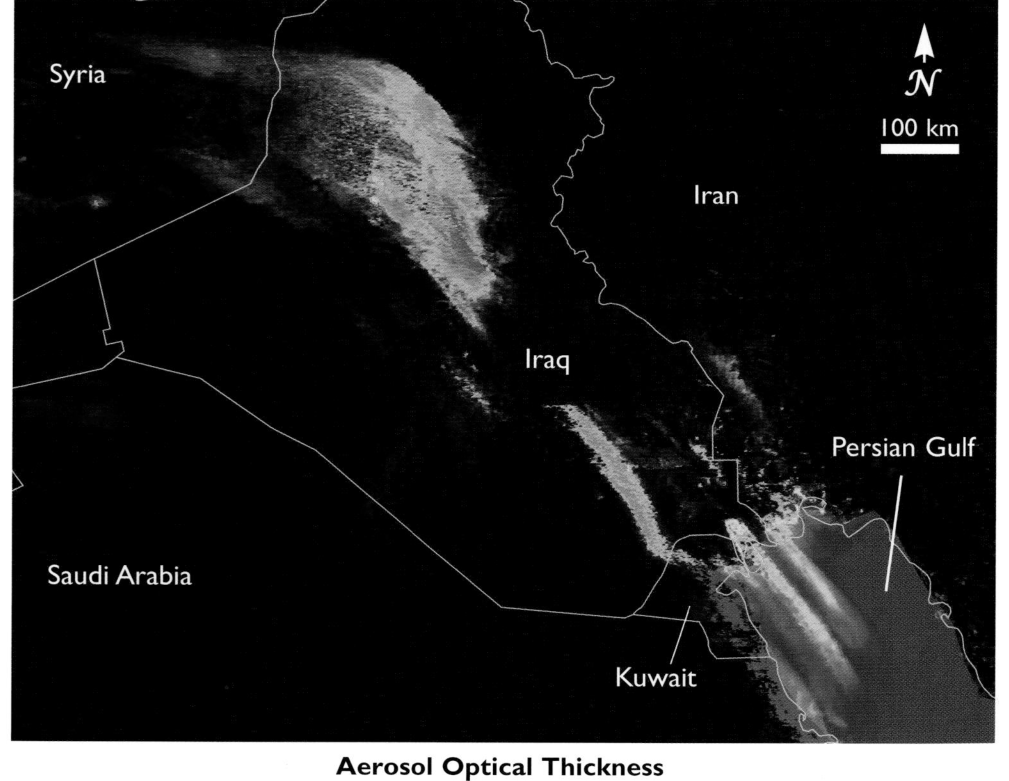


Dust storms frequently occur over places like northern Africa and southwest Asia all year round. Extensive dust clouds were observed from space over Iraq, Syria, and Kuwait, as well as over the Persian Gulf on August 7, 2005 (top panel). The corresponding intensity of the dust plume is also depicted using a unitless quantity called 'optical thickness' (bottom panel). The higher the dust optical thickness, the more dust is in the atmosphere. (Data from the MODIS instrument on the Aqua satellite.)

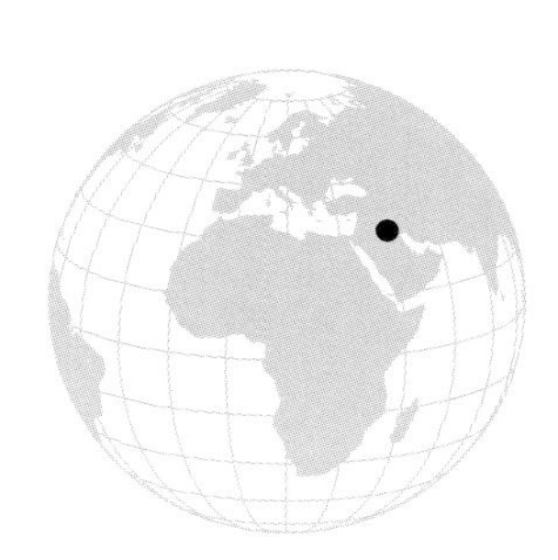

# Atmospheric Pollution: A Global Problem

David P. Edwards

Two potential sources of carbon monoxide (CO) pollution are shown, industrial activity and agricultural burning and wildfire. The grassland fire is from South Africa in September 2000. (Left photograph by Jim Ross, NASA Dryden Flight Research Center; right photograph © 2006 Rinderart, courtesy ImageVortex.com.)

Data from NASA's Terra satellite is adding to our understanding of how pollution spreads around the globe. The information will help scientists protect and understand the Earth.

The Measurements of Pollution in the Troposphere (MOPITT) instrument is able to detect carbon monoxide (CO) gas in the lower part of the atmosphere. This is a good indication of high pollution. It is produced when fossil fuels are burnt in industrial processes and also by the domestic heating and motor vehicle traffic associated with urban areas. Vegetation fires, either for agricultural purposes and deforestation, or wild fires, also produce large amounts of CO.

Accumulation of CO is most pronounced in the Northern Hemisphere where most human activity occurs and happens in winter when the photochemical destruction of CO is decreased by lack of sunlight. This is particularly clear in eastern Asia. Later in the year, CO levels in the Northern Hemisphere decrease while agricultural fires, which are common practice in developing countries, lead to large amounts of pollution in tropical Africa, South America, and Indonesia.

*Our Changing Planet*, ed. King, Parkinson, Partington and Williams
Published by Cambridge University Press 

December 2003, January 2004, February 2004

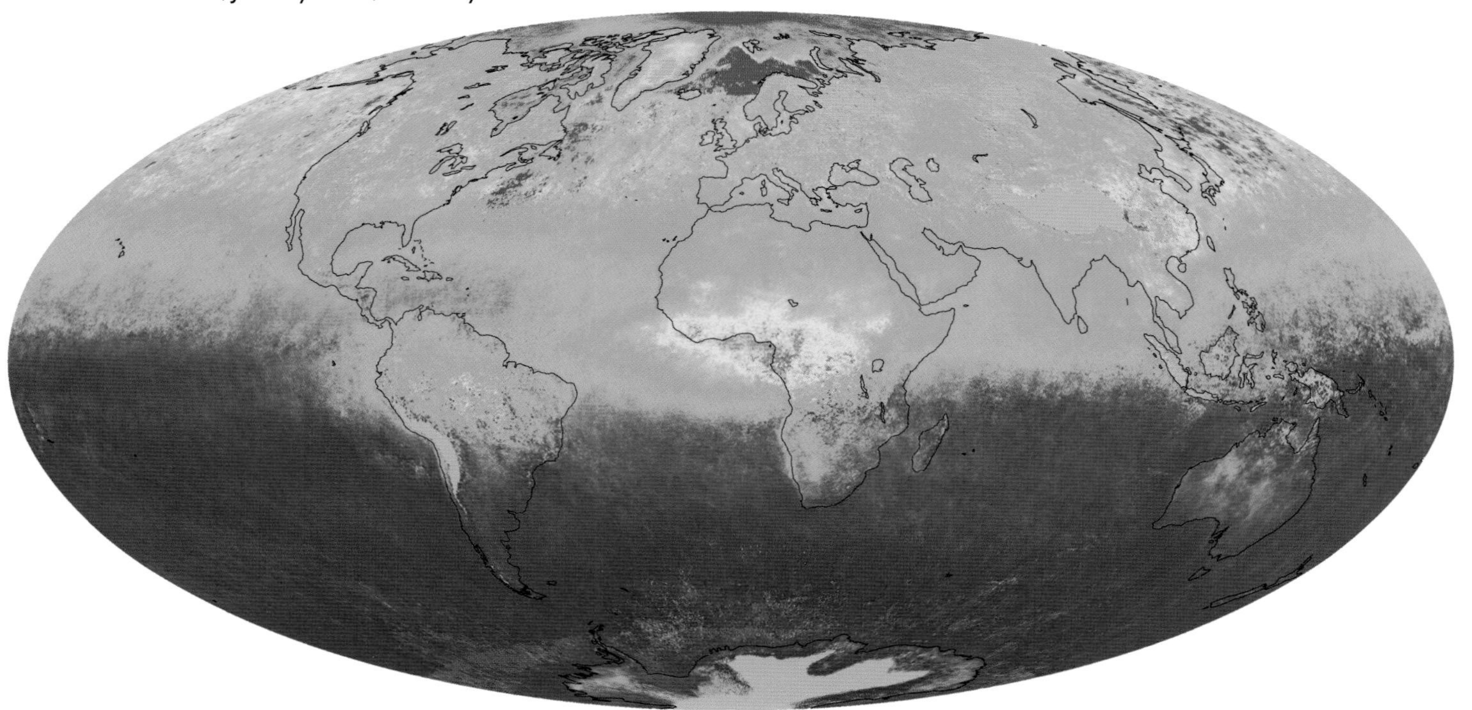

March, April, May

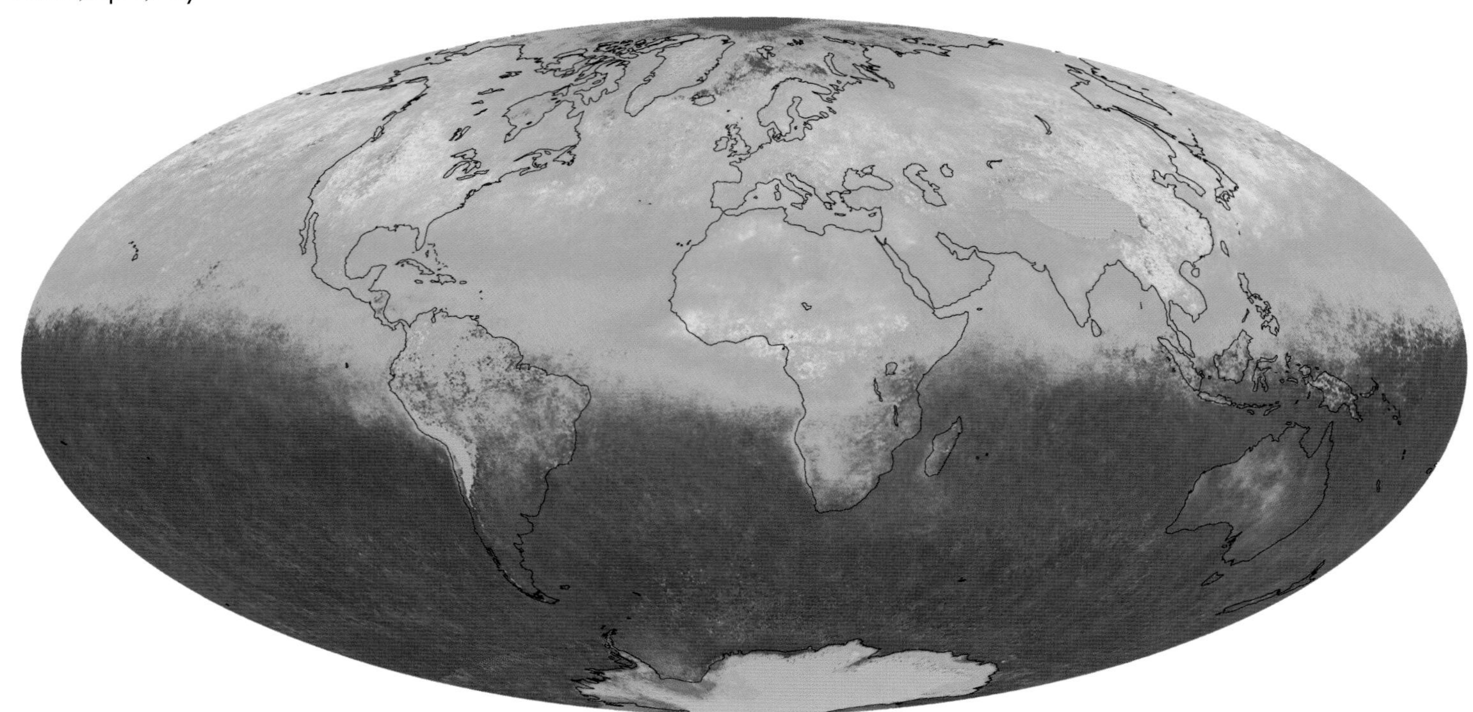

June, July, August

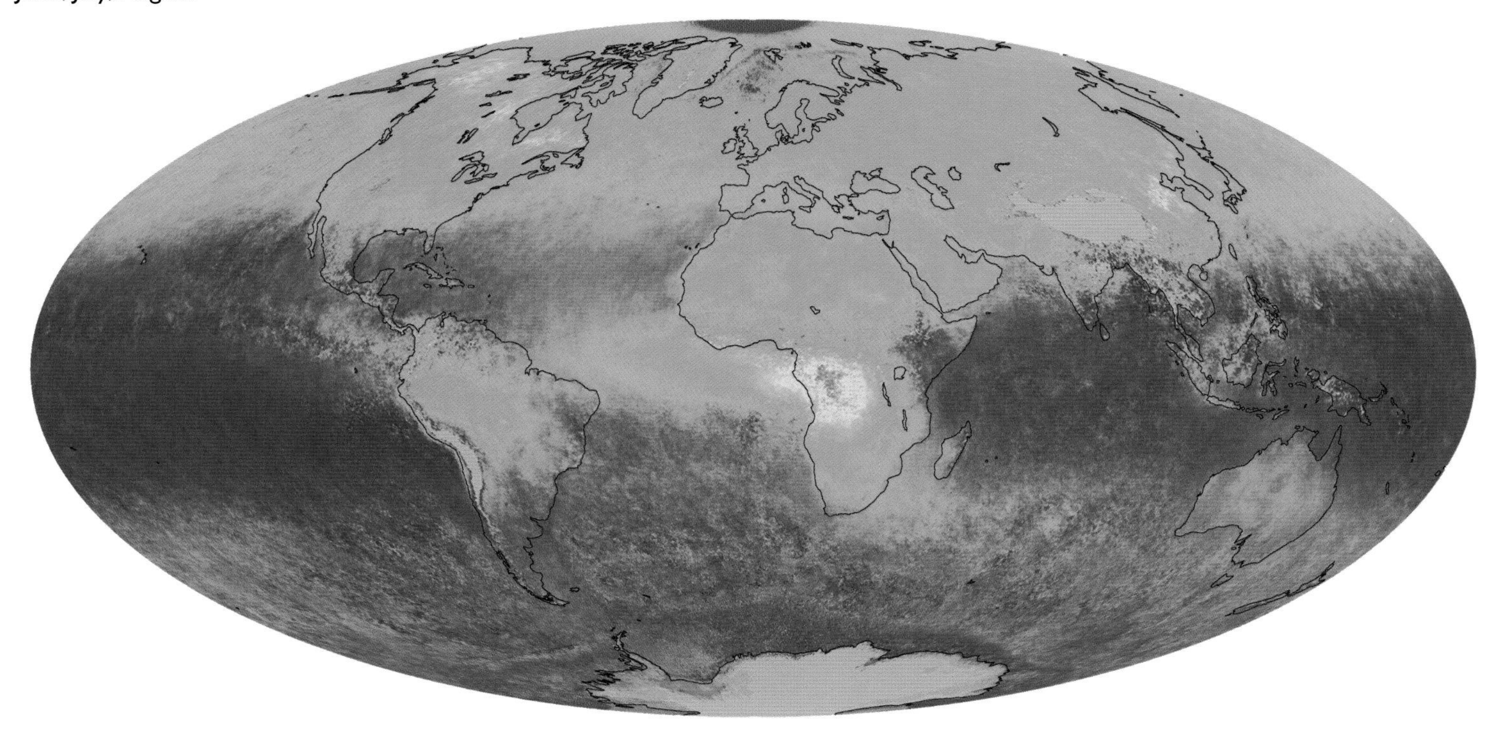

September, October, November

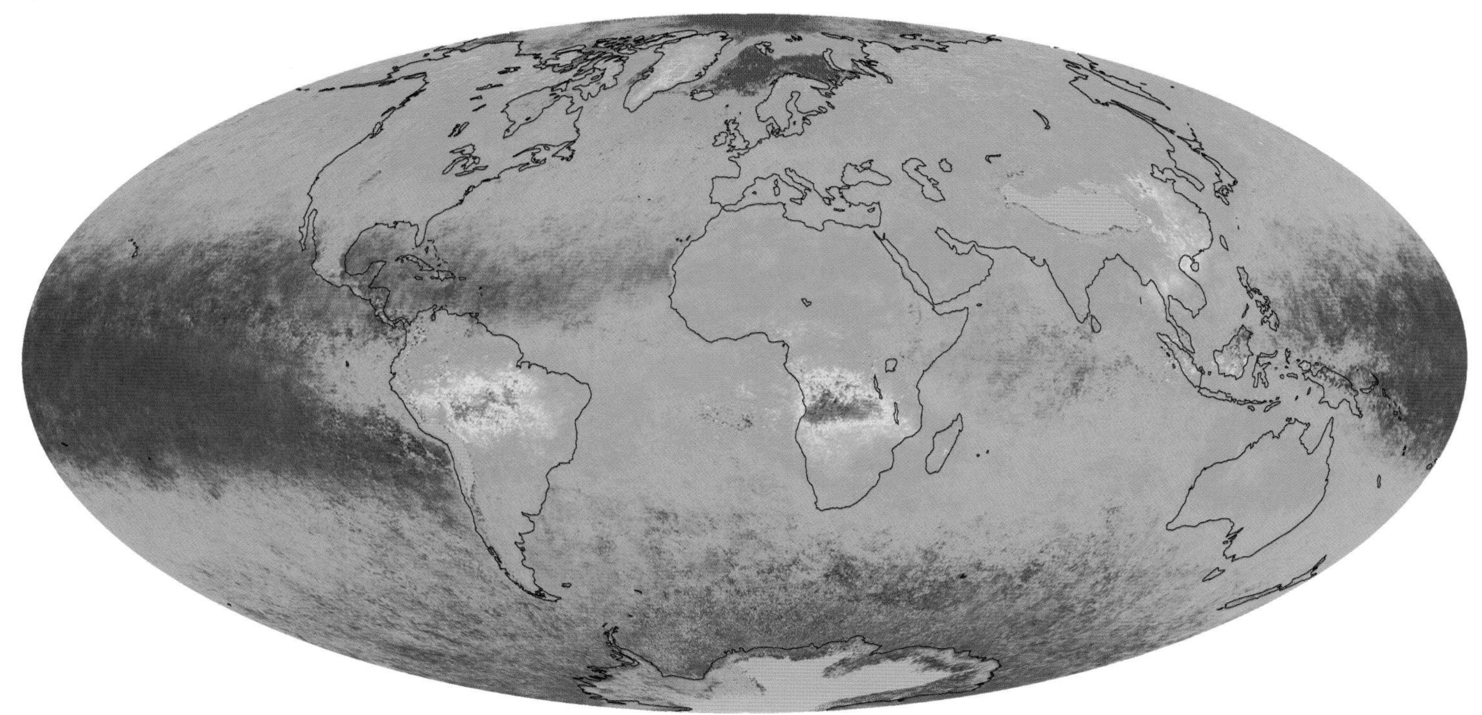

The seasonally changing global distribution of CO pollution at an altitude of 700 hPa (about 3 km). Averages are shown separately for December 2003 to February 2004, March to May, June to August, and September to November 2004. High CO pollution levels are shown in red. Northern Hemisphere pollution sources are predominantly urban and industrial, while high CO in the tropics and Southern Hemisphere often result from agricultural biomass burning. (Data from the MOPITT instrument on the Terra satellite.)

Fires

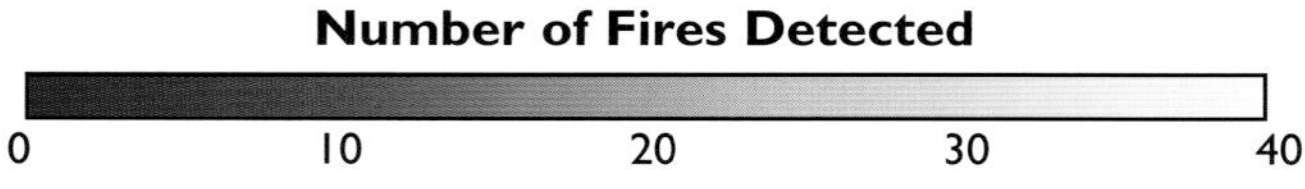


Pollution outflow from Spring 2003 fires in Siberia. In the Spring of 2003 there were a large numbers of fires in Siberia, especially in the Lake Baikal region. The left image shows the number of fires detected during the month of May 2003. These fires produced large amounts of carbon monoxide gas (right). This pollutant can last for over a month. The long lifetime allows the gas to cross the Pacific Ocean and reduce air quality over North America before continuing on around the globe. (Data from the MODIS instrument (left image) and MOPITT instrument (right image) on the Terra satellite.)

In late Summer 2002 and Spring 2003, Terra observed big fires in western Russia and Siberia. CO molecules can last for several weeks in the atmosphere, allowing them to travel long distances and impact air quality far from the point of emission. The fires led to a 'dirty' 2002/2003 winter atmosphere in the Northern Hemisphere with high amounts of CO and other pollutants. Peak levels of CO hung over the United States. The satellite observations showed Russian fires have a huge impact on air quality on a global scale and indicated how pollution must be thought of as a global problem.

Terra and other NASA Earth-observing satellites provide vital tools for monitoring global levels, sources, and destinations of CO and other pollutants. In addition to indicating seasonal changes, the growing data record is beginning to show large interannual changes in atmospheric pollution. These variations often depend on whether meteorological and climatic conditions result in widespread vegetation fires in a particular year. The continuing satellite measurements of these pollutants is providing clues about how our planet may be changing.

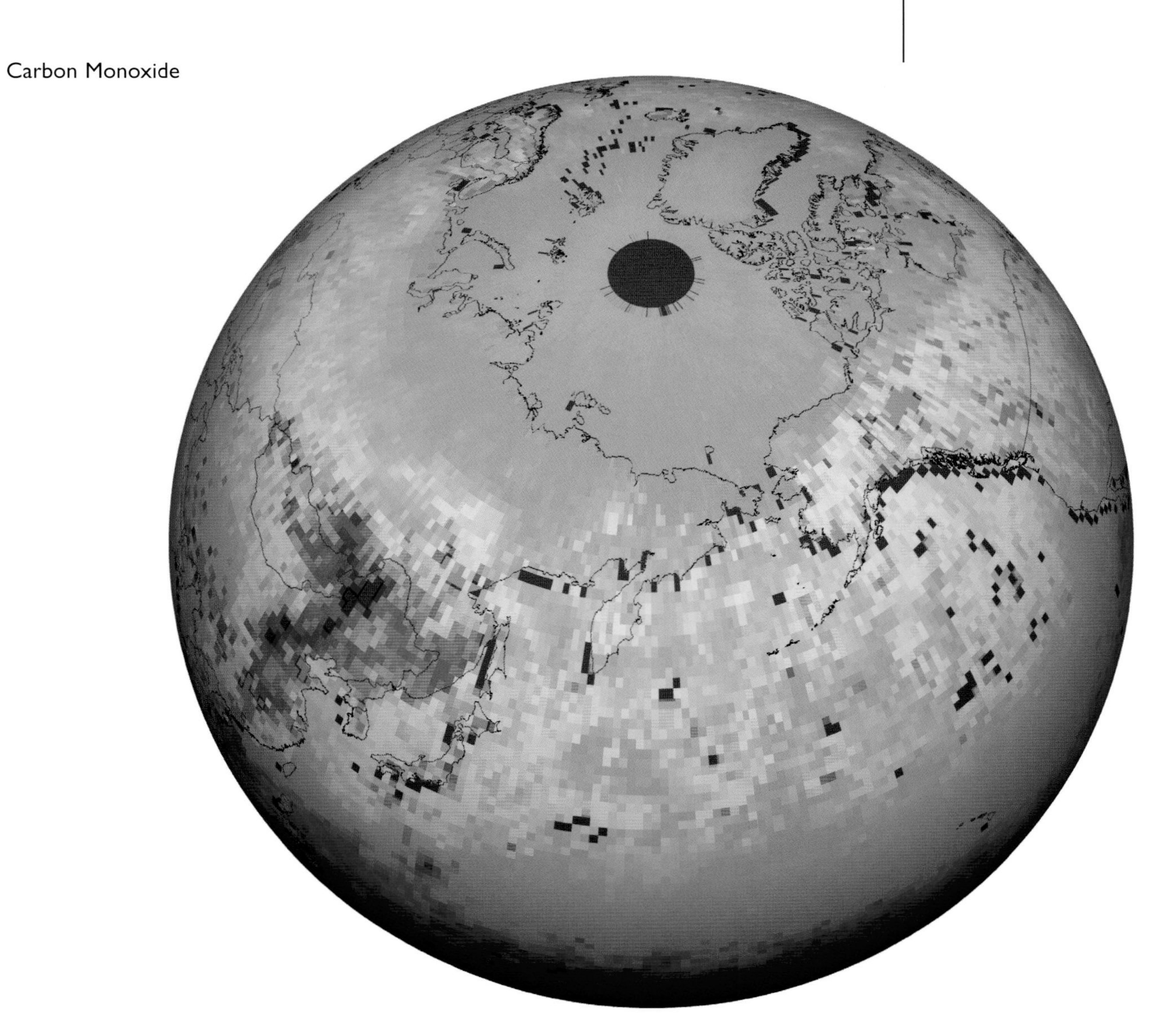


**Carbon Monoxide Concentration** (parts per billion)

0.0 67.5 135.0 102.5 270.0

Steven Platnick

# Ship Tracks

January 27, 2003

An unusually high number of ship tracks were observed in low clouds off the coasts of France and Spain in this true-color satellite image on January 27, 2003. As the ships moved about the eastern Atlantic, they left an impermanent 'track' of their recent movement. (Data from the MODIS instrument on the Terra satellite.)

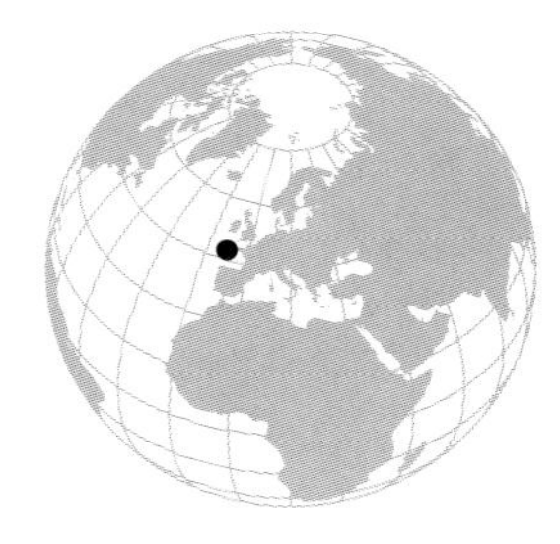

*Our Changing Planet*, ed. King, Parkinson, Partington and Williams
Published by Cambridge University Press 

France
Spain

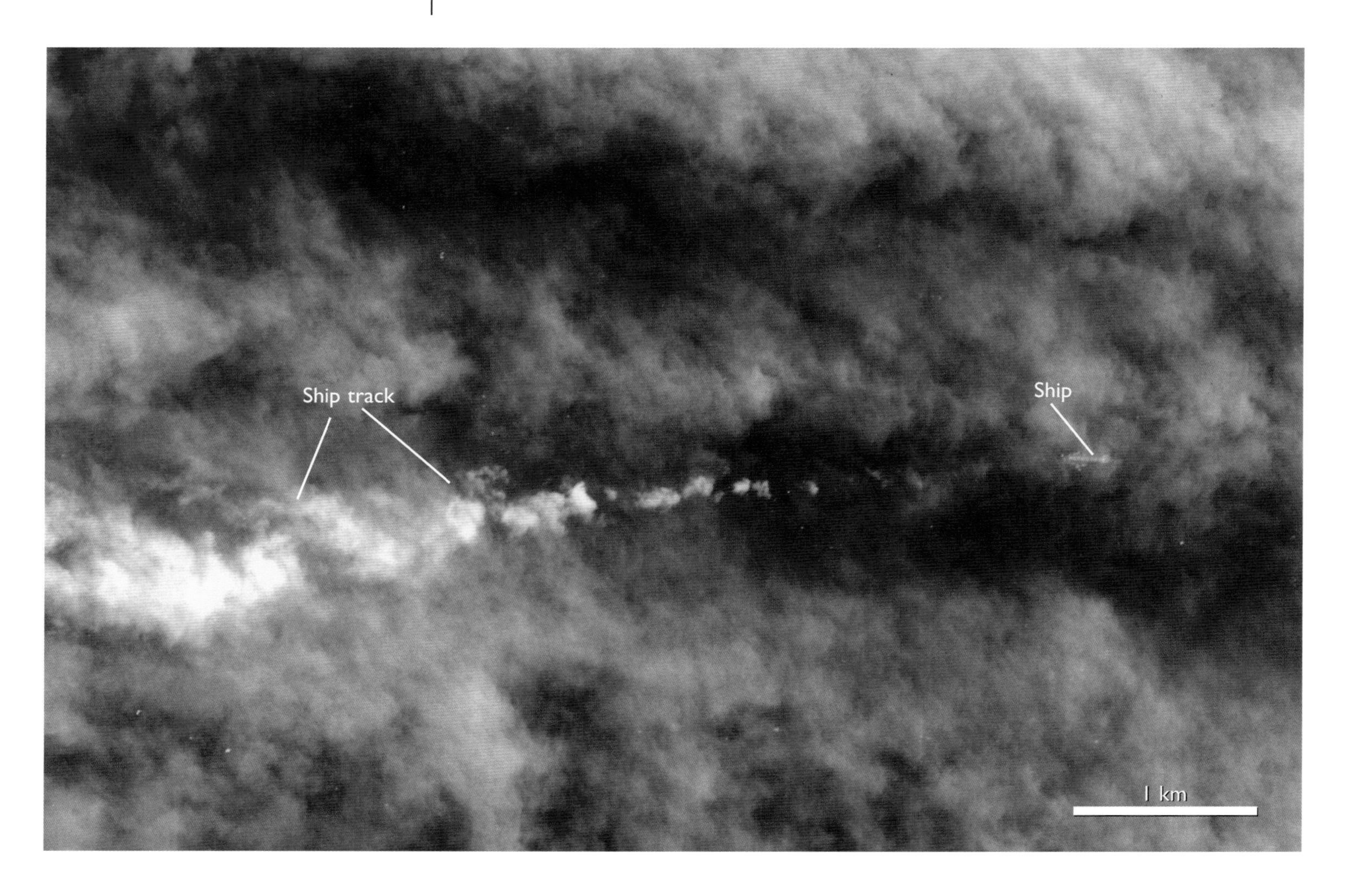


Photograph of a container ship off Monterey, California on June 13, 1994 and the resulting ship track produced downwind of the ship from condensation onto particles produced by the ship exhaust. An approximate length scale is shown. (Photograph from the RC-10 camera on the NASA ER-2 aircraft.)

Ship tracks are linear features of enhanced cloud reflectance in marine stratiform (layered) clouds caused by pollution in ship exhaust. Example tracks, seen in visible or nearby wavelengths of light, are seen in high spatial resolution aircraft photography and large-scale satellite imagery.

Ship tracks form when aerosols (very small suspended airborne particles) are directly emitted by the exhaust of ships and/or form out of the exhaust gases. A subset of these aerosol particles are soluble in water and act as 'seeds' for the condensation of water vapor into liquid droplets. This process of condensation is referred to as droplet nucleation and the aerosol particles are referred to as cloud condensation nuclei (CCN).

The extensive stratiform clouds seen in these images exist in the boundary layer or lowest level of the atmosphere, just above the ocean surface, so that ship-produced CCN are readily incorporated into the cloud. The greater number of CCN in these portions of the cloud result in two modifications to the cloud droplets: first, a larger number of droplets are produced since more nuclei are present, and second, the resulting droplets are smaller in size since the cloud water is distributed among more droplets. In effect, the larger number of smaller droplets reflects more sunlight, thereby making the ship-modified portion of the cloud brighter. The bright linear features seen in photographic and satellite images are a result of that droplet modification.

This mechanism, whereby ship pollution increases cloud reflectance, is analogous to the potential modification of clouds by widespread continental aerosol sources (industrial pollution, fires, etc.) and so is of great climatic interest. As such, ship tracks are an example of

Schematic illustration of the effect of cloud nucleation particles from ship exhaust resulting in a larger number of cloud droplets in a given volume of cloud, but with smaller droplet sizes.

what is often referred to as the 'indirect effect' of aerosols on climate, in the sense that the aerosols affect solar energy indirectly through the modification of cloud properties. However, potential global and regional changes in cloud reflectance due to aerosol sources that are often a large distance away from unpolluted clouds makes the detection and quantification of the process difficult. The 'direct effect' of aerosols (scattering and absorption of sunlight directly by aerosols) is also of great interest. Generally speaking, the narrower and less diffuse ship tracks are associated with situations where the relative difference between the ship- and wind-speed vectors is large (for example, a fast ship moving in the opposite direction of a strong wind). Tracks formed when the ship and wind speeds are similar and in the same direction tend to leave wider and more diffuse ship tracks. Satellite images show examples of this movement for selected tracks between the time of the Terra satellite overpass at 1200 UTC and the Aqua satellite overpass on the same day just 100 minutes later.

Ship tracks can be found throughout most of the world's oceans, especially in regions with relatively shallow atmospheric boundary layers where stratiform and cumulus cloud layers are common and ship exhaust can be easily incorporated into developing clouds. These preferred regions tend to correspond to oceanic sub-tropical high-pressure regions, especially off the west coast of the major continents. Of course, the presence of commercial shipping lanes is also important; ships powered by diesel engines (such as container and tanker ships) are efficient sources of CCN particles and therefore more likely to produce tracks. Though the satellite example discussed above is of ship tracks off northern Europe, this is a relatively rare place to see such a widespread occurrence of tracks. The west coast of North America is a more common location for extensive ship tracks. Tracks are also commonly found off the west coast of South America (Peru, Chile), the Kamchatka Peninsula in the northwest Pacific, and off southern Africa.

High resolution details of ship tracks in the lower middle portion of the image shown at beginning of this chapter. The observations are at 1200 UTC (top) and 1340 UTC (bottom). During these 100 minutes, the ship tracks have moved from the tail of each arrow (top) to the head of each arrow (bottom), due to the combined effect of the ship movement and the local wind vector during this time period. (Data from the MODIS instrument on the Terra satellite at top and Aqua satellite at bottom.)

1200 UTC

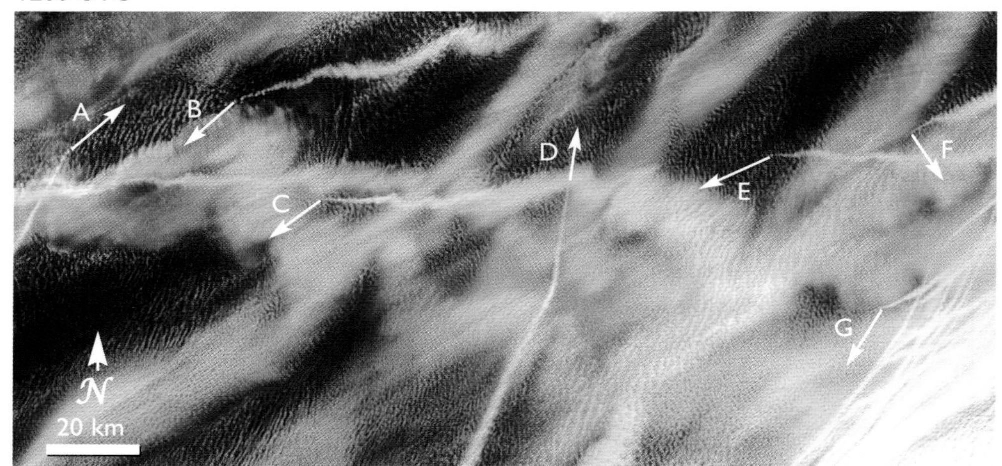


1340 UTC

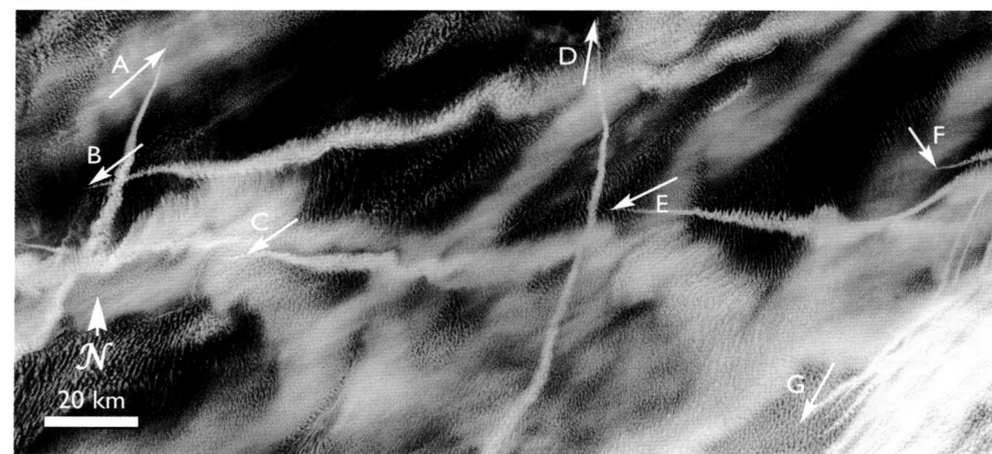


Below: Cloud droplet radii in micrometers for ship track clouds and the background liquid water clouds in which they form. The individual ship tracks correspond to those shown in the 1340 UTC image on the right. Relative to the unpolluted background clouds, substantially smaller droplet sizes are found within the tracks as expected. (Data from the MODIS instrument on the Aqua satellite.)

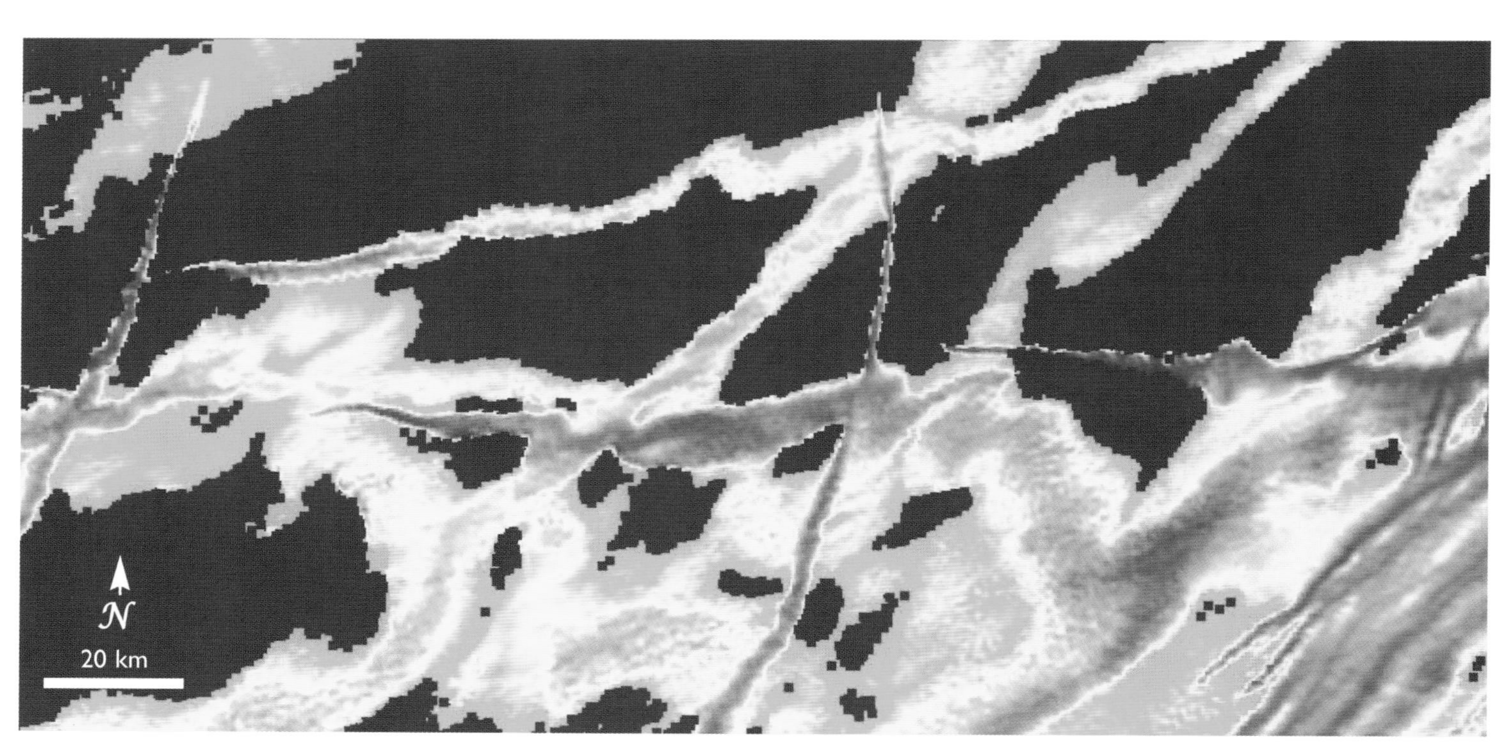


**Cloud Droplet Effective Radius** (μm)

4 8 12 16 20

# Airplane Contrails

Patrick Minnis

Persistent spreading contrails at various stages of growth and dissipation over Hampton, Virginia on October 10, 2001. (Photograph by Patrick Minnis.)

One of the most visible signs of human influence on the atmosphere is the condensation trail, or contrail, formed behind high altitude aircraft. This anthropogenic cirrus cloud can occur as a single line or in imposing geometrical formations as clusters of crisscrossing or parallel lines. Like natural cirrus clouds, contrails are composed of ice crystals and can produce the same dramatic optical displays, especially around sunrise or sunset. Persistent contrails also play a role in climate because they reflect sunlight and trap infrared radiation just like their naturally formed cousins. Thus, the presence of a contrail cluster in an otherwise clear sky can diminish the amount of solar energy reaching the surface during the daytime and increase the amount of infrared radiation absorbed in the atmosphere at all times of day. These opposing effects can simultaneously cool the surface and warm the air within the troposphere. Currently, the overall impact appears to be a warming effect, but research is continuing to unravel the role of this phenomenon in climate change.

The only difference between natural cirrus clouds and contrails is their origin. Natural cirrus clouds typically require an excessive amount of humidity to form out of thin air. Amorphous, liquid water droplets readily form when the relative humidity slightly exceeds

*Our Changing Planet*, ed. King, Parkinson, Partington and Williams
Published by Cambridge University Press 

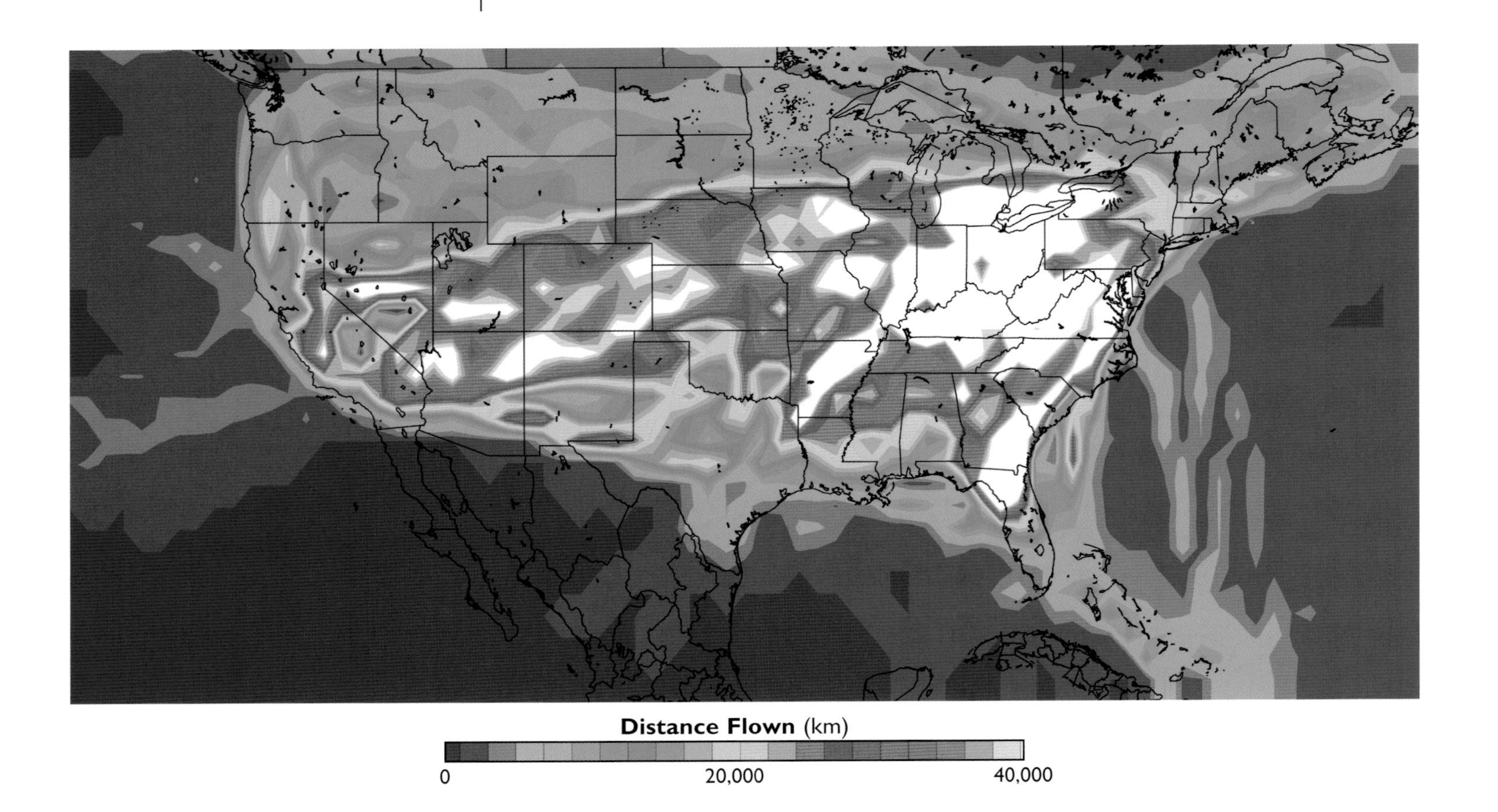


Commercial air traffic density over the United States on September 10, 2001. The cumulative length of all flights between 7 and 15 km altitude within each 1° latitude-longitude box provide a means to estimate the potential coverage of contrails over the United States.

100%, the saturation or dewpoint. At temperatures below freezing, ice clouds can exist at lower humidities because the thermodynamic relationship between the solid and gas phase differs from that between the liquid and gas phase of water. However, extra water vapor is needed to enable the gas molecules to line up in the regular formation of an ice crystal lattice. Most natural cirrus clouds form at humidities near or above the dewpoint, usually as a result of forming a water droplet first. Once formed, the ice crystal can grow, even at lower humidities because less water vapor is needed. Thus, the cold atmosphere at high altitude can often support the existence of a cirrus cloud, but is not moist enough to form one.

When the temperature is less than about −39°C and the conditions are right for cirrus to exist, aircraft exhaust can short circuit the natural cirrus formation process. The mixture of the ambient air with the water vapor in the exhaust temporarily raises the humidity above the dewpoint and causes initiation of a multitude of tiny ice crystals. If the humidity is high enough, the crystals continue growing and the contrail spreads. Otherwise, it dissipates rapidly or gradually depending on the relative humidity. Persistent contrails can often grow into natural-looking cirrus clouds within a few hours, a phenomenon that is best observed from space. Although they typically last for only 4–6 hours , some clusters have been observed to last more than 14 hours and travel thousands of kilometers before dissipating. These persistent contrails are estimated to have caused cirrus cloud cover to rise by 3% between 1971 and 1996 over the United States and are well-correlated with rising temperatures, though not the only cause of them.

Increasing air traffic in nearly every country of the world will cause a rise in global cirrus coverage unless the upper troposphere dries out or advances in air traffic management and weather forecasting or in aircraft propulsion systems can be used to minimize contrail

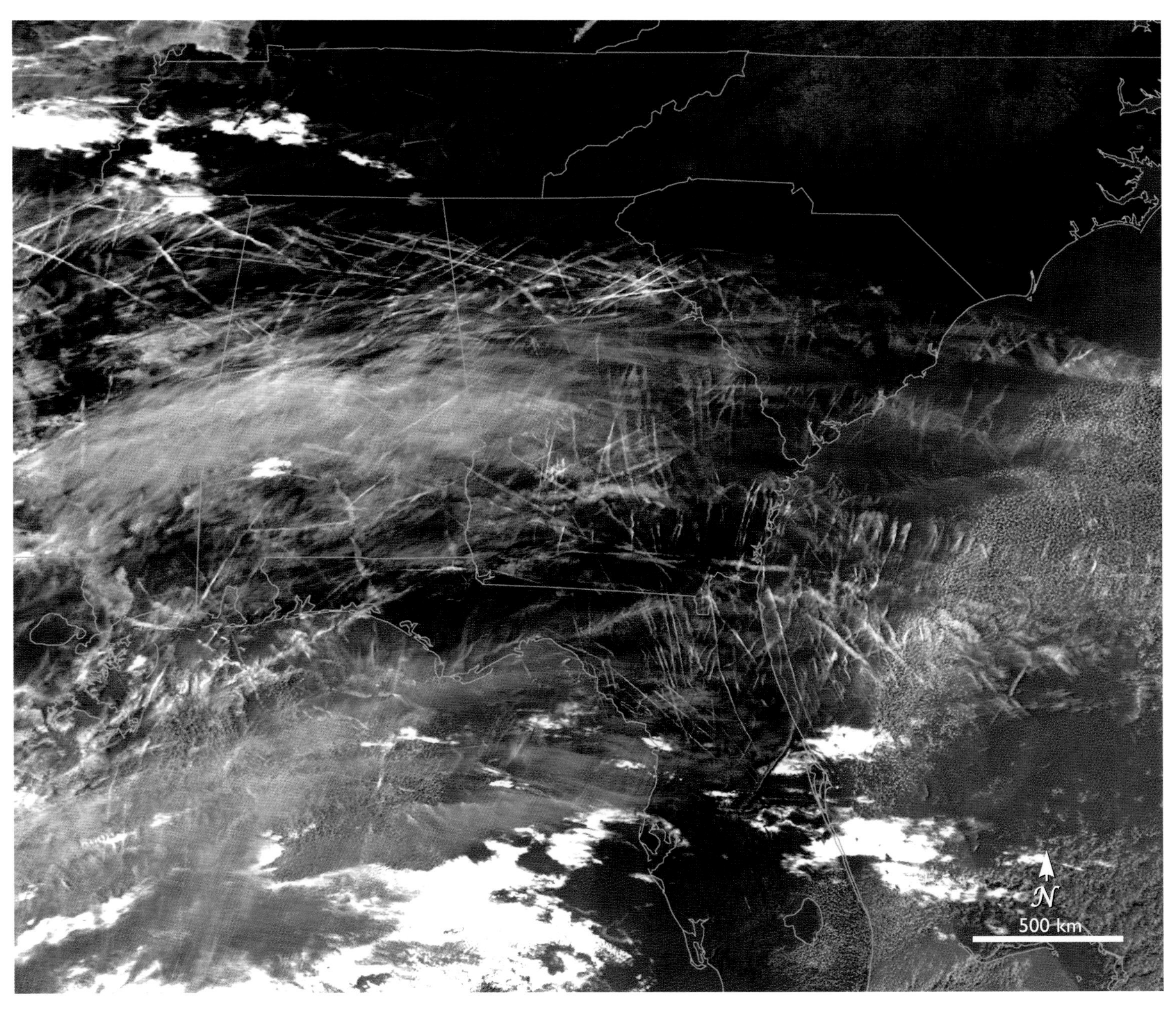


formation. Scientists are using satellite data and detailed models of the atmosphere to better understand contrail formation and dissipation in order to predict future cirrus coverage and climatic effects. The same models and data are also critical for designing ways to mitigate the climate effects of increased air travel. Perhaps, in the future, vivid atmospheric optical displays will be reserved for natural cirrus clouds and man's flight will be barely detectable from the ground and from space.

Enhanced infrared image of widespread contrails over the southeastern United States on January 29, 2004. (Data from the MODIS instrument on the Terra satellite.)

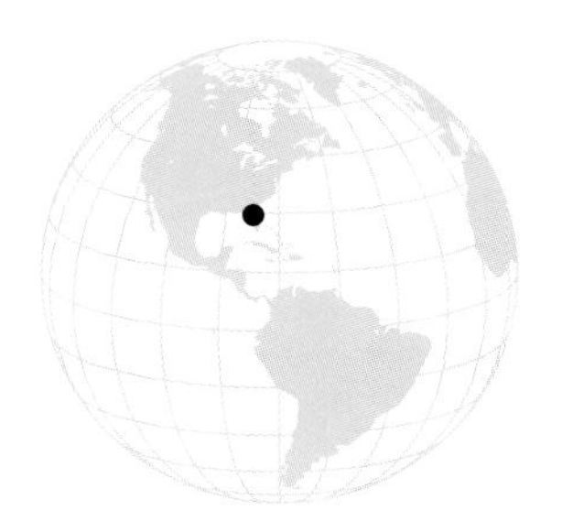

Steffen Beirle
Ulrich Platt
Thomas Wagner
Mark O. Wenig
James F. Gleason

# Weekly Cycle of Nitrogen Dioxide Pollution from Space

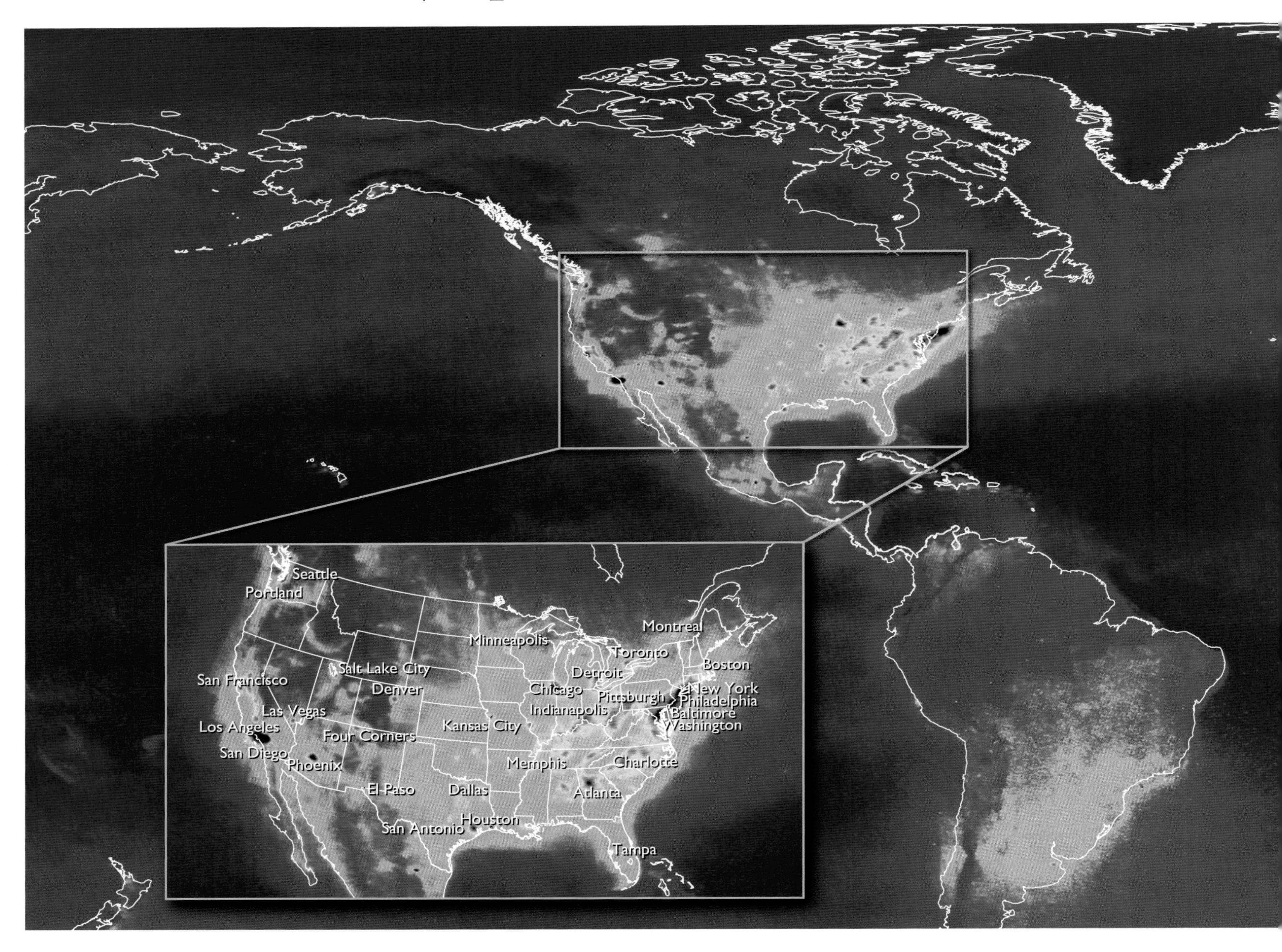


Amount of nitrogen dioxide ($NO_2$) in the troposphere for 2006. The principal sources of $NO_2$ occur over large populated regions, heavily industrialized areas, and individual power plants. (Data from the OMI instrument on the Aura satellite.)

Since the onset of the industrial revolution about 150 years ago, the atmospheric composition has changed dramatically. Amongst the various pollutants emitted by mankind, nitrogen oxides ($NO+NO_2$) are highly important. $NO_2$ is formed from NO in the atmosphere. NO is produced in high-temperature processes (fossil-fuel combustion and lightning) and by the oxidation of naturally produced nitrogen compounds. Nitrogen oxides play a key role in many atmospheric reactions, such as the formation of ozone smog, and when they are removed from the atmosphere they form acid rain.

*Our Changing Planet*, ed. King, Parkinson, Partington and Williams
Published by Cambridge University Press 

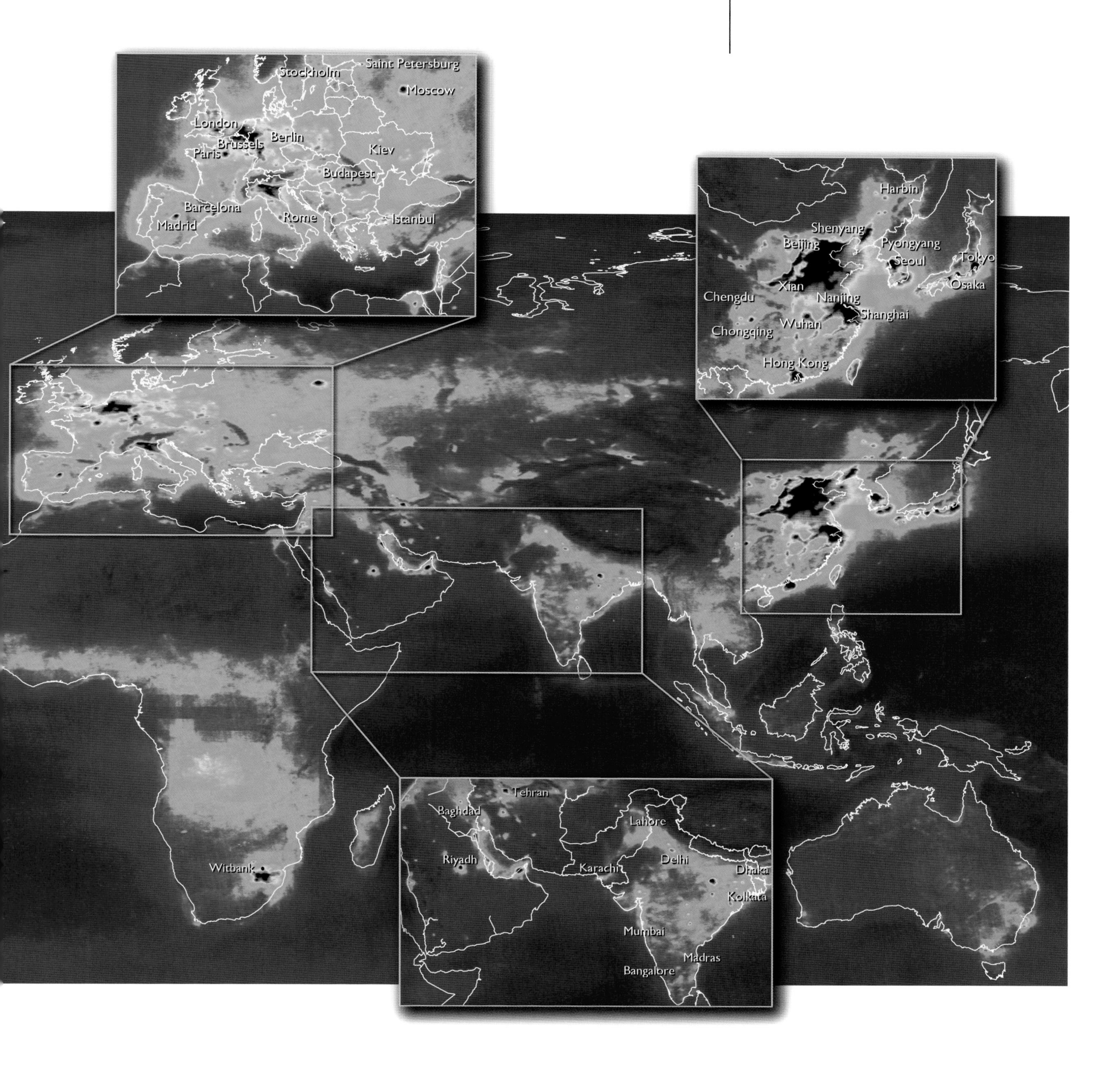


**Nitrogen Dioxide** ($10^{15}$ molecules/$cm^2$)

0 2 4 6 8 10

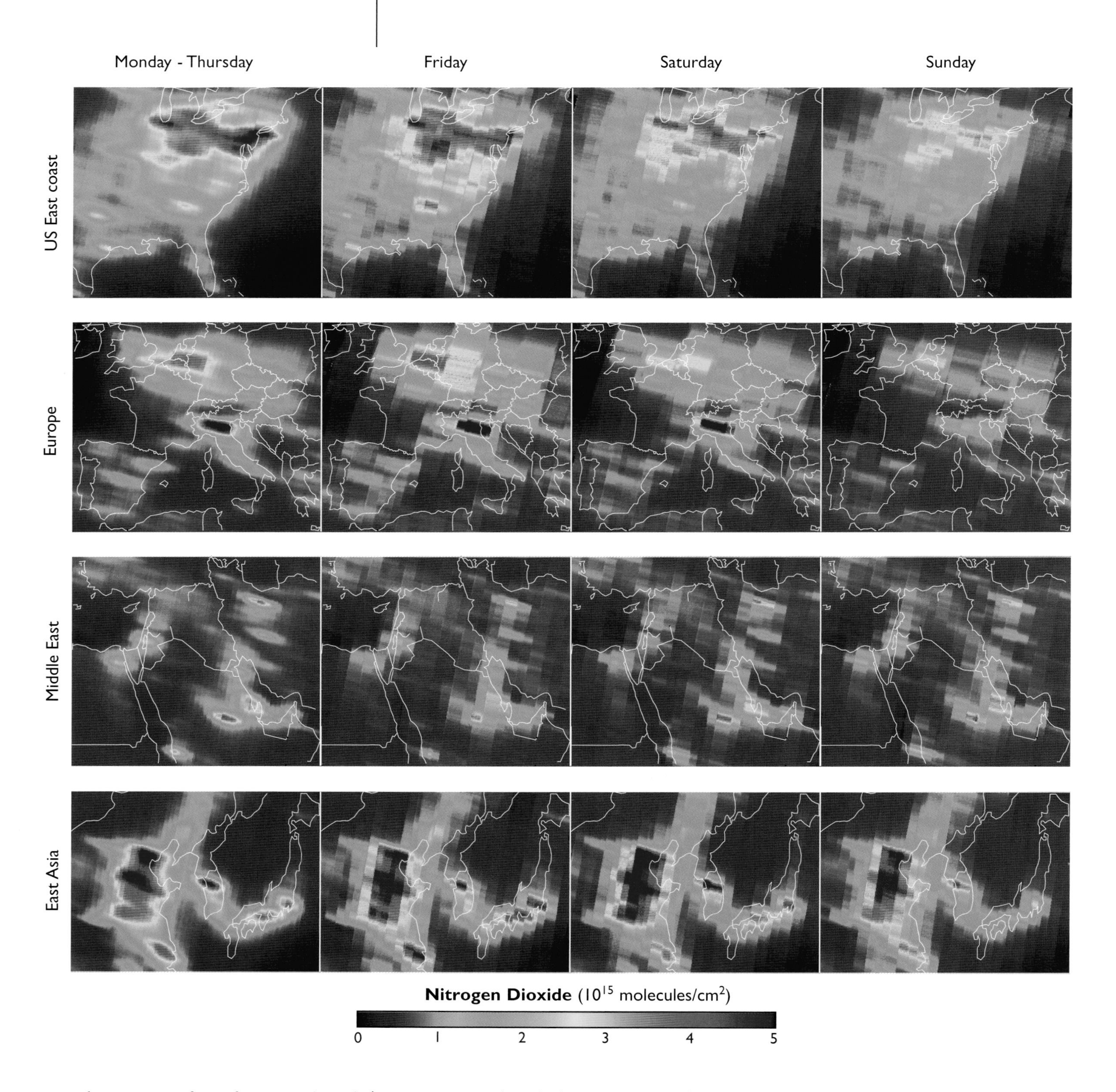


Column amount of $NO_2$ for various days of the week in the eastern United States, Europe, the Middle East, and eastern Asia from 1996–2001. (Data from the GOME instrument on ESA's ERS-2 satellite.)

Nitrogen dioxide, like every atmospheric trace gas, absorbs light of characteristic wavelengths. Instruments on Earth-orbiting satellites measure the sunlight reflected by the Earth and its atmosphere. Analyzing the reflected sunlight enables the measurement of $NO_2$ pollution levels on a global scale.

High amounts of $NO_2$ are observed in urban population areas with significant human activities, primarily transportation, and from the emissions from large coal-fired power plants. Other sources of $NO_2$ arise from biomass burning and lightning that account for the enhanced column densities observed over the Congo basin of central Africa and Indochina. Regions with high industrial activity, such as Mexico City, Hong Kong, Beijing, Brussels, and

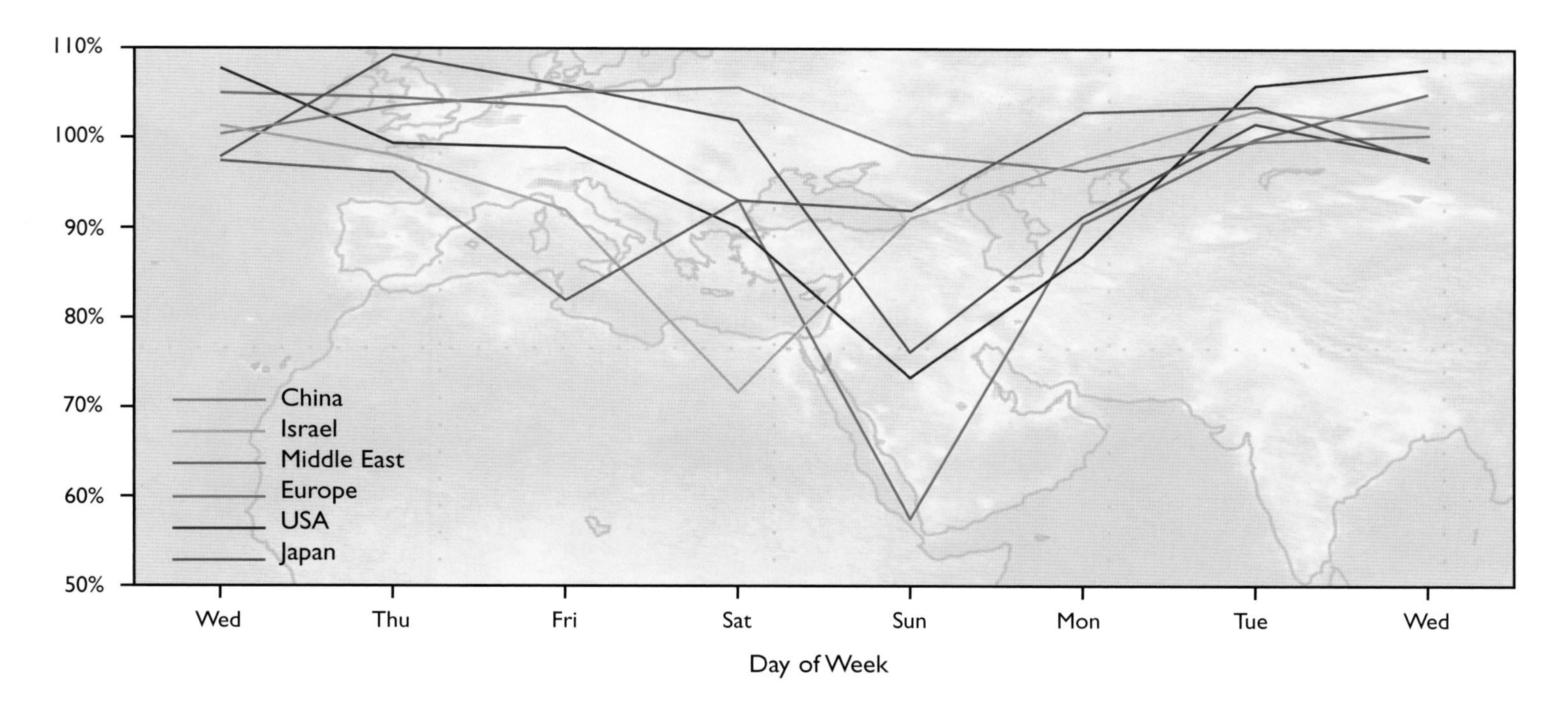


Weekend cycle of $NO_2$ for selected regions. All plots are normalized with respect to the median value for Monday-Thursday. (Data from the GOME instrument on the ERS-2 satellite.)

Pittsburgh clearly show up on a global column composite of $NO_2$. In addition to these and other metropolises, coal-fired power plants in the Four Corners region of the western United States and Witbank, South Africa, are clearly identified.

Human transportation practices in western countries follow the seven day cycle. Consequently, many emissions are reduced over weekends. We have actually observed this 'weekend effect' from space.

When these results (averaged over 6 years) are shown for different polluted regions, separately for working days and the weekend, a strong weekly cycle in nitrogen dioxide is clearly evident. In the industrialized regions of the United States, Europe, and Japan, a pronounced Sunday minimum of $NO_2$ pollution can be found: the levels of $NO_2$ in the atmosphere on Sunday are about 70% of the levels on working days of the week. In contrast, the weekly cycle in the Middle East is shifted: in Israel, the $NO_2$ is lowest on Saturday, the Sabbath being the religious day of rest, whereas Islamic countries show a minimum on Friday. In China, however, weekend levels are not significantly reduced compared to working days.

The detection of a weekly cycle of $NO_2$ from space means more than nice evidence of cultural and religious habits: as the different sources of nitrogen oxides like traffic, power plants, or industry have different weekly patterns, they can be discriminated by analyzing the weekend effect. Thus the satellite data help us learn more about the regional influence of different sources of nitrogen oxides.

Richard D. McPeters

# The Ozone Hole

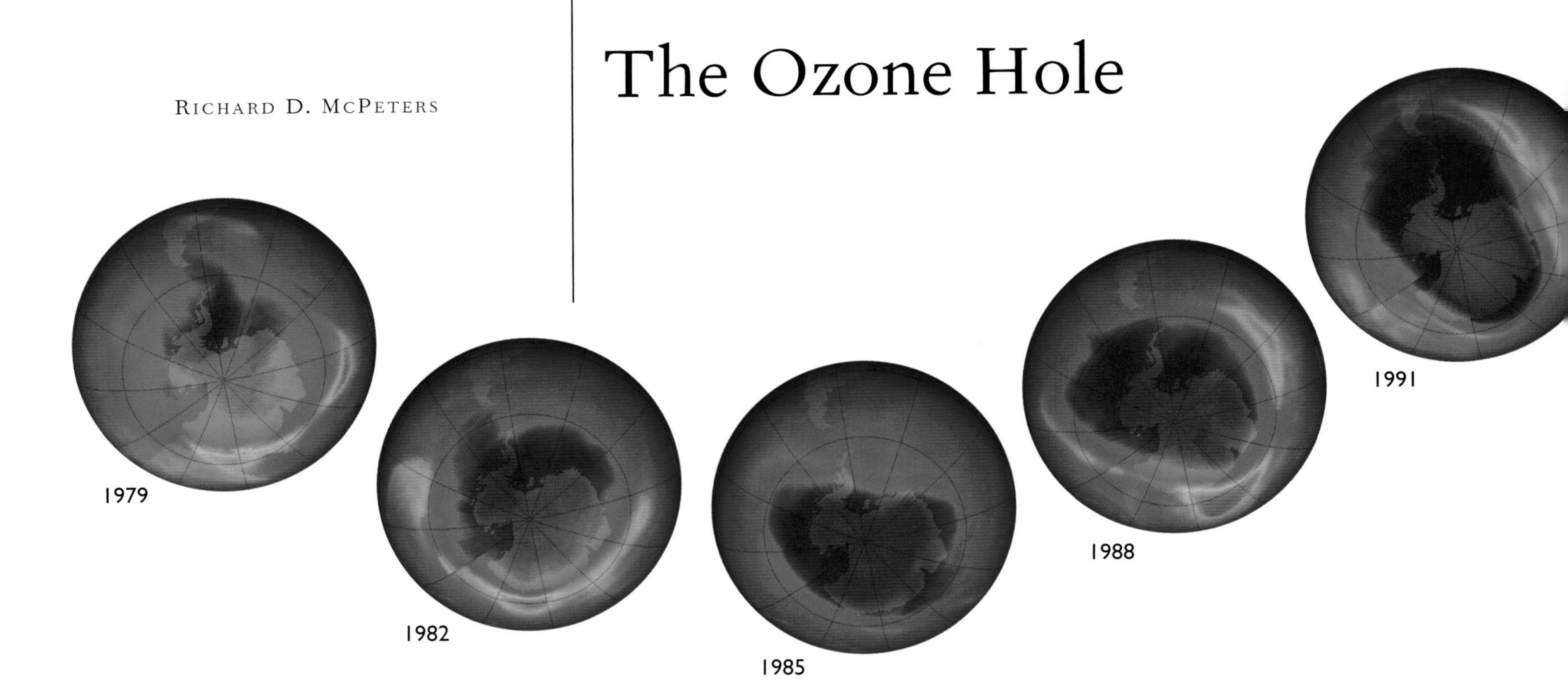


The minimum ozone content in the Southern Hemisphere occurs over Antarctica during September and October. This sequence shows the minimum total ozone content of the Southern Hemisphere for selected years from 1979 to 2006. (Data from the TOMS instrument on the Nimbus 7, Meteor 3, and Earth Probe satellites, and the OMI instrument on Aura.)

In the early 1980s scientists at the British Halley Bay station in Antarctica noticed that each October total ozone was getting lower and lower. After skeptically checking their instruments they finally realized that the results were real—a hole in the ozone layer really was developing each October, but they could not tell at what level in the atmosphere. Joseph Farman and colleagues finally published their surprising findings in 1985. It turned out that NASA scientists analyzing data from the Total Ozone Mapping Spectrometer (TOMS) on the Nimbus 7 spacecraft had also noticed the low total ozone values and were trying to convince themselves that they were real. They quickly confirmed that the ozone hole was large, covering almost the entire continent of Antarctica, and was not just over Halley Bay.

The stratospheric ozone layer, which is located at about 20 km in height, protects the Earth's surface and lower atmosphere from the Sun's harmful ultraviolet radiation. If the

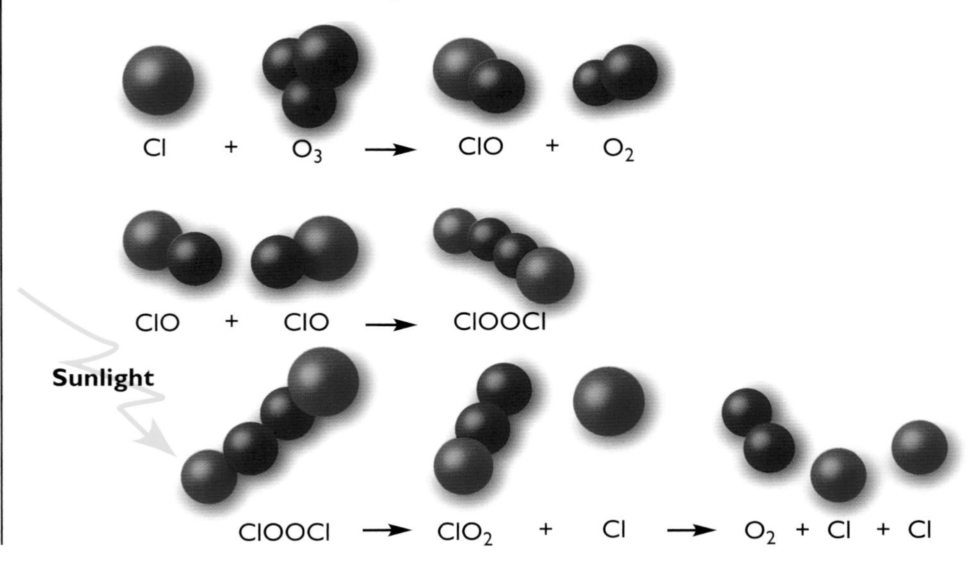


In the ozone hole, reactions on the surface of ice crystals create large concentrations of chlorine monoxide (ClO). ClO participates in a multi-step set of catalytic reactions that convert ozone into oxygen.

*Our Changing Planet*, ed. King, Parkinson, Partington and Williams
Published by Cambridge University Press 

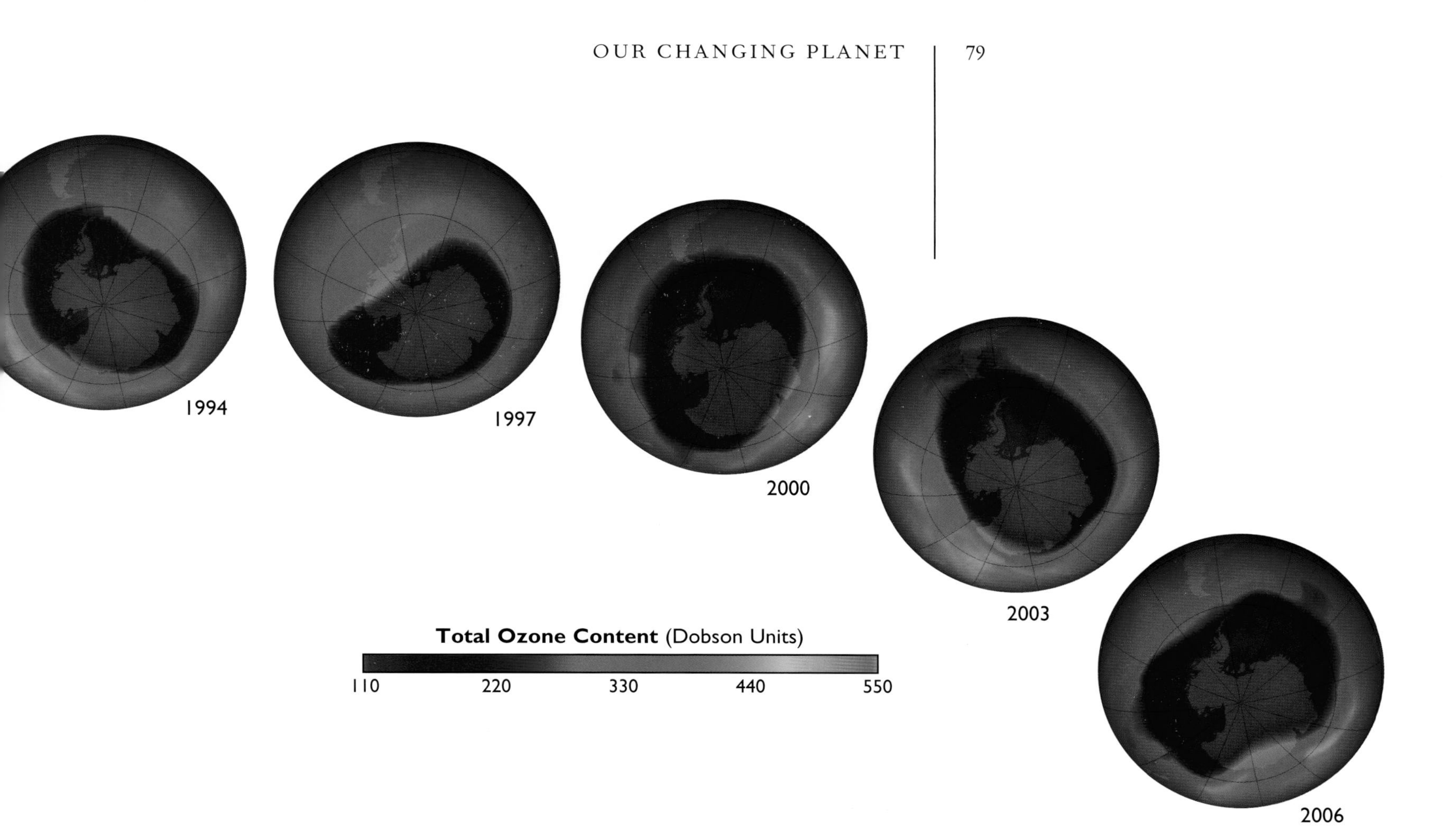


ozone hole were to spread throughout the world, the health effects could be serious. But it turns out that the ozone hole is very much confined to the South Pole. Ozone elsewhere in the world did decrease in the 1980s and 1990s, but only by 5% or so.

Conditions over Antarctica are unique. There is a whirlpool of air over the South Pole, known as the polar vortex. Each winter, during the long polar night, this rotating mass of air

Average area of the Antarctic ozone hole as a function of time, from 1980 to 2006, as compared to the area of North America and Antarctica. Vertical bars indicate maximum and minimum single day areas.

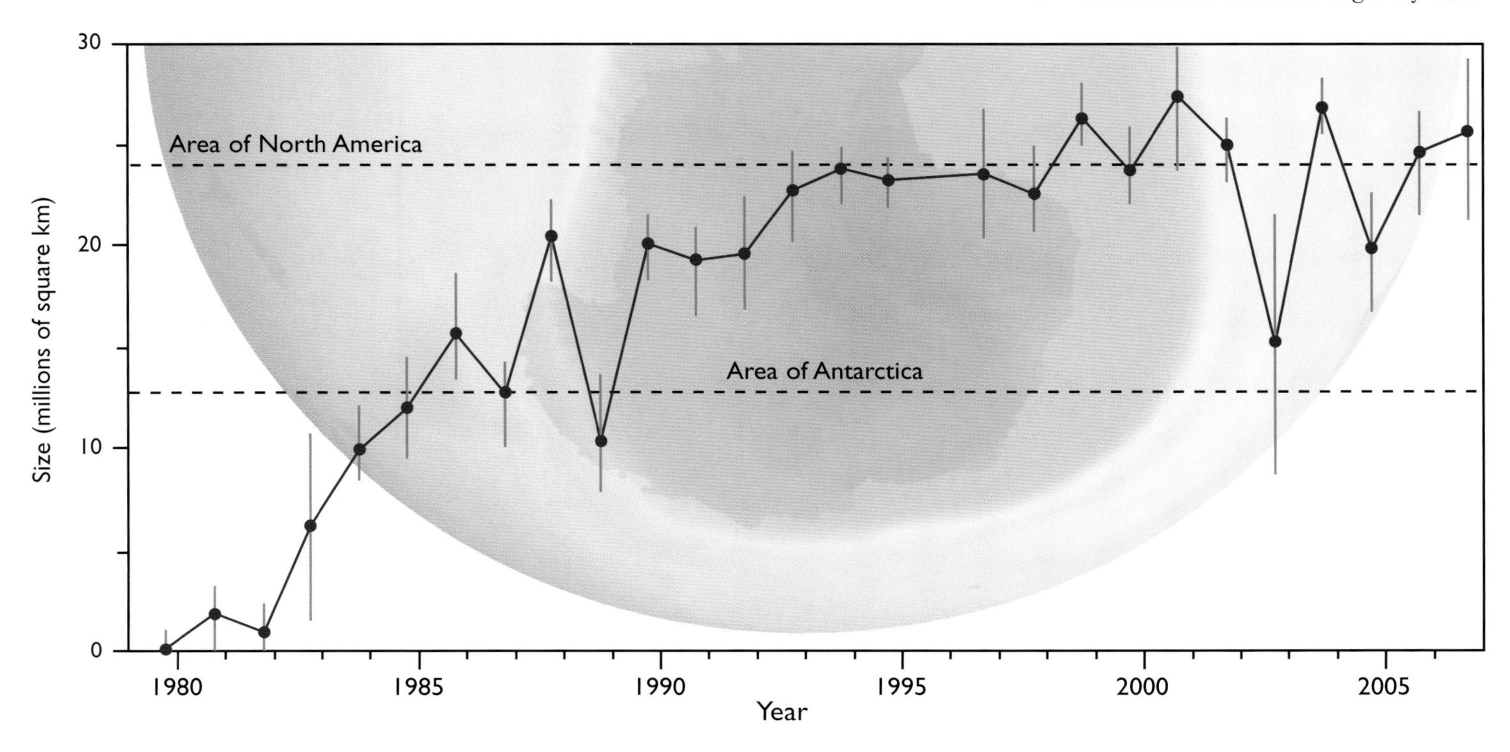

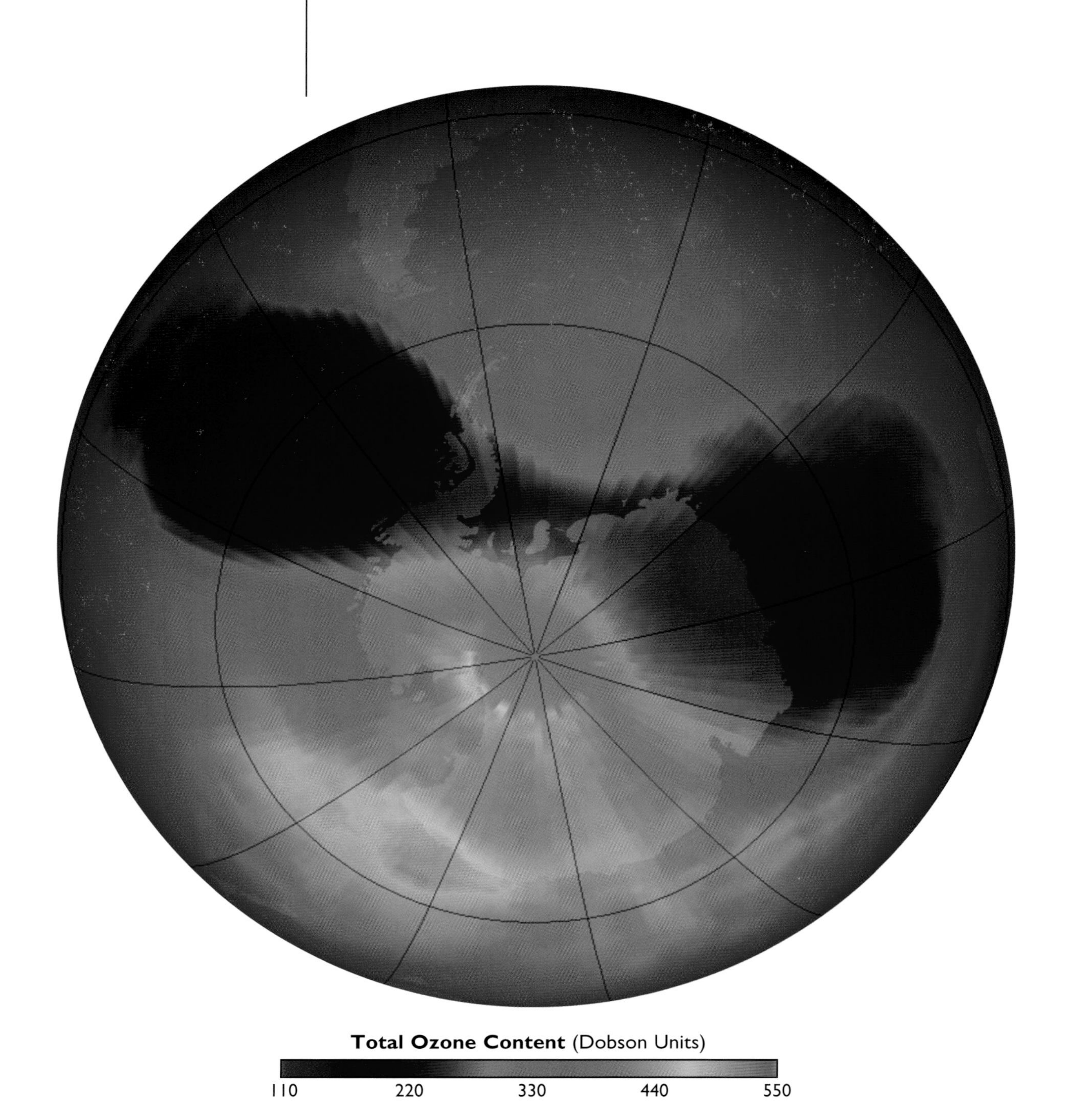


Total ozone content over Antarctica on September 24, 2002 split into two lobes, not previously observed since satellite observation began in late 1978. (Data from the TOMS instrument on the Earth Probe satellite.)

becomes extremely cold and forms a very stable vortex. This has been known since the 1950s when the first intensive scientific expeditions went to Antarctica. Nobody anticipated it, but if you add chlorine to the vortex you get an ozone hole. Chlorine levels in the atmosphere have been rising for the last half century as a result of increasing use of chlorofluorocarbons (or CFCs). CFCs are very stable, innocuous molecules until they rise to the top of the stratosphere where ultraviolet sunlight breaks them apart, releasing chlorine. Chlorine destroys ozone in a catalytic cycle—a series of reactions that has the net result of turning two ozone molecules into three oxygen molecules and returning the chlorine to repeat the cycle over and over.

Sunlight drives these reactions, so as sunlight returns to the Antarctic in August and September, the concentration of ozone begins to drop rapidly. In the extreme cold of the winter polar stratosphere (about −79°C), fine ice crystals form such that even faster reactions between chlorine and ozone take place. Aircraft flights into the ozone hole in 1987 clearly established this link between chlorine and ozone loss. If ozone-rich air from lower latitudes could mix in, this ozone destruction would not be very noticeable, but the polar vortex isolates this air as more and more ozone is destroyed. At the peak of the ozone hole in Antarctica, in late September and early October, essentially all the ozone between about 12 and 20 km altitude is destroyed.

Why is there not an ozone hole at the North Pole? The temperatures there are not quite as cold and fewer ice crystals are formed. But more importantly, the polar vortex in the north is not as stable and usually breaks apart at just about the time ozone destruction is becoming important.

From the early observations of ozone in the Antarctic starting in the 1950s until about 1980, ozone in the polar vortex was low, but only about 270 Dobson Units (DU), which means that the total column of ozone in the atmosphere, if brought down to the Earth's surface, would be 0.27 cm thick. Normal global average ozone is about 300 DU, ranging from a low of about 220 DU in some places to as high as 600 DU in others. 1984 was the first year in which ozone dropped well below 200 DU and there was clearly an ozone hole over Antarctica. Using 220 DU as the edge of the ozone hole, the average area covered by the ozone hole increased throughout the 1980s and in recent years has covered an area about equal to that of all of North America.

But then, in September of 2002, something unusual happened, something not seen in the 50 years of Antarctic observations—the ozone hole split in two. No one knows why, but there were a greater number of large planetary waves than usual, which warmed the upper atmosphere at the poles and cut ozone loss. However, scientists say that these large-scale weather patterns in the Earth's atmosphere are not an indication that the ozone layer is recovering.

We do expect to see the recovery of the ozone hole in the next several decades as the regulation of CFCs that was implemented worldwide in the late 1980s leads to reduced chlorine levels. But the lesson of the 2002 ozone hole is that we should also expect to be surprised from time to time.

Michelle L. Santee

# The Chlorine Threat to Earth's Ozone Shield

Thin clouds made of ice, nitric acid, and sulfuric acid mixtures form in the polar stratosphere when temperatures drop below −78°C (−108°F). In such polar stratospheric clouds, active forms of chlorine are released from their reservoirs. (Photograph over Iceland on February 4, 2003 by Mark Schoeberl.)

We now know that the severe depletion of the stratospheric ozone layer over Antarctica — the 'ozone hole'—is caused by chlorine chemistry. But where does the chlorine come from, and how does it get converted into the form that destroys ozone?

The primary source of stratospheric chlorine is chlorofluorocarbons, or CFCs, chemical compounds composed of chlorine, fluorine, and carbon. CFCs were introduced in the 1930s and used in a variety of industrial applications, such as refrigerants, solvents, aerosol propellants, foam packaging, and insulation. CFCs were used so extensively because they are stable and unreactive in the lower atmosphere and are therefore nontoxic. The very properties that made them so useful, however, turned out to pose a grave threat to the Earth's protective ozone layer.

Because they are inert and insoluble in water, CFCs are carried along intact as winds loft air from the troposphere, the lowest layer of the atmosphere, into the stratosphere, the region between about 12 and 50 km altitude. There, high-energy ultraviolet radiation breaks down the CFCs, releasing chlorine atoms (Cl). In the mid-1970s, it was recognized that chlorine

*Our Changing Planet*, ed. King, Parkinson, Partington and Williams
Published by Cambridge University Press 

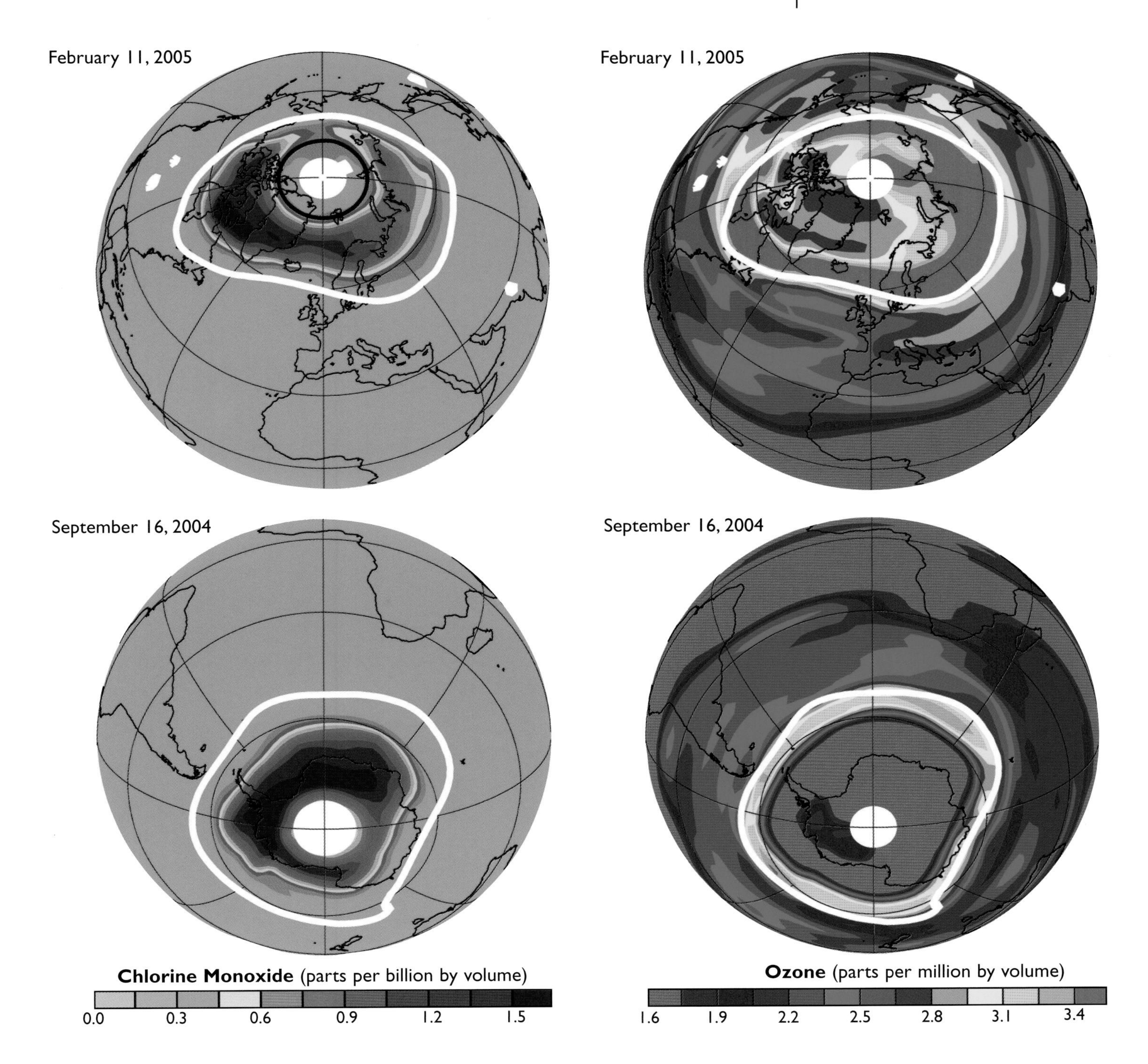


Maps of chlorine monoxide (ClO, left) and ozone ($O_3$, right) for representative days during late winter in the Northern (top) and Southern (bottom) Hemispheres. The data correspond to ~50 hPa in pressure or ~18 km in altitude. The white lines demark the approximate edge of the winter polar vortex in each hemisphere. Because ClO is only enhanced in sunlight, only daytime measurements are shown for this species; the black circle on the ClO maps identifies the edge of polar night, poleward of which is in continuous darkness at this season. (Data from the MLS instrument on the Aura satellite.)

from CFCs could destroy ozone in the stratosphere without being consumed itself. The net result is the conversion of ozone molecules ($O_3$) into oxygen molecules ($O_2$), with the Cl atom freed to attack more $O_3$. In this way a single Cl atom can destroy thousands of $O_3$ molecules. For their work in helping to elucidate these processes, Mario Molina and F. Sherwood Rowland were awarded the 1995 Nobel Prize in Chemistry.

Even after the connection with chlorine from CFCs was made, however, calculations indicated that ozone loss would be gradual and moderate. So scientists and the public alike were shocked by the news in 1985 that British scientists had discovered a dramatic downturn in ozone levels over Antarctica. Why hadn't this phenomenon been predicted?

The problem was that early models failed to account for polar stratospheric clouds (PSCs). In the very low temperatures that prevail in the polar regions during winter, water

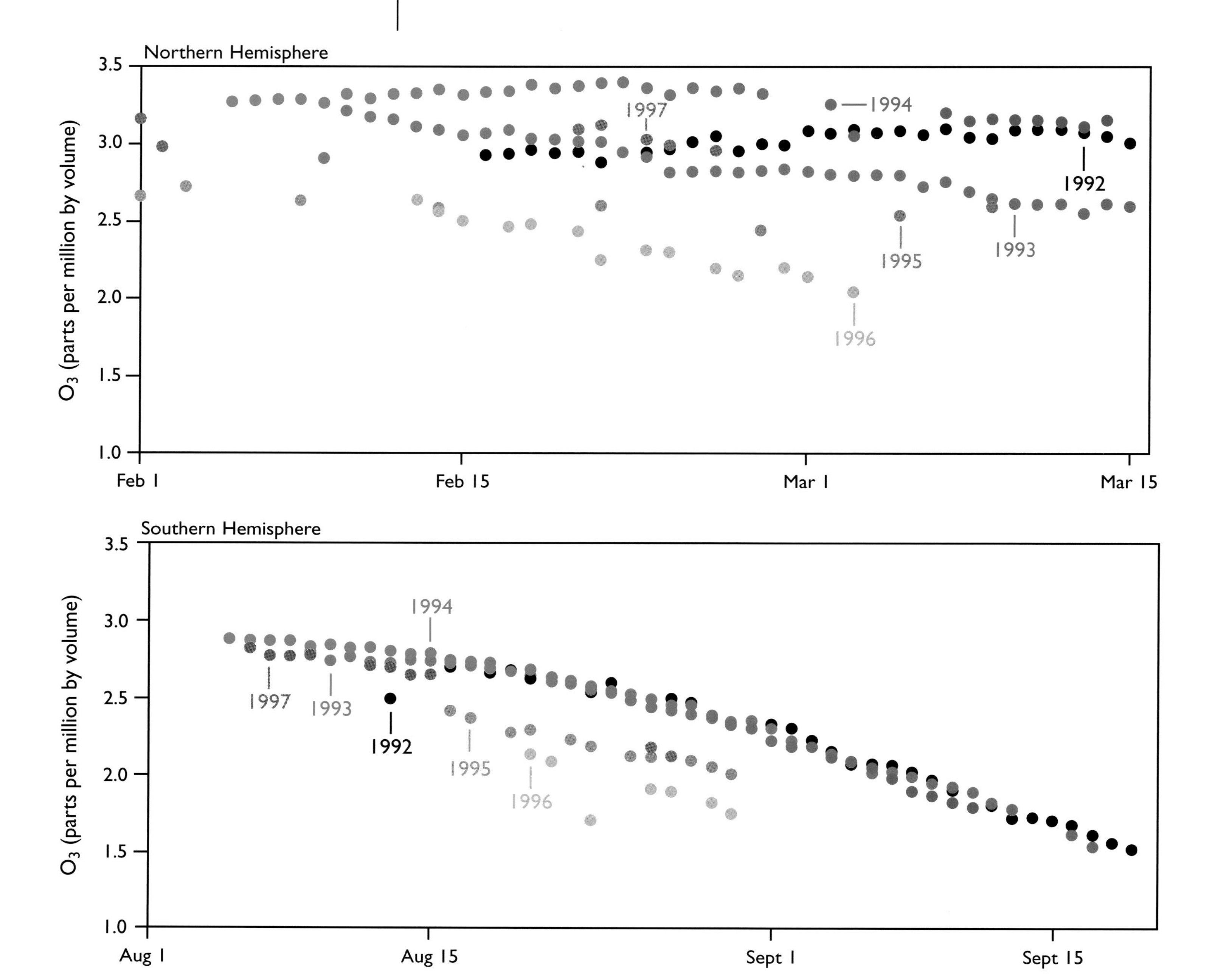


Daily average ozone values during late winter in the Arctic (top) and Antarctic (bottom), calculated within the ozone loss region in each hemisphere. The measurements correspond to ~70 hPa in pressure or ~17 km in altitude. Data from 6 years are plotted in different colors. In general, the Arctic experiences much greater year-to-year variability in ozone abundances and much less severe chemical ozone destruction. (Data from the MLS instrument on the UARS satellite.)

vapor and nitric acid condense to form clouds. PSCs were long known to occur—for example, they were described in 1911 by noted Antarctic explorer Robert Falcon Scott—but they remained little more than scientific curiosities until the discovery of the Antarctic ozone hole. It turns out that PSCs facilitate the conversion of chlorine to active forms that destroy $O_3$ by providing surfaces on which certain reactions can occur very rapidly.

The predominant form of reactive chlorine in the stratosphere—the 'smoking gun' that signals chlorine-catalyzed $O_3$ destruction—is the chlorine monoxide radical, ClO. Measurements made in 1987, using an instrument on a converted U2 spy plane, revealed an abrupt increase in ClO closely correlated with a dramatic decline in $O_3$ inside the Antarctic polar vortex, a region of air kept isolated from its surroundings by strong encircling winds. These results cemented the link between elevated levels of ClO and $O_3$ destruction.

It was not until the launch of NASA's Upper Atmosphere Research Satellite (UARS) in September 1991 that scientists were able to map the full three-dimensional distribution of ClO in the stratosphere and track its seasonal evolution around the globe. UARS found an

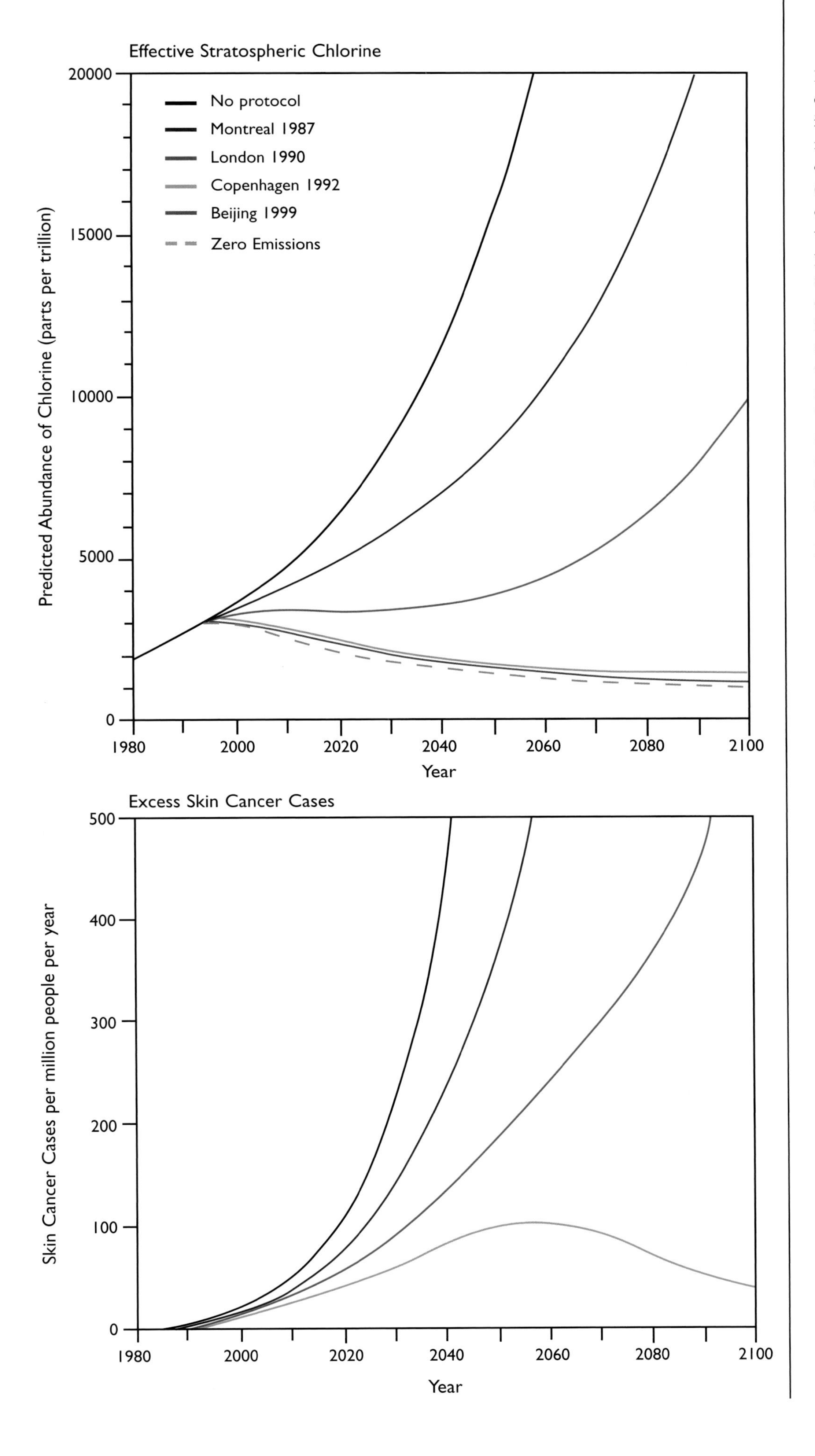


Predictions for the future abundance of effective stratospheric chlorine are shown in the top panel assuming (1) no Protocol regulations, (2) only the regulations in the original 1987 Montreal Protocol, and (3) additional regulations from the subsequent Amendments and Adjustments that were agreed upon at the city names and years indicated. The 'zero emissions' line shows stratospheric abundances if all emissions were reduced to zero beginning in 2003. The lower panel shows how excess skin cancer cases would increase with no regulations and how they will be reduced under the Protocol provisions. These increases are projected skin cancer cases above what occurred in 1980 before ozone depletion was observed, which was about 2000 per million population. (Adapted from *Scientific Assessment of Ozone Depletion: 2002*, World Meteorological Organization.)

almost exact coincidence between enhanced ClO and depleted $O_3$ throughout the Antarctic vortex.

UARS measurements made during Northern Hemisphere winter led to another startling discovery: vast stretches of the Arctic can be covered by ClO abundances as large as those observed in the Southern Hemisphere. So why has the Arctic been spared an ozone hole?

Winter in the Arctic stratosphere is not as cold nor long-lasting as that in Antarctica, leading to fewer, less persistent PSCs. Consequently, the horizontal and vertical extent, duration, and magnitude of chlorine activation are typically much smaller in the Arctic than they are in the Antarctic, and ozone loss, while it does occur, is correspondingly smaller. The Arctic also exhibits much more year-to-year variability than the Antarctic. Throughout the mid-1990s, the Arctic winters were unusually cold, and observations showed substantial ClO enhancement and $O_3$ loss. These years were followed by several warm winters, with very little $O_3$ destruction. Then the 1999–2000 and 2004–2005 Arctic winters were once again exceptionally cold.

At its peak (around 2000), chlorine in the stratosphere reached levels more than 6 times its natural abundance. In recent years, however, stratospheric chlorine has started to decline in response to regulations enacted under an international agreement known as the 'Montreal Protocol on Substances that Deplete the Ozone Layer' and its subsequent amendments. Because of the long residence times of CFCs, it will take several more decades to completely cleanse the stratosphere of excess chlorine and curtail ozone depletion. Despite reductions in chlorine, global climate change, which tends to warm the lower atmosphere, but cool the stratosphere, may lead to more PSC formation and exacerbate $O_3$ loss. In July 2004, NASA launched a new satellite, named Aura, to study the health of the Earth's atmosphere and look for signs of ozone layer recovery.

The ozone hole serves as a good example of how an apparently harmless class of chemical compounds can prove to be a serious danger to life on Earth and thus illustrates how human activities can change the natural state of our atmosphere in unexpected ways. But it also shows how national governments, industry, and society can identify a global environmental problem and work together to solve it.

Polar stratospheric clouds near Kiruna, Sweden. (Photograph by Lamont Poole, NASA Langley Research Center.)

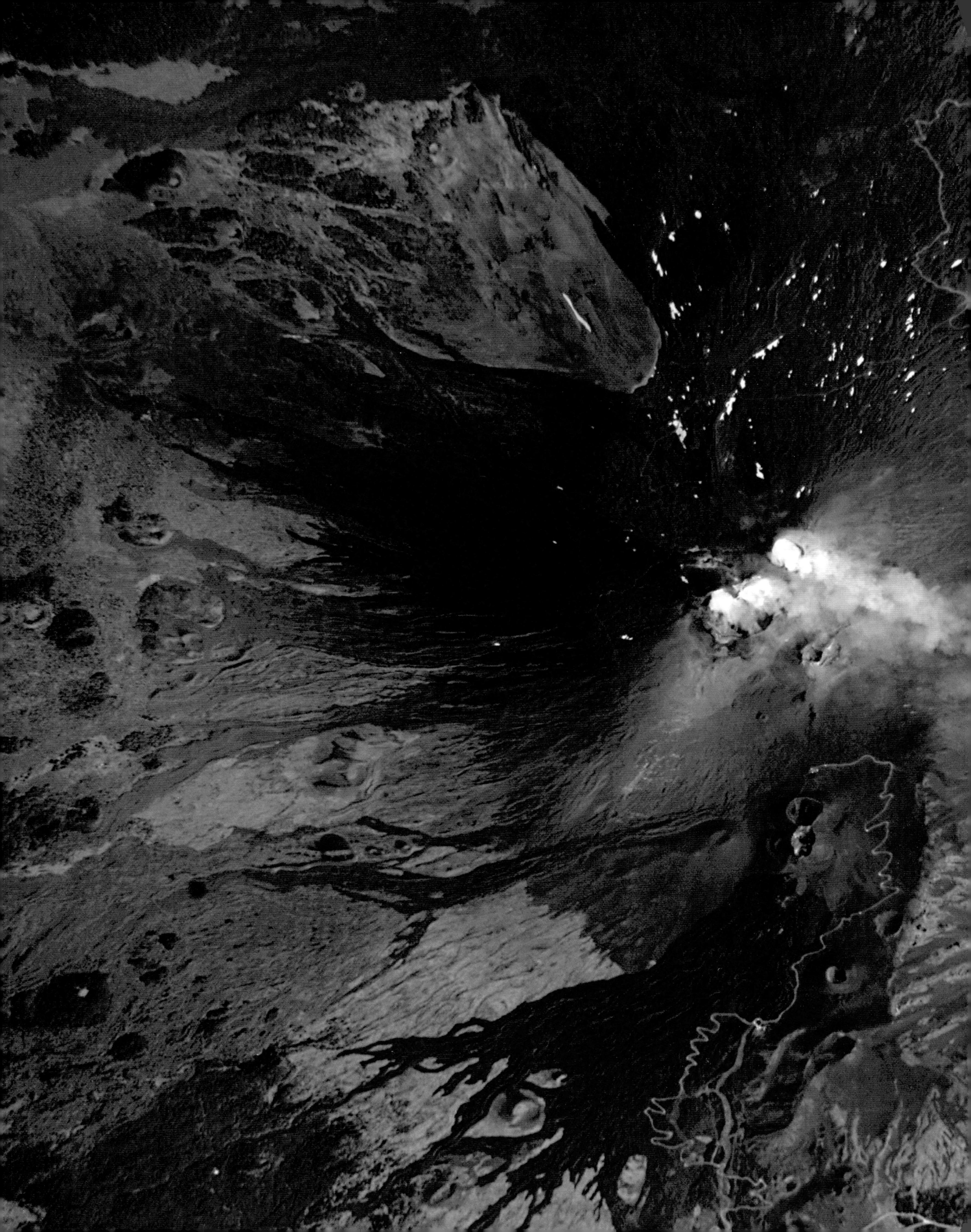

Mount Etna, Sicily, photographed from the International Space Station on August 2, 2006. One of the most consistently active volcanoes in the world, Mount Etna has a record of eruptions extending back to 1500 B.C. Locals reported hearing explosions on July 26, 2006 and this photograph, taken by an astronaut, captures plumes of steam and possibly ash originating from summit craters a few days later. (Photograph from the International Space Station, NASA Image Exchange; image number ISS013-E-62714.)

# The Vital Land: Introduction

Kim C. Partington

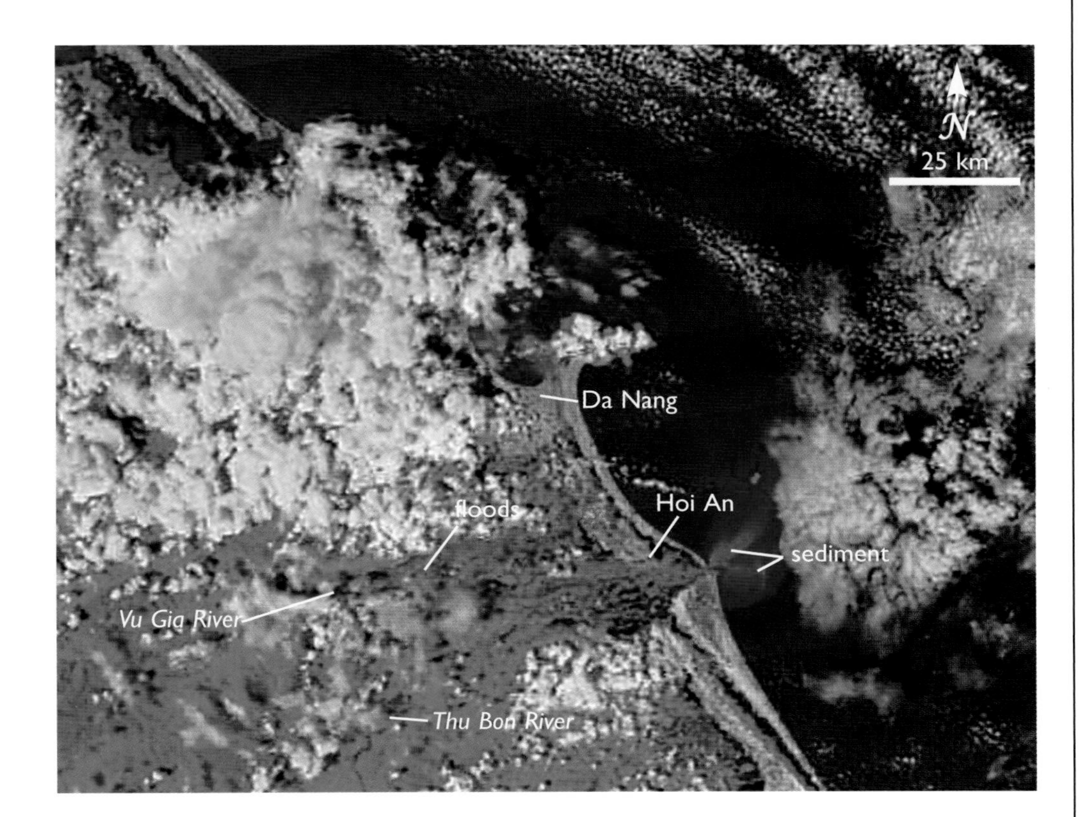


Flooding on the coast of Vietnam, October 3, 2006. Typhoon Xangsane pounded the coastal city of Da Nang with sustained winds of 150 km per hour and heavy rain on October 1. When this image was taken, two days later, the clouds had cleared enough to enable a view of the hard-hit Vietnam coast. The land between the Vu Gia River and the Thu Bon River is covered with water. Mud gives the water on land its pale blue color while a similar pale blue indicates sediment-laden water emptying into the ocean. By the time that this image was taken, the storm had killed 41 people in Vietnam, and damage to infrastructure was extensive. The red box indicates an active fire. (Data from the MODIS instrument on the Aqua satellite; image courtesy of the MODIS Land Rapid Response Team.)

Our species, *Homo sapiens*, has spent the last 150,000 years or more adapting to life on land, yet it is a mark of our success that many of us feel removed from its control over our lives. This success has been sustained by the development of advanced technologies, but these technologies do not protect us from some of the more violent natural events that take place. The land itself is not as stable as it seems, and we are passengers on a planetary-scale process of convection that has its origin deep within the planet. This process can often be violent, brought about by a gradual build-up and sudden release of stress in the form of earthquakes and volcanic eruptions, and our rapidly increasing population is often concentrated in areas of particular vulnerability. The chapters entitled "The San Andreas Fault: Adjustments in the Earth's Crust" and "Mount Pinatubo: An Enduring Volcanic Hazard," describe cases in point. Floods are also violent phenomena that can appear without warning, most often from sustained or very intense rainfall, and the chapter entitled "Extreme Floods" describes how the number of reported floods has been increasing over the last few decades. Catastrophic events can also take the form of epidemics of disease, in part encouraged by our ever-increasing population. Perhaps surprisingly, satellites can be used to monitor the seasonal development of environmental conditions conducive to outbreaks of diseases and "Satellite Monitoring of Ebola Virus Hemorrhagic Fever Epidemics" shows how satellites can be used to pinpoint when particular locations in central Africa are vulnerable to a particularly deadly disease called Ebola Hemorrhagic Fever. Satellites are being used to monitor these and other planetary danger zones in unprecedented detail, which may yet assist in predicting catastrophic events as well as assessing the aftermath.

Opposite: Mount Pinatubo, Philippines, imaged on March 6, 2001. The image features the volcano located 90 km north of Manila. Once standing 1,780 m (5,840 ft) above sea level, Mount Pinatubo was dormant for 600 years until 1991 when the volcano erupted twice in June of that year. The volcano spewed millions of tons of ash high into the atmosphere and covered nearby Clark Air Base. The result was the destruction of the volcano's summit, reducing the elevation to 1,485 m and leaving behind a caldera (a huge depression). Following the eruption there were massive mudslides triggered by heavy tropical rains that mixed with volcanic ash. The homes of more than 100,000 people were destroyed. (Data from the IKONOS satellite; image © 2005 Space Imaging.)

*Our Changing Planet*, ed. King, Parkinson, Partington and Williams
Published by Cambridge University Press 

N
75 km

While satellites can be used to watch for signs of major disruptive events and catastrophes, they can be used to detect more subtle changes too. Our greater reliance on technology makes many of us less sensitive to some of the more gradual changes in our environment that are taking place, as we become removed from the sources of production that sustain our lives and lifestyles. Even for those of us involved in primary production, slow changes can be masked by the passage of the seasons and the changing weather and vegetation patterns that these bring. Monitoring of these subtle changes requires long-term observations using instruments that are finely calibrated. In some cases, these instruments are required to observe planetary-scale conditions such as the temperature and reflectivity of the Earth's surface, with direct implications for the energy balance of the planet as a whole. The ability of satellites to monitor these two conditions is demonstrated dramatically in the chapters entitled "Temperature of the Land Surface" and "The Sunlit Earth," respectively. In other cases, the focus is on broad regions of the planet which play a particular planetary role in terms of climate or biology. The Earth's many glaciers, described in "Glaciers: Scribes of Climate; Harbingers of Change," provide invaluable indicators of decadal and longer changes in regional climate. In contrast, seasonal snow, described in "Snow Cover: The Most Dynamic Feature on the Earth's Surface," responds sensitively to changes in temperature and precipitation from year to year. Both snow and glaciers have been showing a general pattern of retreat. A general pattern of warming in the Northern Hemisphere is also implicated in changes taking place across the great swath of coniferous forest that encircles the Earth at sub-Arctic latitudes, as described in "Boreal Forests: A Lengthening Growing Season." Finally, in "Soil Moisture: A Critical Underlying Role," soil conditions required for successful agricultural production are shown to be sensitive to climate behavior many thousands of kilometers distant, demonstrating that although satellites can be used to observe changes taking place, unraveling of the mechanisms involved in creating these changes is, in general, much harder. Nevertheless, monitoring the conditions of these planetary features provides an important indication of how the Earth is changing in terms of its various land-based elements—the lithosphere (solid earth), the cryosphere (snow, glaciers, floating ice, and permafrost), the hydrosphere (water), and the biosphere (plants, animals, and other life-forms).

Over the last few decades, the observations provided by satellites have enabled insight into all of the events described in the following chapters, from the sudden unexpected incidents that create such havoc to the more gradual changes that may otherwise go unnoticed. Satellites are providing information needed to help us to adapt for the future; they are helping to identify areas at high risk from catastrophic natural events, and they are supporting relief efforts in the aftermath of disasters. Satellites will continue to provide a constant health check on our environment.

Opposite: The dependence of large populations on limited natural resources is illustrated dramatically in this June 3, 2002 image of the Nile Valley in northeast Africa. All of Egypt's large cities lie along the Nile, which supports fertile land amid the desert landscape. (Data from the MODIS instrument on the Terra satellite; image courtesy of the MODIS Land Rapid Response Team.)

David T. Sandwell
Bridget R. Smith

# The San Andreas Fault: Adjustments in the Earth's Crust

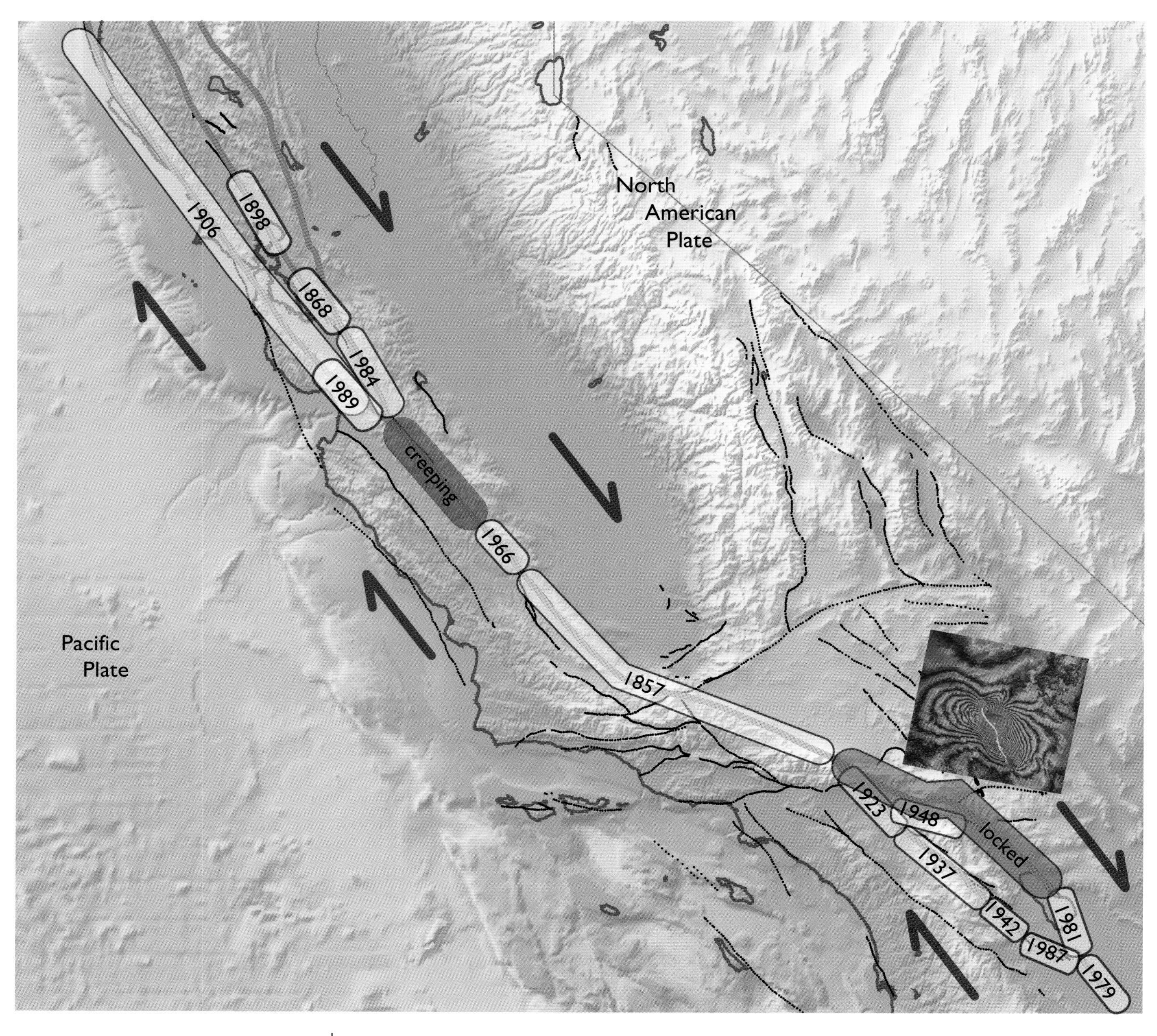


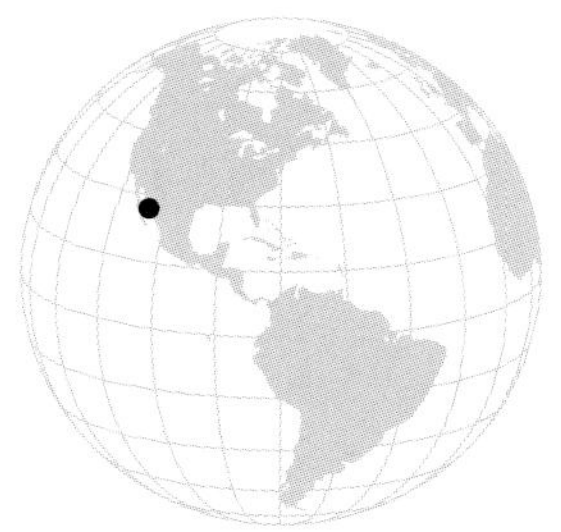

The sudden release of energy along a major earthquake fault is one of the most destructive forces of Nature. During the 1906 San Francisco earthquake, poorly constructed buildings collapsed in a matter of seconds killing thousands of occupants. This was followed by fires that destroyed 28,000 buildings and left more than half of the residents of the city (225,000) homeless.

Destructive earthquakes, such as the 1906 event, almost always occur along the boundaries of the tectonic plates, which are well mapped globally. While most plate boundaries

*Our Changing Planet*, ed. King, Parkinson, Partington and Williams

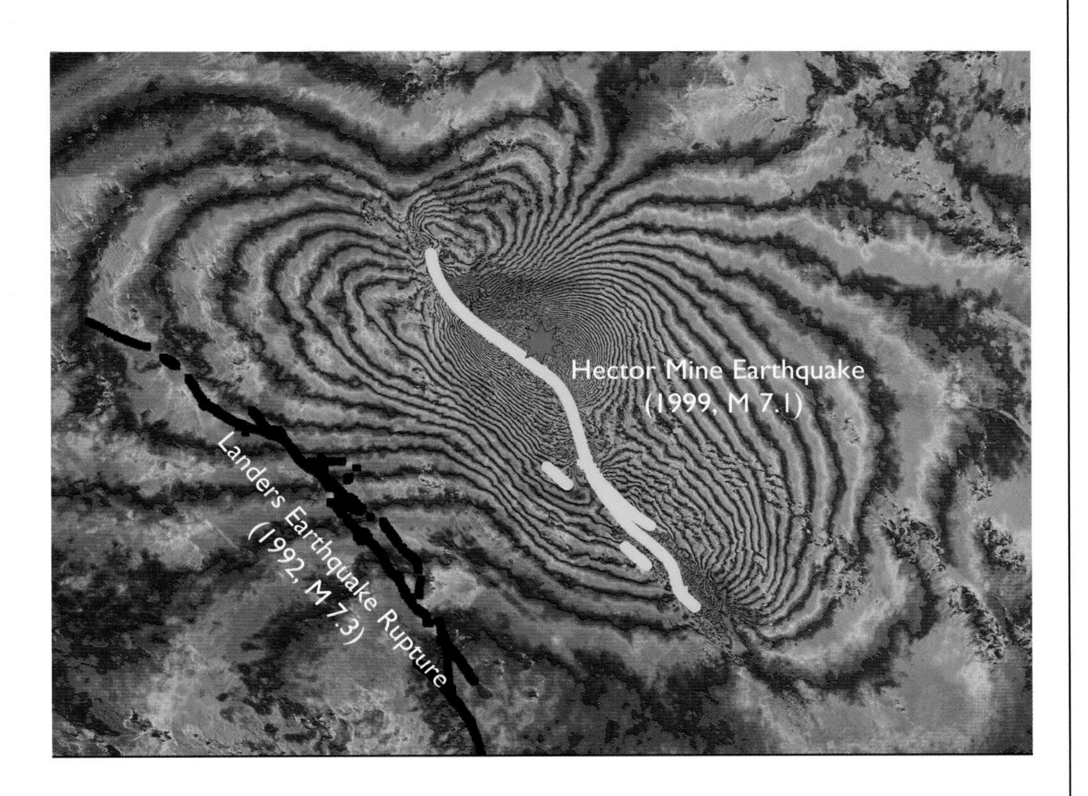


Radar interferogram. Since space-based geodetic measurements have become available, there has not been a major earthquake on the San Andreas Fault. However, two major earthquakes have occurred in the Mojave Desert in 1992 (Landers) and 1999 (Hector Mine). This radar interferogram shows the ground motion associated with the 1999 Hector Mine earthquake. The ERS-2 satellite radar imaged the land before and after the earthquake. The difference in radar phase between the two acquisitions, shown here, reveals the rupture in great detail; one color cycle (or 'fringe') represents 28 mm of ground motion. The technique of radar interferometry offers the ability to monitor fault zones on continental scales and it is highly complementary to the 400 continuously operating Global Positioning System receivers that have been deployed along the San Andreas Fault zone. (Data from the Synthetic Aperture Radar (SAR) instrument on the European Space Agency ERS-2 satellite.)

occur in the deep ocean and thus far from cities, a major transform fault (the San Andreas Fault) cuts across the most populated areas in California. Over a period of 500 years or so, every section of the San Andreas Fault will rupture to release tectonic stress, suggesting that residents need to be prepared for the inevitable.

At depths of greater than 30 km, the ductile nature of the rock results in plates sliding freely past one another. However, shallower fault zones are colder and more brittle and undergo stick-slip behavior. Initially, the shallow surfaces of the fault remain locked because of friction and surface imperfections; this interseismic period can last hundreds of years. Eventually the growing stress exceeds the strength of the rocks and the fault ruptures causing a co-seismic slip event. Recently scientists have discovered that slip can either be fast and destructive or slow and largely unnoticeable. Fast ruptures generate elastic waves that propagate outward and destroy buildings. Slow ruptures, which have only recently been detected using the Global Positioning System (GPS), can release the tectonic stress over several days. Of course, it is important to monitor all faults in populated areas to establish the rate of stress buildup and release. Following the co-seismic event, the fault continues to slip generating aftershocks and dissipating the stress concentrations at the margins of the rupture. This post-seismic deformation can lasts for tens of years. Some scientists believe there is also a short period of concentrated deformation just prior to a major earthquake, although this period of 'pre-seismic deformation' is poorly documented.

Over geologic timescales (millions of years), the earthquake cycle is constantly repeated resulting in nearly continuous motion. In terms of global plate tectonics, earthquakes are relatively unimportant since they are just minor squeaks along the plate boundaries. However, on human timescales, the earthquake cycle is highly irregular and only partially observed on a single fault segment. A full understanding of the process will require observations over many cycles by many generations of people or by observing many active fault systems in various stages of their earthquake

Opposite: The major sections of the San Andreas Fault zone. These sections undergo repeated earthquake activity except along the creeping section where the plates slide smoothly at all depths. Recent major earthquakes are dominated by the 1857 Fort Tejon Earthquake and the 1906 San Francisco Earthquake. The southernmost locked section of the San Andreas Fault has not experienced a major earthquake in at least 300 years. The next event along this section could release more than 7 m of accumulated slip; typically large California earthquakes have a maximum slip of 6 m.

Images taken from Knob Hill after the earthquake in San Francisco, California on April 18, 1906. This view is from the corner of Van Ness Street and Washington Street. (Photograph credit: W. C. Mendenhall and US Geological Survey.)

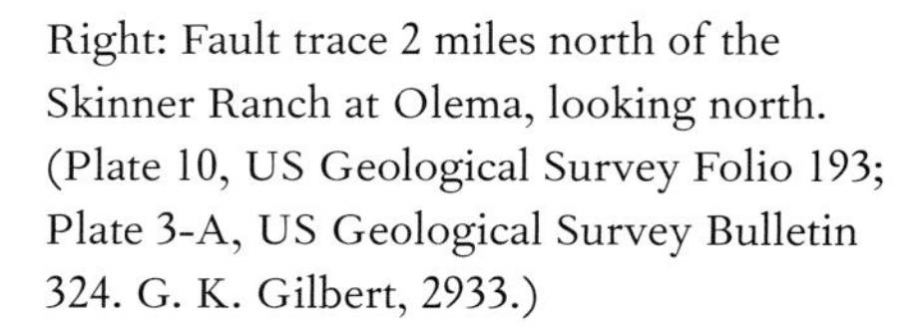

Right: Fault trace 2 miles north of the Skinner Ranch at Olema, looking north. (Plate 10, US Geological Survey Folio 193; Plate 3-A, US Geological Survey Bulletin 324. G. K. Gilbert, 2933.)

Far right: A fence offset 2.6 m (8.5 ft) by a main fault one half mile northwest of Woodville, looking northeast. The fault fracture accompanying the earthquake is inconspicuous, although the horizontal displacement is considerable. Marin County, California. (Photograph submitted by G.K. Gilbert from the Earth Science Photographs from the US Geological Survey Library, by Joseph K. McGregor and Carl Abston, USGS Digital Data Series DDS-21, 1995.)

Modern reconstruction of the 1906 fence on the same site and in the same position. Note the displacement of approximately 7.6 m (25 ft) by 2006. (Photograph courtesy Tammi Grant.)

cycle. Modern space-based geodetic measurements such as GPS and radar interferometry have recorded all but the pre-seismic element of the earthquake cycle. With these new tools, scientists are beginning to understand the earthquake process and may someday be able to provide useful earthquake forecasts or at least assess the danger of individual faults.

# Mt. Pinatubo: An Enduring Volcanic Hazard

Peter J. Mouginis-Mark

Eruptions of Mt. Pinatubo from Clark Air Base in the Philippines, June 12, 1991. (Photograph by Dave Harlow, courtesy USGS.)

The June 1991 eruptions of Mt. Pinatubo on the island of Luzon, in the Philippines, were the second largest of the 20th century (only an eruption in 1912 in Alaska was larger). Material from Mt. Pinatubo had a major impact on the global atmosphere as about 20 megatons of sulfur dioxide were injected into the stratosphere. A large amount of volcanic aerosols was produced, circling the Earth at an altitude of about 25 km in 21 days, and covering 42% of the Earth's surface within 2 months. The effects of these aerosols on the short-term climate were observed by NOAA and NASA satellites, and a cooling of up to 0.5°C to 0.6°C was recorded in the Northern Hemisphere for the period 1992 to 1993.

The Mt. Pinatubo eruption also created 5 to 6 $km^3$ of volcanic material called 'ignimbrite' which formed from fragmented volcanic rocks produced by the violent eruption. This was deposited on the surrounding landscape to depths that locally exceeded 200 m and it continued to act as a great hazard to local people for a decade after the eruption. Most dramatically, tropical storms and typhoons frequently generated large mudflows known as lahars, by

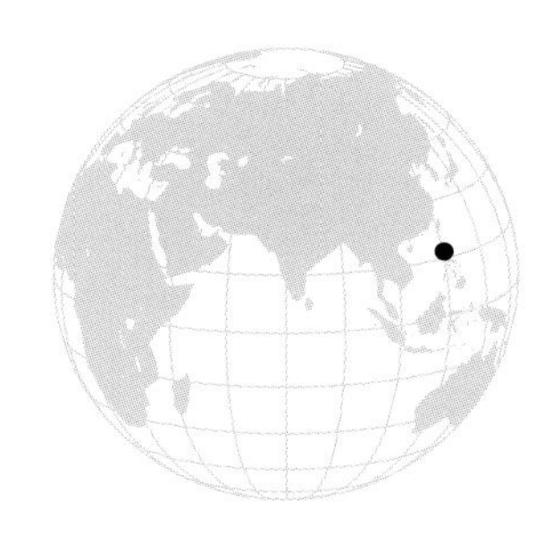

*Our Changing Planet*, ed. King, Parkinson, Partington and Williams
Published by Cambridge University Press 

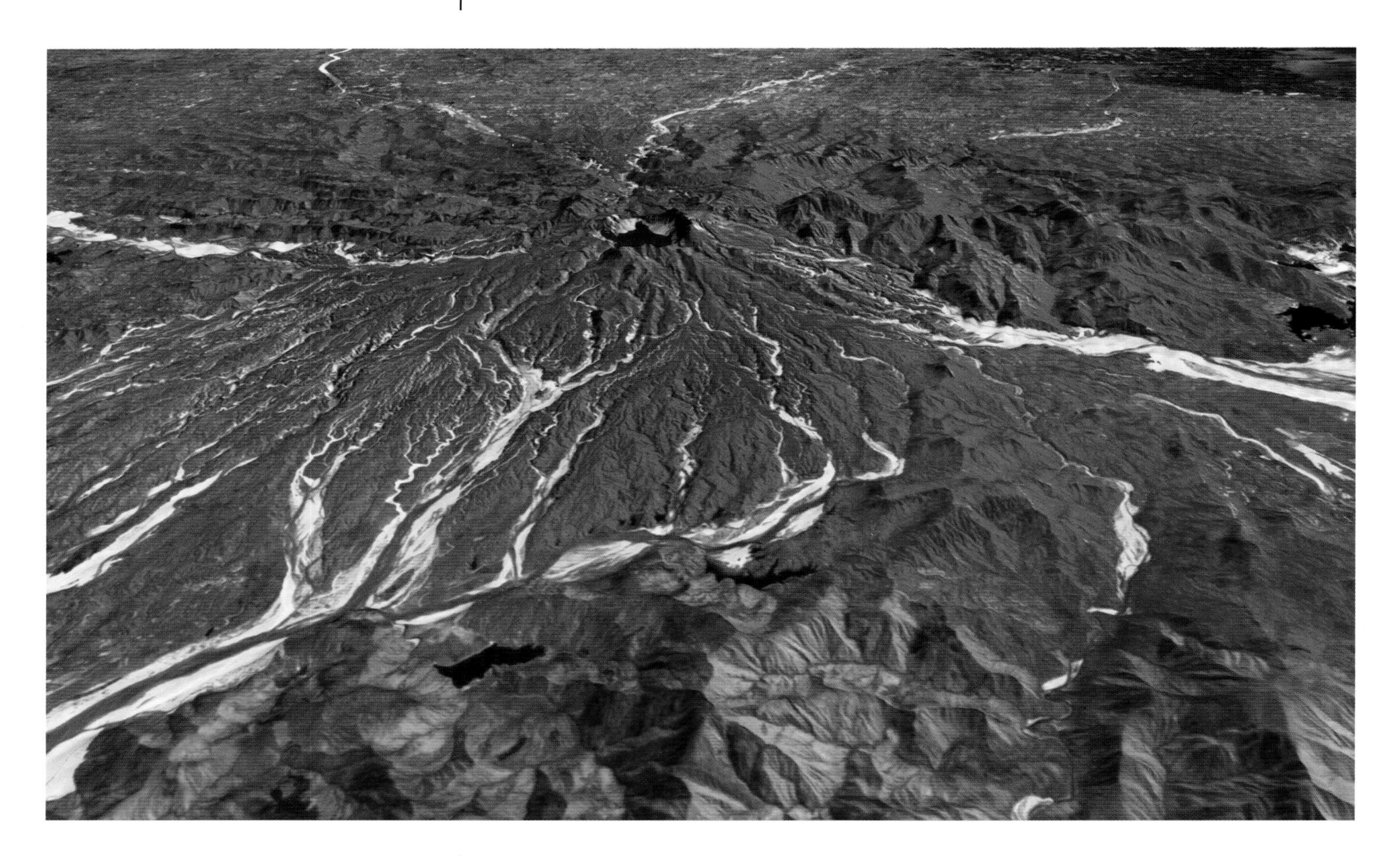

Elevated three-dimensional perspective of Mount Pinatubo on January 29, 2002. The summit crater is just above the image center. (Data from the ASTER instrument on the Terra satellite.)

eroding ignimbrite near the summit and re-depositing the material at the populated lower elevations, destroying buildings and modifying the landscape in a dramatic and dangerous fashion.

This process takes place in areas that are frequently difficult or dangerous to study. Satellite images of Mt. Pinatubo over the last decade have proven invaluable in monitoring the changing landscape. The areas of greatest lahar deposition have been the river flood plains including the Pasig-Potrero river system. Here a single lahar event may bury houses and fields under 2–5 m of reworked volcanic material. Closer to the summit, as much as 80 m of erosion has been recorded in a single day. Vegetation growth helps to stabilize the ignimbrite and provides a good measure of how much unstable material remains to be transported downslope. As a result, the rate of re-vegetation of the ignimbrite is also monitored using satellite data. Satellites have proven invaluable in monitoring Mt. Pinatubo during the evolution of the landscape from the initial eruption through gradual stabilization, providing input to risk assessment studies and an improved understanding of the process.

The great effect of lahars on the many towns downstream from Mt. Pinatubo is illustrated in this photograph. Many buildings like the house at right, have been almost totally buried. Residents who have stayed in the area raise their homes on tall pillars to help mitigate against the risk of future lahars. (Photograph by P. Mouginis-Mark, University of Hawaii.)

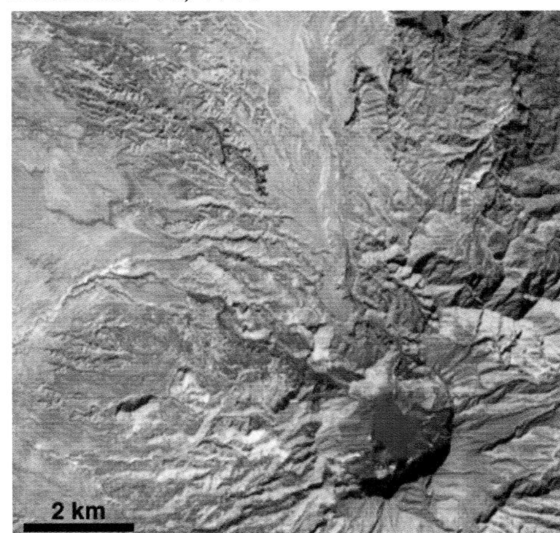


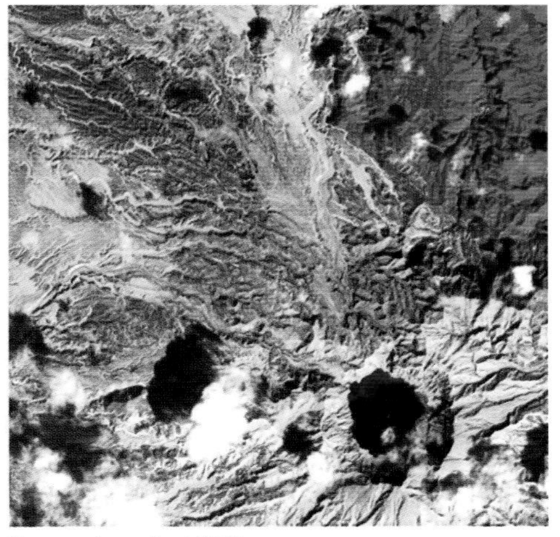


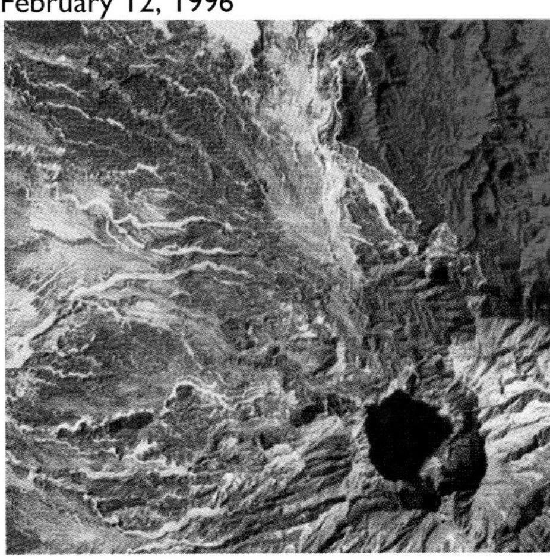

Time sequence showing the rate of growth of vegetation to the northwest of the summit of Mt. Pinatubo. Images were obtained on December 18, 1991; December 11, 1994; February 12, 1996, and December 5, 1998. The summit crater is at the lower right of each view and vegetation is indicated by red in this false-color image. (Data obtained from the SPOT satellite.)

Below: Spectacular scenery found close to the volcano summit. Erosion has cut deep canyons into the layers of ignimbrite that were formed during the 1991 eruptions. Note people at bottom center for scale. (Photograph by P. Mouginis-Mark, University of Hawaii.)

# Extreme Floods

Sebastien Caquard
G. Robert Brakenridge

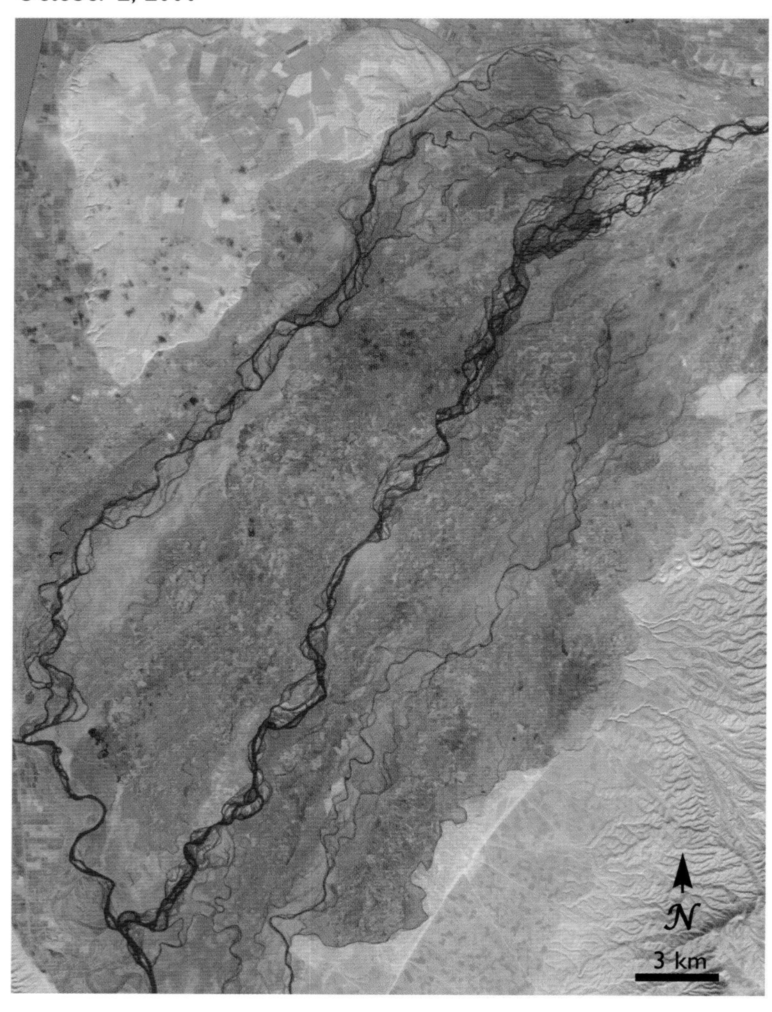


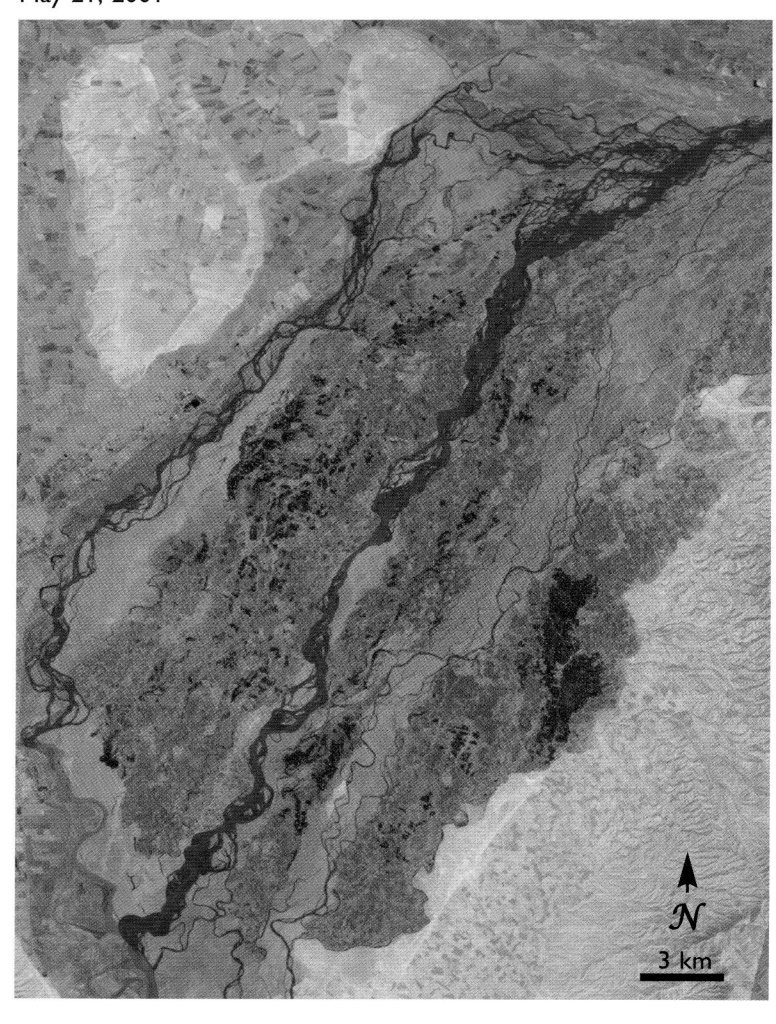


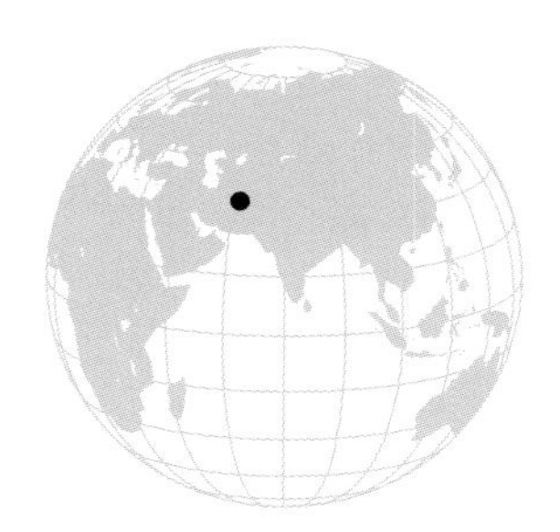

Historically, floods have always had a tremendous impact on human civilization. In some areas such as the Nile valley or Mesopotamia, annual floods were essential for subsistence crops. However, large floods can have a catastrophic impact on modern economies, on the environment, and on human beings caught in harm's way.

Extreme floods are rare within a specific area and are not likely to occur during any particular year. For example, for those living along the Mississippi River, the '100 year recurrence interval flood' is estimated to have a 1% chance to occur in any one year. It is quite possible to have 2 extreme floods in the same area within a year or two, then no similar event for many decades. In part because of how rare they are, the climatic circumstances that produce these extreme events are not well modeled and their occurrences cannot yet be predicted.

During the 20th century, floods were the leading natural disaster in terms of the number of lives lost and property damage. Since the late 19th century, developed countries and

*Our Changing Planet*, ed. King, Parkinson, Partington and Williams
Published by Cambridge University Press 

October 18, 2003

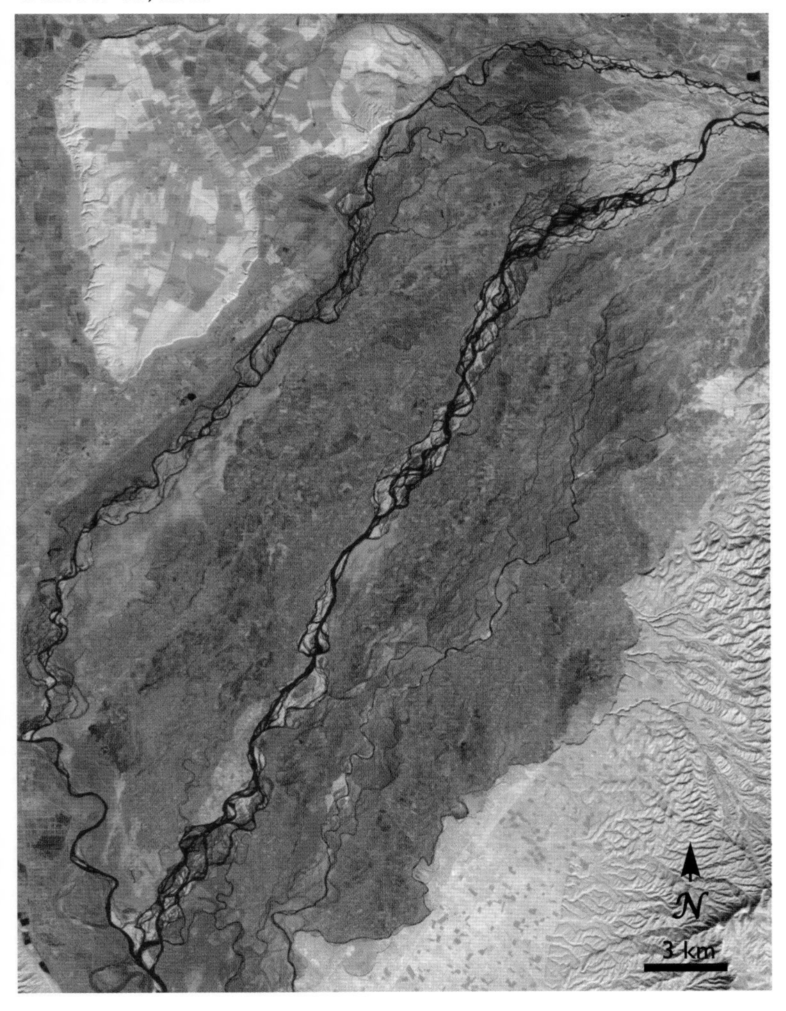


July 19, 2005

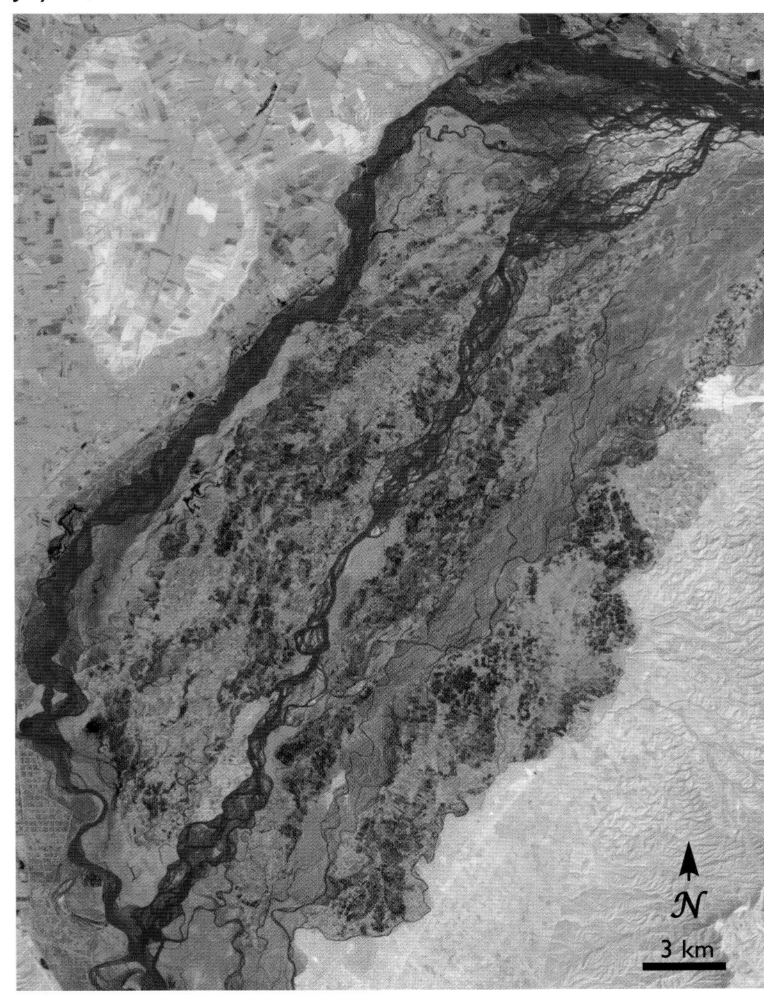


The Panj River on the border between Afghanistan and Tajikistan between October 2000 and July 2005. A better understanding and analysis of flood events is enhanced by the use of satellite technologies in the global monitoring of floods and local measurement and analysis. (Data from the ASTER instrument on the Terra satellite, obtained courtesy of the NASA/GSFC/METI/ERSDAC/JAROS and the US/Japan ASTER Science Team.)

more recently developing countries have put tremendous investments in engineering projects to reduce the consequences of floods. Despite this effort, the frequency and the magnitude of extreme flood events have significantly increased since the 1950s, concentrated in several regions of the world.

It is now widely acknowledged that such changes are partially due to variability in the Earth's climate. Other factors such as changes in watershed land use due to agriculture, forest clear-cutting, or human colonization have also increased the magnitude and consequences of floods. Reducing these consequences presents a complex challenge that may involve international watershed management along with better knowledge and awareness of flood-generating processes. Floods are natural phenomena that cannot be eradicated; societies must learn to adjust to such events.

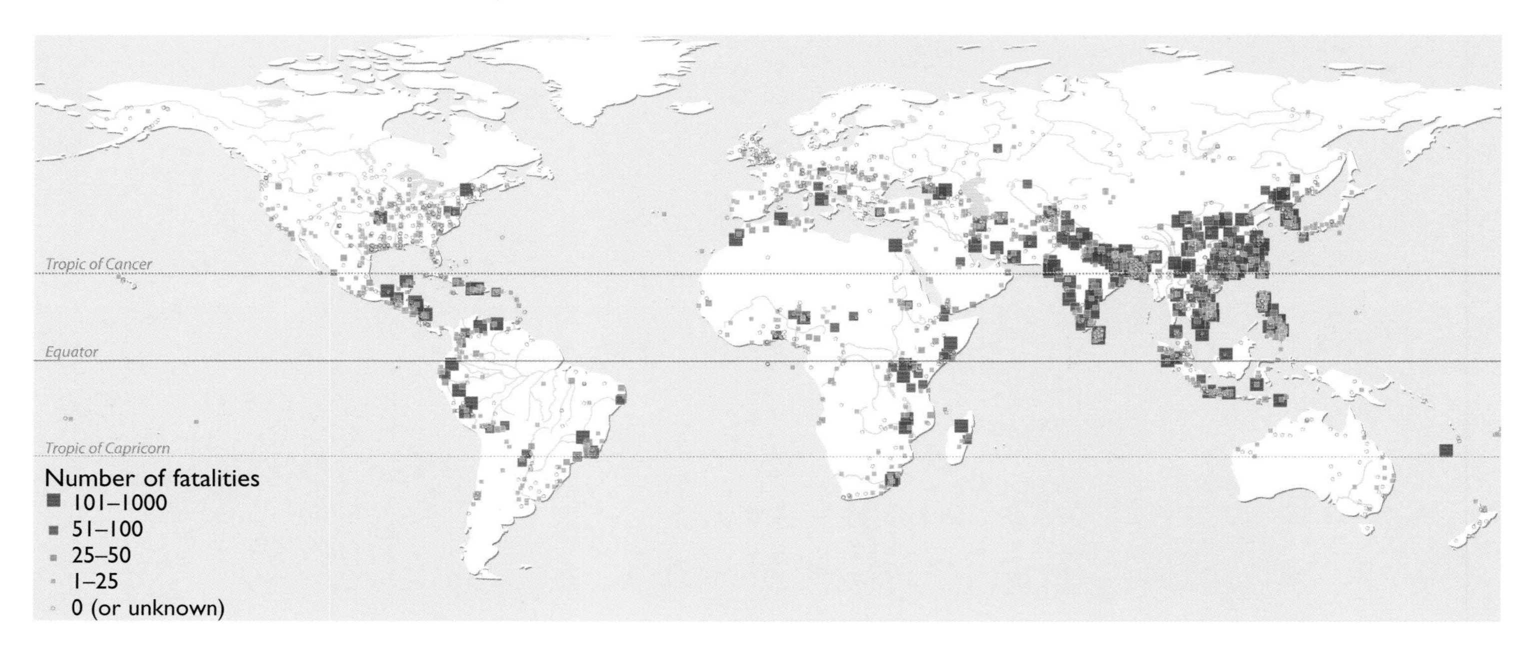


The devastating aftermath of floods shown by a map of fatalities covering the period 1985 to 2003. As a result of floods, more than 300,000 people died and 4 million were displaced between 1985 and 2003. Because of high population density and the consequences of the Monsoon season (from June to September), floods have a particularly catastrophic impact in South Asia. (Source: © Dartmouth Flood Observatory.)

Flooded neighborhoods in New Orleans, Louisiana as a result of Hurricane Katrina, September 2005. (Photograph courtesy Jocelyn Augustino, FEMA.)

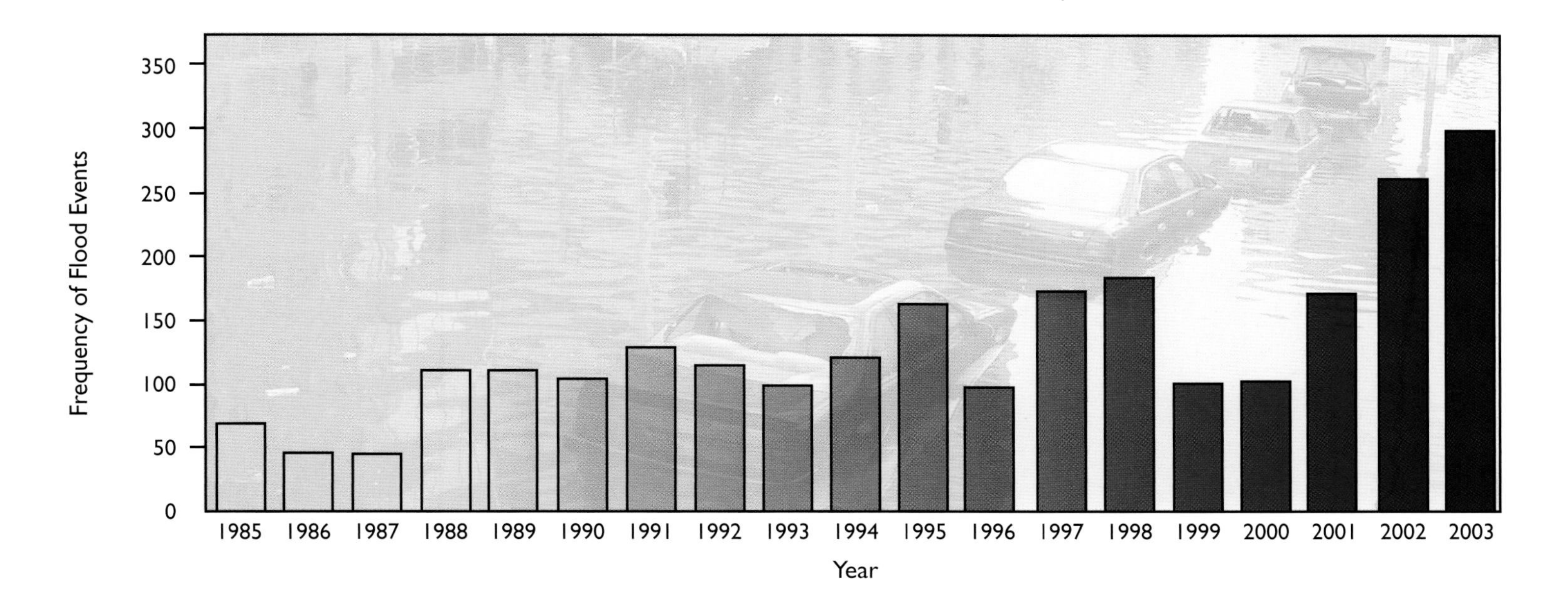


Frequency of flood events around the world from 1985 to 2003. The darker the blue, the better the data reliability.

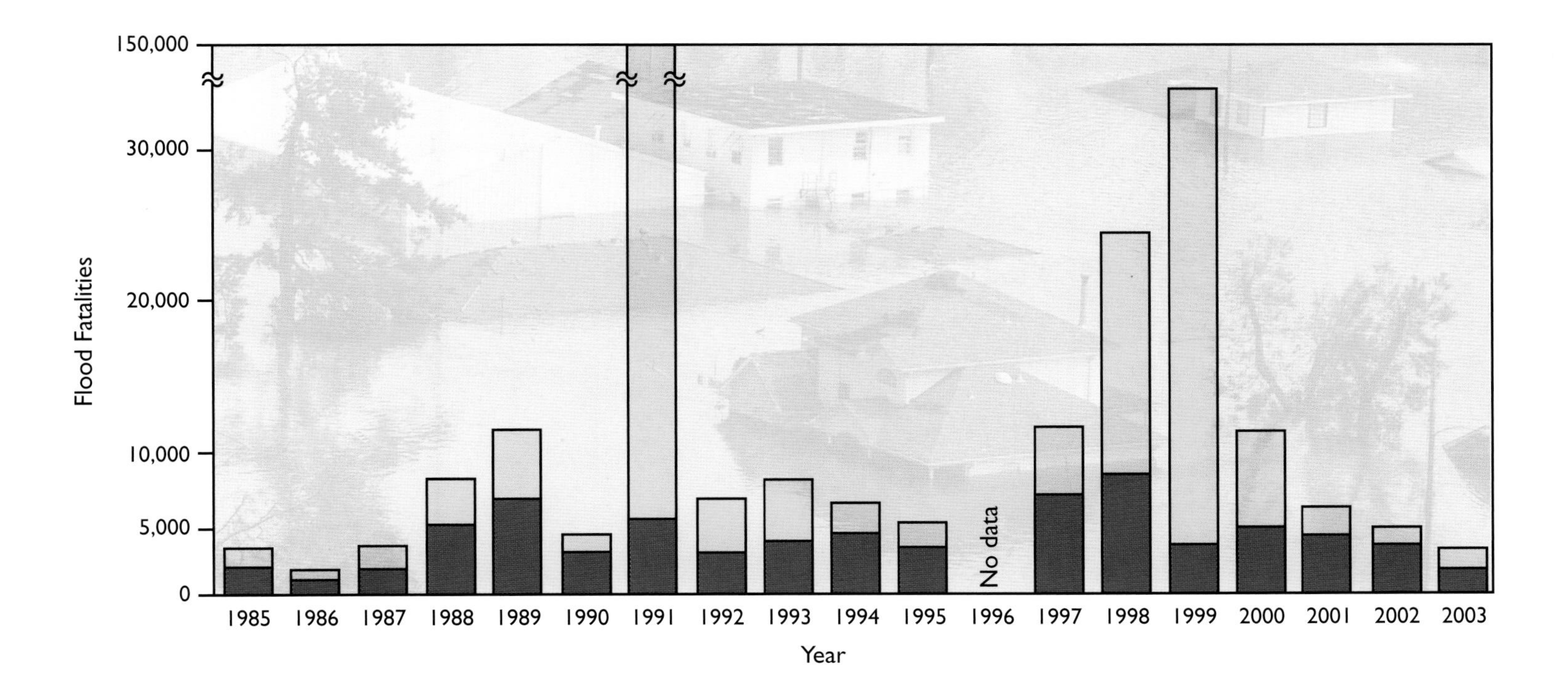


The number of fatalities caused by floods from 1985 to 2003. The vertical blue bars represent the fatality totals, with the lighter blue part of the bar representing the number of fatalities during the worst three floods of the year. For example, the worst flood in 1991 was caused by a tropical cyclone in Bangladesh that killed 138,000 people. The number of fatalities decreased between 2000 and 2003, while flood frequency was increasing. (Source: Dartmouth Flood Observatory.)

Jorge E. Pinzon
Compton J. Tucker

# Satellite Monitoring of Ebola Virus Hemorrhagic Fever Epidemics

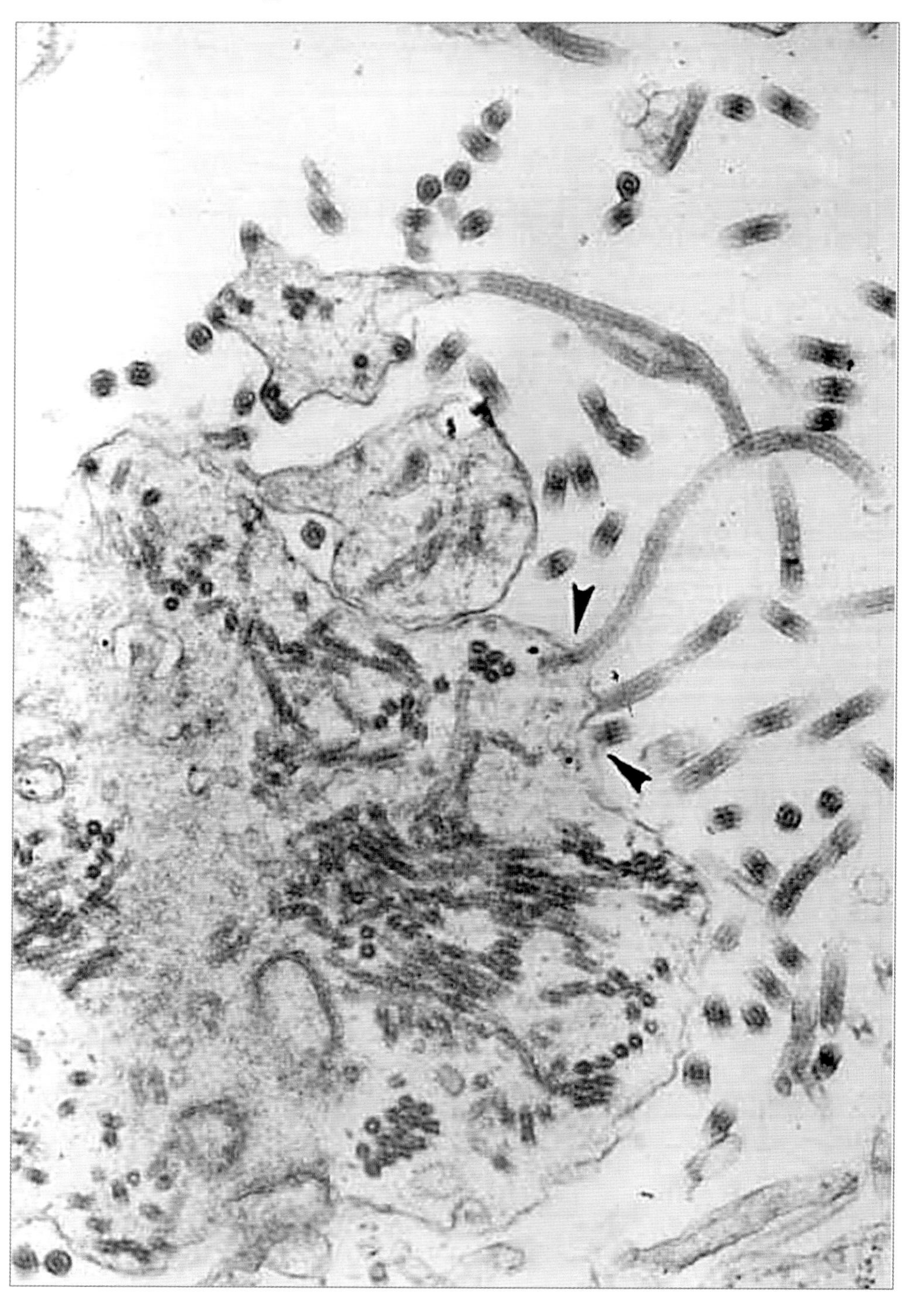

The first isolation and visualization of Ebola virus, 1976. The Ebola virus consists of four distinct subtypes: Zaire, Côte d'Ivoire, Sudan, and Reston. The first 3 subtypes have been identified to cause illness in humans. The Reston strain is highly pathogenic for non-human primates, but it has not to date caused illness in humans. (Micrograph courtesy F. A. Murphy, Department of Pathology, University of Texas Medical Branch, Galveston.)

Ebola virus hemorrhagic fever, named after the Ebola River in Central Africa, was identified for the first time in June 1976, during an outbreak in Nzara and Maridi, Sudan. In September 1976, a separate outbreak was recognized in Yambuku, Democratic Republic of the Congo (DRC). One fatal case was identified in Tandala, DRC in June 1977, followed

*Our Changing Planet*, ed. King, Parkinson, Partington and Williams
Published by Cambridge University Press 

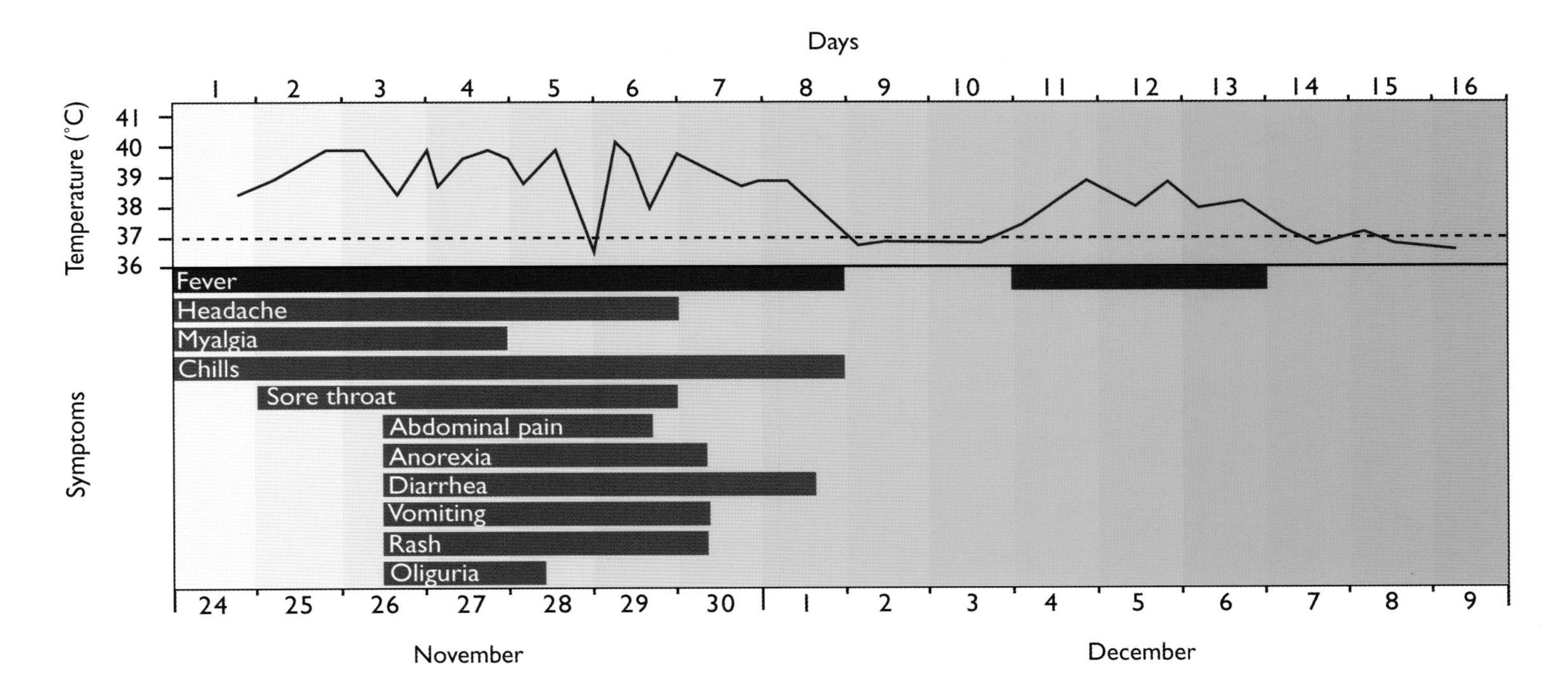


Clinical course of Ebola hemorrhagic fever in a patient presumably infected during necropsy of infected chimpanzee. Health care workers have frequently been infected while treating Ebola patients without correct infection precautions. The majority of deaths occur on the 8th day, although death can occur from 2 days to 21 days after the onset.

by another outbreak in Nzara, Sudan, in July 1979. Ebola causes severe hemorrhagic fever in humans and non-human primates. Its high mortality rate, from 50% to 90%, the ignorance of the mechanisms of its sporadic emergence, and a lack of vaccine or treatment make Ebola one of the most dangerous diseases for humans. Virtually all of the fatal cases had a severe manifestation of gastrointestinal hemorrhage with visible blood loss.

The disease reappeared in Africa at the end of 1994, when 3 outbreaks occurred almost simultaneously. In October, a new strain of the virus designated Ebola-Ivory Coast was isolated from a biologist who became infected during an autopsy of a chimpanzee in Taï, Cote d'Ivoire. The following month, multiple cases were reported in northeast Gabon in the gold panning camps of Mekouka, Andock, and Minkebe, where residents noted the simultaneous occurrence of deaths among gorillas in the surrounding forest. This was the first of three epidemics to occur in Gabon from 1994–1997, each associated with deaths of wild animals and caused by the Ebola-Zaire virus. Later that same month, there had been a large outbreak of Ebola-Zaire virus in Kikwit, DRC, where the epidemic was established through interhuman transmission in a charcoal pit. However, knowledge of the natural reservoir of the disease, its potential vector(s), and mechanism(s) of transmission remained elusive. In Gabon, 2 additional outbreaks were reported in February and July 1996, respectively, in Mayibout II, a village 40 km south of the original outbreak in the gold panning camps, and a logging camp between Ovan and Koumameyong, near Booue.

The largest Ebola hemorrhagic fever epidemic occurred in Gulu District, Uganda from August 2000 to January 2001. In December 2001, Ebola reappeared in the Ogooue-Ivindo Province, Gabon with extension into Mbomo District, DRC, lasting until July 2002. During 2002 to 2005, 4 more epidemics occurred in Gabon and DRC, in the same area, with a high mortality rate of 80% to 90%.

Satellites play a potential role in investigating the seasonal context and occasional temporal clustering of Ebola hemorrhagic fever outbreaks. During a 2-month period in 1976, near simultaneous appearances of Ebola epidemics in Nzara, Sudan, and Yambuku, DRC,

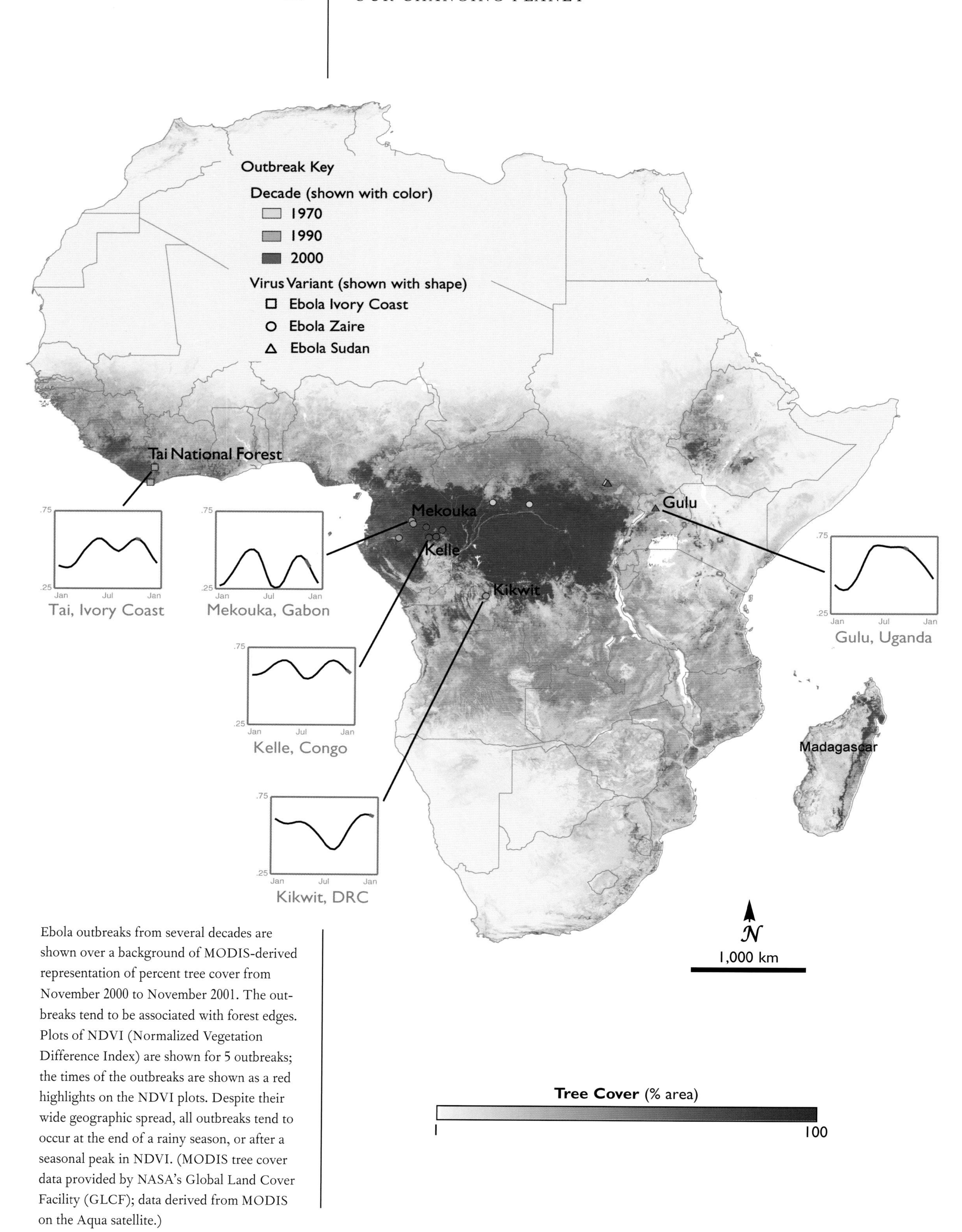


Ebola outbreaks from several decades are shown over a background of MODIS-derived representation of percent tree cover from November 2000 to November 2001. The outbreaks tend to be associated with forest edges. Plots of NDVI (Normalized Vegetation Difference Index) are shown for 5 outbreaks; the times of the outbreaks are shown as a red highlights on the NDVI plots. Despite their wide geographic spread, all outbreaks tend to occur at the end of a rainy season, or after a seasonal peak in NDVI. (MODIS tree cover data provided by NASA's Global Land Cover Facility (GLCF); data derived from MODIS on the Aqua satellite.)

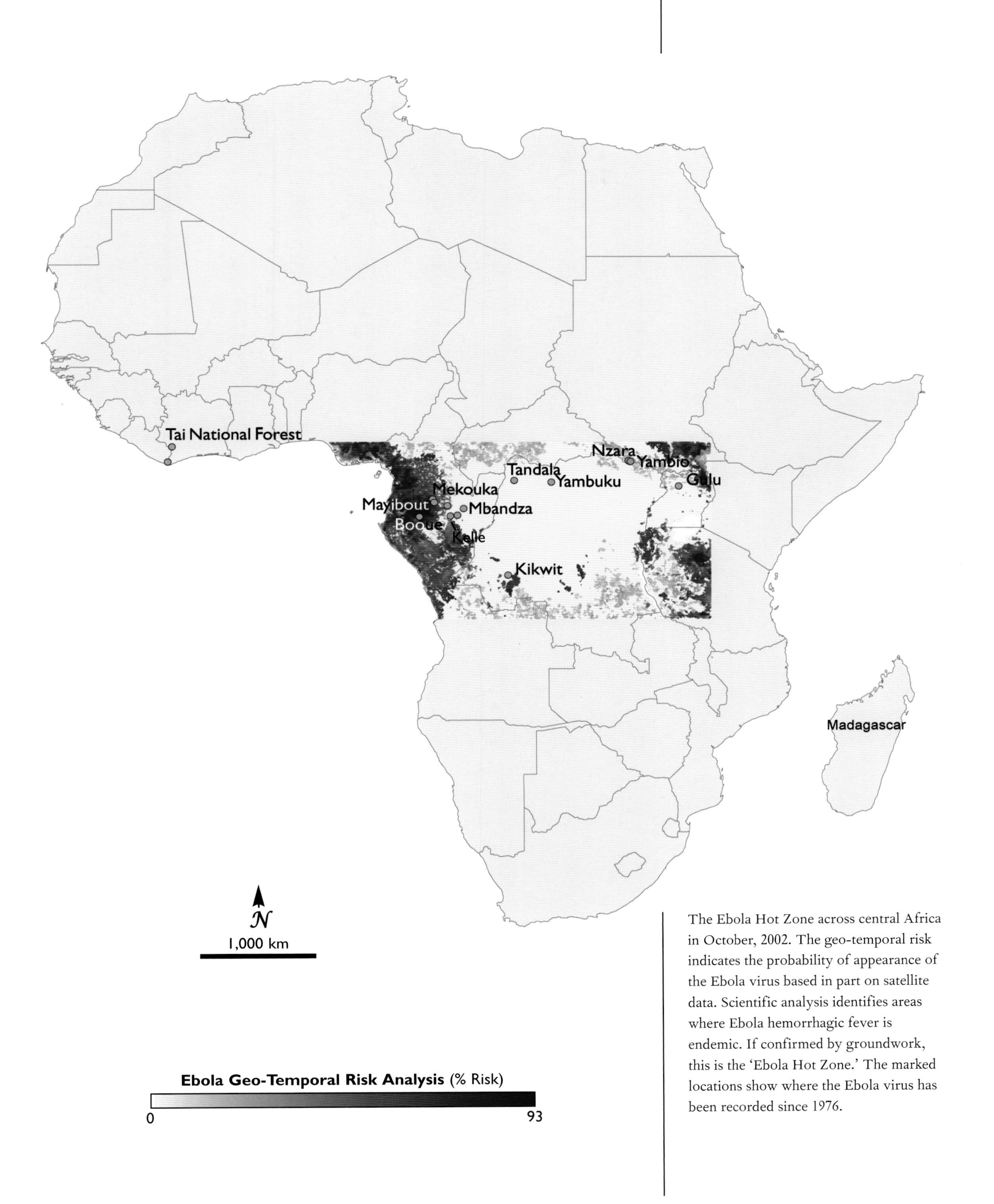


The Ebola Hot Zone across central Africa in October, 2002. The geo-temporal risk indicates the probability of appearance of the Ebola virus based in part on satellite data. Scientific analysis identifies areas where Ebola hemorrhagic fever is endemic. If confirmed by groundwork, this is the 'Ebola Hot Zone.' The marked locations show where the Ebola virus has been recorded since 1976.

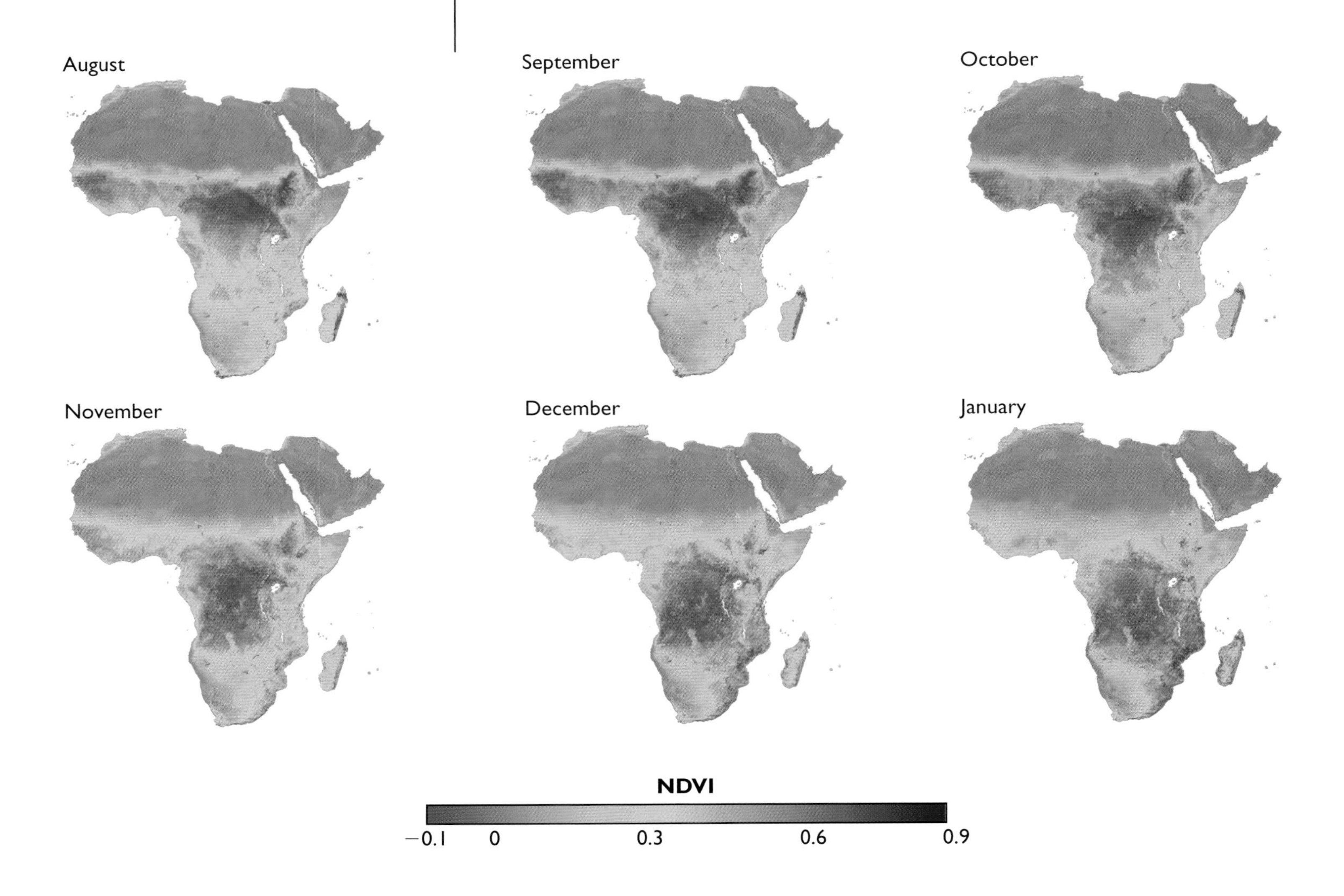


Average seasonal profiles of Normalized Difference Vegetation Index (NDVI) derived from 25 years of monthly AVHRR data covering the period 1981–2005. NDVI, which shows the amount of green vegetation present, can be used as a proxy (substitute) for rainfall and is a factor in Ebola outbreaks.

occurred in 2 geographic locations separated by hundreds of kilometers and involving 2 separate viral strains (Sudan and Zaire EBOV strains). During a period of a few months in late 1994, outbreaks in Taï, Cote d'Ivoire; Mekouka, Gabon; and Kikwit, DRC, also occurred in 3 different geographic regions and involved 2 different viral strains (Cote d'Ivoire and Zaire EBOV strains). Fifteen years passed between the 1976–1979 and 1994–1996 temporal clusters of Ebola cases without identification of additional cases.

Despite extensive field investigations to define the natural history of the Ebola hemorrhagic fever virus, the origin and mechanism of disease transmission, from reservoir to humans, remains a mystery. Nevertheless, Ebola hemorrhagic fever and several other infectious diseases, e.g., Rift Valley fever, cholera and hantavirus, have been studied using satellite data that suggest a role of climate in the onset of Ebola epidemics. The satellite imagery is used firstly to identify proxy environmental and climate indicators of Ebola risk, and secondly to characterize areas subject to this infectious disease.

The geography of Ebola. All Ebola hemorrhagic fever outbreaks occurred in either tropical moist forest or gallery tropical forest in a matrix of savanna. (Photograph by Compton Tucker.)

Normalized difference vegetation index (NDVI) time series data (a measure of the productivity and

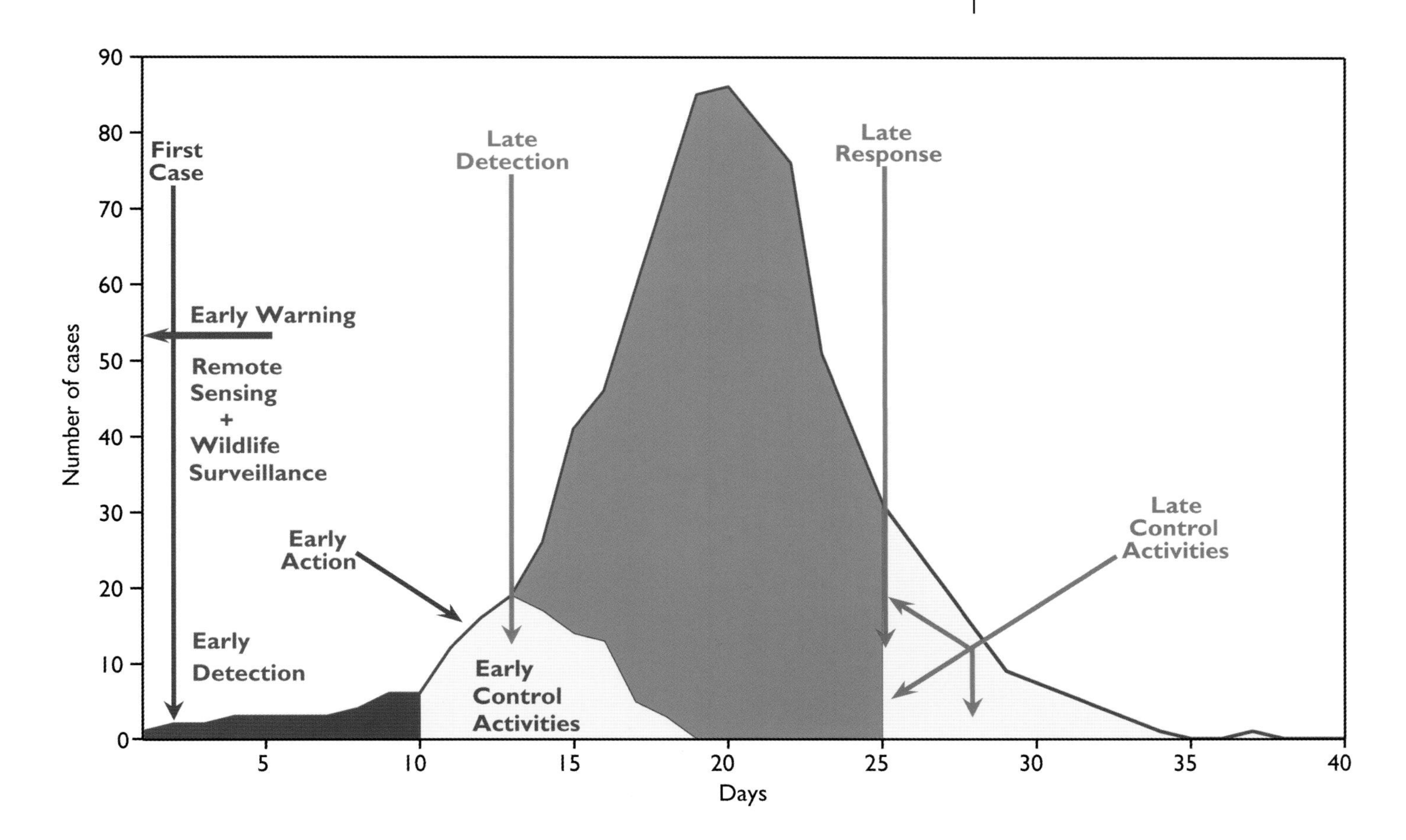


health of vegetation) can be used as a proxy for rainfall. These data are derived from measurements made by the AVHRR instruments carried on the NOAA series of meteorological satellites. The NDVI is computed from the red and near infrared channels of the AVHRR.

The NDVI is sensitive to different types of vegetation, showing, for example, changes that take place in the wet and dry seasons. Relevant changes in vegetation over time, derived using robust statistical techniques, can be used to relate changes in NDVI to documented outbreak sites of Ebola hemorrhagic fever. Landsat data confirms that all Ebola hemorrhagic fever outbreaks occurred in either tropical moist forest or gallery tropical forest in a matrix of savanna. More specifically, these data have enabled satellite identification of the 'Hot Zone' of Ebola hemorrhagic fever in Africa. This is the area where Ebola hemorrhagic fever appears endemic, pinpointing locations for more detailed study of these areas.

The Ebola virus hemorrhagic fever has brought into focus the very real threats posed by emerging diseases and the need to use and develop methods to prevent their spread. A similar remote sensing protocol, as that developed for the Ebola virus, is adaptable to examination of other emerging infectious diseases through the investigation of environmental and climatic events linked to disease occurrences, and how these vary in time. This information is important for public health communities in the 4 phases of preparedness, response, recovery, and mitigation of the epidemic management. The most significant impact of this field of research is the design of remote sensing systems for infectious disease forecasting, which has implications for proactive versus reactive measures of epidemic control.

Ebola control activities indicated in the context of a graph which shows the average time profile of all Ebola-Zaire strain outbreaks (a total of 6 epidemics, including 3 from 1994–1996 and 3 from 2002–2004). The use of spatial and temporal models based on remote sensing data could lead to the formulation of specific plans to manage or control Ebola epidemics, thus facilitating early interventions to prevent its spread.

Dark green indicates an outbreak that has been detected early, within 10 days of the first case. With early, proactive measures, an outbreak can be contained after just the first few cases.

Bright green indicates that the outbreak has been detected only after many cases have been identified. By this time, the disease has already passed a threshold whereby the outbreak follows a rapidly escalating course. The only control is reactive, that is, to contain the spread of the infection through more sanitary measures employed in the region until the outbreak has run its course.

Dorothy K. Hall
James L. Foster

# Snow Cover: The Most Dynamic Feature on the Earth's Surface

Snow cover and thaw lakes in northern Alaska, October 15, 2002. The Brooks Range is also visible on the image. (Data from the MODIS instrument on the Terra satellite; image courtesy of the MODIS Rapid Response Team, NASA Goddard Space Flight Center.)

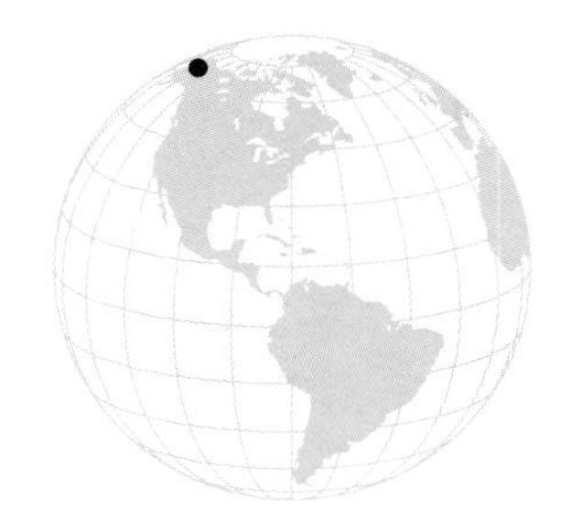

*Our Changing Planet*, ed. King, Parkinson, Partington and Williams
Published by Cambridge University Press 

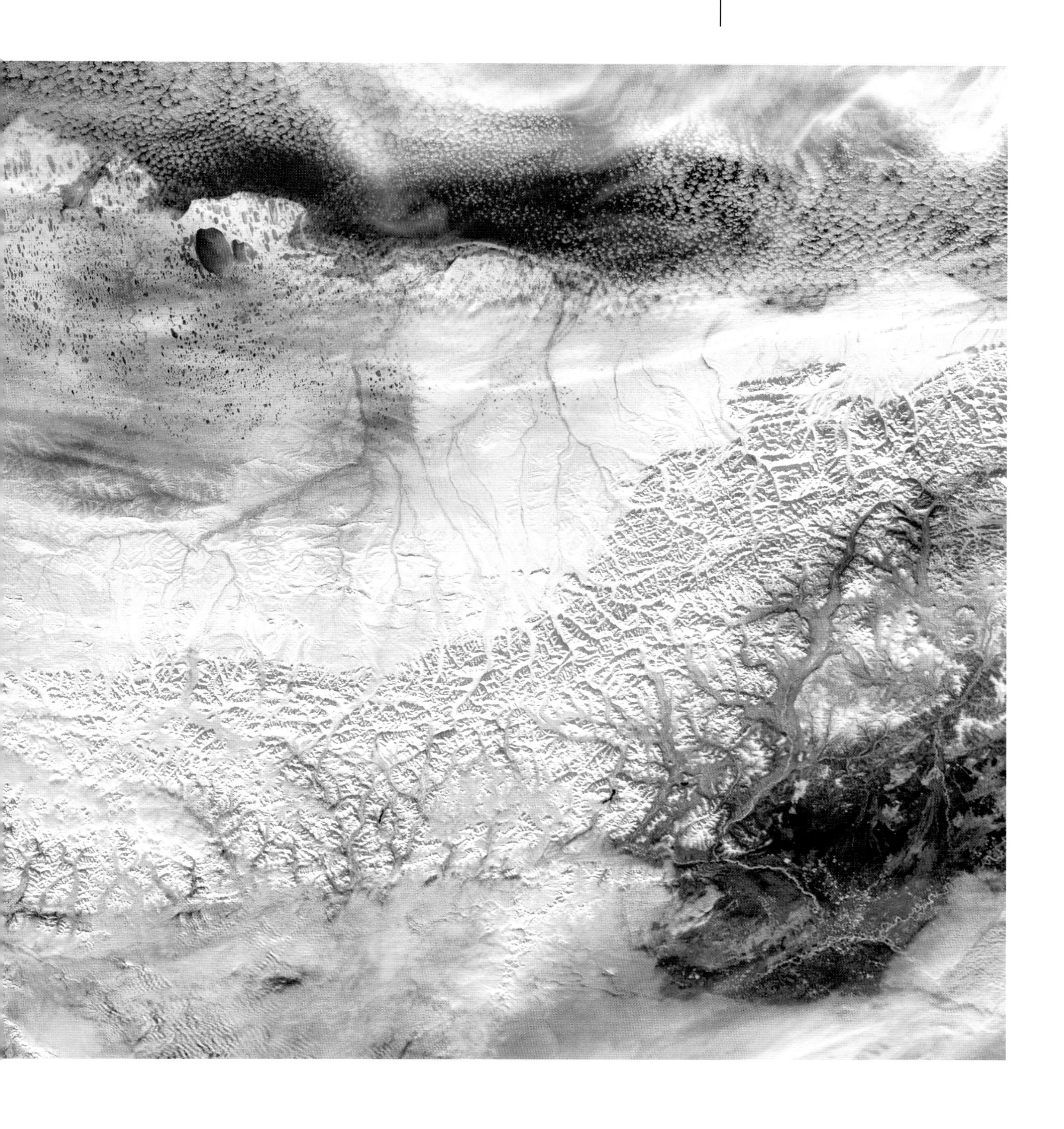

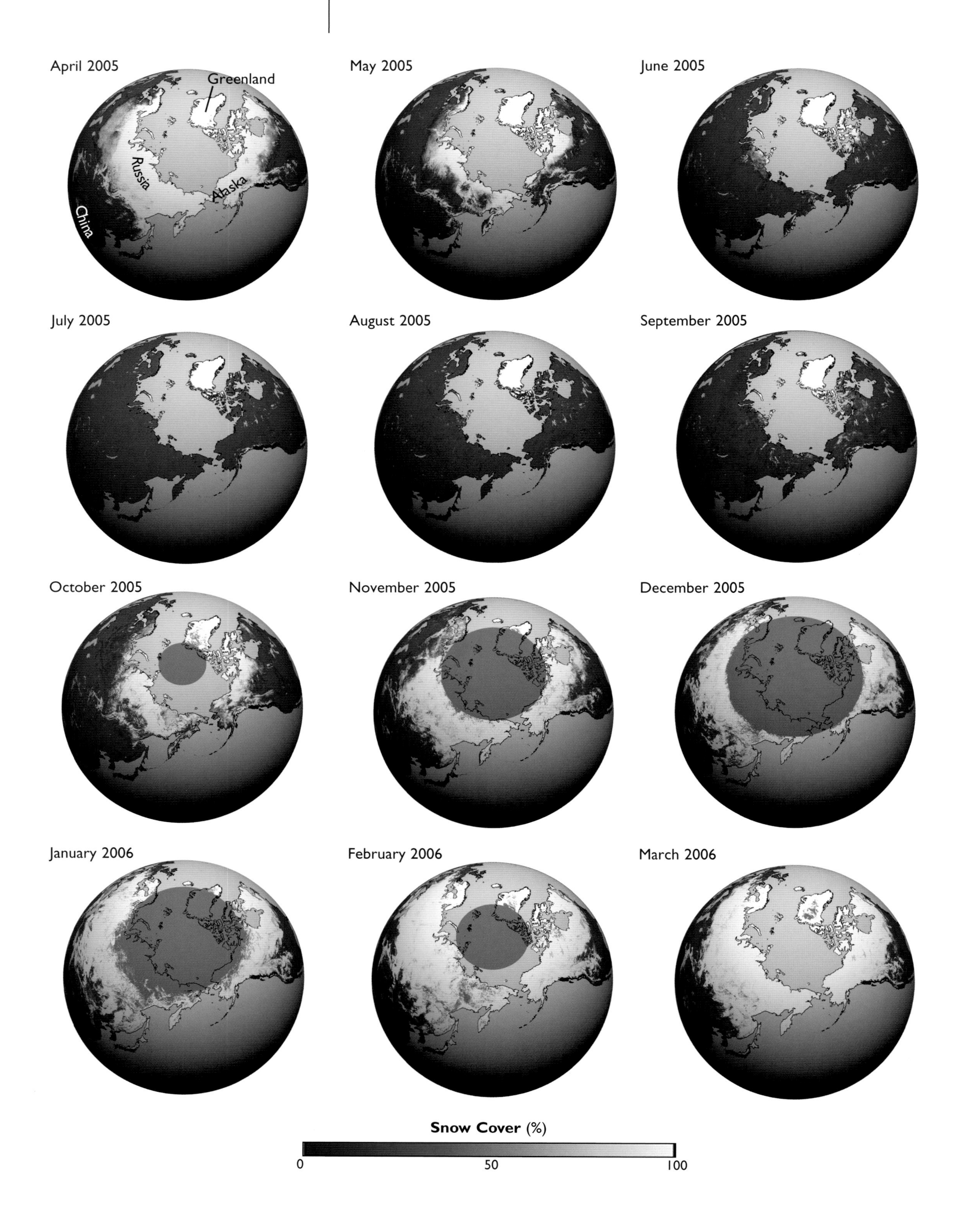
April 2005
Greenland
Russia
Alaska
China
May 2005
June 2005
July 2005
August 2005
September 2005
October 2005
November 2005
December 2005
January 2006
February 2006
March 2006
Snow Cover (%)
0
50
100

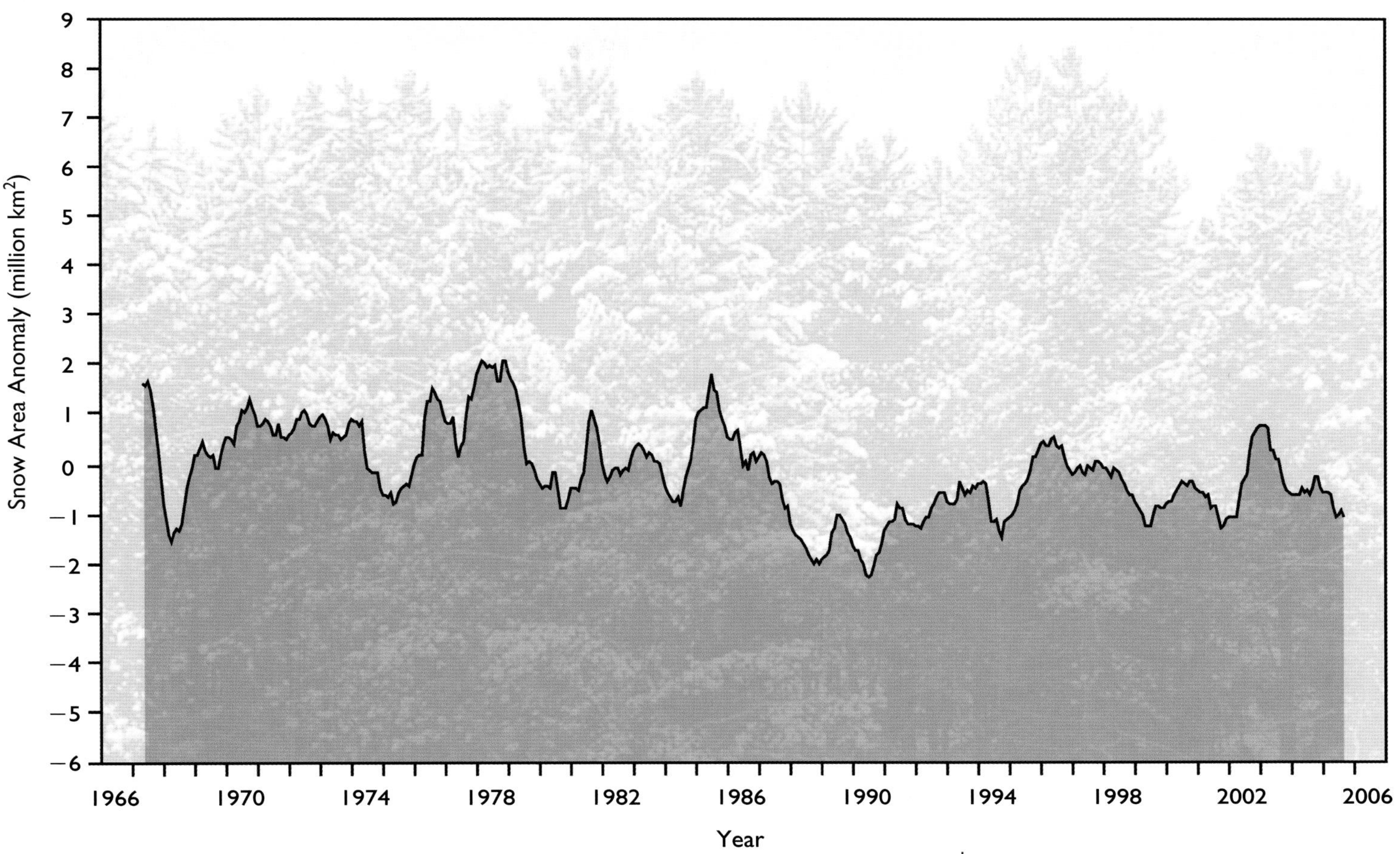


Northern Hemisphere snow cover anomalies, or departure from 30-year mean snow cover. The plot above is based on monthly snow maps from the National Oceanographic and Atmospheric Administration, displaying snow-covered area from 1967–2006. Note the stepwise decrease in snow extent in the mid- to late-1980s, and the contrast in snow extent between the 1970s through the mid-1980s, with the period from about 1987–2005. (Data courtesy Dave Robinson and Tom Estilow, Global Snow Lab, Department of Geography, Rutgers University; photograph by Gracey Stinson, courtesy morgueFile, Inc.)

Few of us ever think about the importance of snow in our daily lives except as a nuisance or a facilitator for recreation. But snow has major economic and climatological significance for our planet both in obvious and obscure ways.

Most of the Earth's seasonal snow cover—that is, snow that falls and melts away each year—is found in the Northern Hemisphere, and it can cover more than 40% of the Earth's land surface. Since there is much more land in the Northern Hemisphere than in the Southern Hemisphere (at mid and high latitudes), there is more potential for seasonal snow cover in the Northern Hemisphere. Seasonal snow cover in the Southern Hemisphere is generally located in relatively high mountain areas, such as in the Andes in eastern South America. In the Northern Hemisphere, the snow cover from one year to the next at the same time of year can vary enormously, by as much as 3 million $km^2$ (approximately equivalent to the size of India).

Ground- and aircraft-based photographs have long been used to map snow cover. With the advent of weather satellites in 1960 came a major step forward in snow mapping, when snow was first detected on a satellite image. Its high albedo allowed the snowline to be delineated, albeit at a very coarse resolution at first. Satellite instruments now map snow cover on a global basis at least once a day. Visible satellite sensors are well suited for measuring snow cover extent with great detail, while remote sensing instruments operating in the microwave portion of the electromagnetic spectrum are used by scientists to estimate the amount of water stored in snowpacks (snow water equivalent), especially over large geographic areas. And microwave sensors have the added advantage that they can measure snow cover through clouds and darkness.

Opposite: Monthly snow cover for the Northern Hemisphere from April 2005 to March 2006. The various colors represent percentages of snow cover. The gray circle in October through February indicates no data available. Portions of continents without snow cover are shown in dark gray. (Data from the MODIS instrument on the Terra satellite.)

Image of snow crystals obtained using low temperature scanning electron microscopy at the United States Department of Agriculture (USDA)/Beltsville Agricultural Research Center in the Electron Microscopy Unit. (Courtesy USDA/Scanning Electron Microscopy Unit, Beltsville, Maryland.)

Snow cover regulates the amount of solar radiation that is absorbed by the ground, by reflecting much of the incident solar radiation back to space. Economically, a snowpack is crucial for storing water to be released as spring meltwater (snowmelt runoff); however, if meltwater is released too quickly, flooding can result.

Satellite records show that the extent of the Northern Hemisphere annual snow cover has decreased by about 10% since 1966, mainly due to earlier spring snowmelt. Decreases in spring and summer snow extent have been measured since the mid-1980s over both the Eurasian and American continents. This is both a result of, and has an impact on, the Earth's climate. During the Northern Hemisphere winter, the Earth is much brighter because of the high reflectivity of the snow cover. This prevents the radiation from being absorbed into the ground where it would be re-emitted as long-wave radiation or heat. Thus the presence of snow cover directly influences the amount of the Sun's radiation that is available to warm the Earth. When snow cover melts, it permits more solar radiation to be absorbed thus heating the lower atmosphere. So when snow cover retreats, local or regional warming is enhanced.

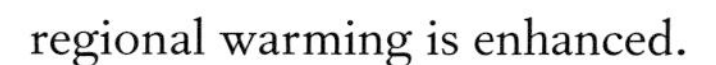

The first exhaustive studies with photography of snowflakes were performed by Wilson A. Bentley and presented in his 1931 book. Bentley's studies revealed graphically and in great detail that there are thousands of different shapes of snowflakes. Today, snowflakes can be studied in unprecedented detail using electron microscopy, providing detailed information on the crystal structure that is useful for improving our interpretation of satellite data of snow cover—and thus improving our ability to measure the amount of water in a snowpack.

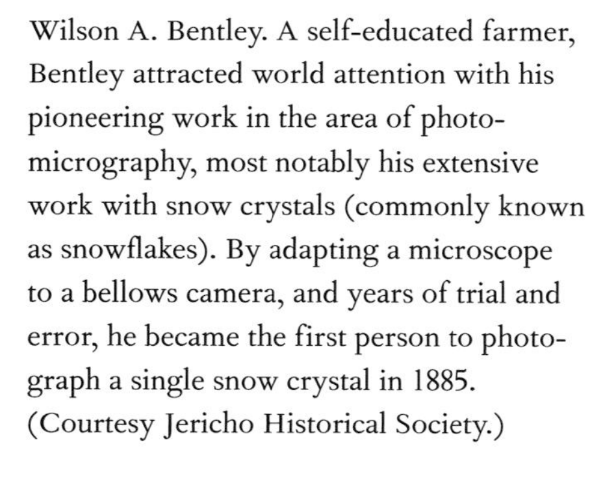

Wilson A. Bentley. A self-educated farmer, Bentley attracted world attention with his pioneering work in the area of photomicrography, most notably his extensive work with snow crystals (commonly known as snowflakes). By adapting a microscope to a bellows camera, and years of trial and error, he became the first person to photograph a single snow crystal in 1885. (Courtesy Jericho Historical Society.)

Glacier National Park, Montana. (Photograph by D. K. Hall.)

Lack of snowfall can have insidious consequences; for example, it can contribute to low lake levels in the Great Lakes. Heavily laden cargo ships cannot carry as much cargo when lake levels are low because the water has to be a certain depth so the ships will not hit the lake bottom. As an example, for every inch (2.5 cm) of water that Lake Michigan loses, a cargo ship must reduce its load by approximately 100 metric tons.

Many regions of the world, the western United States for instance, rely on melting snowpacks for most of their water supply. A dearth of winter snow can contribute to spring and summer drought and an increase in wildfires (and problems associated with low lake levels), while an overabundance can cause flooding, especially if a snowpack melts rapidly. Our ability to predict the extent, thickness, and timing of snowmelt is of critical importance in helping to manage regional water resources. Government agencies rely on satellite data to aid in predicting snowmelt runoff. Satellite imagery, acquired in various parts of the electromagnetic spectrum, allows us to improve forecasts of spring snowmelt and flooding potential.

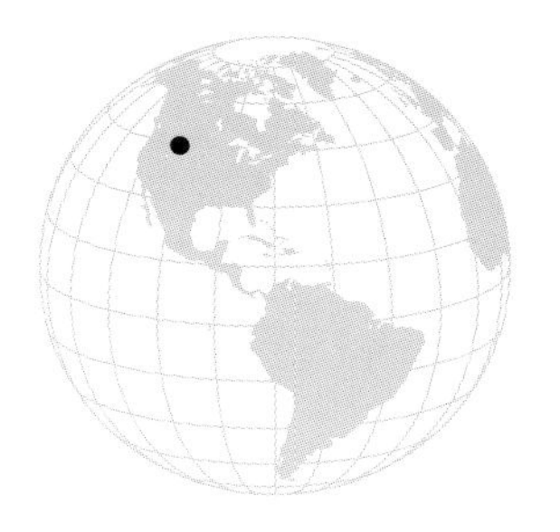

Dorothy K. Hall
Richard S. Williams, Jr.

# Glaciers: Scribes of Climate; Harbingers of Change

The Pasterze Glacier in western Austria, October 3, 2001. Higher summer temperatures and lower winter snowfall have been causing the glacier to recede since 1856. (Data from the IKONOS satellite; image © Space Imaging.)

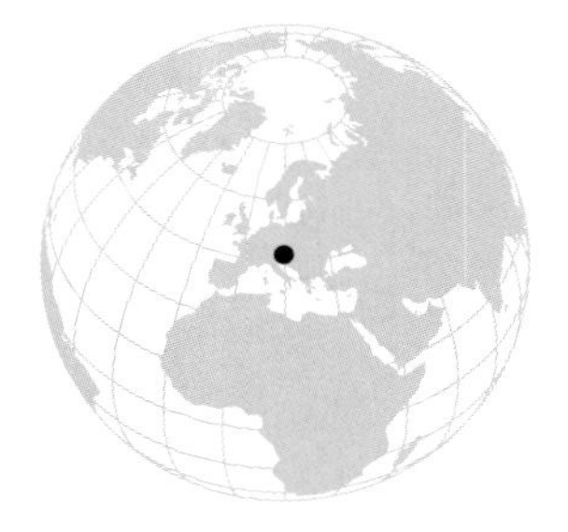

Often slowly and inexorably, but sometimes rapidly and unexpectedly, the Earth's large and small glaciers flow in response to both internal and external forces. Glaciers form from snowfall, and snow that lasts from one year to the next can turn slowly into ice. Over time, glacier ice begins to flow in response to gravity, forming reservoirs of freshwater ice on all of the Earth's continents except Australia, ranging from the size of a football field, to the size of the Antarctic ice sheet (13.6 million km$^2$ in area). Smaller glaciers exist mainly in high mountain areas—even near the Equator—and at high latitudes.

Though formidable in appearance, the Earth's smaller glaciers and ice caps pale by comparison with the area and volume of the huge Antarctic and Greenland ice sheets. More than 99% of the volume of glacier ice on Earth is stored in the Greenland and

*Our Changing Planet*, ed. King, Parkinson, Partington and Williams
Published by Cambridge University Press 

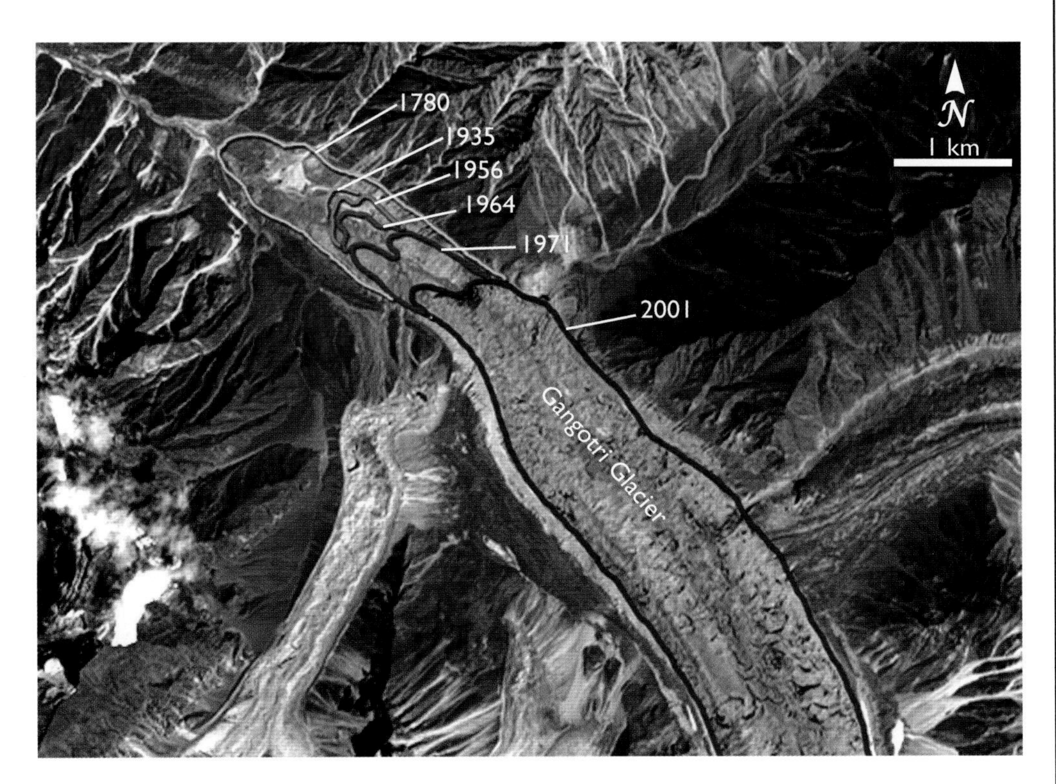


False-color image showing the Gangotri Glacier, situated in the Uttarkashi District of Garhwal Himalaya, September 9, 2001. Currently 30.2 km long and between 0.5 and 2.5 km wide, Gangotri Glacier is one of the largest in the Himalayas. Gangotri has been receding since 1780, although studies show its retreat quickened after 1971. (Please note that the blue contour lines drawn here to show the recession of the glacier's terminus over time are approximate.) Over the last 25 years, Gangotri Glacier has retreated more than 850 m, with a recession of 76 m from 1996 to 1999 alone. (Data from the ASTER instrument on the Terra satellite; measurements from the ASTER science team.)

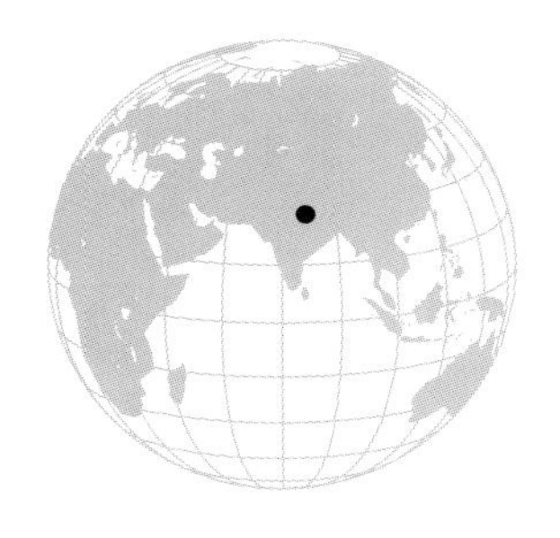

Antarctic ice sheets. The ice sheet covering Antarctica is more than 4,200 m thick in some places.

As they expand and shrink, glaciers exchange water with the atmosphere and ocean through evaporative processes and melting. About half of the 30 cm rise in global sea level measured during the past 100 years has resulted from the shrinking of the Earth's smaller glaciers in response to the warmer climate of the 20th century, which followed several centuries of a colder climate, referred to as the 'Little Ice Age.' Much of the other half of the observed sea level rise is likely due to expansion of the ocean water as the upper layers become warmer. The total potential contribution to sea level rise of all of the glacier ice on Earth is about 80 m, while the potential contribution from the smaller glaciers and ice caps (all glaciers exclusive of the Antarctic and Greenland ice sheets) is about 60 cm. While 60 cm may seem small, it is very significant as densely populated coastal cities and deltas are highly vulnerable to the ongoing sea level rise, at present 3–4 mm per year.

Glaciers hold clues to past climate and atmospheric conditions; as the snow settles on a glacier, it traps minute particles and atmospheric gases that become incorporated into the ice. Trapped gases, including some 'pollutants' generated by humans, can be identified and their abundance measured, for example to study atmospheric constituents before and after the Industrial Revolution. In addition, annual layers of snowfall in ice cores can be counted, allowing dating of events such as volcanic eruptions. Ice cores containing ice that is 900,000 years old have been obtained from the Antarctic ice sheet, revealing the composition and temperature of the atmosphere at the time that the snow fell.

William O. Field with a theodolite at Columbia Glacier Station 9, September 9, 1931. (Photograph from the WOF Collection, Archives, University of Alaska Fairbanks, number F-31-163.)

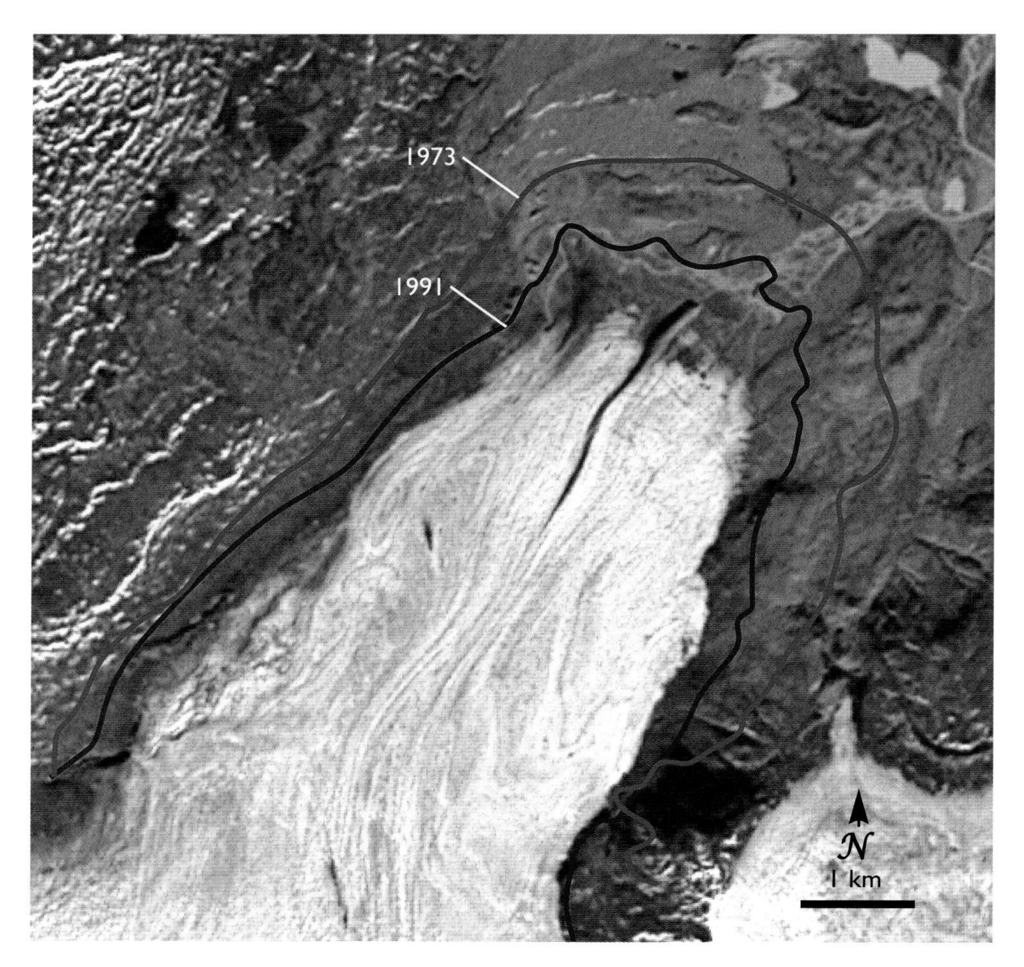


Eyjabakkajökull, an outlet glacier of the Vatnajökull ice cap in Iceland. This glacier has been retreating since a 2.8 km surge occurred in 1972–1973. The true-color image here shows the glacier terminus in September 2000, by which time it had retreated 1.8 km. The light- and dark-blue outlines show the terminus extent in 1973 and 1991, respectively. (Data from the Landsat ETM+ instrument; outlines derived by D.K. Hall, R.S. Williams, Jr., O. Sigurðsson, and J.Y.L. Chien from Landsat MSS and TM data.)

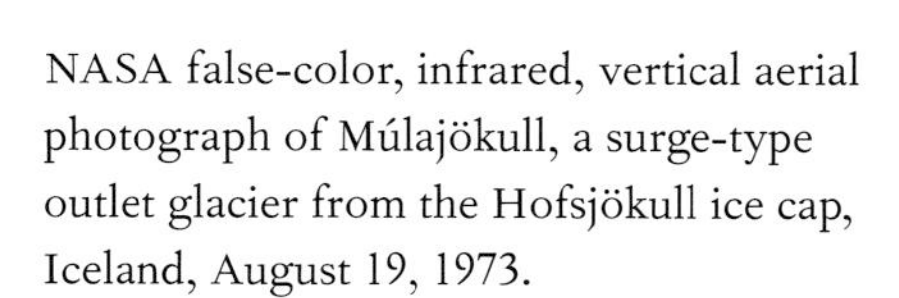

NASA false-color, infrared, vertical aerial photograph of Múlajökull, a surge-type outlet glacier from the Hofsjökull ice cap, Iceland, August 19, 1973.

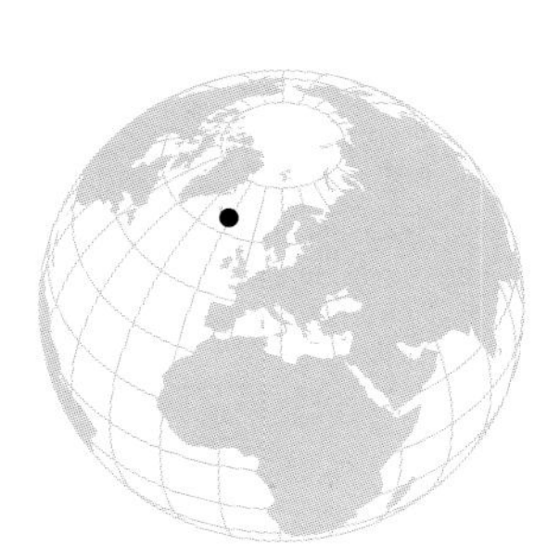

1977

2002

Quelccaya ice cap, Peru, the largest single glacier in the Peruvian Andes. The photograph on the left was taken in 1977, and the photograph on the right was taken in 2002, from the same place, demonstrating a dramatic recession. (Photographs by Lonnie G. Thompson, Byrd Polar Research Center, Ohio State University.)

Globally, most small glaciers and ice caps are receding, especially those in the lower latitudes where temperatures are already closer to the freezing point than they are in high-Arctic regions. The smaller glaciers and ice caps are excellent indicators of regional and global climate change; they can respond to climate change on a relatively short timescale (years to decades). The larger glaciers and ice masses actually influence regional climate because the highly reflective snow surface reflects much of the incoming solar radiation back to space where it is not available to heat the Earth's surface.

Throughout most of the 20th century, and previous to that, the only way to study glaciers was to visit them, often trekking through dangerous territory and in harsh conditions, and spending countless days and weeks making measurements. Research is still being conducted in this way, although the use of helicopters has made many glaciers more accessible. However in the last 30 years or so, satellite measurements have augmented the field measurements revealing a wealth of information, much of which is not even possible to obtain in the field. Together, field and satellite-based measurements allow us to measure precise changes in the areal extent and even the thickness of small glaciers, ice caps, and ice sheets, and to study these changes in the context of the Earth's climate and climate change.

Party at Muir Glacier in 1897. Lila Vanderbilt Sloane (second from the left) was the mother of William O. Field, and her parents are on either side of her. (LaRouche photo, Seattle, photograph courtesy J. O. Field.)

Kyle C. McDonald
John S. Kimball

# Boreal Forests: A Lengthening Growing Season

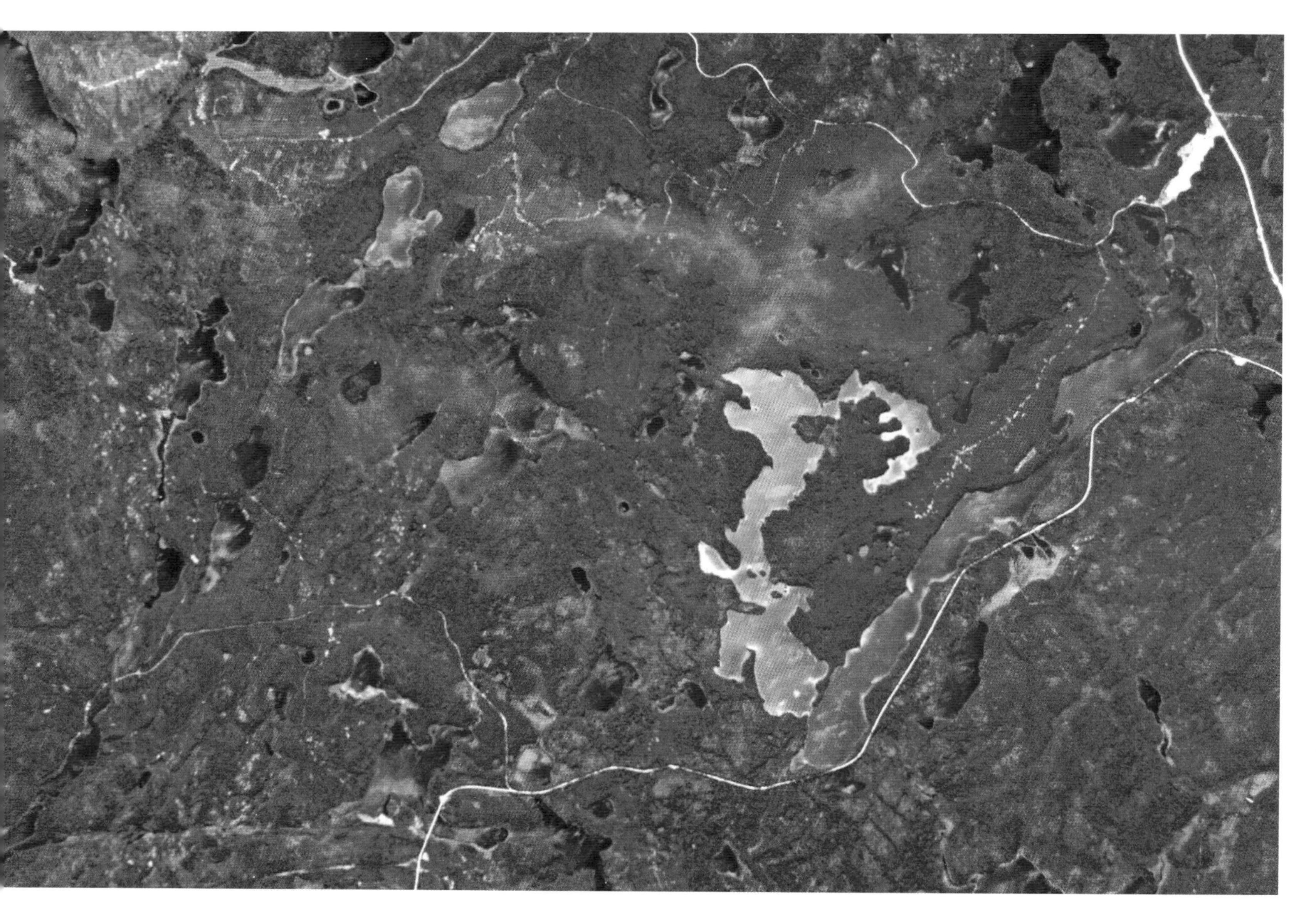

Boreal forest in central Ontario, Canada in the early afternoon of July 25, 2005. (Photograph from the International Space Station, NASA Image Exchange, image number ISS011-E-10893.

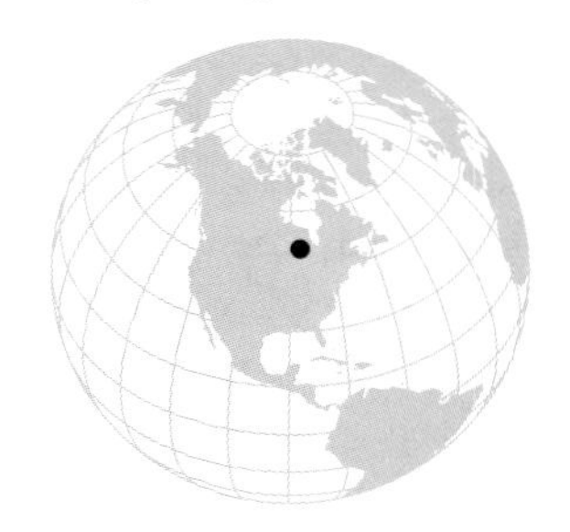

The boreal forest stretches from Alaska, through Canada, and across northern Europe and Siberia, comprising a band of predominantly coniferous trees interspersed with a smaller component of hardy broadleaf deciduous species. Conifers include spruce, pine, fir, and larch while broadleaf deciduous species include, birch, aspen, and poplar. Because cold winters, brief summers and dry climate conditions prevail, the boreal forest is often referred to as a green desert. Cold, short growing seasons have allowed the development of permanently frozen soil, or permafrost, which underlies much of the region's shallow soils. The presence of permafrost leads to shallow rooting depths and pooling of surface water creating a deceptively moist environment in this otherwise desert-like region.

*Our Changing Planet*, ed. King, Parkinson, Partington and Williams
Published by Cambridge University Press 

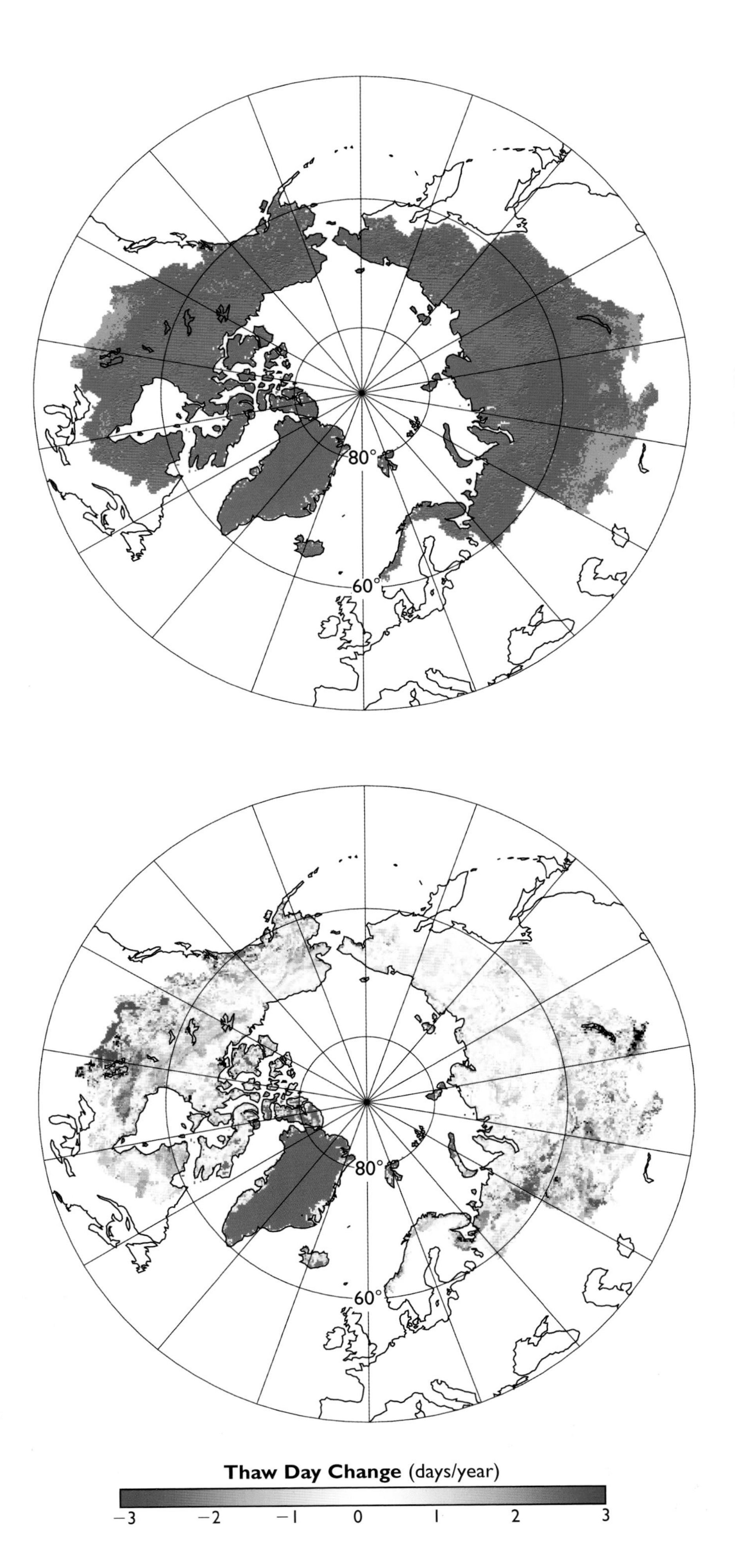


The boreal forest. In this graphic, it is delineated as coniferous and broadleaf forest and consists of a circumpolar band of predominantly coniferous needleleaf trees covering an area of approximately 3 million km$^2$. Comprising about 17% of the global land area, it is one of the world's two largest forest belts.

Map showing the trend in springtime thaw day across the pan-boreal region from 1988 to 2001, excluding permanent ice and snow, barren, and sparsely vegetated areas (indicated in gray). Areas experiencing advance toward earlier thaw are indicated in red and orange, while light and dark blue delineate regions tend toward later thaw. The timing of springtime thaw governs the growing season length and other critical processes in the boreal forest. Multi-year brightness temperature measurements have been used to derive these changes in boreal forest growing season across multiple years. (Data from the Special Sensor Microwave Imager (SSMI) instrument on a DMSP satellite.)

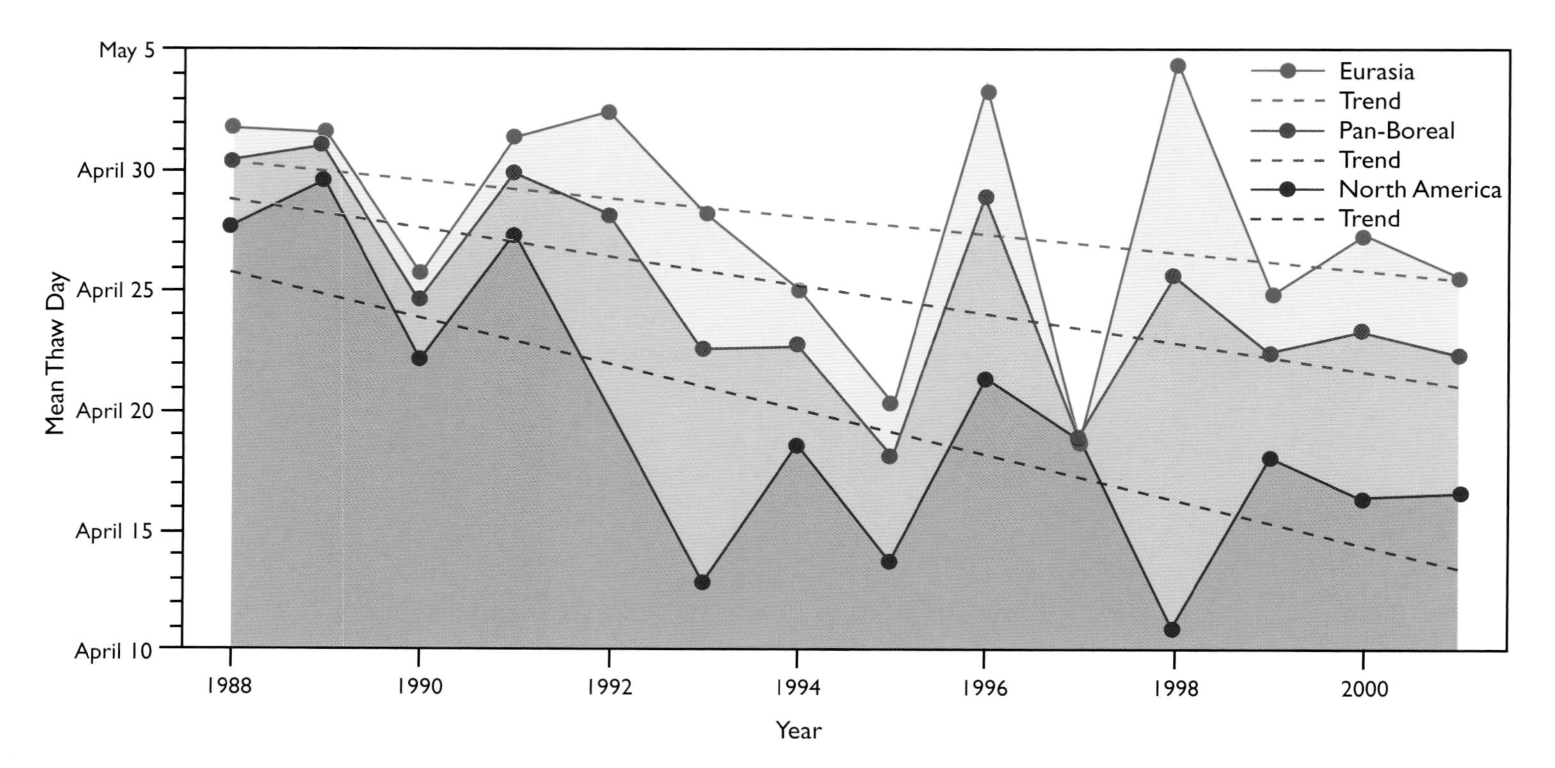


Trend in mean thaw day for the entire pan-boreal region and for North America and Eurasia. North America is experiencing a more pronounced advance in thaw day than is Eurasia. (Data from the SSMI instrument on a DMSP satellite.)

*No! There's the land. (Have you seen it?)*
*It's the cussedest land that I know,*
*From the big, dizzy mountains that screen it*
*To the deep, deathlike valleys below.*
*Some say God was tired when He made it;*
*Some say it's a fine land to shun;*
*Maybe; but there's some as would trade it*
*For no land on earth—and I'm one.*

Excerpt from "The Spell of the Yukon" by Robert Service

Life in the boreal forest is dominated by seasonal cycles of dark cold winters, and brief warm summers. The timing of spring thaw and length of the growing season govern annual cycles of forest growth, fire, and other successional processes. An integral component of the boreal landscape, fire, unlocks nutrients from vegetation and cold soils, releasing them for new plant growth. Animal and insect life cycles are also controlled by the seasons.

Climate models predict a rise in global temperatures of 1.4°C to 5.8°C over the next century, with higher increases expected at high northern latitudes. Evidence indicates that warming temperatures are resulting in earlier springs and longer growing seasons, enhancing boreal forest productivity. Continuation of warming temperatures and longer non-frozen periods may also create drier soils, leading to increased frequency of fire.

NASA scientists are using satellite-borne radar to map seasonal cycles of the boreal forest. The all-weather capability and sensitivity to landscape freeze-thaw state render radars particularly useful for monitoring seasonal processes at high latitudes. This

The foothills of Alaska's Brooks Range, taken on September 1, 2005. The transition from dense boreal forest through treeline, alpine tundra, and snow-capped peaks is apparent in this image. (Photograph by Kyle McDonald.)

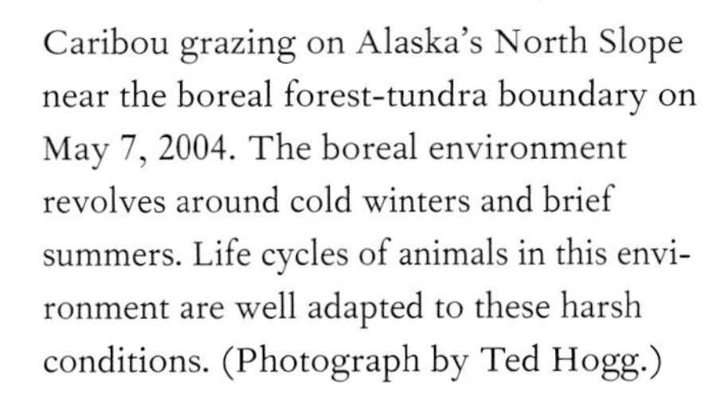

Caribou grazing on Alaska's North Slope near the boreal forest-tundra boundary on May 7, 2004. The boreal environment revolves around cold winters and brief summers. Life cycles of animals in this environment are well adapted to these harsh conditions. (Photograph by Ted Hogg.)

The trans-Alaska pipeline on May 7, 2004. The 1280 km long pipeline, that winds through Alaska's boreal forest, carries crude oil from Prudhoe Bay on the North Slope, to Valdez on Prince William Sound, North America's northernmost ice-free port. (Photograph by Ted Hogg.)

capability extends surface meteorological station-based growing season measurements to a broad view across the vast boreal landscape and allows investigation of the extent and variability of these changes. Changes are being observed with the mean thaw date across the boreal region becoming earlier by more than 5 days per decade. Satellite microwave remote sensing is now being used to assess the implications of these changes, for surface moisture regimes, standing vegetation biomass, and landscape changes caused by fire and other disturbances.

Michael F. Jasinski
Thomas J. Jackson
Manfred Owe
Brian A. Cosgrove

# Soil Moisture: A Critical Underlying Role

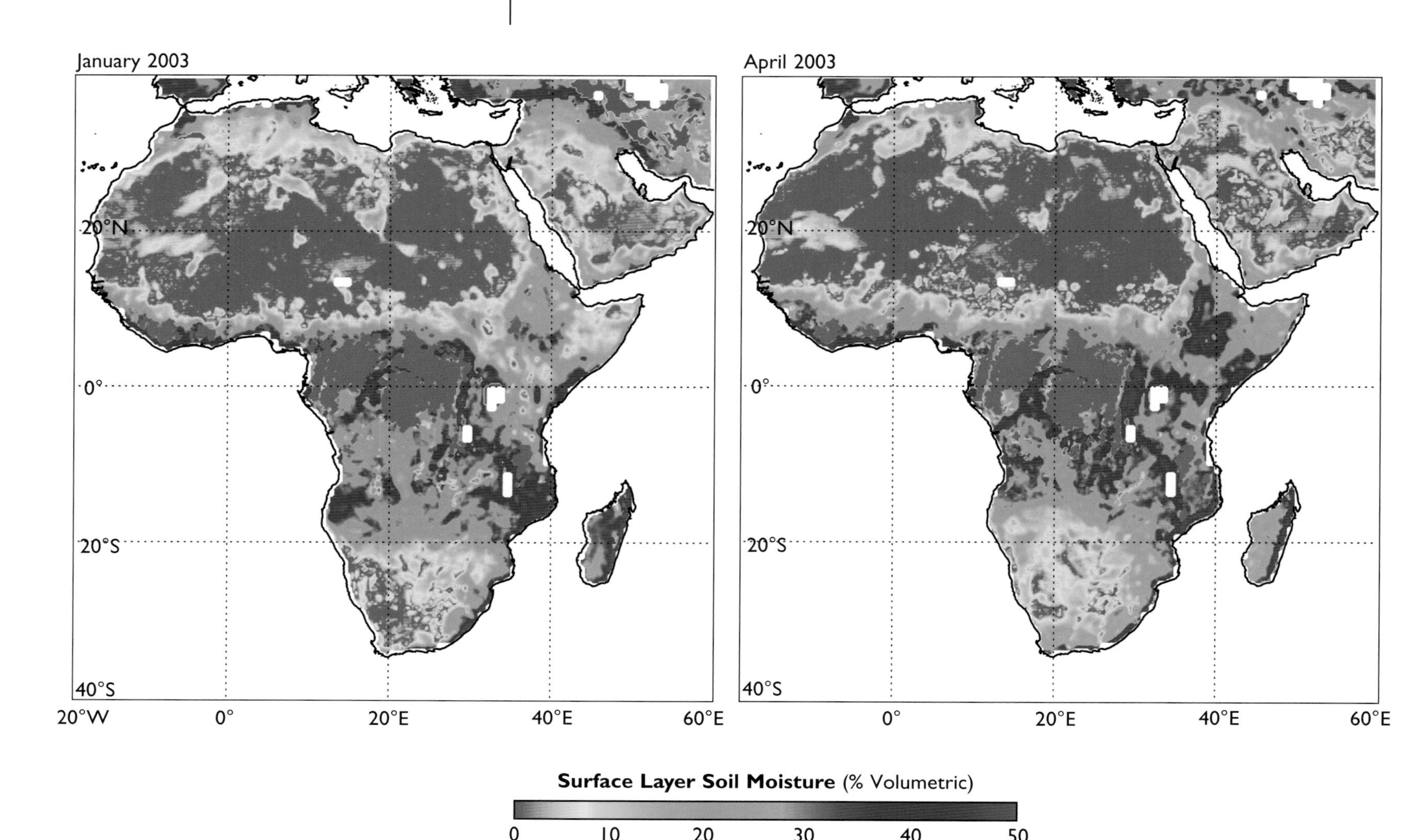


Seasonal patterns of surface soil moisture across Africa for January, April, July, and October 2003. Areas depicted in gray indicate sufficiently heavy vegetation, where sensors are unable to penetrate. (Data from the AMSR-E instrument on the Aqua satellite; images courtesy Manfred Owe, NASA Goddard Space Flight Center and Richard de Jeu, Vrije Universiteit, Amsterdam.)

Soil moisture is a fundamental natural resource of our planet. Located within the top few meters of the land surface, it constitutes only about 0.001% of all the water in the world. However, its repeated fluctuation from wet to dry plays an important role in the Earth's constantly changing environment, influencing the daily activities of humans, if not their very survival.

All natural soils consist of a complex combination of solid mineral particles, organic matter, water, and air. Water occupies the numerous minute spaces between the solid particles forming a soil moisture reservoir. Its volume varies greatly, generally ranging from about 10% to 40% of the total soil volume. When all the spaces are filled due to precipitation, snowmelt, or flooding, the soil is referred to as saturated. Soil moisture is then depleted through evaporation at the land surface, withdrawal by plant transpiration, and gravity drainage into deeper groundwater reservoirs and nearby streams.

*Our Changing Planet*, ed. King, Parkinson, Partington and Williams
Published by Cambridge University Press 

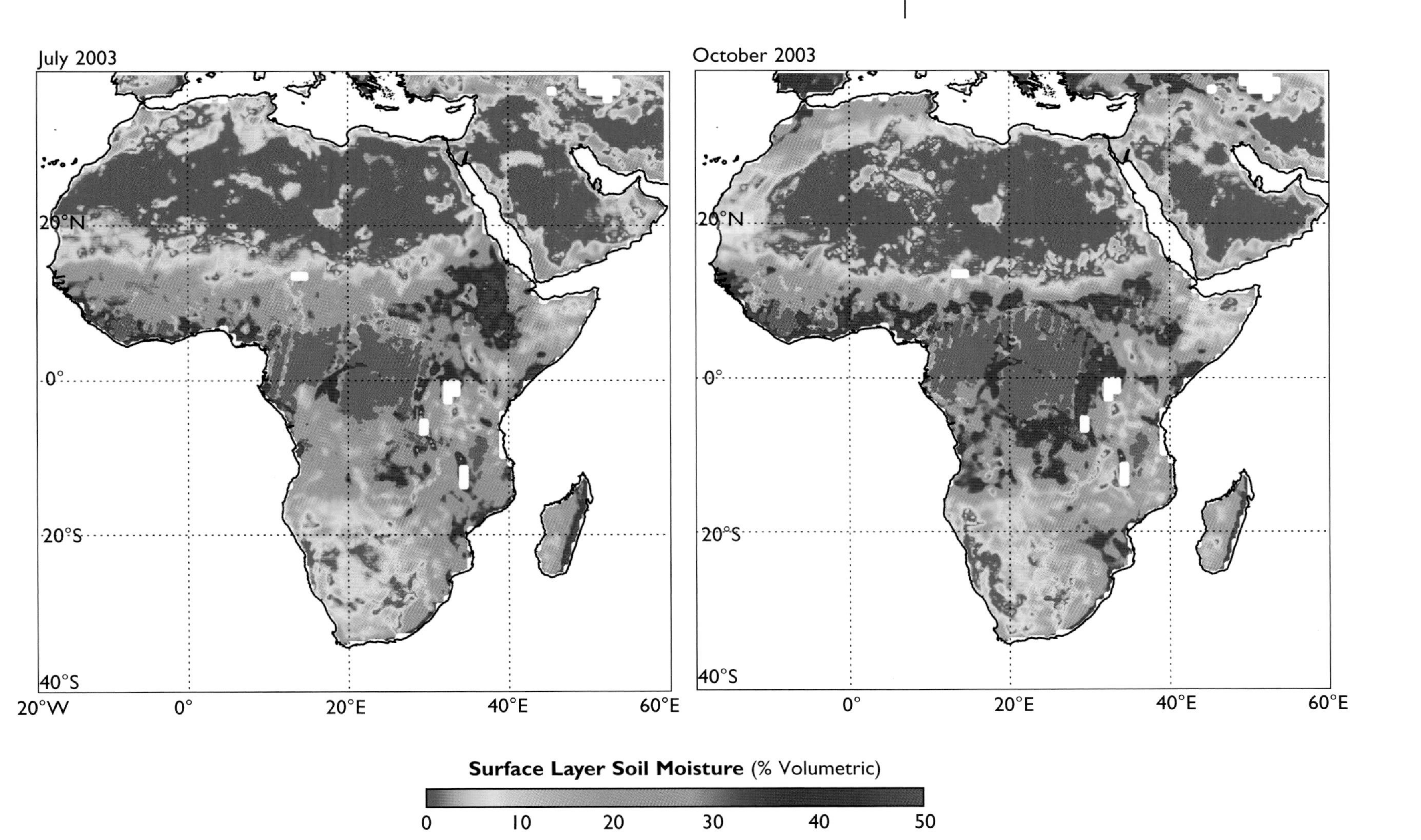


The continuous filling and emptying of the soil moisture reservoir allow it to play a critical role for the Earth's terrestrial environment, with impacts on plant growth and sustainability, water resources, regional weather and climate, and human health and habitat. For example, almost all plant species rely on uptake from the soil moisture reservoir through their rooting system for photosynthesis, chemical nutrient supply, and temperature regulation. When the re-supply of soil moisture through precipitation and irrigation fails to meet plant needs, then land surface temperature increases, growth is limited and crop yields decline. Deficits affect not only the individual plant, but also the surrounding flora and fauna that depend on that plant for food or shelter. Persistent deficits increase the likelihood of forest fires.

Variations in soil moisture can affect flood intensity. For example, unsaturated soils mitigate flood peaks by absorbing heavy precipitation or snowmelt thereby reducing and delaying the runoff from the land to the stream. In contrast, under extended periods of intense

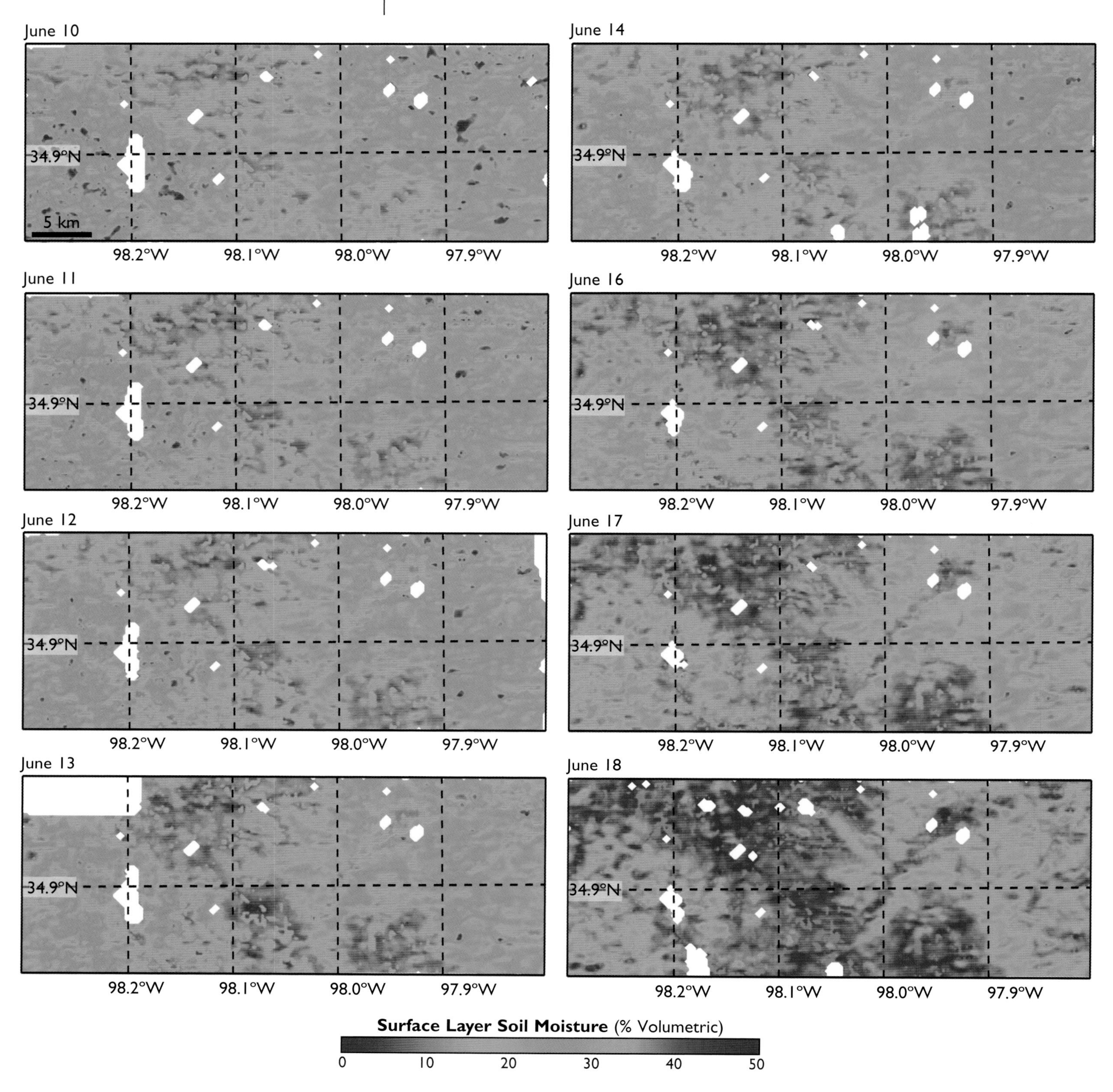


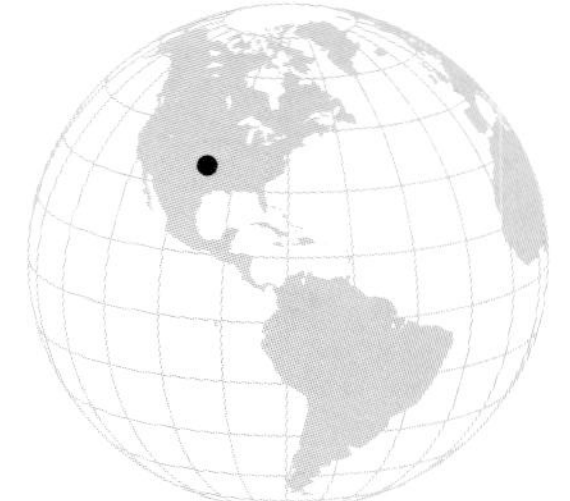

Aircraft-based observations of near-surface soil moisture over a 46 × 19 km$^2$ region in central Oklahoma during June 10–18, 1992. The pattern exhibits soil drying during the 8-day period following a rainfall event. This type of soil moisture change occurs daily across the globe. (Data from the Electronically Scanned Thinned Array Radiometer (ESTAR) on the C-130 aircraft.)

precipitation or spring snowmelt, soils become saturated. Excess water is forced to flow overland entering adjacent streams and rivers more quickly, leading to potentially dangerous flooding.

Spatially variable soil moisture plays several important roles in weather and climate. Contrasts between wet and dry regions affect the heat and water balance of the Earth's surface, influencing atmospheric circulation patterns. Soil moisture is believed to influence summer precipitation in the central United States and other large mid-latitude continental regions. Seasonal variability in soil moisture is associated with the El Niño climatic events of the Pacific Ocean. If the frequency and strength of those phenomena deviate due to future climate change, shifts in current soil moisture patterns would also occur, impacting vegetation, water resources, and climate.

Because it plays an important role in the variability of our Earth's weather, climate, and water resources, NASA engineers and scientists have undertaken the challenge of developing satellite sensors to view and understand the global soil moisture patterns over a timeframe of days to years. Such views from space, when calibrated with field experiments and incorporated into hydrologic simulation models, provide valuable insight on the global energy and water cycles, and how they change on weekly, seasonal, and annual timescales.

One typical example of short-term soil moisture change is that due to gradual dry down after a precipitation event saturates the soil. The accompanying series of 8 images on the opposite page illustrates the daily progression of soil moisture dry down during the period June 10, 1992 to June 18, 1992 over the Little Washita Watershed in central Oklahoma. The images were taken with a prototype soil moisture sensor mounted on a NASA aircraft. Approximately every 2 years since 1990, NASA and the United States Department of Agriculture, together with other federal agencies and universities, have cooperated in the design and implementation of such large-scale hydrology field experiments in order to provide calibration data for global satellite soil moisture observations.

An example of seasonal to annual changes in soil moisture can be readily observed with space-based microwave satellite sensors. Although there is currently no satellite system dedicated exclusively to the monitoring of soil moisture, a number of multi-purpose sensor systems have been used to map the spatial variability of soil moisture at global scales. A series of figures shown early in this chapter depict typical changes in soil moisture within the African continent throughout an annual cycle during 2003, using the AMSR-E sensor aboard NASA's Aqua satellite. The contrasting images readily illustrate the seasonal moisture cycle, especially within the Saharo-Sahelian and Sudano-Sahelian zones and subzones. The rainy season within this roughly 400–500 km wide belt across Africa typically varies from 1.5 months in the north to 3.5 months in the south. When studied over longer periods, results from these satellite technologies may yield important information on long-term climate variability.

NASA P-3B aircraft. This aircraft collects passive microwave observations over crops and forests near Huntsville, Alabama that will be used for soil moisture mapping. (Image courtesy Stephen Ausmus, USDA-ARS.)

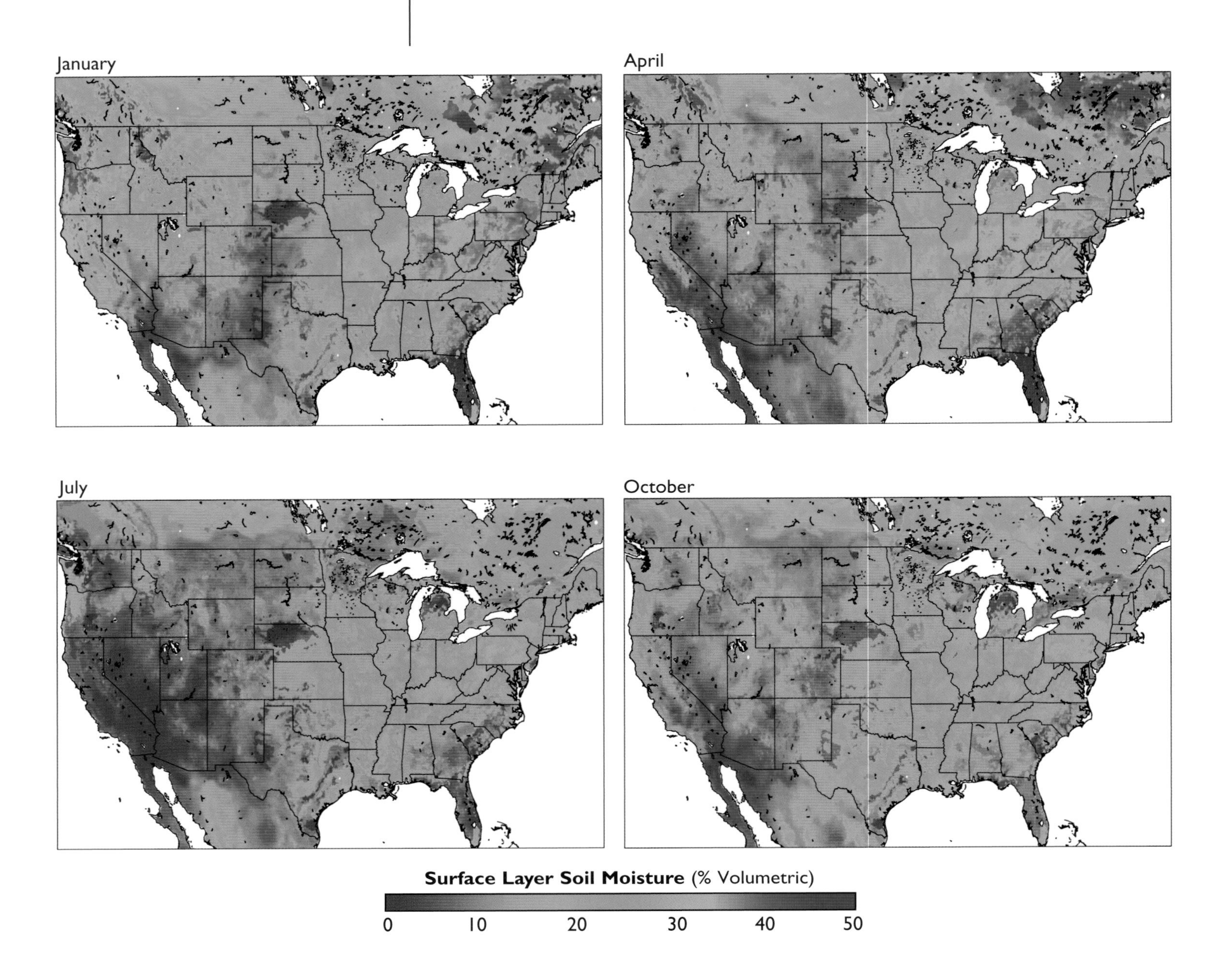


Simulation of average monthly near surface soil moisture over the continental United States during January, April, July, and October 2004. Simulation models that employ satellite imagery as input can be used to understand the consequences of global change. (Data generated using the Mosaic Land Surface Model, courtesy Brian Cosgrove, NASA GSFC.)

Since soil moisture is neither routinely observed by satellites nor measured at weather stations, computer simulation models of the hydrologic cycle are also being developed to assist in the prediction of soil moisture variability. The average soil moisture for winter, spring, summer, and autumn for the continental United States is shown above in a set of maps produced from computer modeling. These models, that use observations of precipitation, net radiation, and other meteorological quantities, are currently being run for all terrestrial regions of the Earth in order to examine the daily, seasonal, and yearly changes in soil moisture.

The increasing pressure of our world's population on natural resources requires that we monitor and manage them with ever increasing care on a sustained basis. Soil moisture, with its quiet but critical role just below the land surface, is as important to sustaining life on Earth as those resources that we can see. Current efforts toward improved satellite observations combined with computer simulation modeling will provide the best opportunity for understanding long-term soil moisture changes and their impact on the global energy and water cycles.

# The Sunlit Earth

Crystal B. Schaaf

January 1–16, 2002

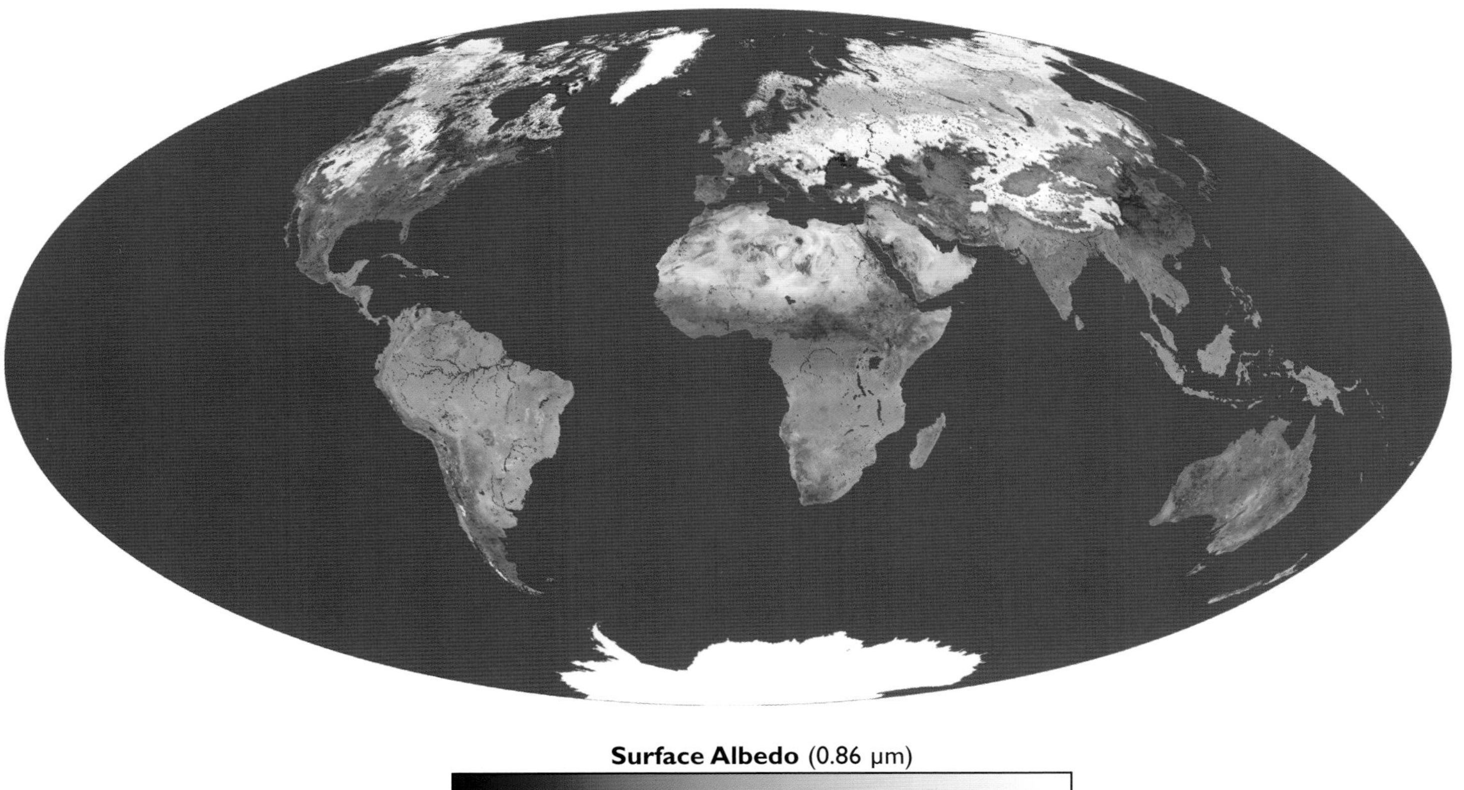


Global surface albedo for the period from January 1–16, 2002. The image shows the proportion of the incoming solar radiation that is reflected from the Earth's surface at a wavelength of 0.86 μm. (Data from the MODIS instrument on the Terra satellite; visualization by Eric Moody, RS Information Systems, Inc.)

From space, the sunlit Earth appears as a fragile planet of white clouds swirling over blue oceans, brilliant snow and ice, dark mountains, bright deserts, and lush vegetation. The way in which sunlight is absorbed or reflected back to space is the engine that drives our entire ecosystem. The ability of vegetation to absorb sunlight and transform it into energy forms the basis of the food chain and allows life to thrive on Earth. Sunlight, as it interacts with the atmosphere and reaches the land and ocean at the surface, also governs our weather and determines both our global and local climate. The reflective character of the surface is known as albedo and monitoring land albedo over the seasons and from year to year helps us better understand the Earth's use of solar energy and how changes are affecting the health of the planet.

Satellite instruments view the Earth from a variety of views and under a variety of illumination conditions. By using multiple cloud-free observations of the same location on Earth over a week or two, it is possible to determine the reflective nature or albedo of that region and then to assess how albedo is changing both in response to the natural annual cycles of vegetation growth and decline, as well as to more abrupt changes caused by snow, catastrophic weather events, fires, and human alteration. This information can

*Our Changing Planet*, ed. King, Parkinson, Partington and Williams
Published by Cambridge University Press 

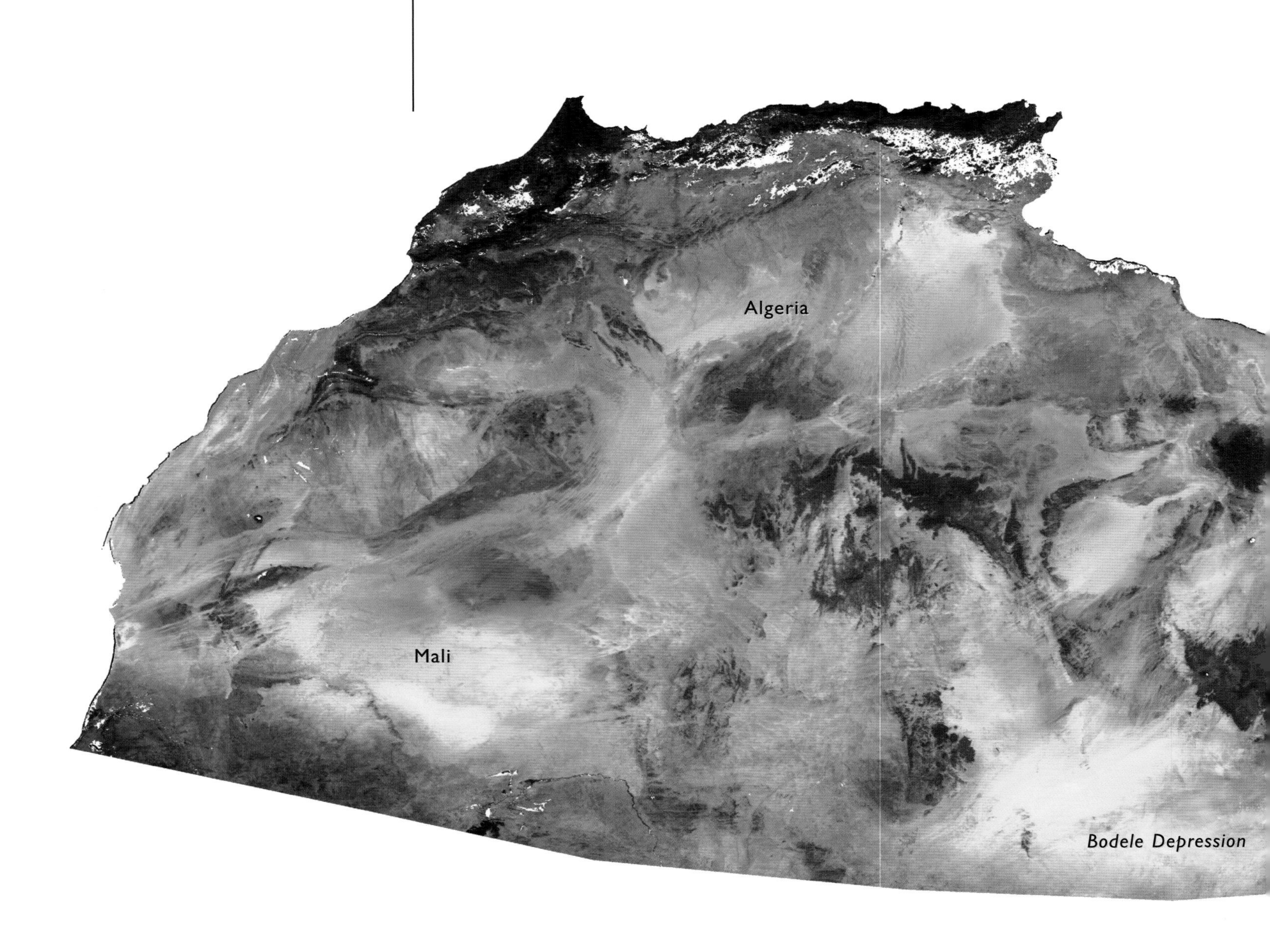


Cloud-free, true-color composite surface albedo of the Sahara Desert and Arabian Peninsula in November 2000. The image shows the large variability in spectral albedo of barren soils and mountains (from dark browns to reds and bright yellows). (Data from the MODIS instrument on the Terra satellite; visualization by Elena Tsvetsinskaya, Boston University.)

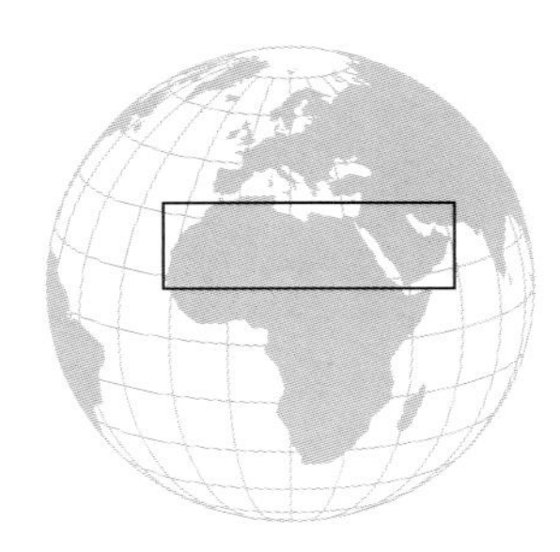

then be used to refine climate models or to drive weather prediction and vegetation growth models.

Bright surfaces (such as snow-covered land, ice, or sandy deserts) reflect the vast majority of the incoming sunlight, thus sending that energy back out to the atmosphere. But even in these bright areas, pools of melted snow, open water exposed in cracks in the ice, local variations in desert soil color, and shadowing by dark, barren mountains or wind sculpted dunes and snowdrifts can lead to vast differences in the albedo. Fires that clear and blacken landscapes also cause extreme variations in albedo. Dark surfaces such as the tropical rain forests of South America and Africa absorb most of the radiation that reaches them and only reflect a small proportion back to the warm tropical atmosphere. The boreal forests of Canada, Europe, and Russia can also be quite dark and absorb a lot of energy during the summer months, but then vary widely from bright to dark during the winter months depending on whether new fallen snow is still lying on the tree boughs or has fallen through and is now obscured by the dark foliage. This combination of snow cover and of snow shadowing by trees determines the timing and amount of snow melt in the region. It is particularly

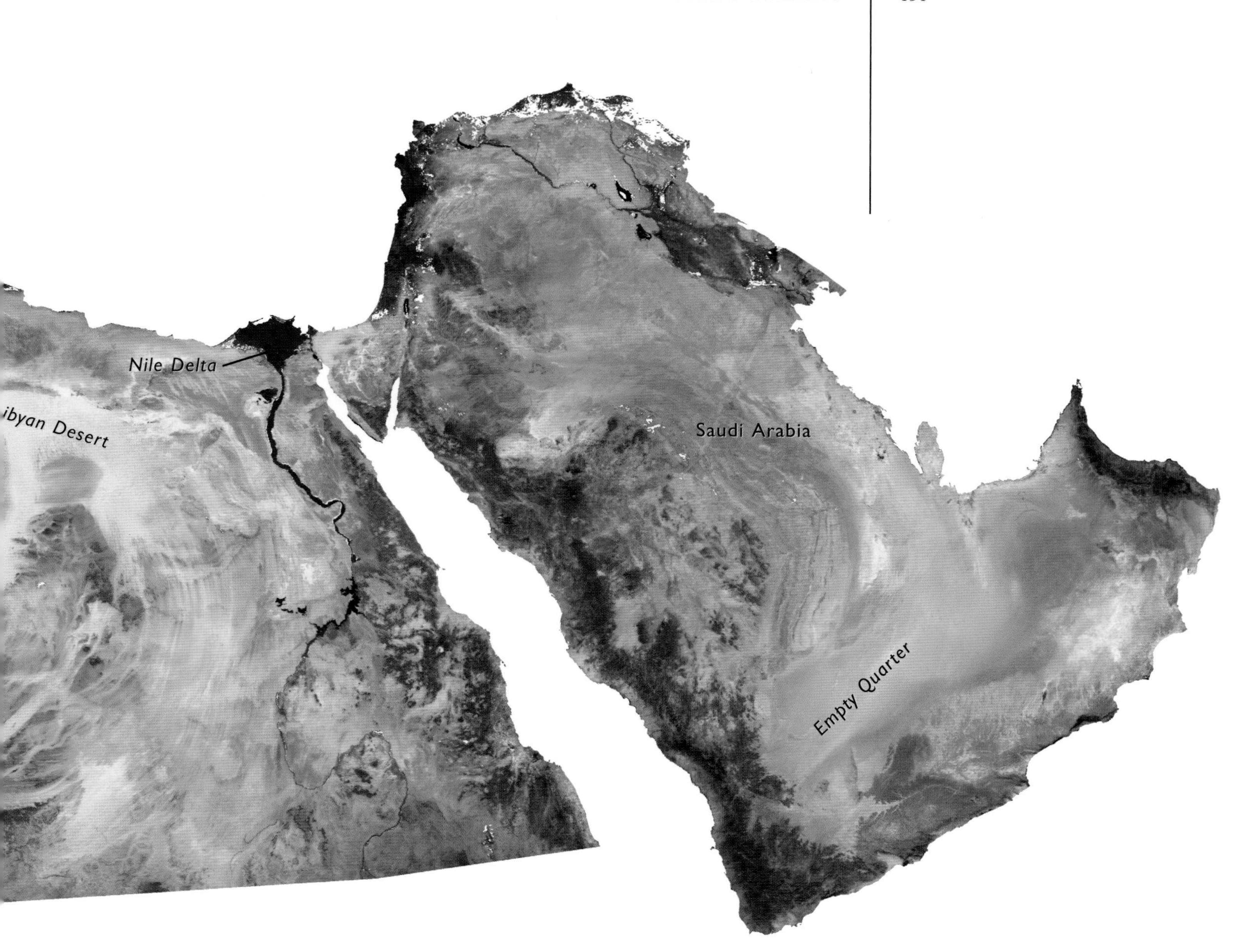

Sand dunes in the Namib Desert located just to the south of the normally dry Kuiseb River and west of the Gobabeb Research Centre, about 40 km inland from the Atlantic coast, on September 27, 2003. This location marks the northern extent of the Namib dune fields, which give way to gravel planes north of the Kuiseb. The photograph was taken looking towards the south. (Photograph by Steven Platnick, NASA GSFC.)

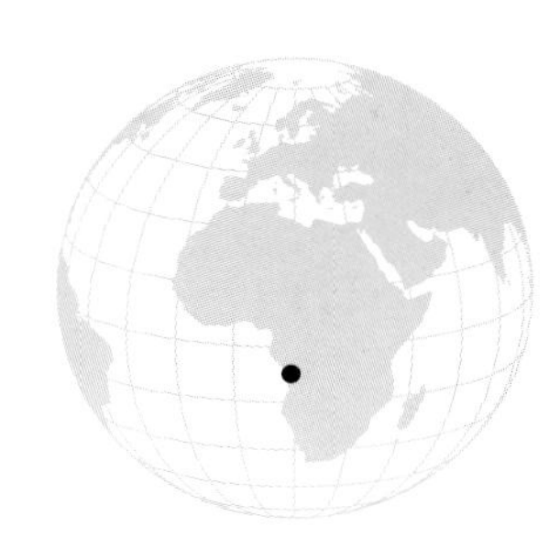

August 28–September 12, 2004

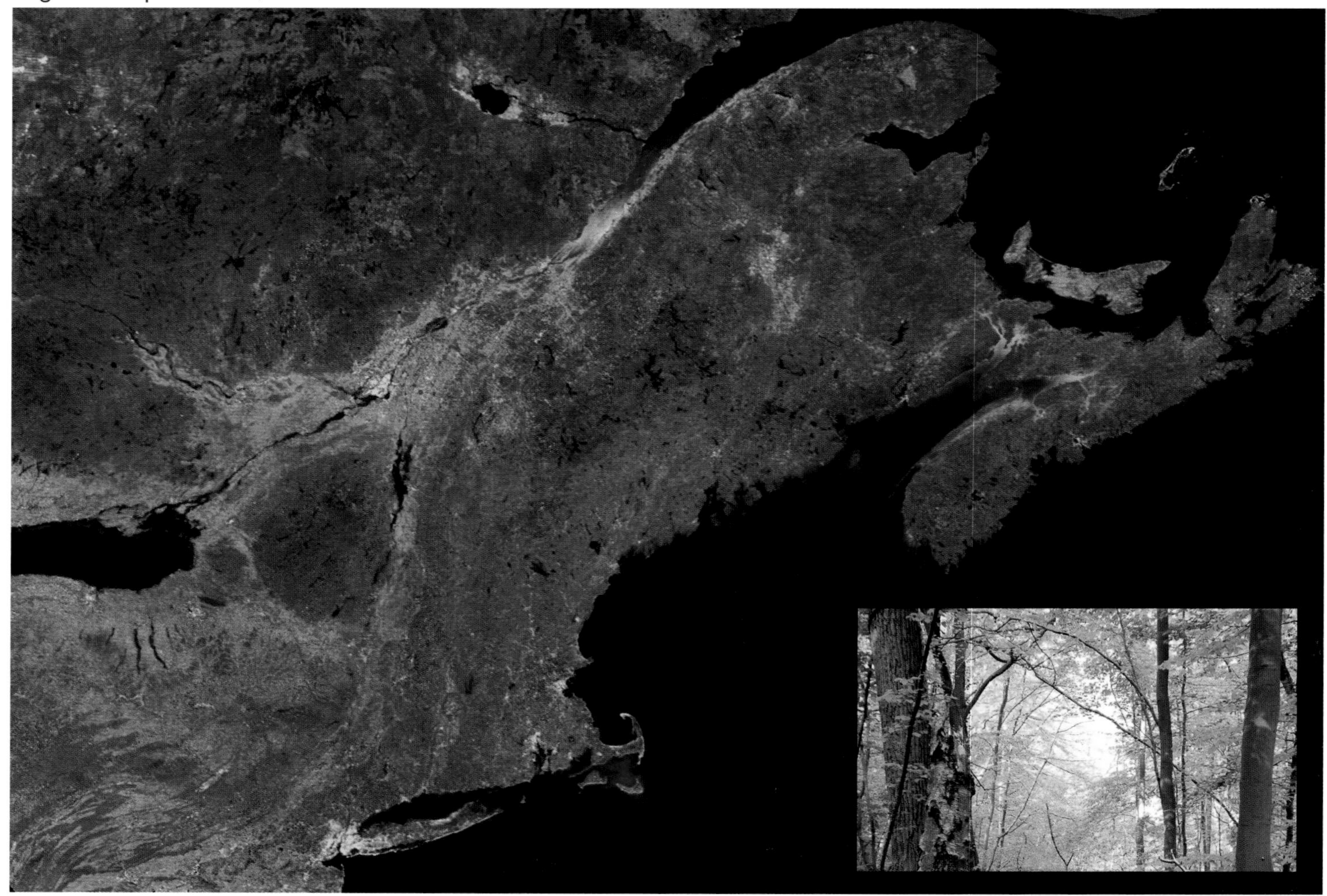

True-color albedos showing the spread of reddish autumn foliage across the northeastern United States and eastern Canada. (Data from the MODIS instrument on the Terra and Aqua satellites; visualization by Jonathan Salomon and John Hodges, Boston University; inset photograph by Crystal Schaaf.)

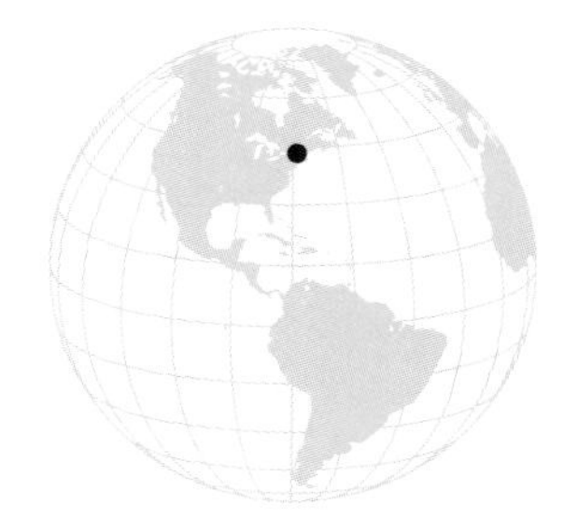

difficult to simulate this cycle correctly and to predict the amount of water that will be released during the spring thaw. Deforestation in these areas can lead to larger areas of exposed snow and less shadowing by trees and this can alter the ability of a region to retain water over the winter as snow pack. Therefore researchers are using satellite observations of the patterns of bright snow and dark trees throughout the winter to help them improve their climate and prediction models.

Satellite images of surface reflectance can also be used to monitor the annual growth cycle of crops and vegetation. In arid lands, overuse of the land can couple with drought to expand barren deserts and reduce the ability of local populations to grow crops and graze their animals. Governments and famine relief organizations carefully monitor the annual extent of vegetated regions to identify potential trouble spots. In more temperate areas, satellite imagery is also used to monitor crop growth so that harvest amounts and revenues can be predicted.

In satellite imagery, healthy vegetation can be detected in the near-infrared wavelengths and highlighted in green. A wave of new growth can be seen sweeping northward across the continental United States in the early spring—yet the farm fields of the Mississippi River Valley have not yet been planted and lag behind the natural vegetation. By midsummer the entire region (both planted and natural) is lush with vegetation, but in the autumn, the harvesting of crops means that the farm fields are barren long before the natural vegetation has gone dormant. Similarly, brilliantly colored autumn leaves lure thousands of tourists to the northeastern United States. Satellite imagery in natural colors can

September 29–October 14, 2004

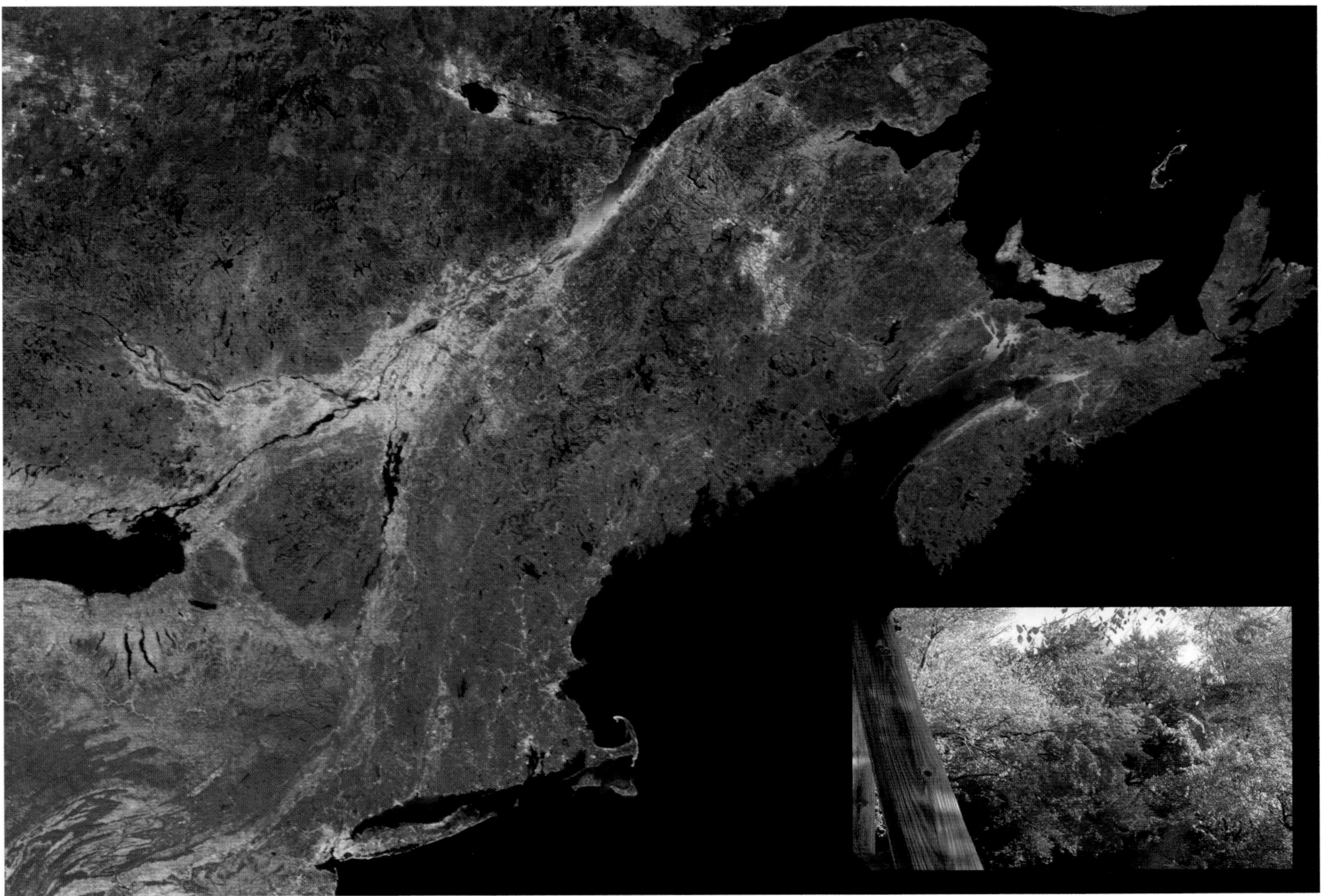

be used to identify the onset and duration of the peak foliage season and also to monitor the impact of drought and insect damage on the extent and brilliance of fall colors.

By using satellite imagery to monitor the Earth's albedo throughout the seasons and from year to year, scientists are now beginning to understand the impact of landscape alterations and the resultant local weather changes. These insights are leading to improved climate and weather prediction models that can simulate these variations more accurately and then can be used to predict the effects of further surface transformations on the health of our sunlit Earth.

Above [inset]: Brilliant fall foliage in late September in Concord, Massachusetts. (Inset photograph by Crystal Schaaf.)

Mixed forest in winter with snow on the ground. The reflectance of this environment is much less than if snow was on open land without a forest canopy cover. (Photograph taken in Stow, Massachusetts by Crystal Schaaf.)

January 1–16, 2001

April 7–22, 2001

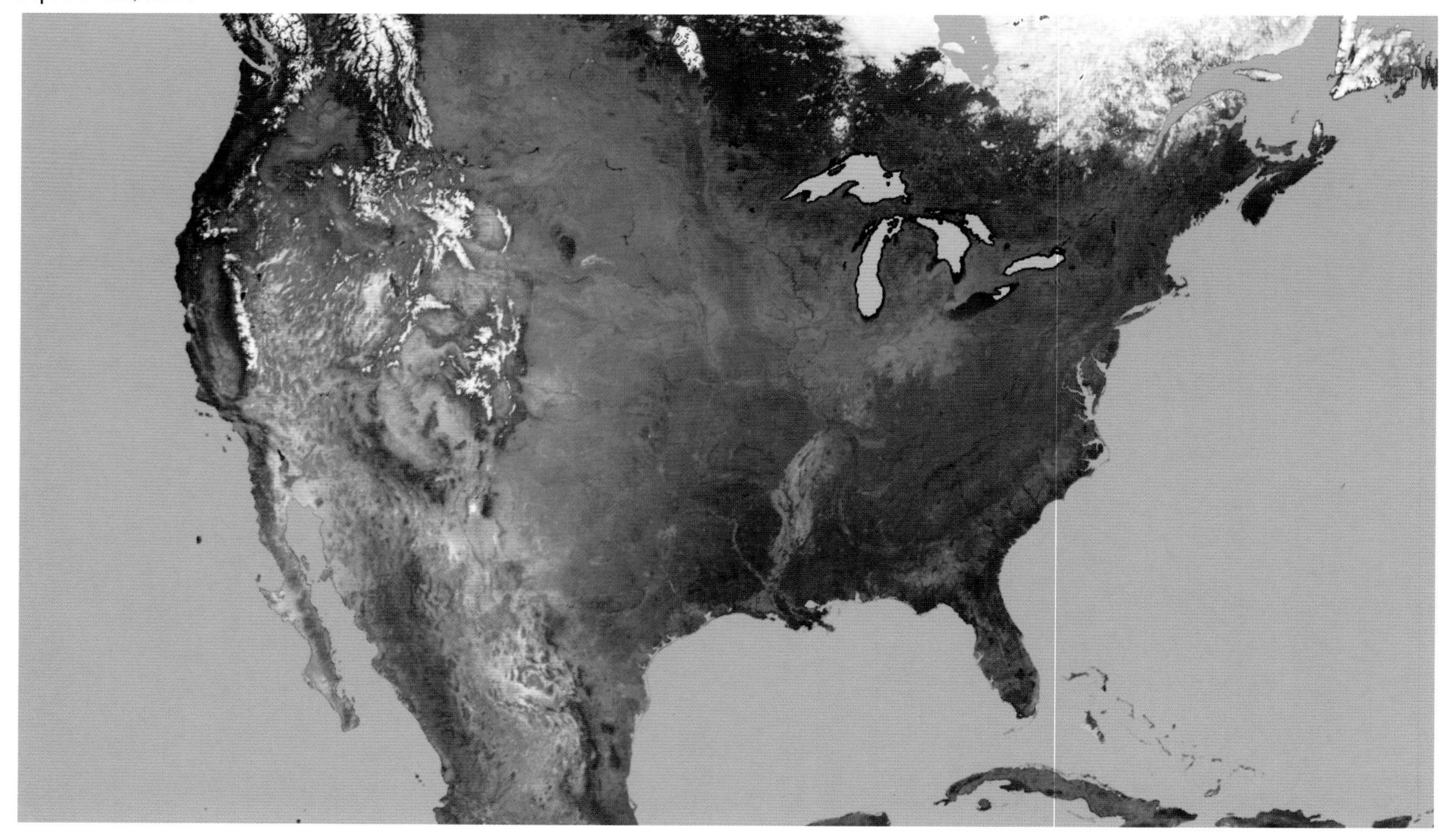

North American surface reflectance where lush vegetation is indicated by the green colors as detected by the near-infrared wavelengths. (Data from the MODIS instrument on the Terra satellite; visualization by Feng Gao, Earth Resources Technology, Inc., MD.)

July 12–27, 2001

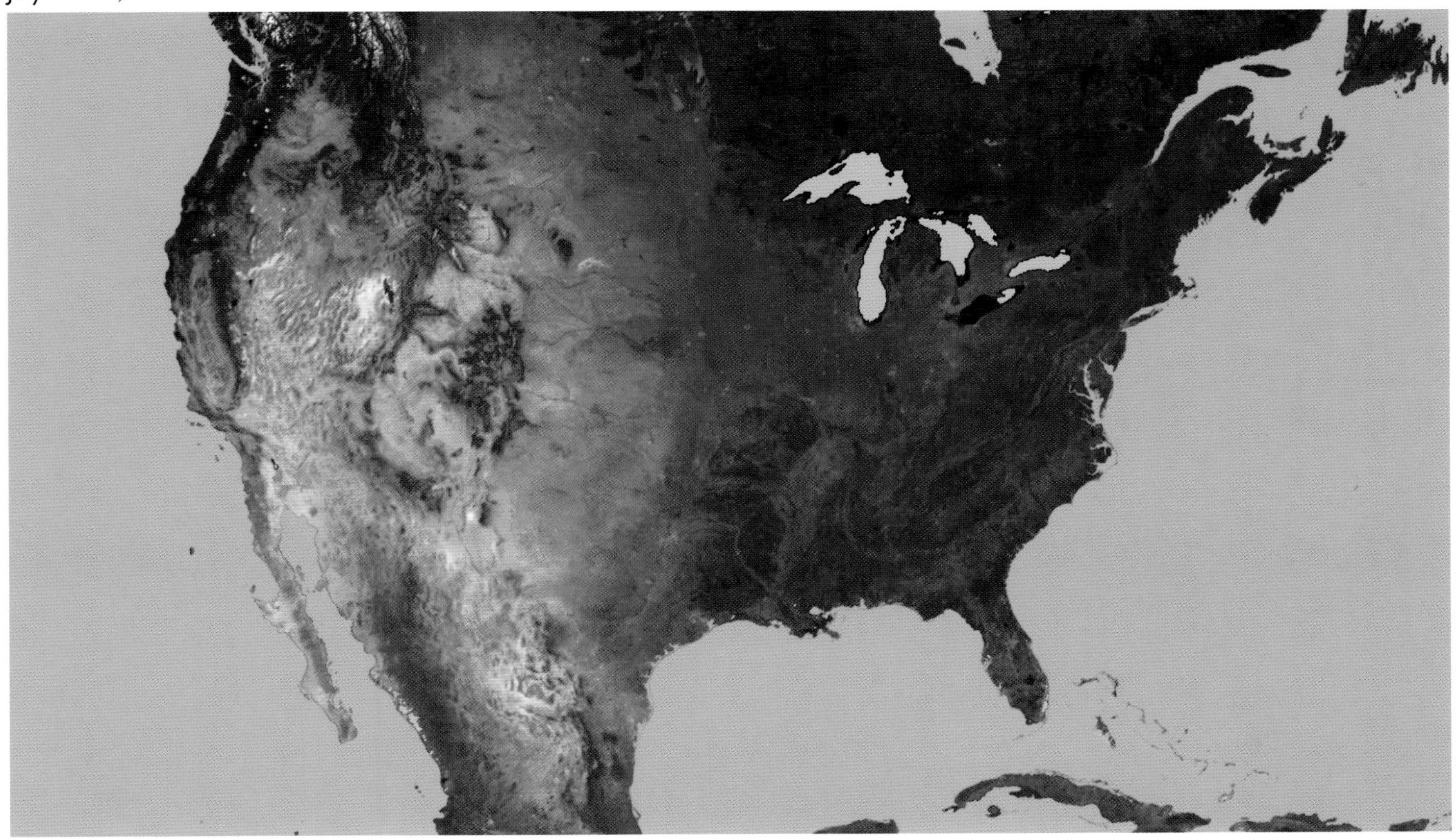

October 16–31, 2001

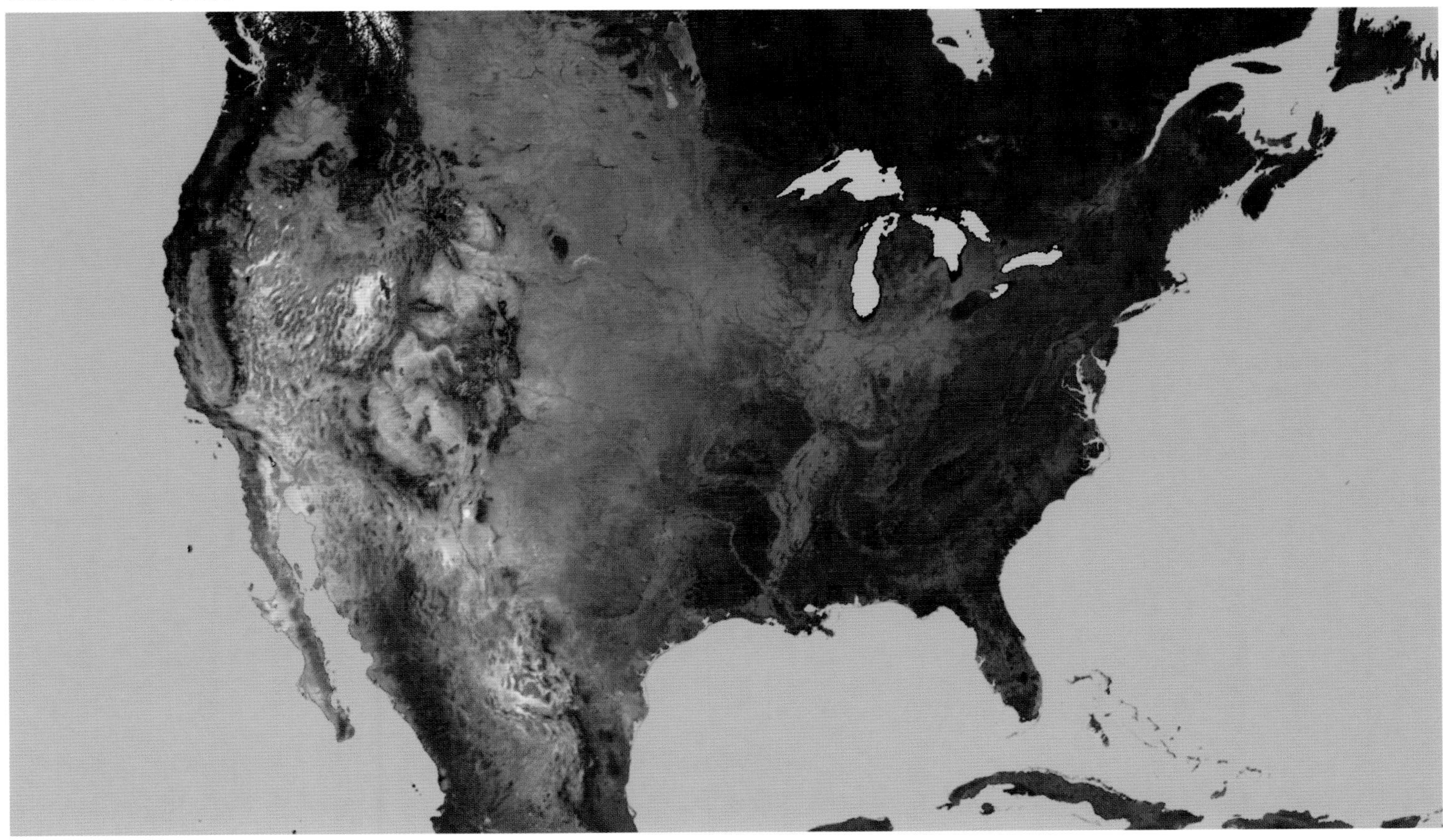

Zhengming Wan

# Temperature of the Land Surface

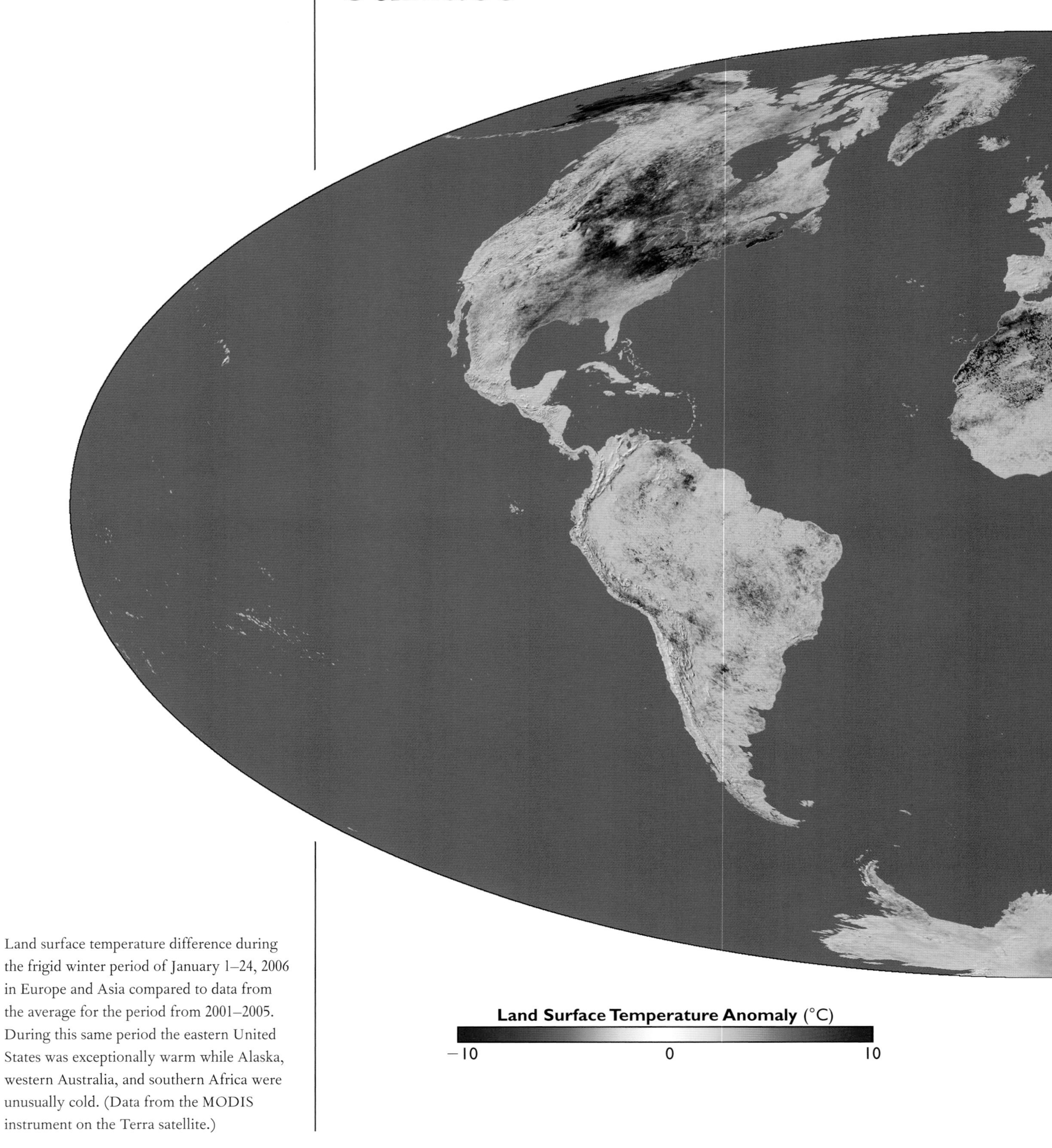


Land surface temperature difference during the frigid winter period of January 1–24, 2006 in Europe and Asia compared to data from the average for the period from 2001–2005. During this same period the eastern United States was exceptionally warm while Alaska, western Australia, and southern Africa were unusually cold. (Data from the MODIS instrument on the Terra satellite.)

*Our Changing Planet*, ed. King, Parkinson, Partington and Williams
Published by Cambridge University Press 

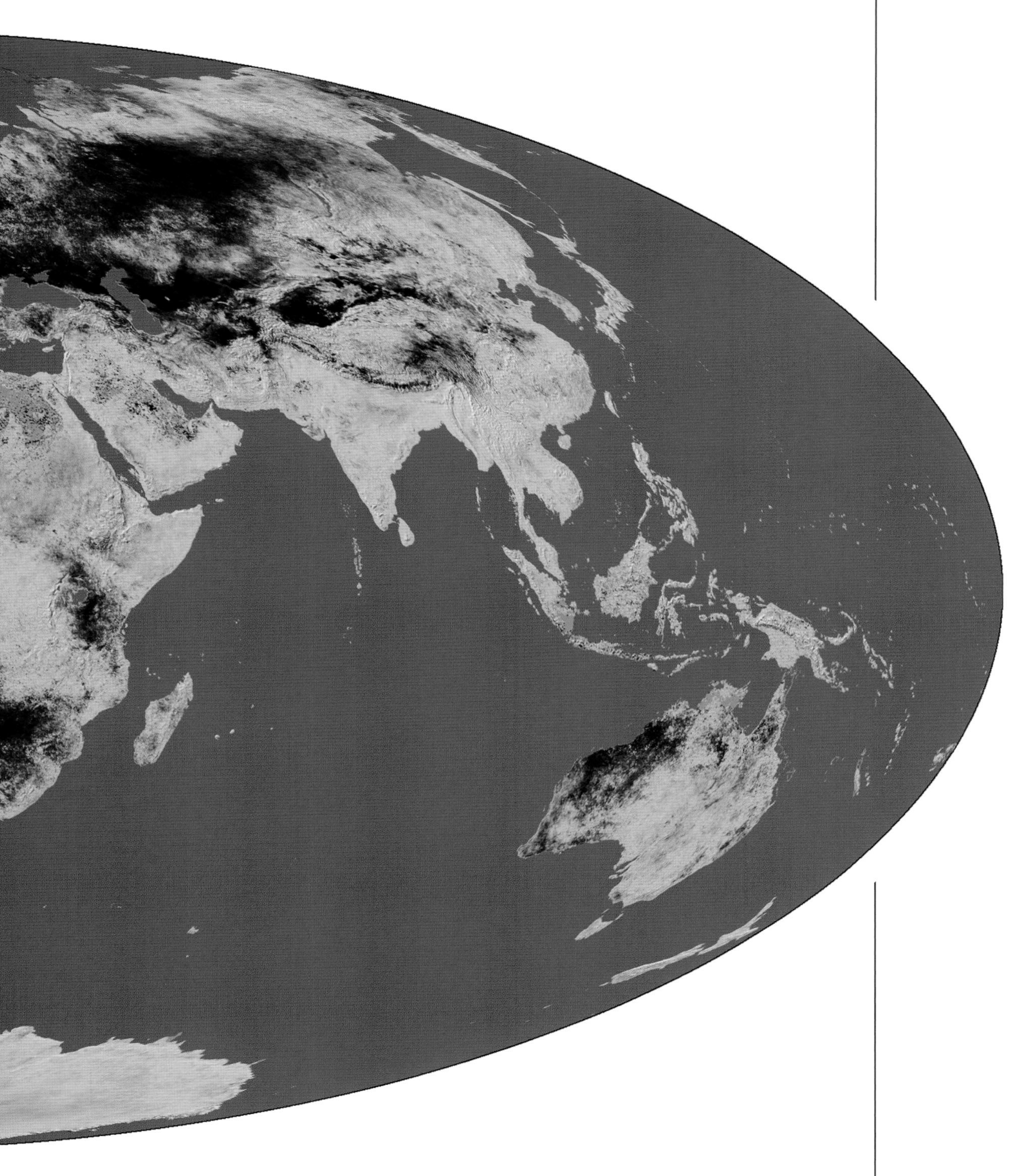

Unlike conventional observations of surface temperature that are actually measurements of air temperature by thermometers 2 m above the ground, satellites measure the thermal radiation emitted from the Earth's surface—whether that surface is bare ground, lakes, treetops, or rooftops. These measurements from space make it possible for farmers to know the temperature of their crops for better estimating water requirements and yields. Since the late

December 2005, January 2006, February 2006

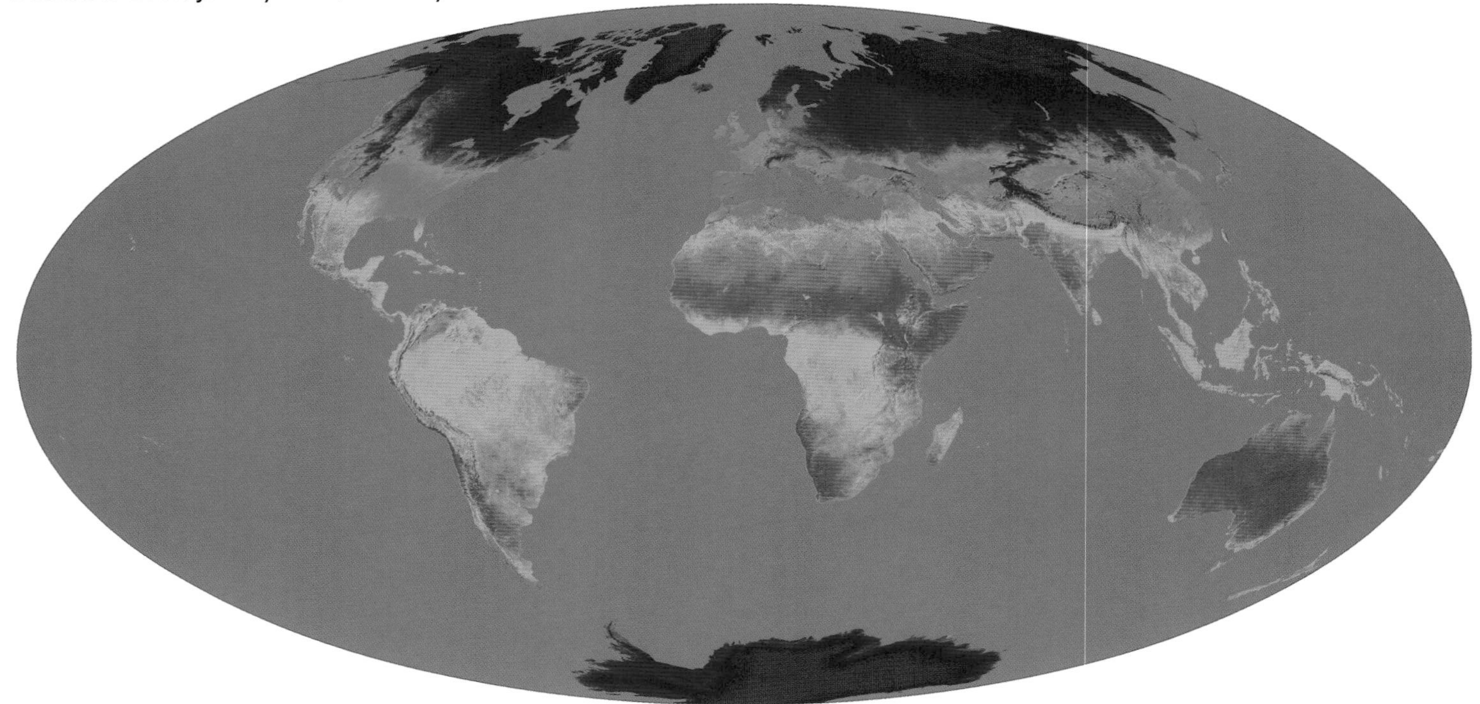

March, April, May

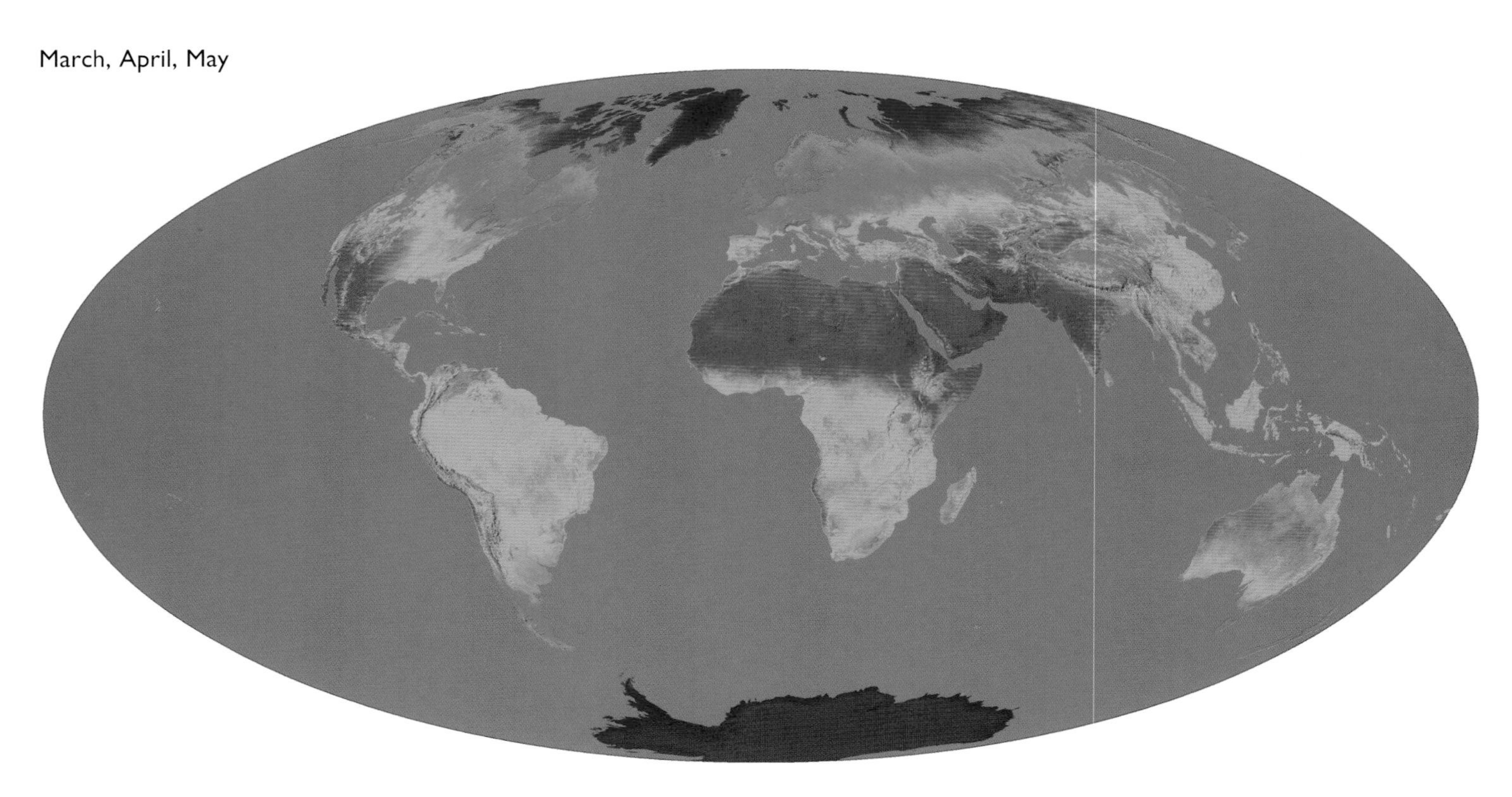

**Land Surface Temperature** (°C)

<−35.0 12.5 >60.0

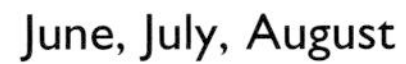
June, July, August

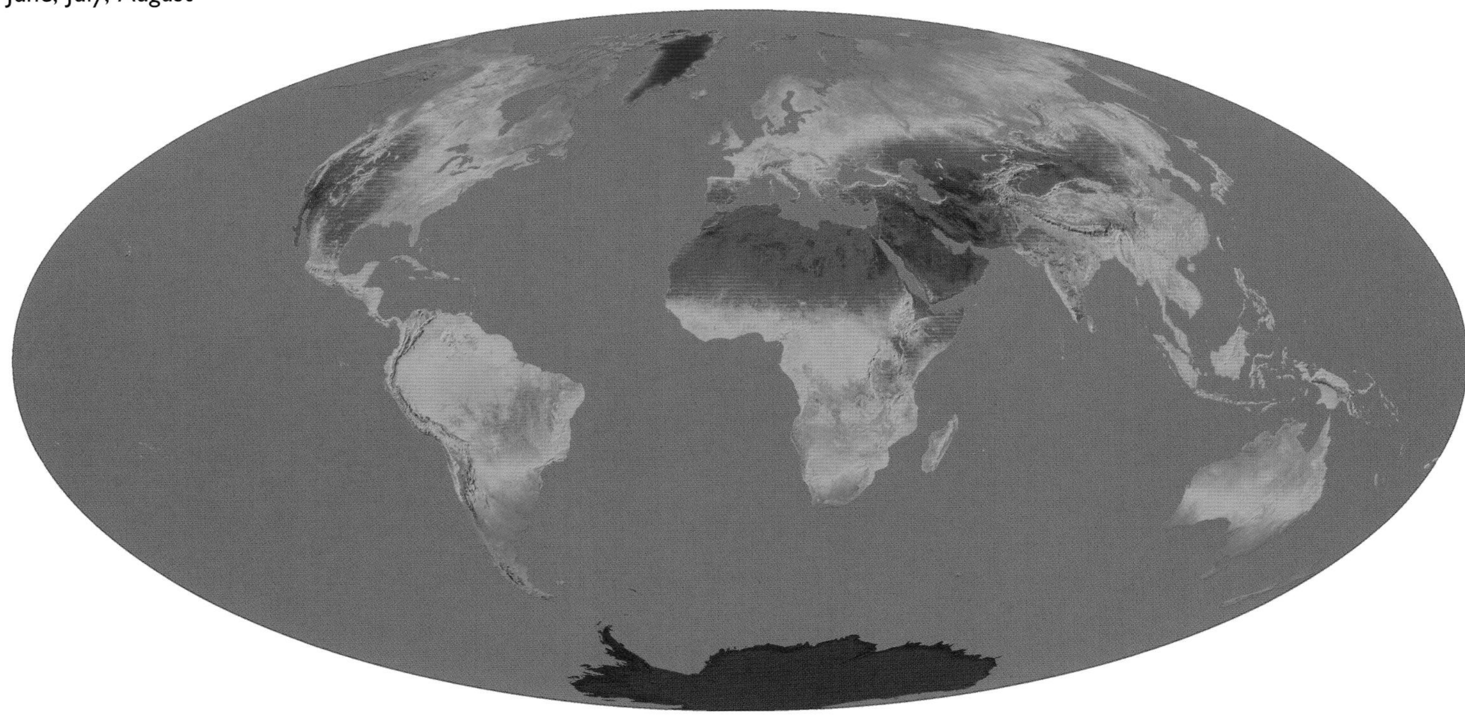

September, October, November

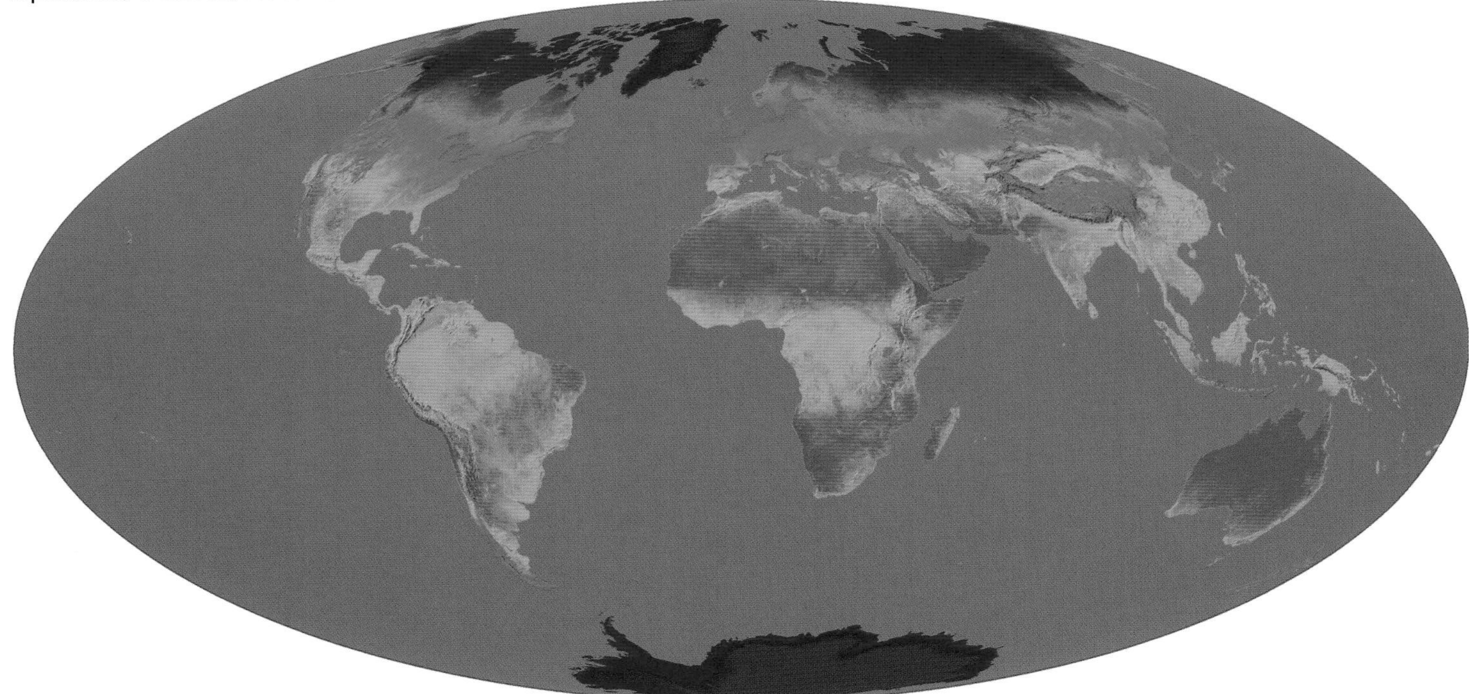

Seasonal pattern of daytime land surface temperature during 2005 and 2006. Averages are shown separately for December 2005 to February 2006, March to May, June to August, and September to November 2006. High land surface temperatures are shown in red. (Data from the MODIS instrument on the Terra satellite.)

1970s, satellites have measured land surface temperatures in clear-sky conditions during both the daytime and nighttime everyday. The multi-year surface temperature data reveal the spatial and temporal distribution of warming and cooling on the land surface, and how these patterns change in response to regional and global change, including the globally significant impact of El Niño events in the tropical western Pacific.

A pre-Monsoonal heat wave left India, Pakistan, Nepal, and Bangladesh baking for much of May and June 2005 (right), in contrast to land surface temperature during the same time period of the previous year (facing page). (Data from the MODIS instrument on the Terra satellite.)

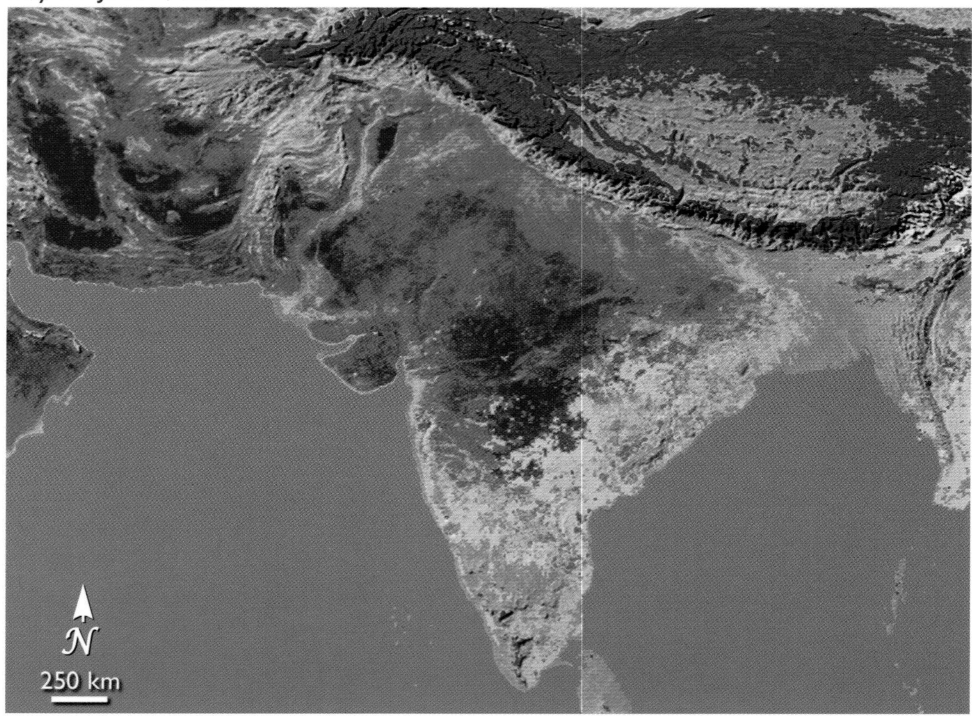


Examining the global distribution of land surface temperature it is apparent that, during northern summer, the temperature in India is nearly as hot as vast areas of Saudi Arabia and the Sahara desert. When contrasting January 2006 with the average temperature of the land surface from 2001–2005 for the same period of time, it is evident that the eastern half of the continental United States was experiencing an unusually mild January while vast areas of Europe and Asia were experiencing temperatures that were nearly 10°C colder than normal. These lows had not been reached in Moscow since the 1920s. There were significant variations in land surface temperature in many other parts of the world, including exceedingly cold temperatures in South Africa, Mozambique, Zambia, Australia, southeastern Greenland, and much of Alaska.

Land surface temperatures may differ by more than 20°C on the same day in different years. In high latitude regions, daytime temperatures are near or below the freezing point (0°C) at the beginning of May, and well above 10°C by the end of May in warm years. Forest fires start earlier in warm years. Fires in the Siberian forests—the largest in the world and vital to the planet's health—have increased 10-fold in the last 20 years. In the discontinuous permafrost regions, permafrost will likely disappear as a result of ground thermal changes associated with global warming of the Earth's climate. This permafrost degradation will have associated physical impacts. Of greatest concern are soils with the potential for instability upon thaw. Such instabilities may have implications for the landscape, ecosystems, and infrastructure. Scientists warn that global warming might be set to accelerate as rising temperatures in sub-Arctic lands melt the permafrost, causing it to release greenhouse gases, most notably methane, into the atmosphere. An estimated 14% of the world's carbon is stored in Arctic lands. This ancient carbon, locked away in these frozen lands, is starting to be released and bacteria are breaking down its organic material.

May 24–June 1, 2004

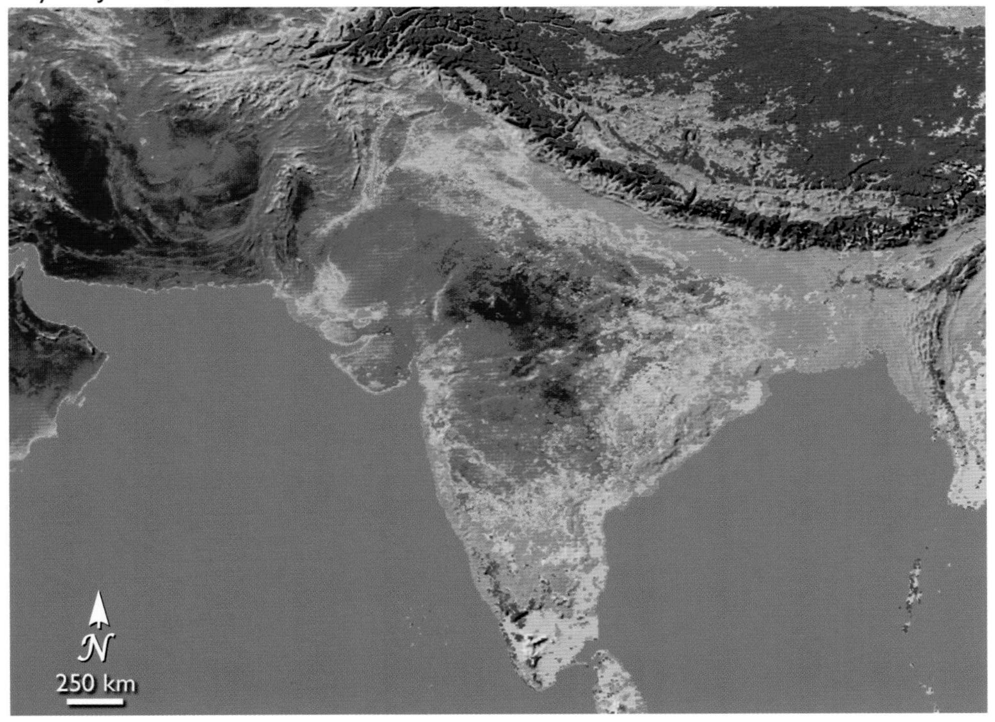


Hundreds of people died from sunstroke and dehydration in a month-long heat wave that swept India, Nepal, Pakistan, and Bangladesh, as south Asia endured one of its hottest summers on record in 2005. The deserts of North Africa are threatening to leap the Mediterranean and creep through Spain. This promoted a national campaign to halt desertification in Spain.

The MODIS data from the Aqua satellite, which flies overhead at about 1:30 pm local solar time, close to the hottest part of the day, have been used to retrieve the daily near-maximum land surface temperature in clear-sky conditions at the global scale. According to the Aqua MODIS land surface temperature data, the hottest place on Earth in 2004 and 2005 was the Lut desert of Iran (which is also shown in the dark red area in the upper left portion of the Terra MODIS land surface temperature images on this page and the previous page). The MODIS data from both the morning (Terra) and afternoon (Aqua) satellites provide a better opportunity for a diurnal series of land surface temperatures. Many applications including global change monitoring and the development of land surface models rely on satellites continuously providing long-term records of the global land surface temperature.

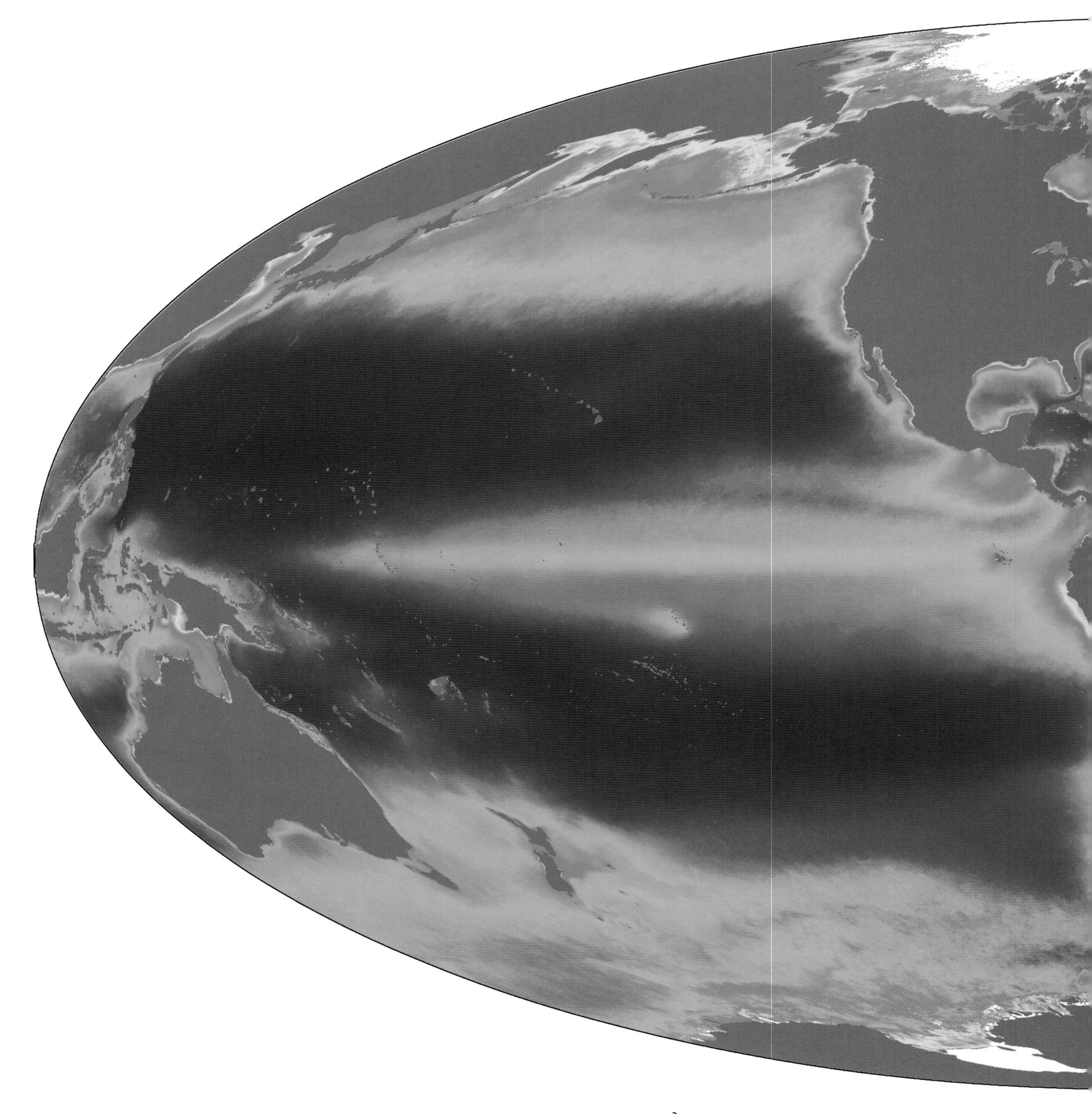

**Chlorophyll-*a* Concentration** (mg/m$^3$)

0.01 0.1 1.0 10 50

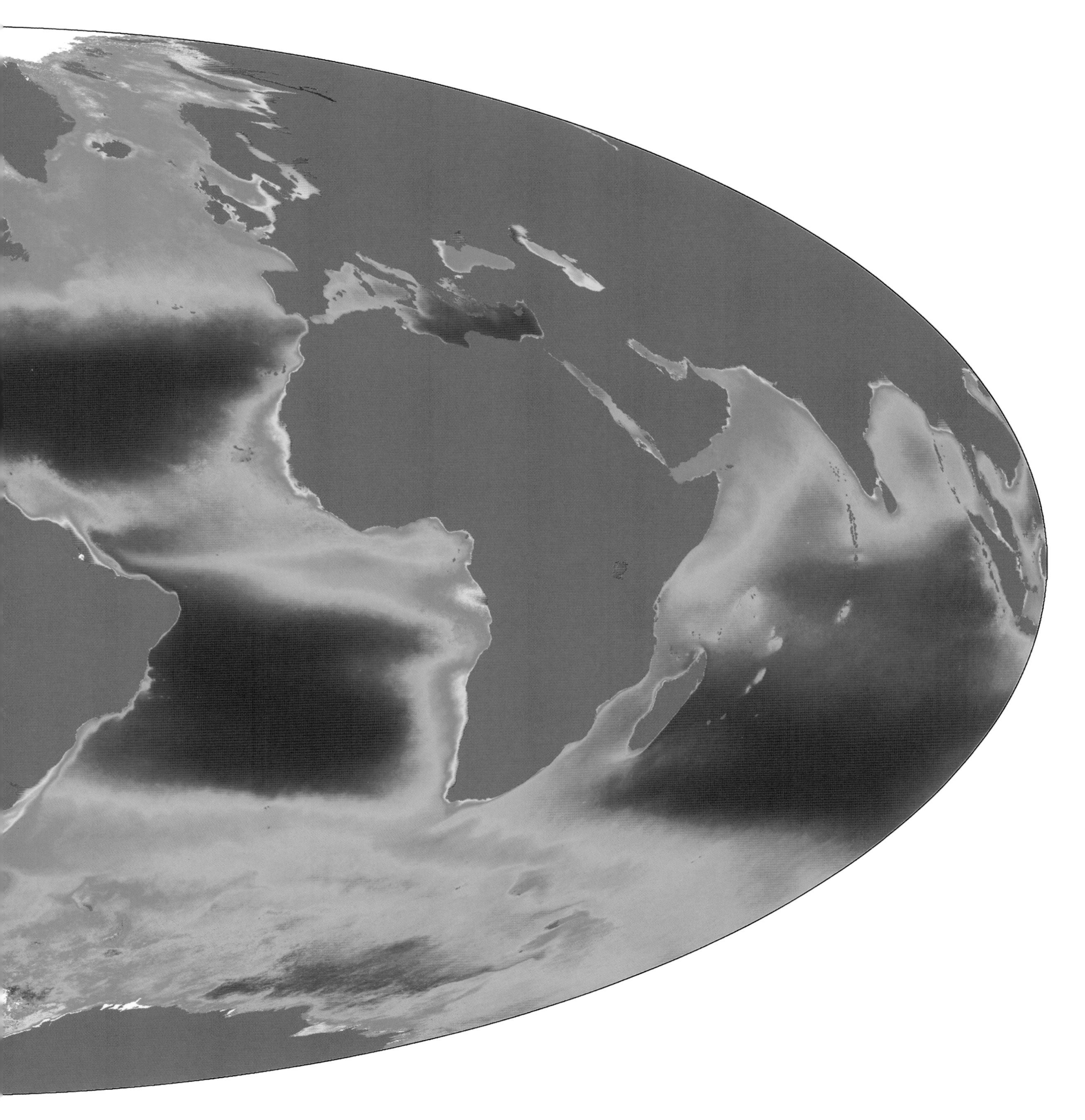

The ocean's long-term average phytoplankton chlorophyll concentration acquired between September 1997 and August 2005. (Courtesy SeaWiFS Project, NASA/GSFC, and GeoEye.)

Rough seas off the coast of Korea, USS Valley Forge in the background. (Photograph courtesy Scripps Institution of Oceanography Archives.)

# Restless Ocean: Introduction

Robin G. Williams

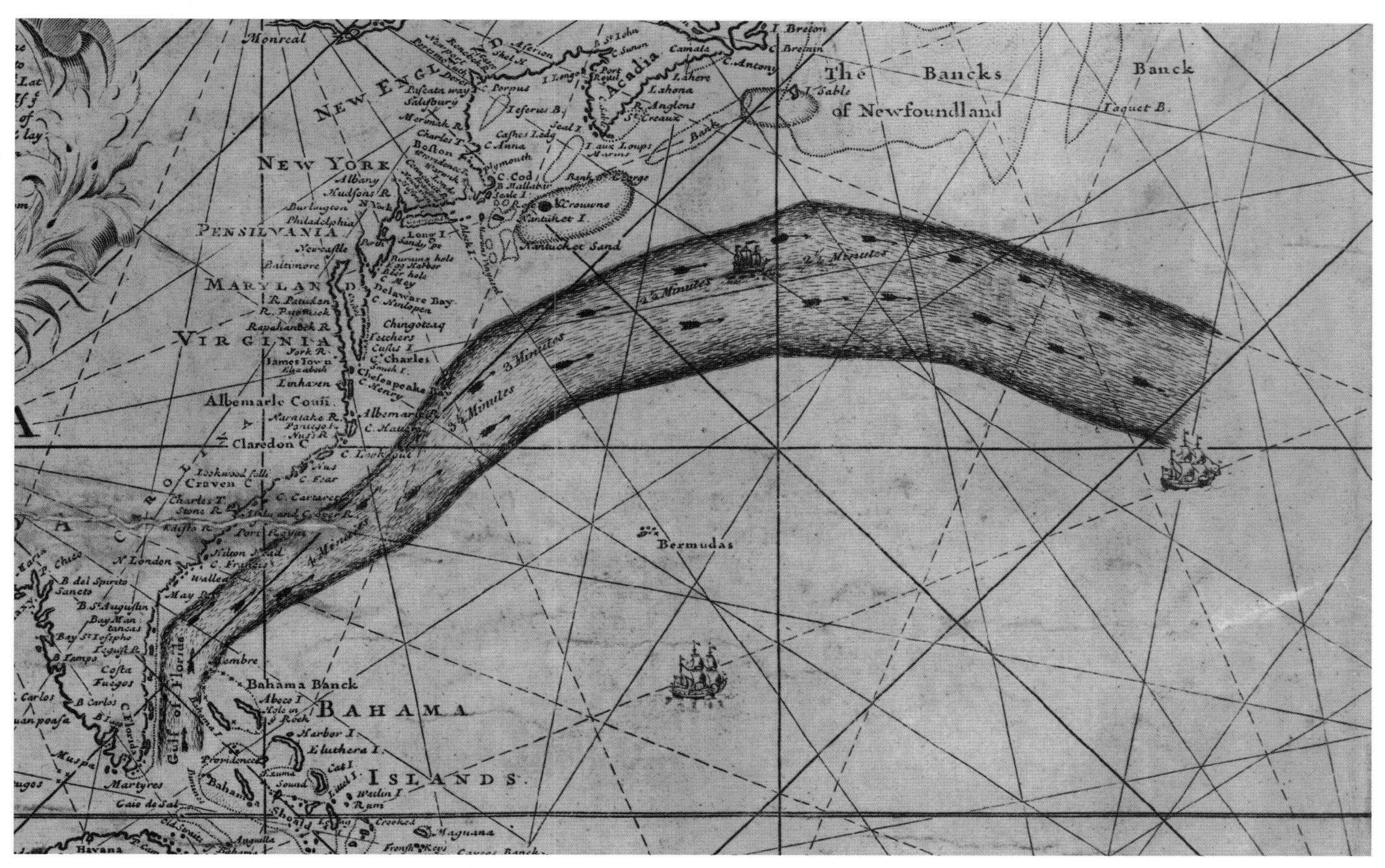


1768 chart of the Gulf Stream by Benjamin Franklin and Timothy Folger. Franklin traveled several times between the newly independent American colonies and Europe. At that time sailing ships were the only means of trans-Atlantic travel. Franklin detected the ships' crossings of the warm waters of the Gulf Stream by the temperature of the wine served at table on the ship. The wine bottles were stored below the waterline against the hull of the ship, causing the wine to be warmer when the ship was in the Gulf Stream, and cooler elsewhere. Franklin also compiled many meticulous temperature measurements that helped delineate the Gulf Stream. (Map image courtesy Library of Congress, Geography and Map Division.)

Despite the fact that the ocean covers 70% of the Earth's surface, we still know very little about this vital and dynamic part of our planet. In large part, this is because mankind has historically been concerned with his immediate environment, the land. However, the other overriding reason is that the investigation of the oceans has always presented great challenges both to seafarers and scientists. In ancient times, civilizations made hazardous voyages across the ocean with the intent of discovering new lands and riches. In the Pacific Ocean, distant ocean voyages were undertaken as long ago as 5,000 BC, which resulted in the peopling of the islands of this great ocean. In the western world, ancient seafaring was dominated by the Phoenicians, the Greeks, and finally the Romans. Along with these voyages of exploration came great scientific discoveries and improved charts, such as that of Ptolemy in AD 150, which acted as a great incentive for explorers to seek out the unknown lands beyond the known seas. In the Middle Ages, scientific enquiry counter to religious teachings was actively suppressed and much of the knowledge gained previously was forgotten. The next significant wave of exploration came with the Vikings and, in the late 9th century, a warming global climate freed the North Atlantic of ice, allowing Vikings to sail

*Our Changing Planet*, ed. King, Parkinson, Partington and Williams
Published by Cambridge University Press 

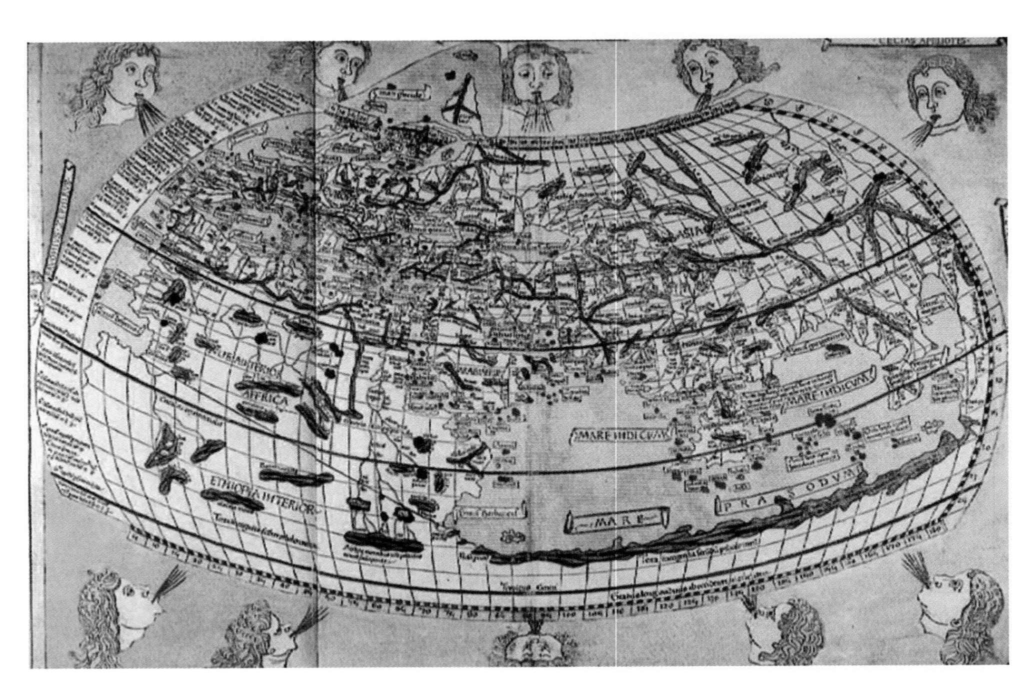


Ptolemy's atlas of the known world, circa AD 150. This atlas depicts a reasonable map of the Mediterranean region, northern Africa, and western and central Asia. The lack of accurate longitudinal measurements gives the map a stretched look in the east-west direction. Also, Ptolemy used a value for the Earth's circumference of 29,000 km, an error that later caused Columbus to believe he had encountered Asia rather than the Americas.

westward, colonizing Iceland and Greenland, and establishing a temporary settlement in Newfoundland. In 1410, Ptolemy's map was republished and, under the leadership of Prince Henry the Navigator of Portugal, a new drive to explore beyond Europe began anew. During this period of discovery, Christopher Columbus sailed to the New World and Magellan/Elcano circumnavigated the globe. Then, in 1588, England defeated the Spanish Armada, thus ending the Spanish dominance of the seas and heralding a new period of exploration led by the British. The discoveries of Captain James Cook in the late 1700s were aided greatly by the marine chronometer, a device that enabled sailors to accurately determine, for the first time, longitude. During the 19th century a more structured effort was made to investigate the oceans from a scientific point of view. An international conference was held in 1853 to establish uniform methods for data collection at sea. In 1859 Charles Darwin published his famous book, *On the Origin of Species*, based upon the data collected during the voyages of HMS Beagle. Then between 1872 and 1876 the British Challenger Expedition took place. This was the first voyage dedicated solely to the scientific study of the oceans. Two of the outstanding contributions of this voyage were the discovery and classification of 4,717 new species of marine life and the measurement of a record water depth of 26,847 feet in the Marianas Trench. At the beginning of the 20th century, the science of oceanography was institutionalized with the establishment of, among others, Scripps Institution of Oceanography in California in 1904. Ironically, the two world wars were of great benefit to oceanography. The German U-boat in World War I led to the invention of the echo sounder and this technology was later used to make depth surveys of the world's oceans. In World War II (WWII), knowledge of currents and other physical properties of the ocean gave navies a significant advantage and the military performed and supported many studies on the transmission of sound in the ocean, waves, currents, and ocean-floor topography. These studies continued after WWII ended as governments perceived a need to maintain strong oceanographic research programs. The Deep Sea Drilling Project was established to collect data about the sea floor, which later confirmed the theory of plate tectonics, continental drift, and sea-floor spreading. Large multi-disciplinary data collection programs paved the way to a detailed understanding of ocean circulation, including how surface and deep circulation are connected in the polar oceans through great vertical movements of water. This discovery led to the conveyor belt theory of ocean circulation, which is key to understanding the future effects of climate change on the ocean. With the scientific and engineering breakthroughs that led to satellite remote sensing in the 1960s, the study of the oceans was revolutionized. It enabled scientists to view, over a large area simultaneously, features in the ocean that previously they were only able to take snapshots of using sampling equipment from ships.

We begin this section with a chapter on ocean bathymetry and plate tectonics and end it with a contribution on a related topic, the devastating Indian Ocean tsunami of 2004. The intervening chapters are concerned with the physical nature of the ocean and the marine organisms that live within it. Throughout, there are many references to the key instruments that have made ocean remote sensing possible. These include the radar altimeter, the infrared sensor, the scatterometer, and the ocean color detector. In "Ocean Bathymetry and Plate Tectonics," an application of radar altimeter data is presented. This instrument, which measures the distance from the ocean surface to the satellite, enables us to create contour maps of ocean surface with all its hills and valleys and from these maps calculate the bottom bathymetry through our knowledge of gravitational effects. The resulting charts have led to advances in our understanding of plate tectonics and sea-floor spreading, estimated to be on average 2 cm per year. "Ocean Surface Topography and Circulation" is the first of the chapters on the physical nature of the ocean. It discusses the small departure of the sea surface from gravitational effects caused by the general ocean circulation. This so-called ocean surface topography reveals the richness of ocean current variability, ranging from ocean current storms (eddies) to basin-wide movements coupled to climate change. In "Heat in the Ocean" we see how data from infrared sensors are used to map heat transport by large currents such as the Gulf Stream, to detect the surface temperature signal of El Niño, and to study hurricanes. In "Sea Level Rise" data are presented that show that global sea level is increasing, but that, interestingly, there are also regional variations. In some parts of the

The global oceanic conveyor belt is a unifying concept that connects the ocean's surface and deep thermohaline circulation regimes, transporting heat and salt on a planetary scale. The system can weaken or shut down entirely if the North Atlantic surface-water salinity drops too low to allow the formation of deep-ocean water masses.

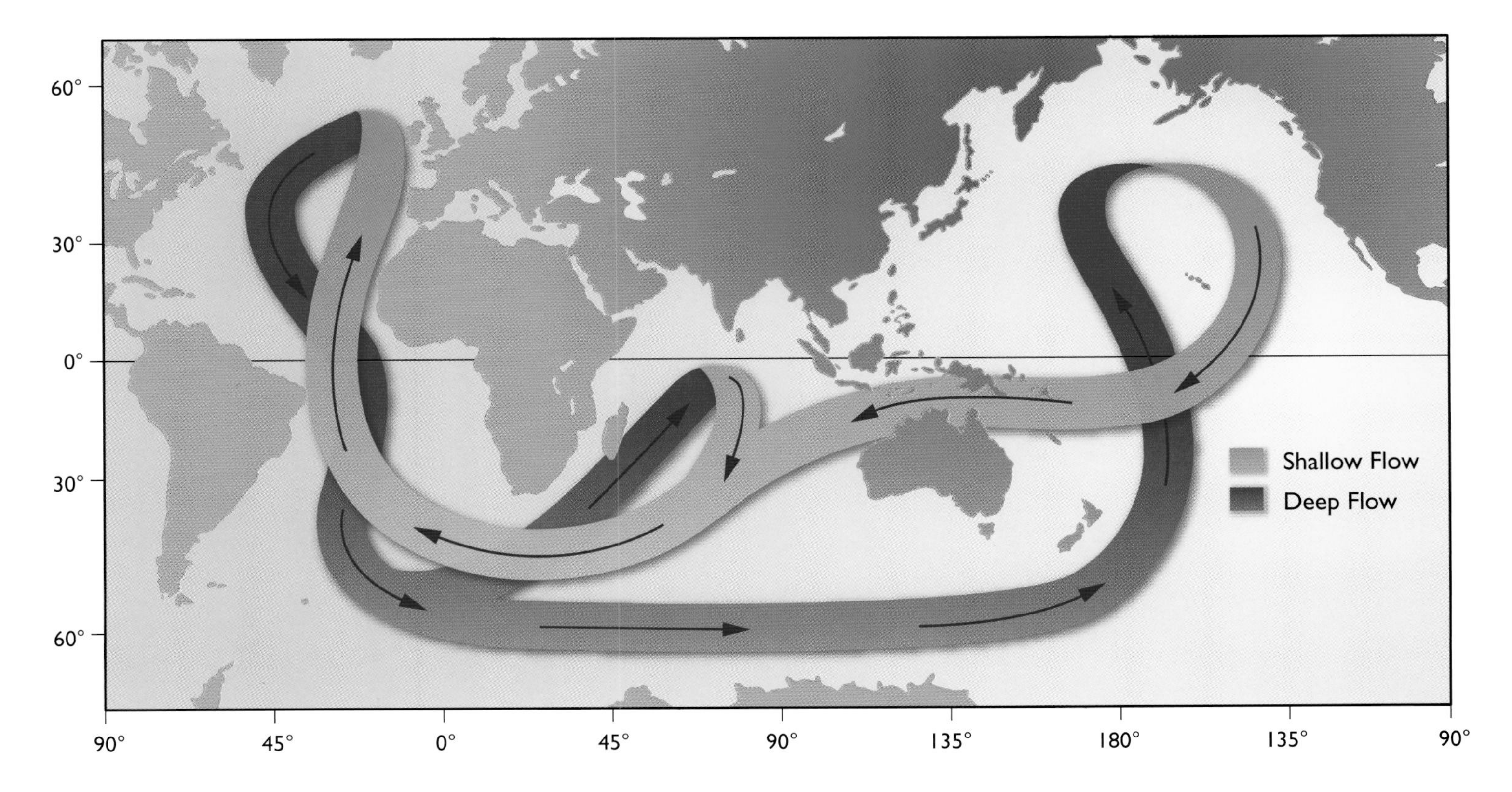

world's oceans, sea level is actually decreasing. In "Tides" another application of altimeter data but over smaller timescales is presented. The main tidal component varies over a period of 12.4 hr and sea-level variations over these scales can also be detected. Improved tidal charts have been generated as a result, with better resolution than those based purely on *in situ* recording instruments. In "Winds Over Ocean" we learn how data from the scatterometer are used to estimate wind speeds and wind direction over the ocean surface (previously not possible), which can be used in forecasting models to improve weather prediction. In "The Stormy Atlantic" the authors show altimeter data that are used to calculate wave height at the ocean surface and that, from the 1960s to the 1990s, the so-called North Atlantic Oscillation phenomenon led to an increase in average wave height over the North Atlantic. "The Ocean Biosphere" chapter, the first on marine organisms, describes how ocean color data are used to produce maps of changes in primary productivity, mainly phytoplankton, microscopic marine plants on which higher life forms depend. Their concentration in the equatorial Pacific is greatly affected by the El Niño phenomenon, for example. In "Coccolithophores and the 'Sea of Milk'" we see how these calcium-rich organisms occasionally appear at the ocean surface. They eventually die and sink to the ocean floor and, over geological time, form sedimentary rocks like the white cliffs of Dover. In "Coral Bleaching" we see how data from ocean color and infrared sensors are combined to identify areas where coral reefs have been damaged by exposure to high ocean temperatures, a phenomenon that is on the increase as sea temperatures rise. This is a serious problem in the tropics and these data enable us to map the affected coral reefs more accurately. "Hunting Red Tides from Space" describes a very interesting teleconnection between dust from the Sahara Desert and the occurrence of red tides in Florida. The dust containing iron-rich minerals is blown across the Atlantic in the atmosphere to the coast of Florida where it falls into the ocean with rain. The iron encourages certain types of algae to multiply, causing so-called red tides near the coasts. The toxins released by these algae are harmful to both fish and humans. In "Marine Sediments" we see how data from ocean color sensors are used to track changes in surface waters caused by marine sediments. Imagery reveals how the Ganges River delivers huge amounts of sediment to its delta in the Bay of Bengal during the Monsoon season. The last chapter, "Tsunamis", describes the Indian Ocean Tsunami that caused so much death and destruction in the coastal areas adjacent to the Indian Ocean in 2004. Fortuitously for oceanographers, a radar altimeter happened to capture the tsunami wave on its journey across the Indian Ocean. This final chapter serves to remind us that we still have much to learn about the ocean, but also that satellite remote sensing can be a valuable tool in the pursuit of new knowledge and understanding about this vitally important part of our planet.

# Ocean Bathymetry and Plate Tectonics

David T. Sandwell

Bathymetry of the Pacific Ocean Basin centered on the Hawaiian Island Chain, which was formed as the Pacific Plate moved over a mantle plume currently beneath the Big Island of Hawaii. Hawaii is the tallest feature on our planet rising 9,700 m (31,800 ft) above the seafloor. (Data derived from sparse ship soundings and dense coverage from the GEOSAT (US Navy 1985) and ERS-1 (ESA 1994) satellite altimeters.)

More than two-thirds of our planet is covered by deep oceans. What would the Earth look like if we could drain the oceans? You would probably ask for a window seat on your next flight to Hawaii. Today we know the topography of Mars 100 times better than the topography of the oceans. Why are the oceans so poorly surveyed? What would you see if we could

*Our Changing Planet*, ed. King, Parkinson, Partington and Williams
Published by Cambridge University Press 

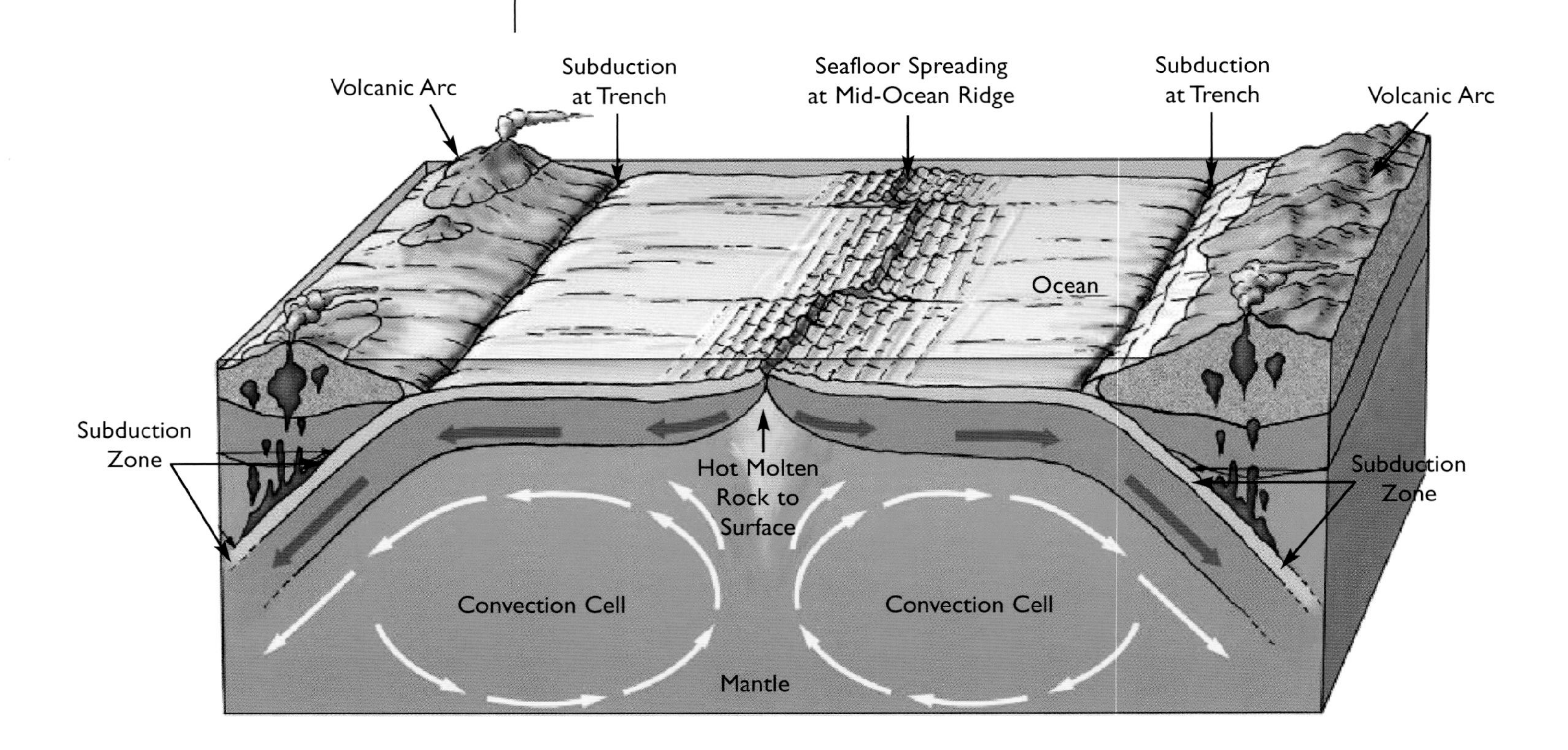


Model of global plate tectonics includes 3 types of plate boundaries that shape the seafloor bathymetry. Seafloor spreading creates an axial rift and corrugated hills. Spreading ridges are offset by transform faults where some of the most destructive earthquakes occur. Subduction of the cooled plate into the mantle creates the deep ocean trenches and produces major earthquakes and tsunami. The plates act as giant radiators of heat. They cool, thicken, and gradually subside as they progress from ridge to trench, forming the broad-scale patterns of ridges and deep ocean basins.

drain the oceans and how is this related to plate tectonics? How has seafloor mapping technology evolved over the past 40 years?

The bathymetry of the ocean floor reflects plate tectonics processes associated with global-scale mantle convection. There are 3 types of plate boundaries, each producing a characteristic type of seafloor bathymetry. The ridge axes represent a singular point in the system where the upwelling magma meets cold seawater. Hydrothermal cooling of magma chambers at the ridge axes recycles the entire volume of the ocean in just a few million years and delivers nutrients for transient biological communities on the seafloor. Scientists believe that life originated in these hydrothermal environments and that hydrothermal circulation is critical for maintaining life on the Earth.

Since one cannot directly map the topography of the ocean basins from space, most seafloor mapping is a tedious process that has been carried out over a 40-year period by research vessels equipped with echo sounders. So far only a few percent of the oceans have been surveyed at 200 m resolution. It has been estimated that 125–200 ship-years of survey time will be needed to map the deep oceans and this would cost a few billion dollars. Mapping the shallow seas would take much more time and funding. Fortunately, such a major mapping program is largely unnecessary because the ocean surface has broad bumps and dips which mimic the topography of the ocean floor. The extra gravitational attraction of features on the seafloor produces minor variations in the pull of gravity that produce tiny variations in ocean surface height. These bumps and dips can be mapped using a very accurate radar altimeter mounted on a satellite.

The ability to infer seafloor bathymetry from space was first demonstrated in 1978 using Seasat altimeter data, but the spatial coverage was incomplete because of the short 3-month lifetime of the satellite. Most ocean altimeters have repeat ground tracks with track spacings

The bathymetry of the Indian Ocean reveals the triple junction between the African Plate, the Indo-Australian Plate, and the Antarctic Plate. Fracture zones record the direction of spreading at these 3 plate boundaries. The large-scale bathymetry is dominated by cooling and subsidence of the plates as they slide away from the ridge axes. (Data derived from sparse ship soundings and dense coverage from the GEOSAT (US Navy 1985) and ERS-1 (ESA 1994) satellite altimeters.)

of hundreds of kilometers so they cannot be used to infer bathymetry. Adequate altimeter coverage became available in 1995 when the United States Navy declassified radar altimeter data from one of their mapping missions flown in 1985. With today's technology, a new altimeter mission could achieve a 5-fold improvement in global ocean floor bathymetry.

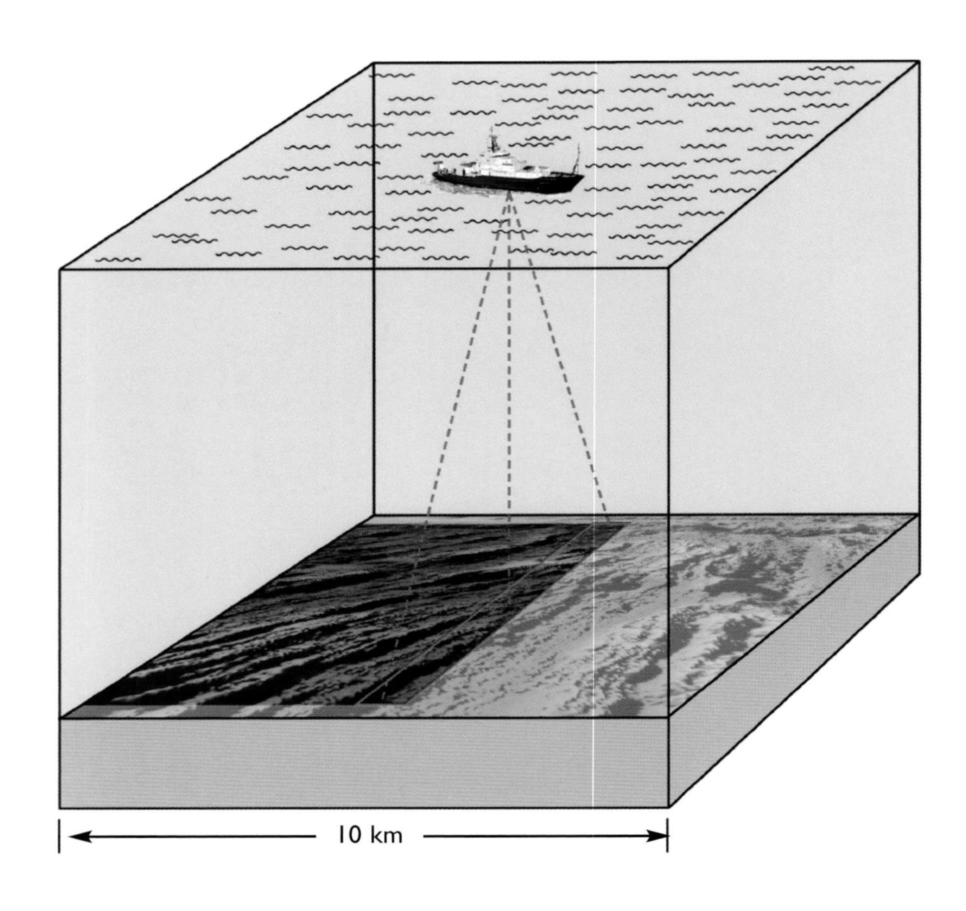


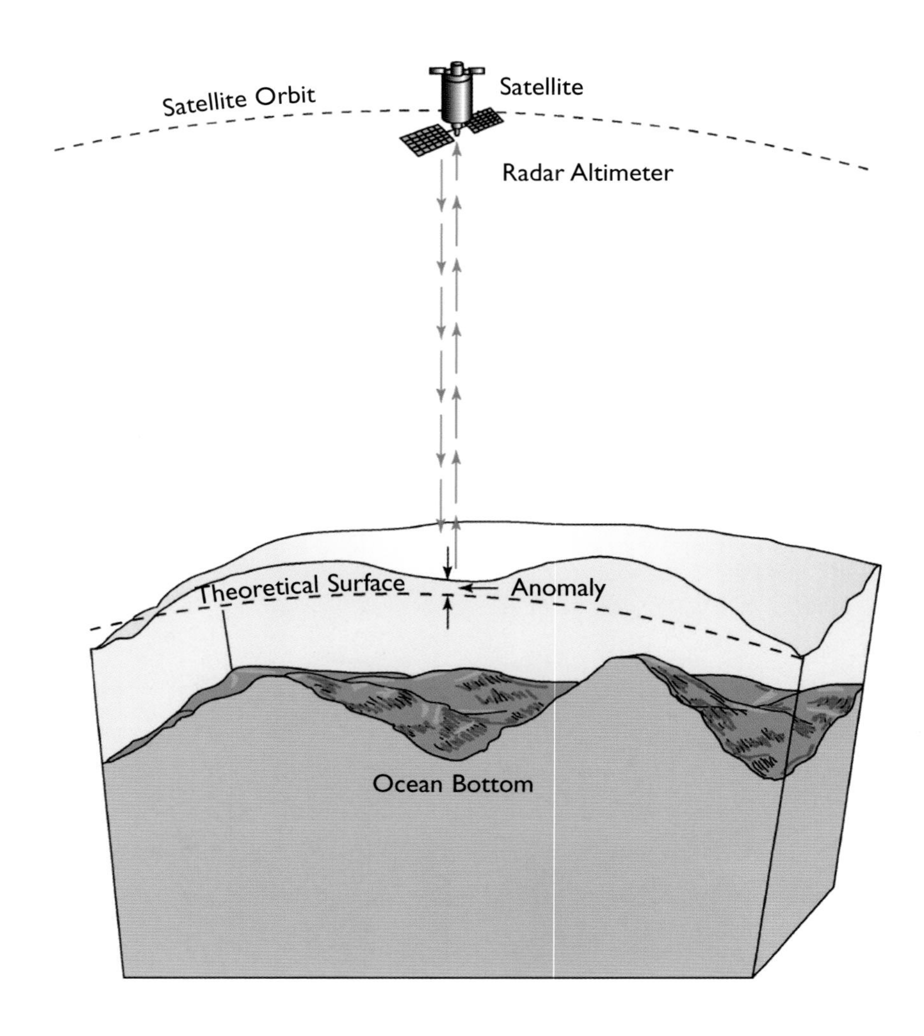


Modern tools for mapping the deep ocean floor. (top) A shipboard multi-beam echo sounder uses sound waves to map 10–20 km wide swaths at ~200 km horizontal resolution. (bottom) An Earth-orbiting radar cannot see the ocean bottom, but it can measure ocean surface height variations induced by ocean floor topography. While the resolution of the echo sounder technique is far superior to the resolution of the satellite altimeter technique, complete mapping of the deep oceans using ships would take 200 ship-years at a cost of billions of dollars. Indeed, the shipboard and altimeter methods are highly complementary. When interesting features are discovered in satellite gravity measurements, these can be surveyed in fine detail by ships.

# Ocean Surface Topography and Circulation

LEE-LUENG FU

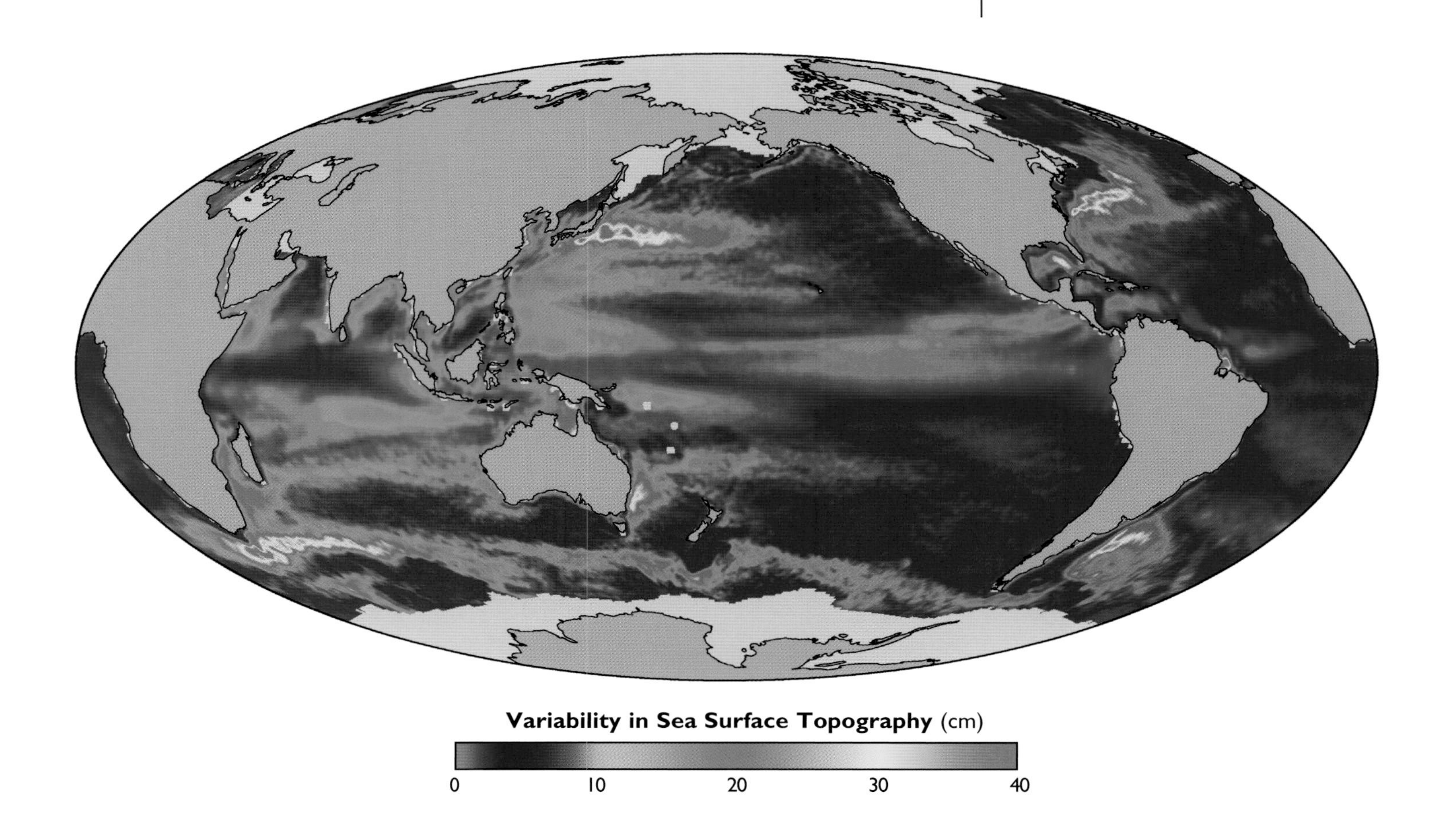


Variability of the ocean surface topography derived from radar altimeter measurements. The variability reflects the presence of ocean eddies. (Data from TOPEX/Poseidon and ERS satellites. Image processing by L.-L. Fu.)

Surfers and sailors are well aware of the ocean's turbulent nature as displayed by breaking waves and swirling currents. Current meters in the ocean are tossed around, making it difficult to measure the speed and direction of currents. Measurements from these instruments are often dominated by random waves rather than by the steady flow that is of interest to navigators and fishermen. Only after averaging a large number of measurements is it possible to detect a steady flow. This steady flow is what we usually refer to as ocean circulation. It transports the heat stored in the ocean around the world and controls weather and climate.

The slow movement of ocean circulation is driven by the differences in pressure created by the variation of sea-surface elevation. Contrary to our intuition, the ocean surface is not flat but has hills and valleys. If the ocean is at rest, the sea-surface elevation is determined by Earth's gravity. This shape is called the geoid. When the ocean is in motion, its surface elevation departs from the geoid, leading to a pressure force that balances the motion. The deviation of the sea-surface elevation from the geoid is called ocean surface topography. Knowledge of the ocean surface topography allows us to calculate the speed and direction of ocean currents. The TOPEX/Poseidon satellite, operated jointly by the United States and France, has been making measurements of ocean surface topography since October 1992.

*Our Changing Planet*, ed. King, Parkinson, Partington and Williams
Published by Cambridge University Press 

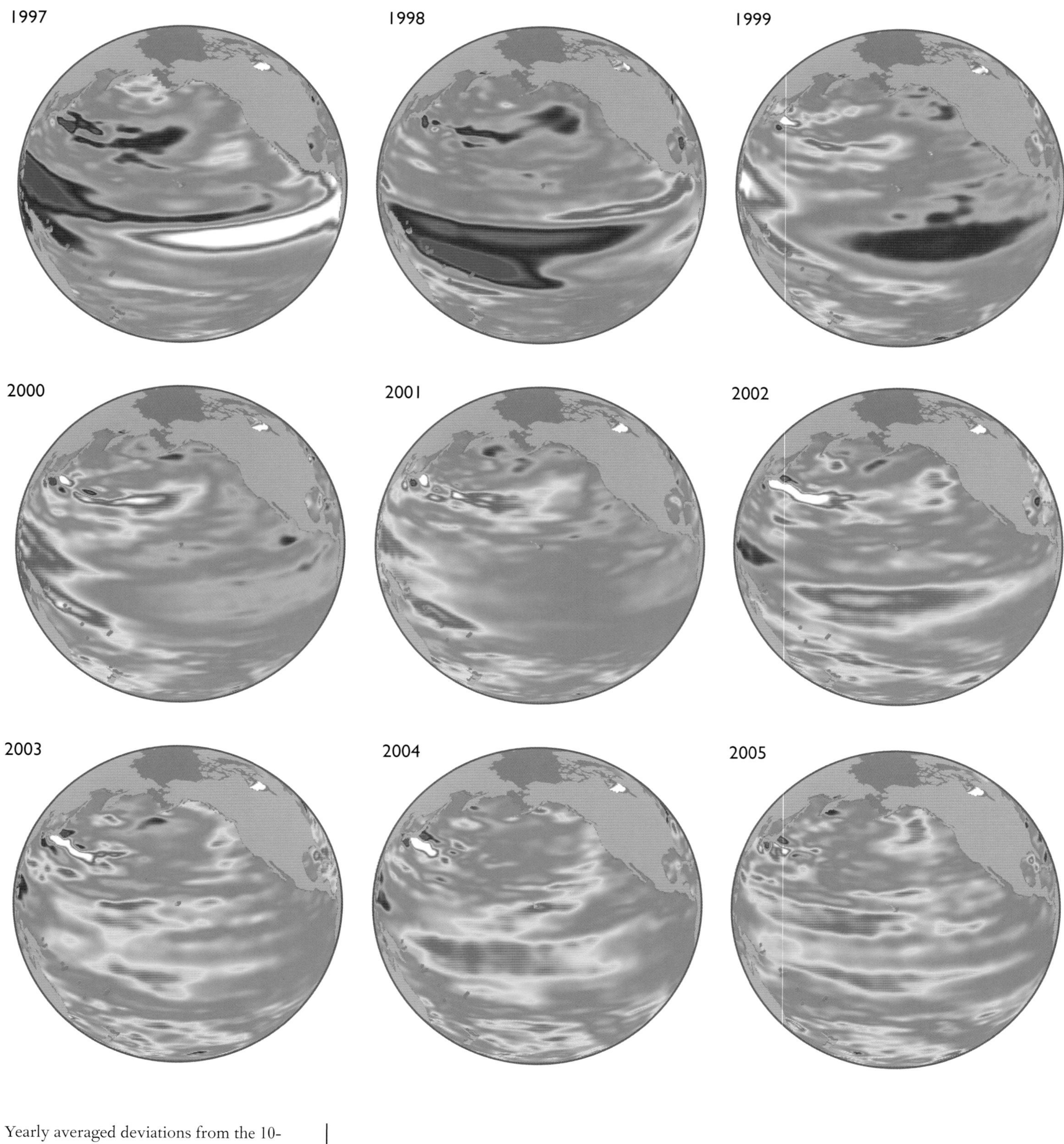


Yearly averaged deviations from the 10-year mean ocean surface topography of the Pacific Ocean, 1997–2005. The color scale indicates deviations from the mean topography computed from the TOPEX/Poseidon data record from late 1992 to 2002. (Data from the radar altimeter instrument on the TOPEX/Poseidon satellite. Image processing by L.-L. Fu.)

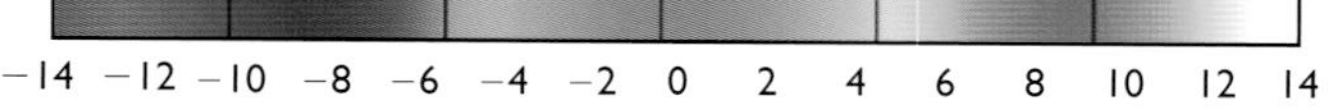

The data provide the first global view of ocean circulation and how it changes on timescales of weeks to years. This ability to map the ocean surface topography marks a milestone in the understanding of ocean circulation and its effects on climate.

The most energetic variability of ocean circulation is in its eddies, which are the 'storms' in the ocean currents. Typically the diameter of these eddies is about 100 km, and they pass by a location in a month with current speeds of 10 to 100 cm per sec. The intensity and tracks of these eddies have been mapped using satellite data. The 'storm' alleys of the ocean are located in the vicinity of the major ocean currents: the Gulf Stream, the Kuroshio to the east of Japan, the East Australian Current, the Brazil-Malvinas Confluence, and the Antarctic Circumpolar Current. Ocean eddies play an important role in ocean circulation, heat transport, and biogeochemical cycles. Our ability to monitor them from space has applications in navigation, in offshore operations, and in fishery, hurricane, and climate forecasting.

The ocean is also changing on timescales from years to decades. The heating and cooling of the ocean by the atmosphere causes the expansion and contraction of the ocean, leading to the seasonal rise and fall of the sea surface. Over timescales of 3–7 years, El Niño and La Niña create massive redistributions of heat in the ocean. After the strong El Niño in 1997, the Pacific Ocean went into a prolonged La Niña cycle until 2001, with lower topography (less heat storage) in the eastern Pacific and higher topography (more heat storage) in the western Pacific. Subsequently, the large-scale pattern of the surface topography of the Pacific Ocean seems to have transformed into a phase of the so-called 'Pacific Decadal Oscillation,' a long-lived El Niño/La Niña-like cycle of Pacific climate variability with a timescale of 20 years. Such long-term changes in the ocean are linked to the variability of the atmosphere and have large-scale, long-lasting impacts on the weather and climate of the world. However, a 10-year record is too short to tell the whole story about global change in the ocean. Jason, the follow-on satellite to TOPEX/Poseidon, was flying side-by-side with TOPEX/Poseidon from 2002 to 2005, thus providing twice the spatial coverage. Together with future satellite altimeters, Jason, which has succeeded TOPEX/Poseidon after its termination in 2005, will continue the measurement of ocean surface topography enabling us to better understand and predict ocean circulation and its effects on climate.

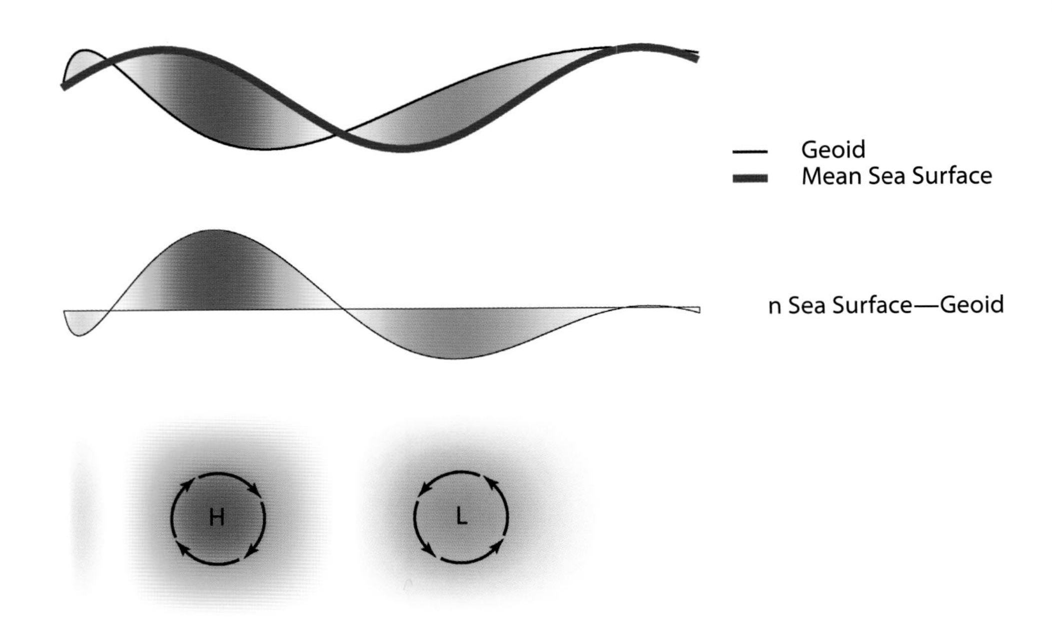


Schematic illustration of ocean surface topography as the difference between the mean sea surface and the geoid. In the Northern Hemisphere, geostrophic currents flow clockwise around the highs of the ocean surface topography and counter-clockwise around the lows. The flow directions, caused by the Earth rotation, are reversed in the Southern Hemisphere.

Peter J. Minnett

# Heat in the Ocean

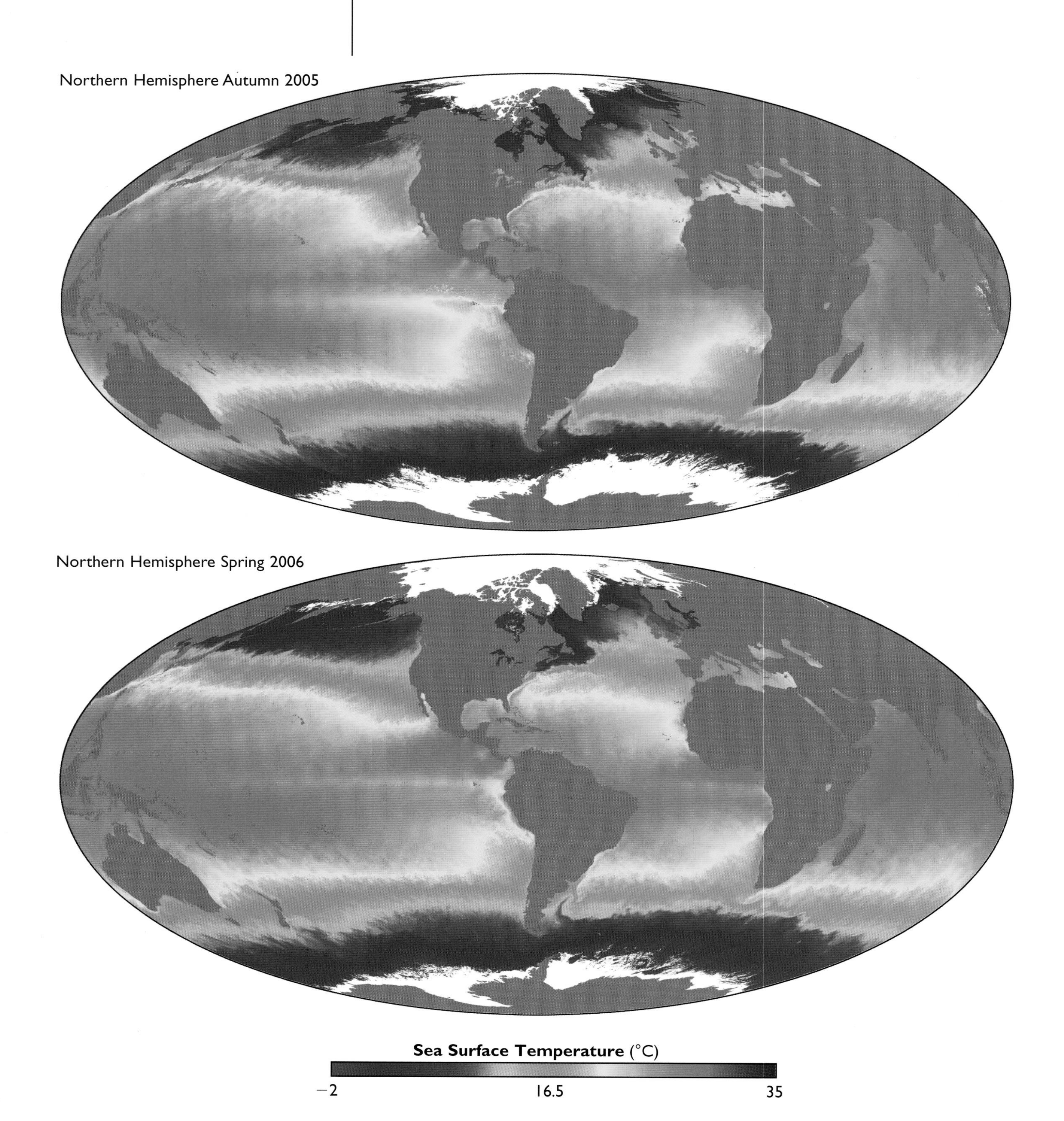


*Our Changing Planet*, ed. King, Parkinson, Partington and Williams
Published by Cambridge University Press 

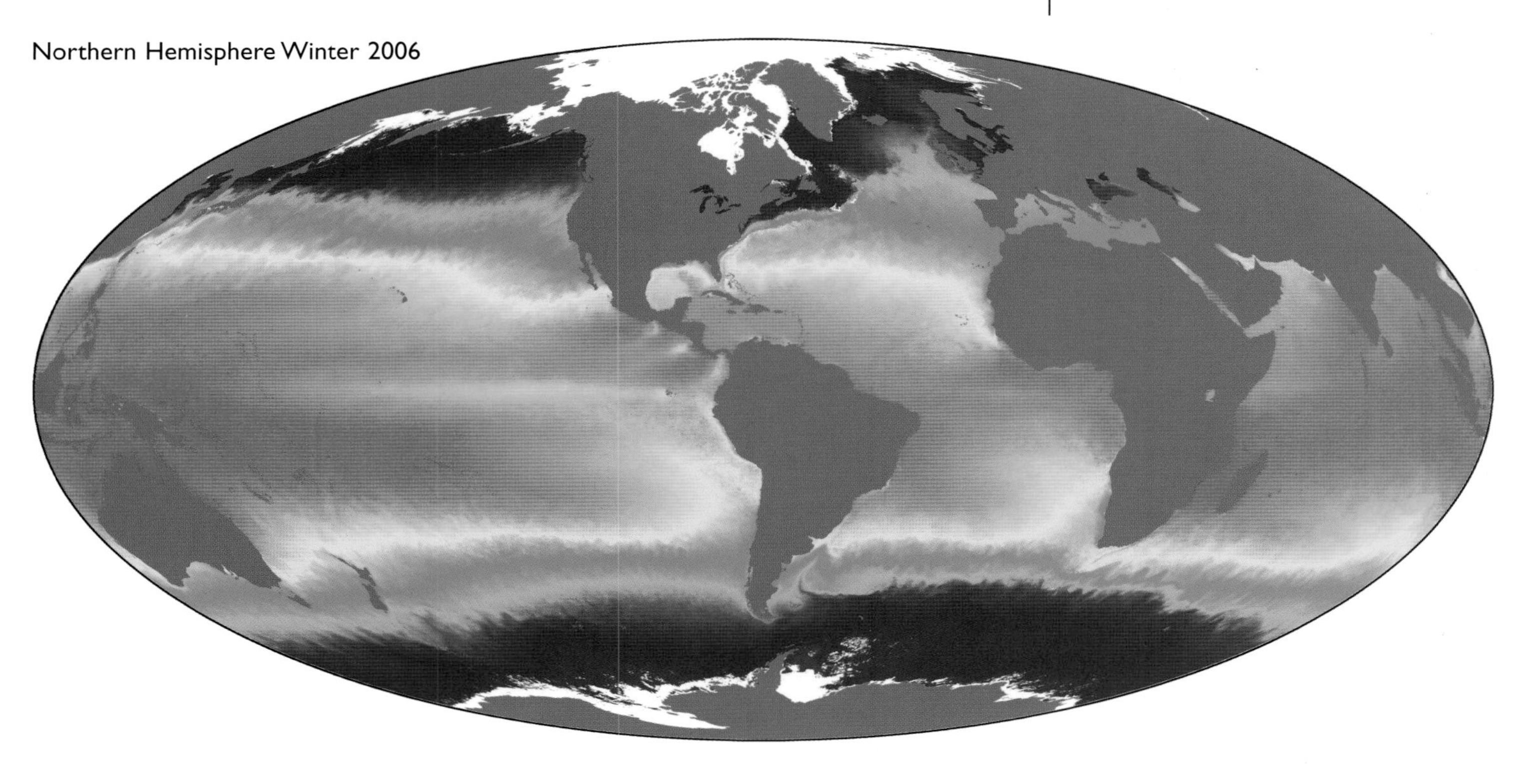


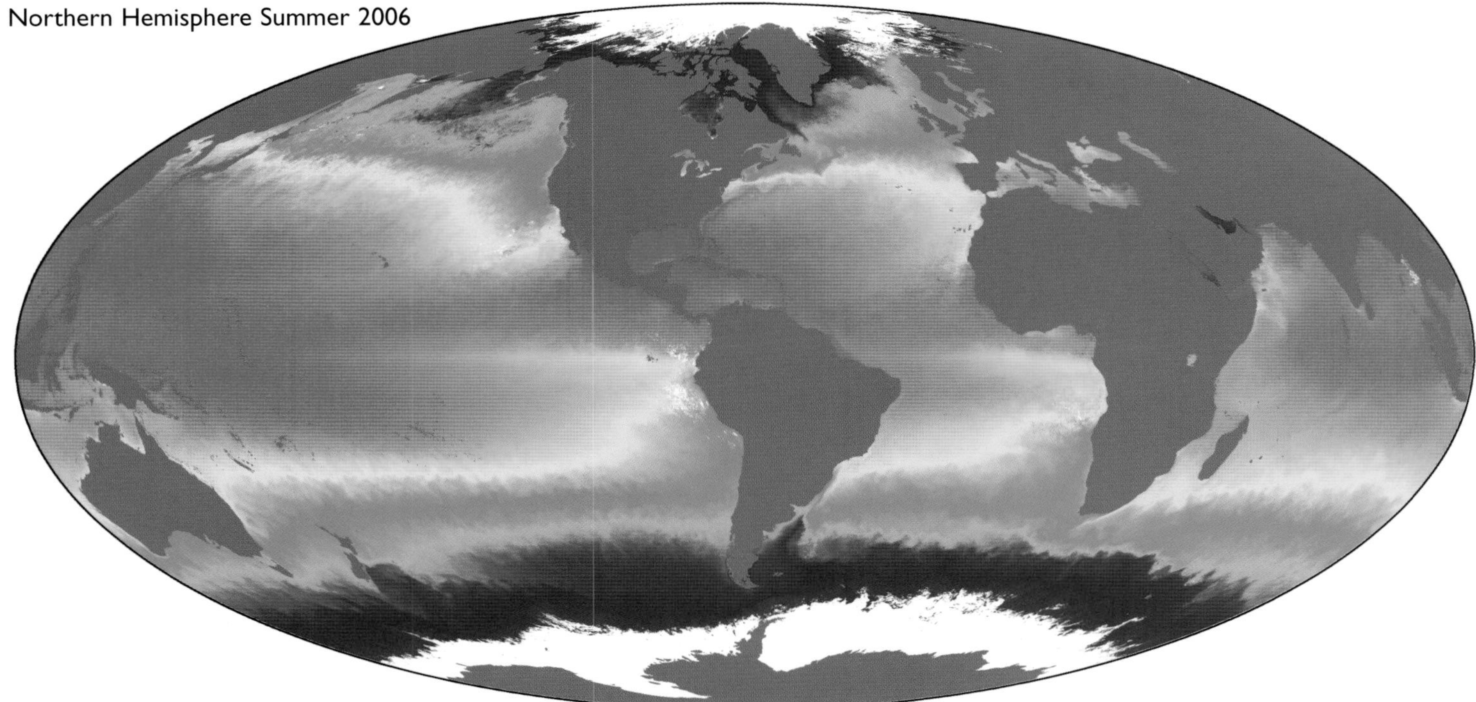


Seasonal global pattern of sea surface temperatures during 2005 and 2006. (Data from the MODIS instrument on the Terra satellite.)

A glance at an atlas reveals pronounced differences in living conditions on opposite sides of the North Atlantic Ocean. Britain is at the same latitude as Labrador; Norway is opposite Greenland. The countries on the eastern side of the Atlantic offer a comfortable

August 27, 2005

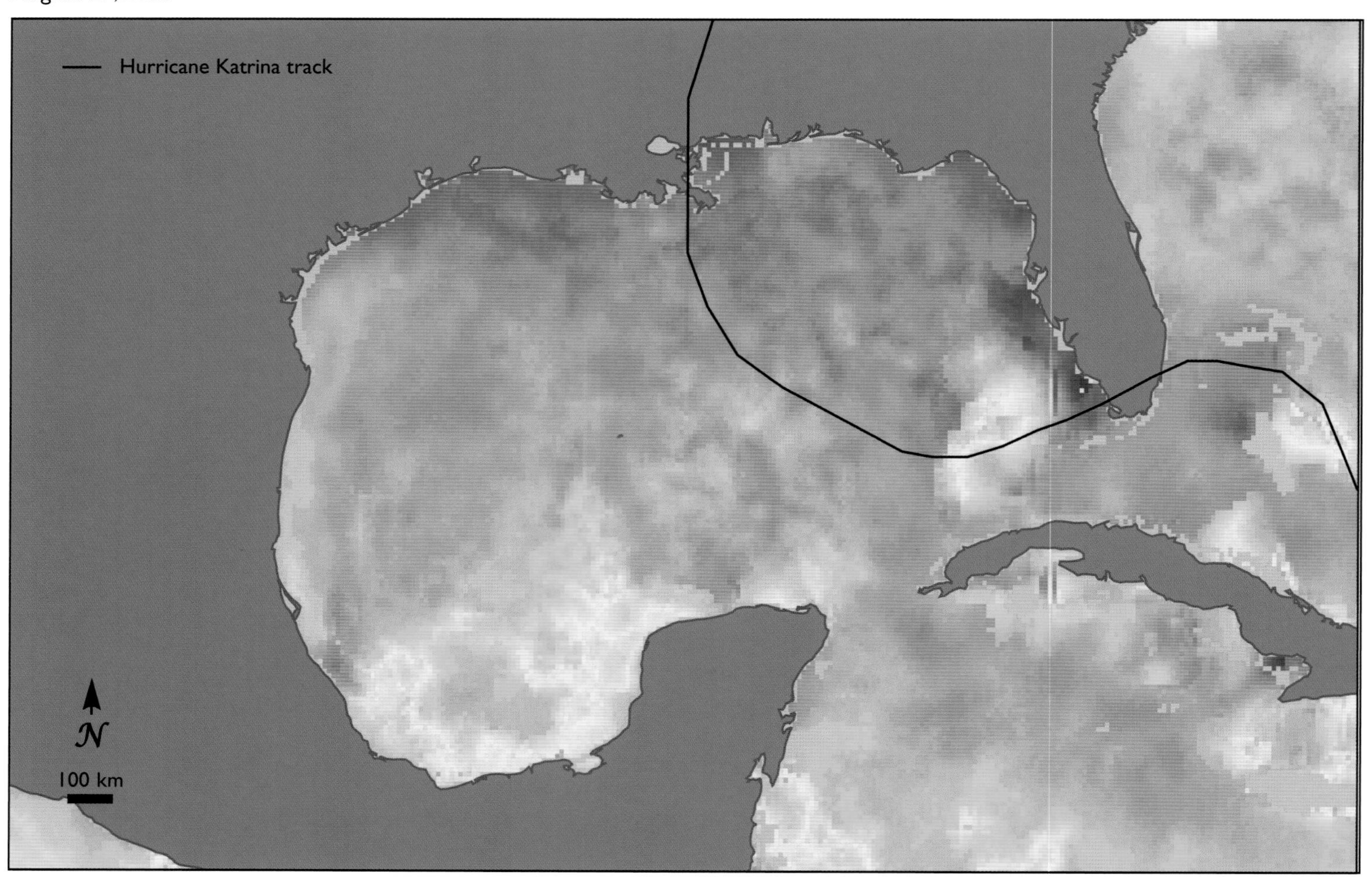


Sea surface temperature in the Gulf of Mexico before (left) and after (right) the passage of Hurricane Katrina, August 27 and 29–30, 2005. The storm underwent intensification after passing over the southern tip of Florida before landfall close to New Orleans, resulting in extensive destruction by flooding as well as winds. The cold water wake along the hurricane track on August 30 (right) marks where Hurricane Katrina extracted sufficient heat from the ocean during its intensification to lower the sea surface temperature by several degrees. The microwave temperature measurements, which are unreliable close to land, have been extrapolated to the coastline. The clouds that have been added to the August 29–30 image are from GOES data from August 29, 2005 1445 UTC. (Data from the AMSR-E instrument on the Aqua satellite, provided by Chelle Gentemann.)

climate whereas those on the western side are frigid and much less hospitable. The reason for this is the temperature of the seas and the currents that move the warm and cold ocean water around the globe. The temperatures of the sea surface have been measured from satellites for about 25 years, and these measurements have provided new insight into the surface currents and have led to improved weather forecasting. The heat in the oceans in the warm, sunny regions of the equator and tropics is transported to higher latitudes by the great ocean currents. In the Atlantic Ocean, the best known feature is the Gulf Stream, which brings warm water from the Gulf of Mexico to Europe. The warm surface waters of the Gulf Stream heat the atmosphere above, warming and moistening the air that the winds bring over the land. By giving up heat to the atmosphere, the ocean currents play a large part in determining the major features of the world's climate. The Gulf Stream is part of the large pattern of ocean currents that links all of the oceans, and the northward flow of water in the Atlantic is compensated on the western side by a southward flow at depth of cold water from the Arctic. These patterns of surface temperature are readily seen in the satellite imagery, which have shown the surface currents to be quite variable, changing in strength and position in periods of weeks to months. Some of these changes, such as the meanders of the Gulf Stream, do not have a very significant impact on the weather, but others, such as the increase in surface temperature across the tropical Pacific Ocean caused by the El Niño phenomenon, have serious consequences.

August 29 and 30, 2005

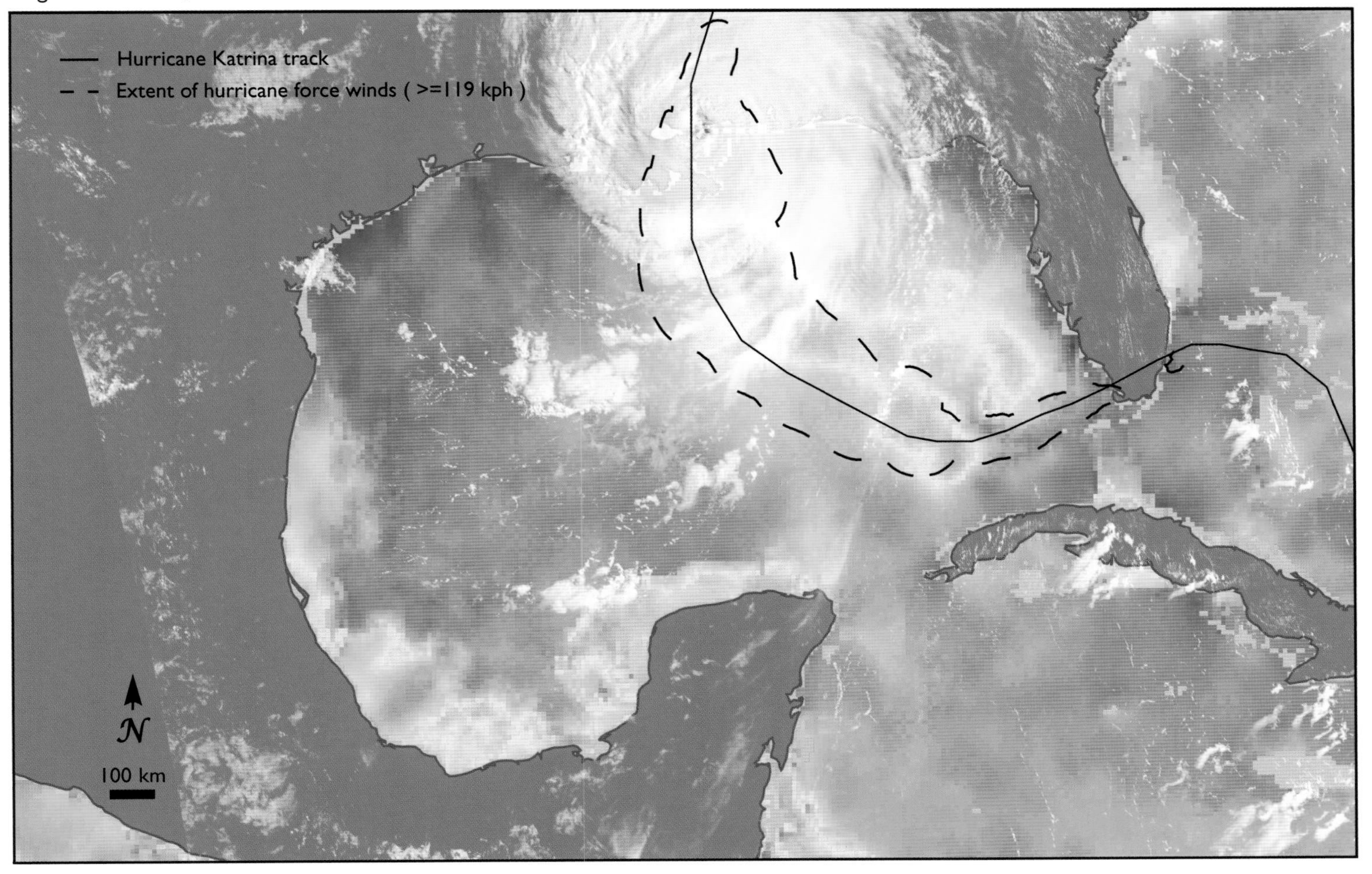


Sea Surface Temperature (°C)

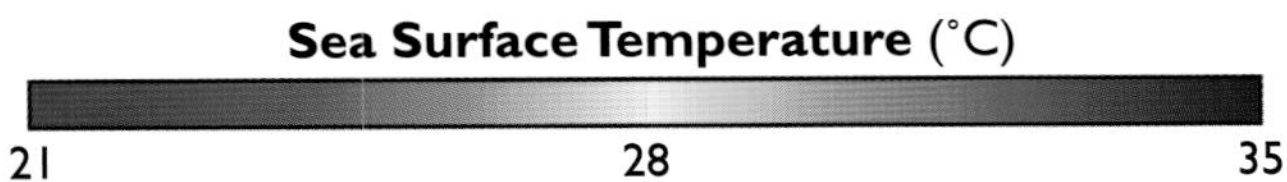


Destructive hurricanes, typhoons, and cyclones grow over the warm seas, feeding on the heat available in the ocean, and satellite measurements of the sea surface temperatures are used to improve the forecasting of intensities and paths of these severe storms. Severe storms can leave a signature in surface temperature in the form of a cold wake, indicating where they have drawn heat out of the ocean to feed their growth. The appearance of a cold hurricane wake is seen in the images of the Gulf of Mexico before and after the passage of Hurricane Katrina in August 2005. On occasion, during a busy hurricane season, the passage of a subsequent storm over such a cold wake can lead to a sudden decay of intensity as the source of energy needed to sustain or intensify the storm is diminished. A clear example of this occurred in August 1998 when Hurricane Danielle rapidly lost strength as she passed over the cold wake of Hurricane Bonnie.

The El Niño phenomenon is characterized by a redistribution of the surface temperature in the tropical Pacific Ocean. The normal situation is for the western Pacific to be warmer at the surface than the eastern, in part driven by the large-scale patterns of winds over the globe, including the trade winds. The large area of warm water in the western Pacific, called the Warm Pool, heats the air above. At unpredictable intervals of several years, the Warm Pool extends to the east, warming the eastern Pacific Ocean and disturbing the atmospheric circulation. This has consequences around most of the globe by changing the positions and intensities of storms and rainfall patterns, and causing drought and floods that disrupt

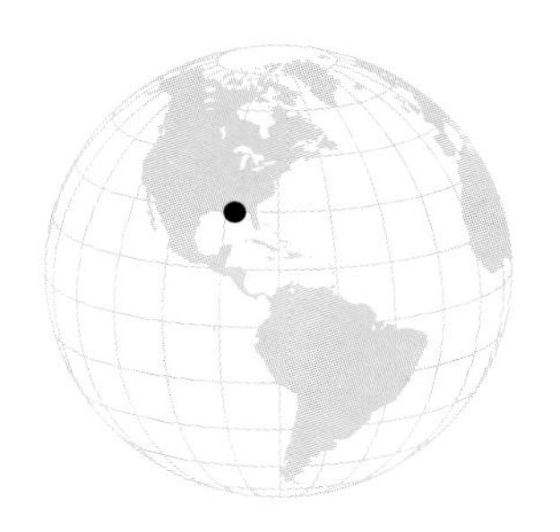

December 1993

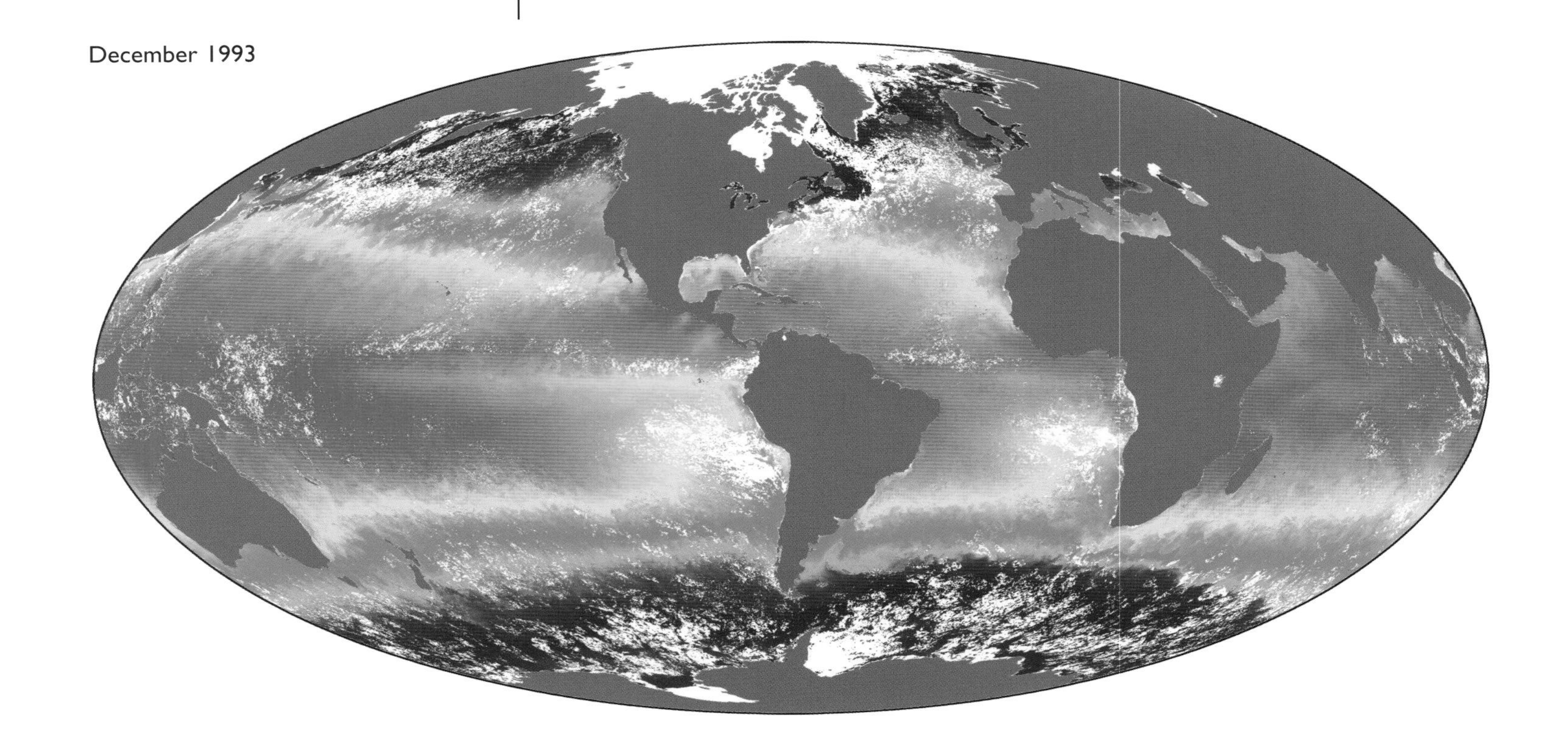

December 1997

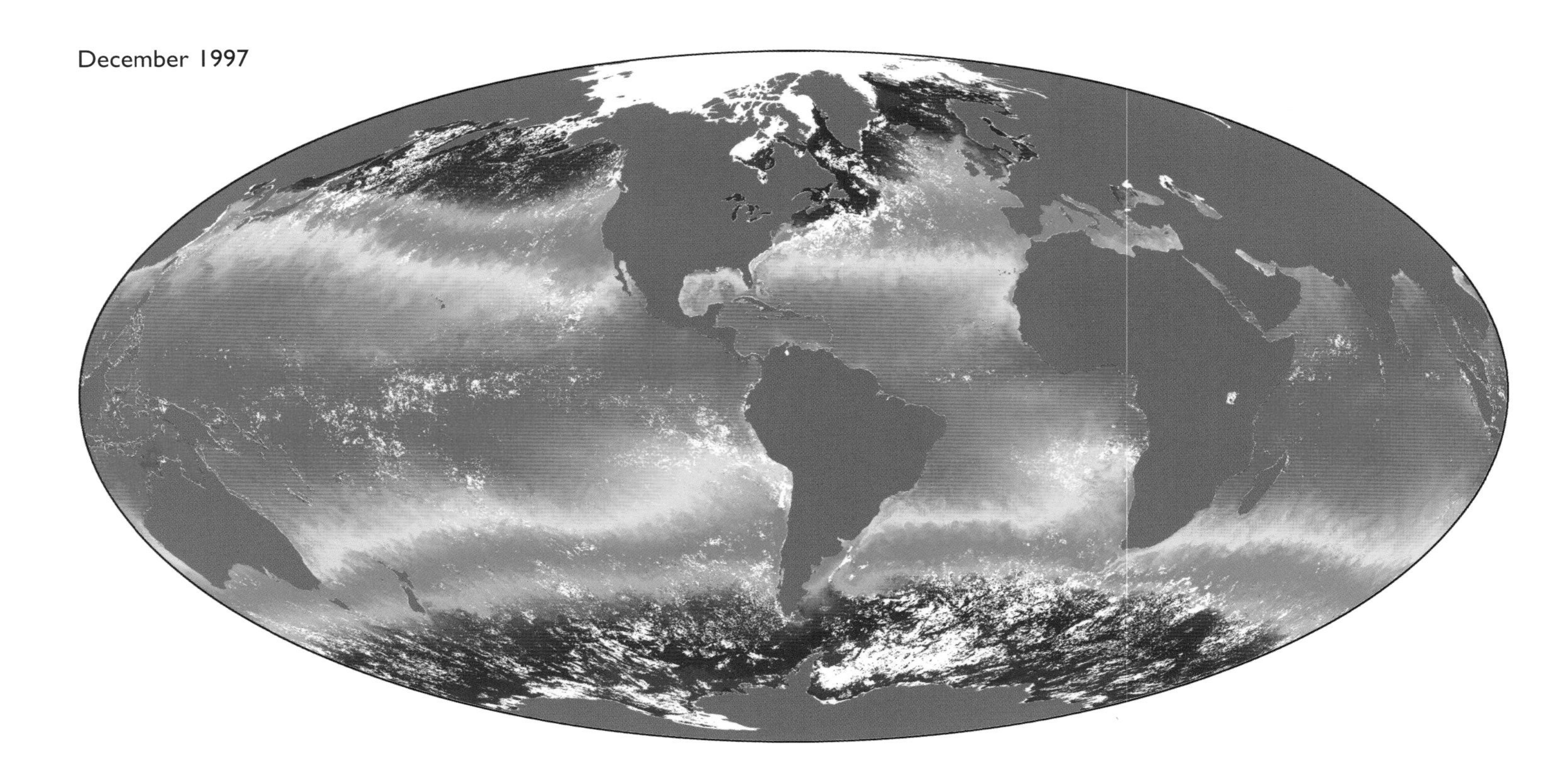

**Sea Surface Temperature** (°C)

−2 16.5 35

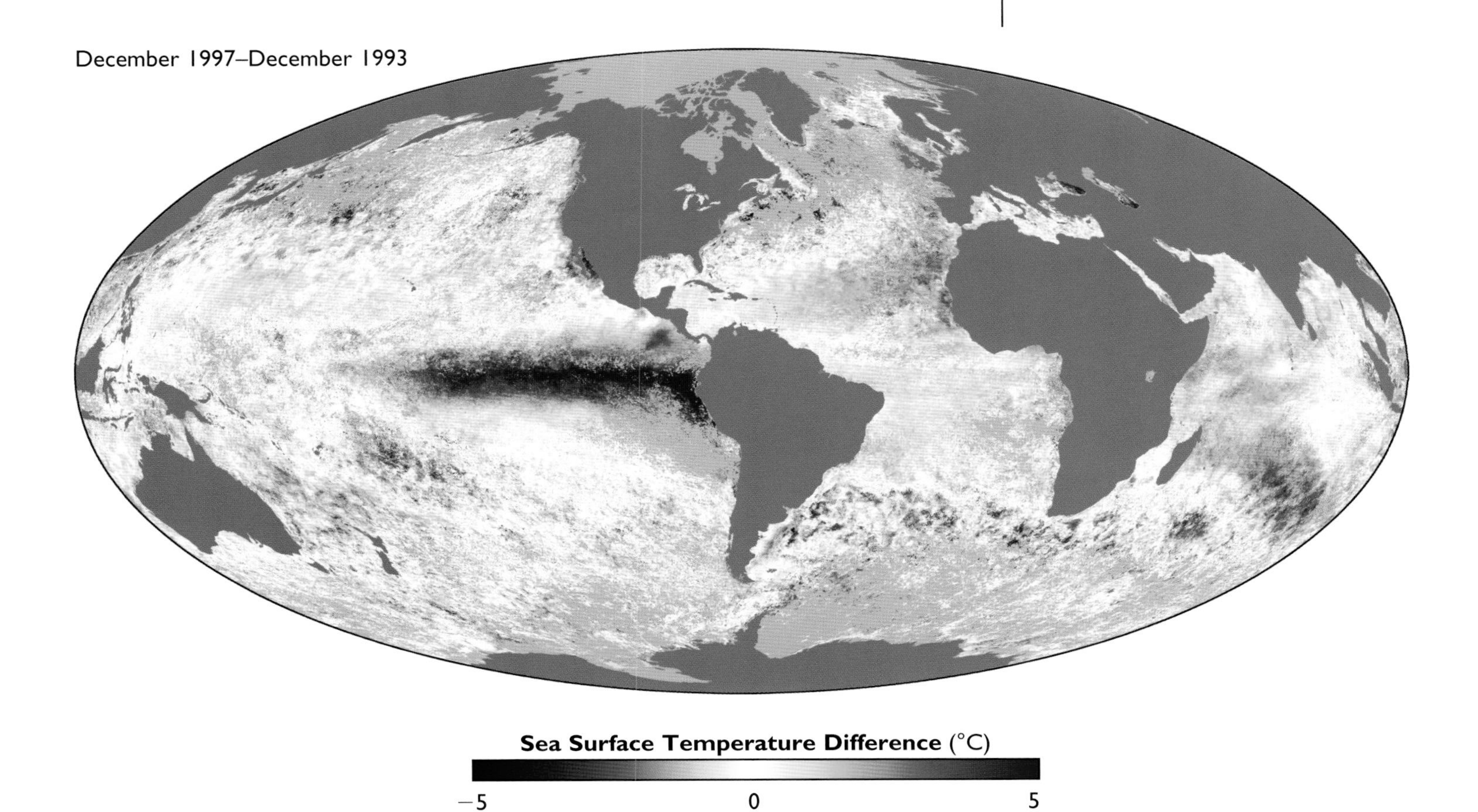


agriculture and lead to loss of life and property. Better measurements of the changes in the sea surface temperatures in critical regions of the Pacific are leading to more accurate seasonal forecasts of El Niño events.

The average surface temperature of the oceans can be thought of as a global thermometer that might indicate the initial effects of climate change. However, because of the complex structure of the ocean currents and the way they influence the surface temperatures, and the large-scale perturbations of the El Niño and other features, the anticipated heating trend expected to result from global warming has not yet been clearly identified. Continued measurements of sea surface temperature from satellites over the coming years and decades will be a key element in detecting and studying changes in the Earth's climate.

Contrasting images of the global ocean-surface temperatures for December 1993, showing the normal situation, and December 1997, during an El Niño event (opposite), and their difference (above). The most significant change between the two cases is that the warm temperatures, usually centered in the western Pacific, extend across the whole Tropical Pacific during the El Niño, and the usually cold waters off Peru, important to the local ecosystem, are missing. (Data from the AVHRR instrument on the NOAA-11 and NOAA-14 satellites.)

# Sea Level Rise

Gary T. Mitchum
R. Steven Nerem

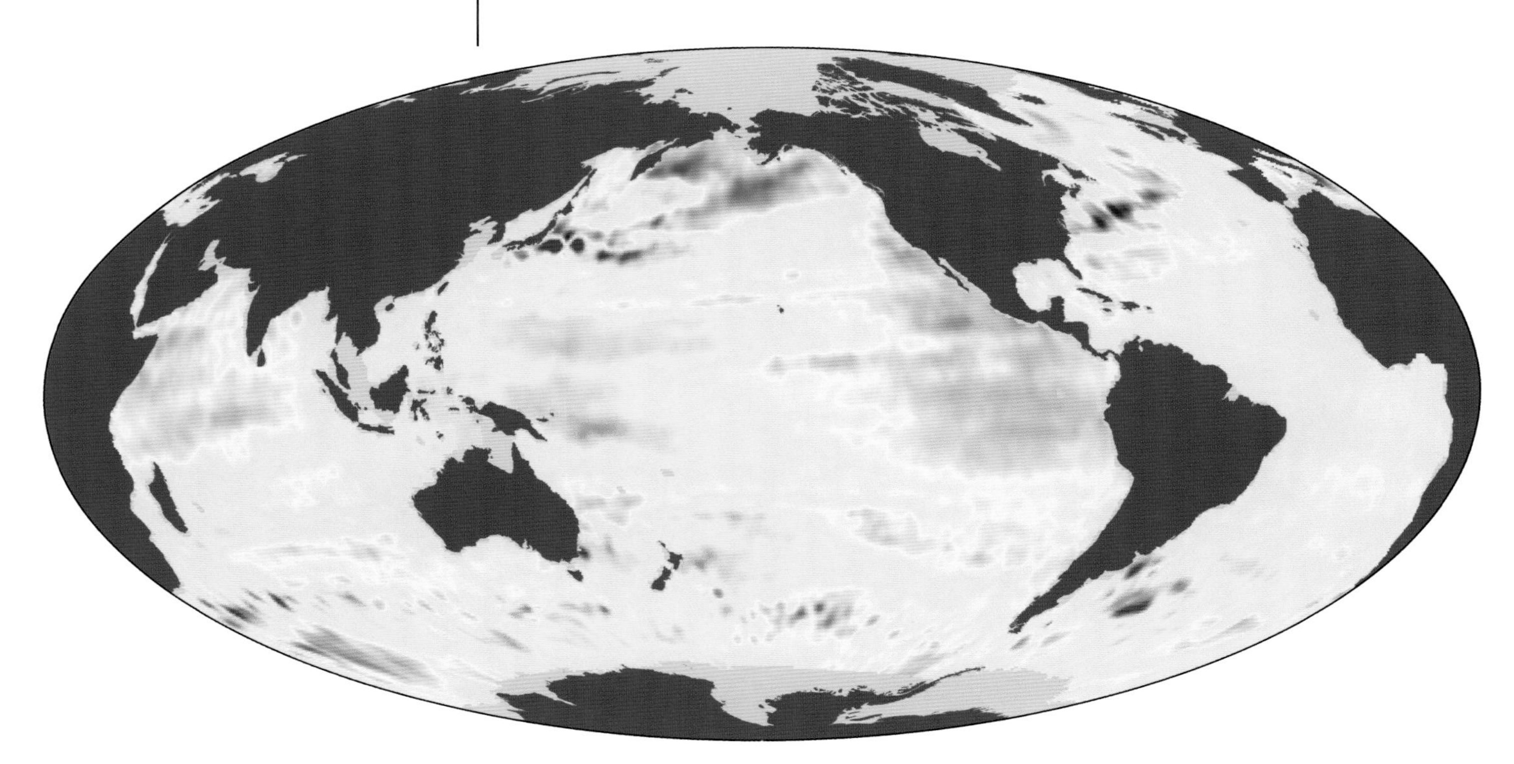

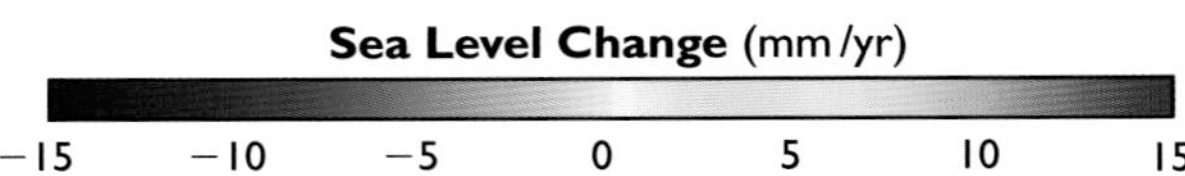


Sea level change during the period 1993–2006. (Data from the TOPEX/Poseidon and Jason-1 altimeter missions.)

Human activities are changing the face of our planet in many ways. We know, for example, that the burning of fossil fuels is increasing carbon dioxide levels in the atmosphere, and we have good evidence that the atmosphere is slowly warming as a consequence. But the long-term ramifications are not entirely clear, and are presently the subject of much debate. If our climate is indeed warming, sea level rise is widely considered to be an unavoidable

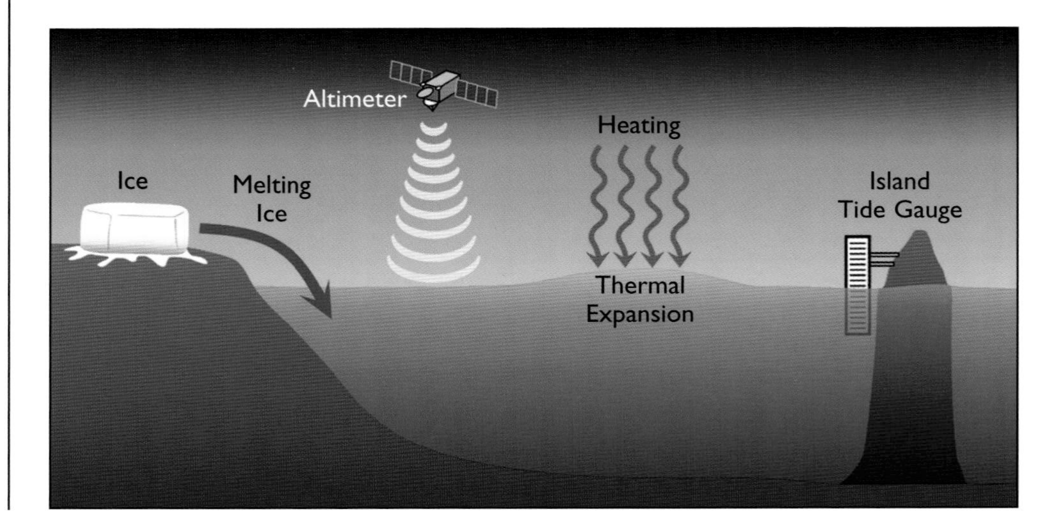


Sea level changes reflect changes in the global ocean volume—higher sea level means that the ocean volume has increased. There are 2 primary ways in which the ocean's volume can be changed. Melting ice that sits on land can increase the amount of water in the ocean. Also, if the waters in the ocean become warmer, the volume will increase because warm water is slightly less dense than cool water and therefore takes up more space. There are now 2 primary ways of observing these sea level, or volume, changes: through tide gauges attached to the land, or through satellite altimeters orbiting the Earth. (Graphic prepared by C. Edmisten, University of South Florida.)

*Our Changing Planet*, ed. King, Parkinson, Partington and Williams
Published by Cambridge University Press 

consequence. This is obviously of direct importance to human beings, as it has been estimated that over one-third of us live within 100 km of the coastline.

Flooding of coastal areas is an important concern, of course, but there is another reason why sea level rise and its observation are important. At present the predictions we have of the consequences of global warming are from highly complex computer models. Unfortunately, because we have little data as of yet, our confidence in these models is less than we would like. Sea level change is actually an indication of ocean volume change, which can happen only if water is added or withdrawn or if heat is added or taken away. According to the predictions of the computer models, sea level rise should occur because of added water from melting ice and from thermal expansion due to increased heat input from a warmer atmosphere or more retention of solar radiation. Accurate observations of volume change, that is, of sea level rise, can therefore serve as a sensitive indicator of the reliability of our prediction models.

In the past we have measured sea level rise with tide gauges located along the continental shorelines and on a relatively few islands in the open ocean. These measurements are very accurate, but a long period of observation is required since so much of the open ocean is not observed. Over the past decade, though, we have seen the development of precise sea surface height measurements of the global ocean from satellite altimeters, most notably from the TOPEX/Poseidon mission that is operated by the United States and France. And remarkably, when one considers that these measurements are made from an altitude of over 1300 km, these data are nearly as accurate as the data from a tide gauge.

Changes in the globally averaged sea level since the beginning of the TOPEX/Poseidon mission in 1992, with seasonal variations removed. The average indicated sea level rise rate is approximately 3.5 mm per yr. Note that El Niño events can temporarily change the global ocean volume, as seen in the large fluctuation in the time series during 1997–1998. With continued measurements from altimetry satellites, we will be able to say whether the sea level rise rate is accelerating and whether computer models of climate changes associated with global warming are accurately predicting the rise rate. (Data from the NASA Radar Altimeter on TOPEX/Poseidon and the Poseidon-2 altimeter on Jason-1.)

Sea level measurement gauge at Male in the Maldives in the Indian Ocean. Sea level rise is of particular concern for inhabitants of low-lying island nations. Sea level measurements are also of special value in estimating sea level rise rates. Historically, tide gauges have been our only source of information about sea level rise, but in recent years satellite altimeters have given us an exciting new tool to apply to this problem. (Photograph courtesy University of Hawaii Sea Level Center.)

With these new data from satellite altimeters, we are essentially measuring the height of the ocean with what amounts to thousands and thousands of tide gauges distributed uniformly over the entire ocean. This uniform coverage allows us to measure sea level rise much more accurately than in the past when only the coastal tide gauges were available. This global coverage might also allow us to examine the spatial patterns of sea level changes, which potentially could make it possible to attribute the observed changes to either heating changes or to water input from melting ice. Our hope is that with continued altimetry we can make definitive tests of the computer models making global change predictions, helping thereby to identify the best models, and in that way to perhaps better assess the future changes that human activities might have on the Earth's climate.

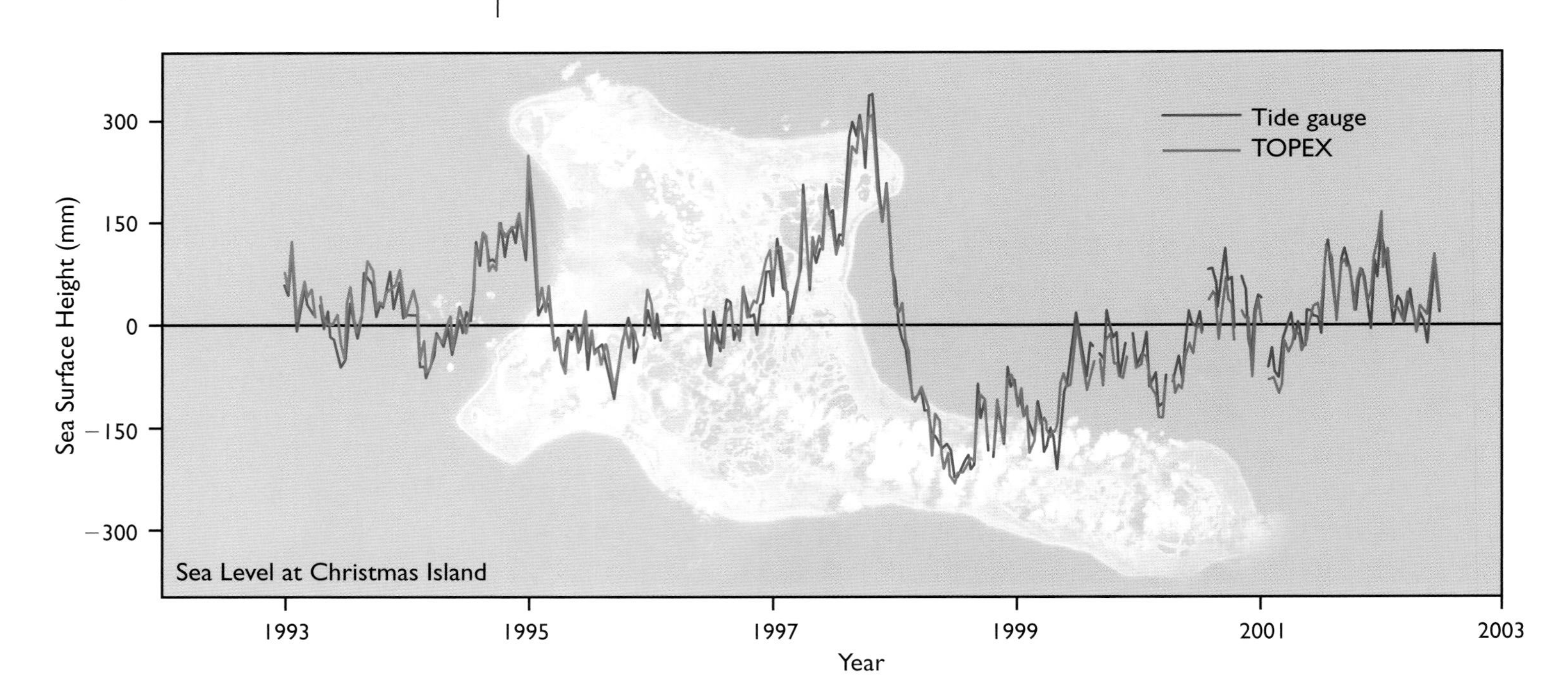


Satellite altimetry is a relatively new tool for measuring global sea surface heights over the ocean. It is important to check that satellite altimetry provides an accurate measurement, and comparisons with the data from tide gauges are used to study this. Shown here are time series of sea level height from Christmas Island in the tropical Pacific Ocean. The large drop in late 1997 is associated with the large El Niño event that occurred that year. The close correspondence of the tide gauge and TOPEX/Poseidon measurements of sea level demonstrates that the altimeter is highly precise and allows us to use the satellite data to estimate sea level rise rates much better than in the past when only the tide gauge data were available. (Data from TOPEX/Poseidon altimeter.)

# Tides

RICHARD D. RAY

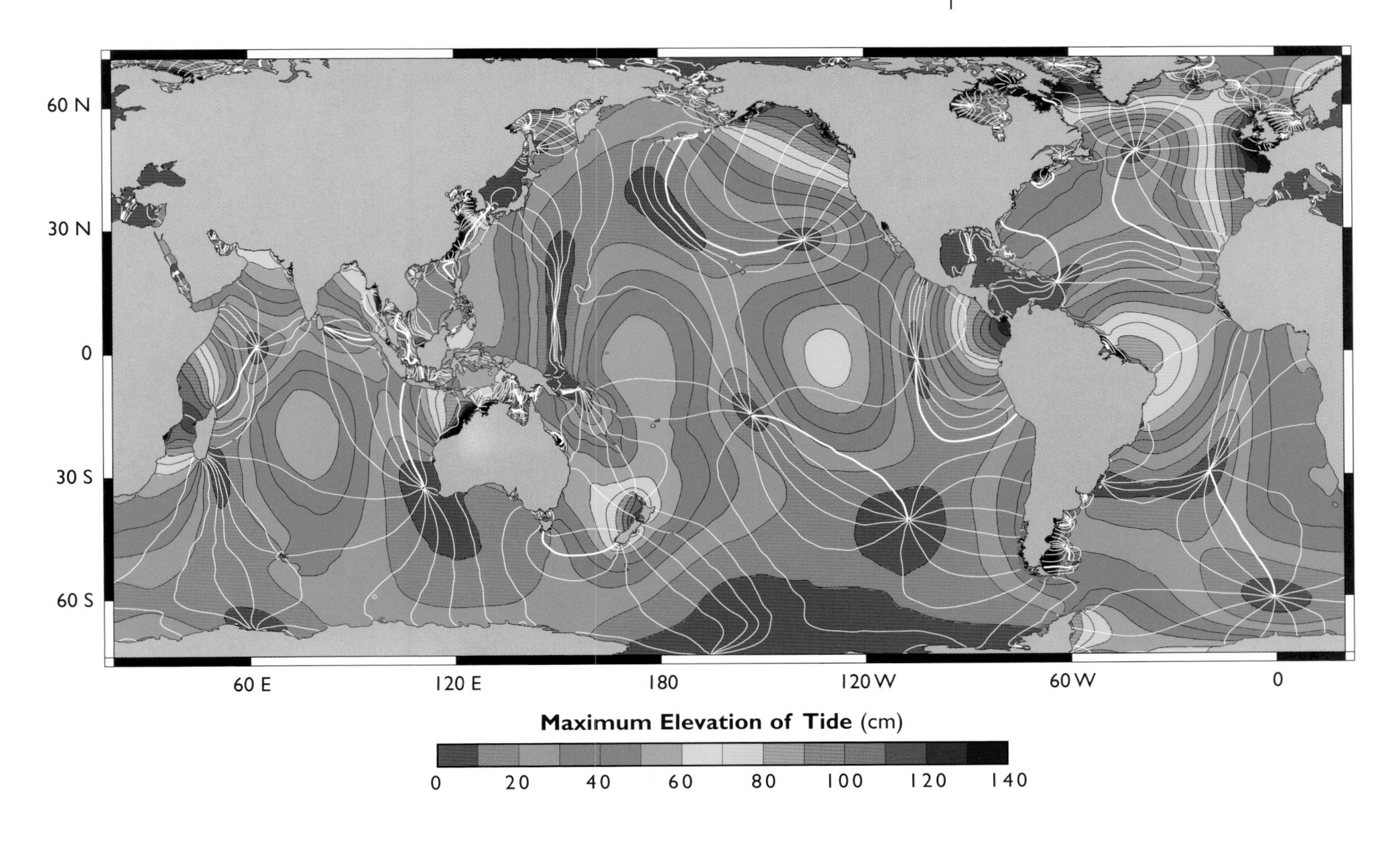


The tide is composed of multiple waves, the largest of which is denoted M2, shown here as determined by TOPEX/Poseidon measurements. This M2 wave cycles once in half a lunar day, or about every 12.4 hours. Colors show maximum elevation at M2 high tide, in cm. White lines connect places that experience high tide simultaneously. To predict the full tide at any place, a dozen or more such waves must be added together.

Pick up a newspaper in almost any coastal city and you find predictions for the day's tides. It seems as easy as clockwork. What is not evident, however, is that these tidal predictions are always based on long series of hourly sea level measurements, taken in close proximity to the city, and usually over many years. These data must then be analyzed by methods at least as sophisticated as the motions of the Sun and Moon, and remember that Sir Isaac Newton claimed that determining the Moon's motion precisely was the one problem that made his head ache. The theory of the Moon's motion and tides is now understood, but accurate tidal predictions still require building upon historical observations.

What if one is living on a small isolated island with no historical tide-gauge data available? Or, what if one needs to know the tide in the middle of the ocean? Scientists have successfully addressed this problem by using altimeter measurements collected by TOPEX/Poseidon, a satellite launched in 1992 by the United States and France. The satellite altimeter acts as a flying tide gauge, repeatedly measuring sea level over the global ocean, and these data have subsequently been analyzed to allow tide predictions in the open sea and, somewhat less accurately, in coastal waters. The predictions are based on adding together many waves, or constituents, each of which is described by a 'cotidal chart', such

*Our Changing Planet*, ed. King, Parkinson, Partington and Williams

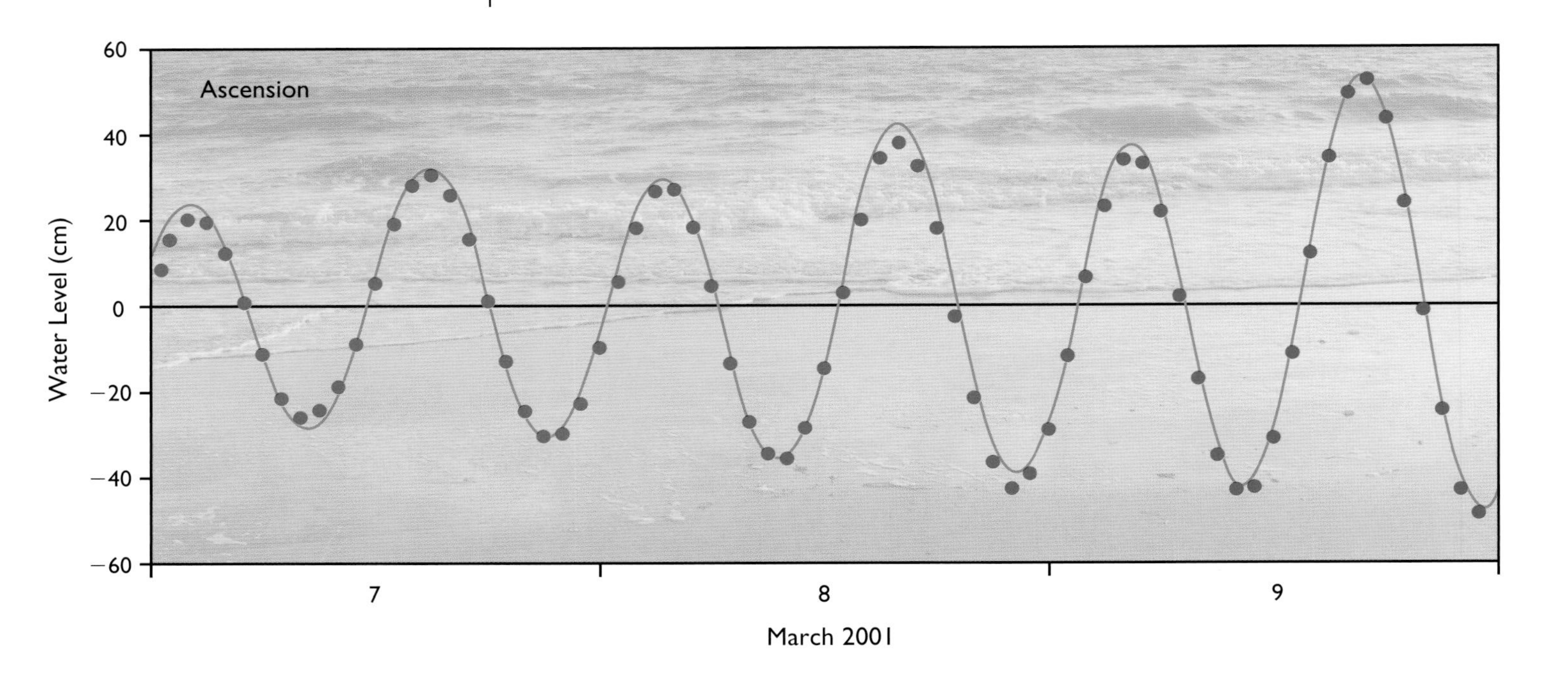


Ground-truth validation of satellite-based tide predictions. A tidal prediction (blue curve), based on analysis of TOPEX/Poseidon satellite altimeter data collected in the vicinity of Ascension Island, in the middle Atlantic Ocean, is found to compare well with actual hourly water level measurements made at the island (red circles.) (Hourly data courtesy Proudman Oceanographic Laboratory, UK.)

as the one shown here for the Moon's principal half-daily tidal wave. At many locations the gravitational tide can now be predicted with an accuracy of a few centimeters or better (although weather effects like storm surges can sometimes spoil the prediction of sea level).

Accurate knowledge of tides is required in countless applications, both military and civilian, coastal and pelagic. Oceanographers are especially keen to predict open-ocean tides with very high accuracies, because the tide can often mask the ocean's smaller, and far more subtle, climate signals. In fact, to use satellite altimetry for any non-tidal science requires first removing the ocean's dominating tide signals from the data.

Organizations like NASA have a special interest in tides, because accurate tidal predictions are critical to all facets of modern space geodesy. Geodesy determines positions, such as a person's position on the Earth's surface or a satellite's position above the surface, and these kinds of measurements, if sufficiently precise, are always affected by tides. For example, a satellite's orbit is perturbed by the gravitational attraction of the ocean's tidal masses; special geodetic satellites like GRACE and LAGEOS are extremely sensitive to such perturbations.

World War II tide disaster. "The tide that failed" during the US Marine amphibious landing at Tarawa, November 20, 1943. Unexpectedly low water stranded United States landing craft on coral reefs more than 500 m from shore, leaving Marines to wade in against withering Japanese defensive fire. The blue curve is a hindcast based on modern tidal measurements, unavailable to military planners during the war.

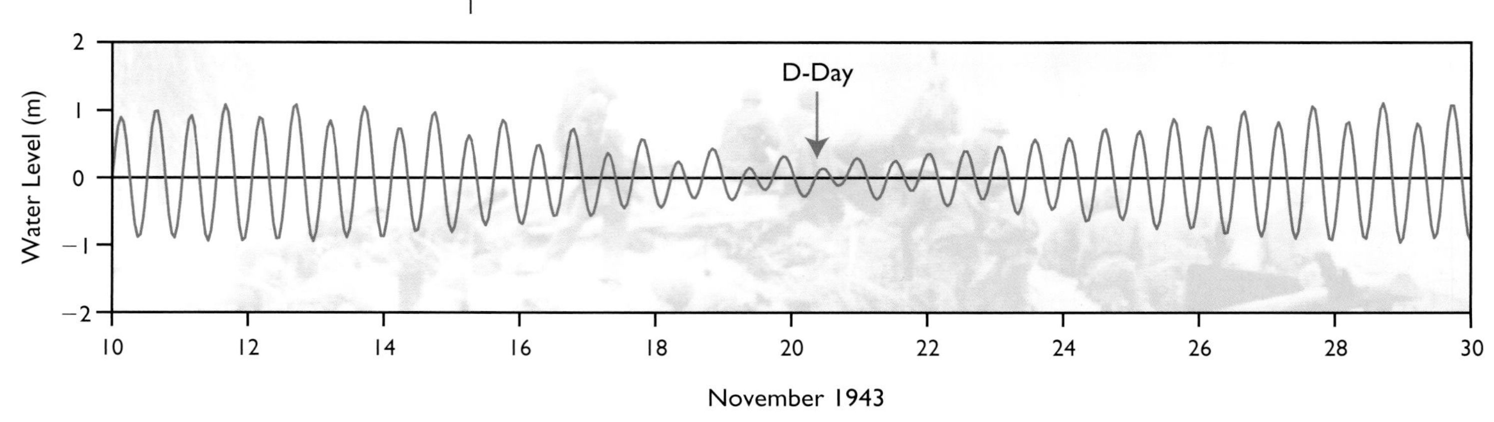

April 20, 2001

September 30, 2002

High and low tides on the Bay of Fundy, on April 20, 2001 and September 30, 2002, respectively. Situated between the Canadian provinces of New Brunswick and Nova Scotia, the Bay of Fundy is famous for having dramatic differences between its high and low tides. Under typical conditions, high tide at the head (the most inland part) of the Bay of Fundy is as much as 17 m (about 56 ft) higher than low tide. (Data from the ASTER instrument on the Terra satellite.)

# Winds Over Ocean

W. Timothy Liu
Xiaosu Xie

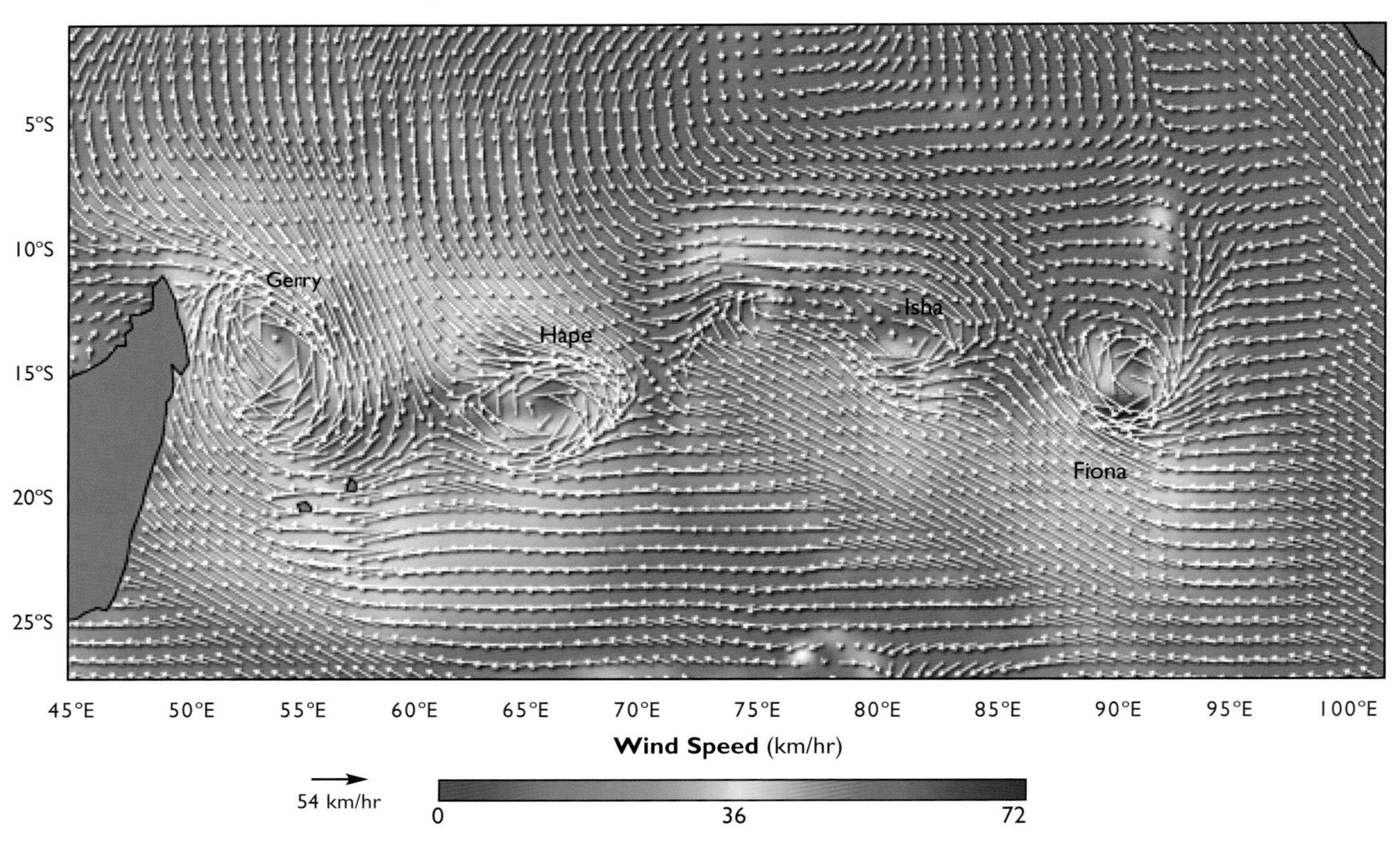


Wind direction and speed in the Indian Ocean, February 11, 2003. Wind direction (white arrows) is superimposed on a color map of wind speed. The figure demonstrates the details of 4 tropical cyclones revealed by 2 scatterometers flying in tandem. (Data from the SeaWinds instruments on the QuikScat and ADEOS II satellites.)

Sailors understand both the importance and the difficulty in getting information on wind over oceans. Textbooks still describe global ocean wind distribution in sailors' terms: doldrums, horse-latitude, trade-wind, and roaring forties. Just a decade ago, almost all ocean wind measurements came from merchant ships and there were no wind measurements in most parts of the ocean, particularly under stormy conditions. Even today, the ability to predict weather accurately by computer models is limited not only by our knowledge of the physical processes, but even more so, by the availability of measurements, such as the winds over ocean.

Spaceborne microwave scatterometers are the proven instruments for measuring ocean surface wind vector (both speed and direction) under clear and cloudy conditions, day and night. They give not only a near-synoptic global view, but also details not yet possible using weather prediction models. The coverage and details are illustrated by the 4 tropical cyclones that were observed at the same time by 2 different scatterometers flying in tandem, as depicted above. Data from the scatterometer on QuikScat have been routinely assimilated in the major operational numerical weather prediction centers of the world, and incorporated in operational analyses by marine weather warning and forecast centers. Space-based measurements of wind over ocean have made a significant impact in weather forecasting and are also used in much weather related research.

*Our Changing Planet*, ed. King, Parkinson, Partington and Williams

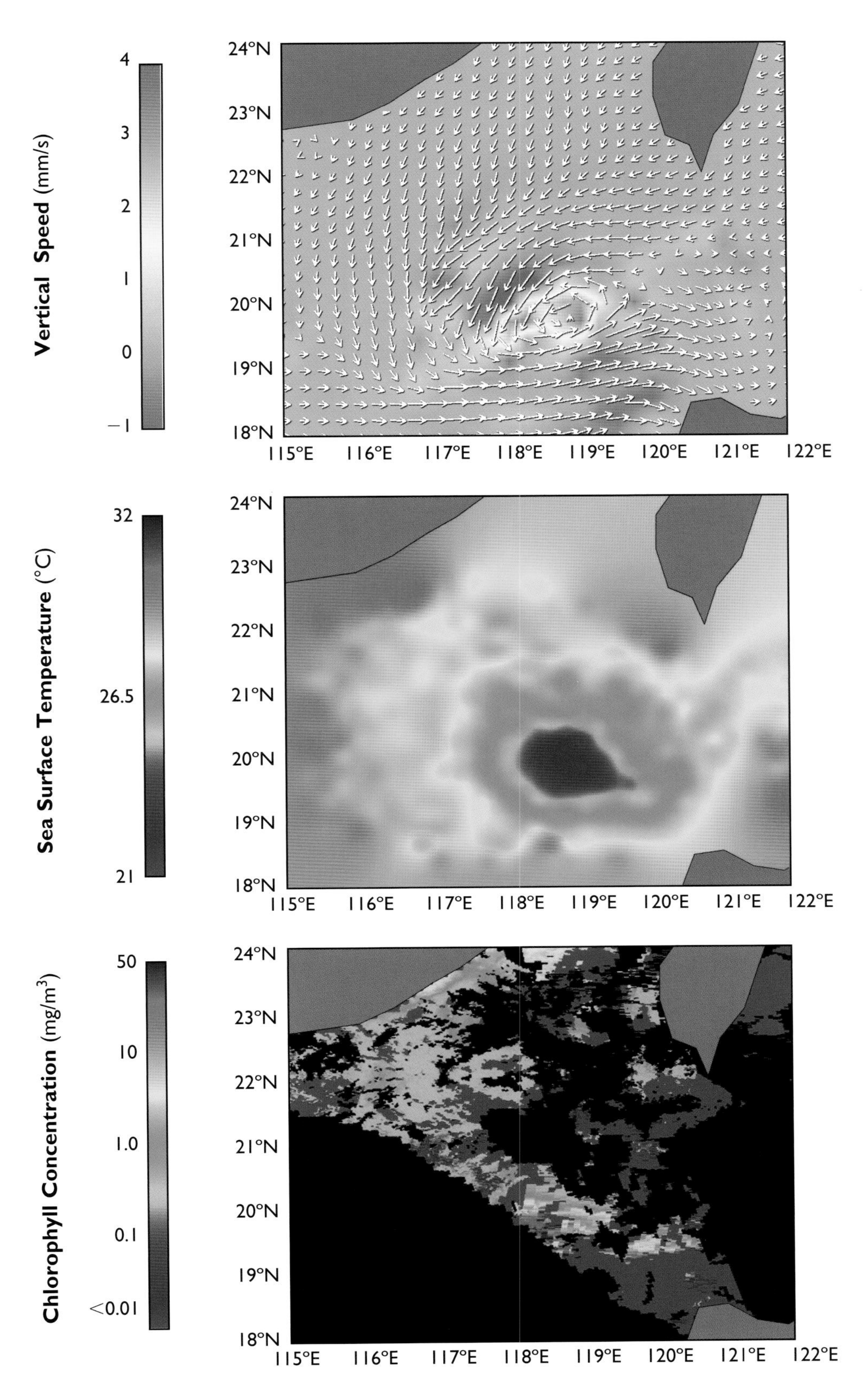


Effects of Typhoon Kai-Tak in the South China Sea in July 2000: (top) wind vector (white arrows) superimposed on a color image of upwelling velocity on July 6, 2000; (middle) sea surface temperature on July 9, 2000; and (bottom) chlorophyll concentration on July 12–15, 2000. Wind-driven ocean mixing and upwelling induce marked cooling and biological productivity, which have dramatic effects on the annual primary carbon productivity. (Wind data from the SeaWinds instrument on the QuikScat satellite; temperature data from the TMI instrument on the TRMM satellite; and chlorophyll data from the SeaWiFS instrument on the OrbView-2 satellite.)

Besides helping us to monitor and predict weather systems, winds measured by the scatterometer are critical to characterize, understand, and predict climate changes. Oceans are the largest reservoir of heat, water, and carbon on Earth, and the redistribution of these quantities through time and space modifies Earth's climate. Wind stress is the single largest source of momentum and energy to the upper ocean. A two-dimensional wind vector field is needed to compute the wind divergence and rotation that control the vertical mixing in the ocean; the

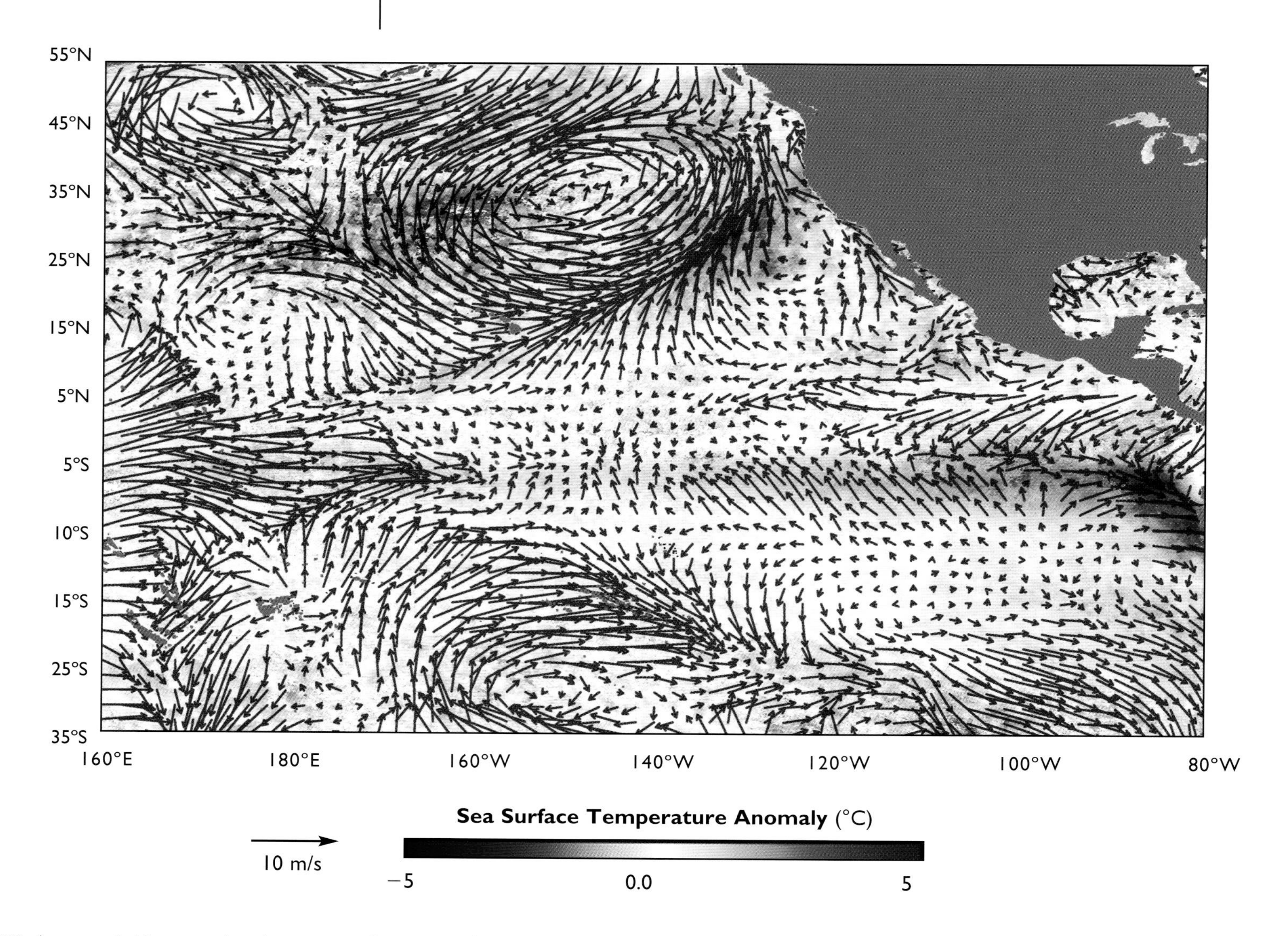


Wind vectors (white arrows) and sea surface temperatures (colors) in the eastern Pacific Ocean, averaged over the last week of May 1997, with the long-term mean removed. The surface winds are related to the interannual ocean warming during an El Niño. (Data from the NASA SeaWinds instrument on the ADEOS II satellite and the AVHRR instrument on a NOAA satellite.)

mixing brings short-term momentum and heat trapped in the surface mixed layer into the deep ocean where they are stored over time, and brings nutrients stored in the deep ocean to the surface where there is sufficient light for photosynthesis. Through photosynthesis, carbon dioxide is transformed into organic compounds. The biological productivity may mitigate greenhouse warming by absorbing carbon dioxide from the atmosphere and offsetting its increased emission from burning of fossil fuels. The stored heat and carbon in the ocean are distributed by horizontal currents, which are driven in part by winds. By moving heat from the warm tropical oceans to the cold high-latitude oceans, the wind-driven currents make Earth a more comfortable habitat.

The effect of wind on ocean heat and carbon storage was clearly demonstrated by I.-I. Lin and her collaborators, in their study of Typhoon Kai-Tak. They computed strong upwelling velocity stirred up by the counterclockwise spinning winds of the typhoon measured by the SeaWinds scatterometer on QuikScat, as the typhoon lingered over the South China Sea on July 6, 2000. In the aftermath, the sea surface temperature dropped by 9°C. Cold water drawn up by the spinning of the winds caused the intense cooling. A few days after the typhoon had moved on, another satellite sensor began to measure a dramatic change in the ocean color, inferring 300-fold increase in phytoplankton. The nutrient-rich water from the deeper part of the ocean sparked massive biological bloom, which provides an important food source for marine life and has been shown to have a significant effect on the carbon cycle.

The atmosphere and ocean are turbulent fluids; processes at one scale affect processes at other scales. Adequate observations at significant temporal and spatial scales can only be achieved from the vantage point of space. The multi-scale interaction connected by surface winds is illustrated on the facing page. The westerly wind events at the equatorial western Pacific have timescales of a few days to a week, but have been shown to be associated with the collapse of the trade winds (steady winds from east to west in the tropical oceans) and the El Niño (equatorial warming in the Pacific), which occurs every few years. The westerly winds have been shown to be connected to the 'Pineapple Express,' the winds that transport moist and warm air from the tropical ocean, through Hawaii, to the west coast of the United States. This branch of the winds is associated with the displacement of a counterclockwise wind circulation during El Niño episodes, which in turn, shifts a warm-cold temperature dipole in the ocean. This ocean dipole had been present in the north Pacific for almost a decade. The approach of warm water induces severe ecological changes at the coast.

Ocean surface wind vectors are essential in monitoring and understanding the oceans' influence on the water cycle over land and ice. The methodology of estimating the advection of moisture over the depth of the atmosphere, using wind vectors measured by scatterometers and water vapor measured by microwave radiometers has been established and validated.

We usually take wind information for granted because it is common in daily weather reports. When a hurricane suddenly intensifies and changes course, when a delay of onset of Monsoon causes drought, and when an El Niño breaks out with the unexpected collapse of trade winds, we then remember the importance of continuous high quality measurements of winds over the ocean. By preserving and extending time series of space-based wind measurements, without sacrificing their research quality, while infusing new technology for improved and expanded applications, we may be able to achieve the routine needs of operational weather forecasting, encouraging new and enabling technology, and advancing scientific investigation in climate and environment changes all at the same time.

David K. Woolf
Ian S. Robinson

# The Stormy Atlantic

Average Significant Wave Height, January 2006

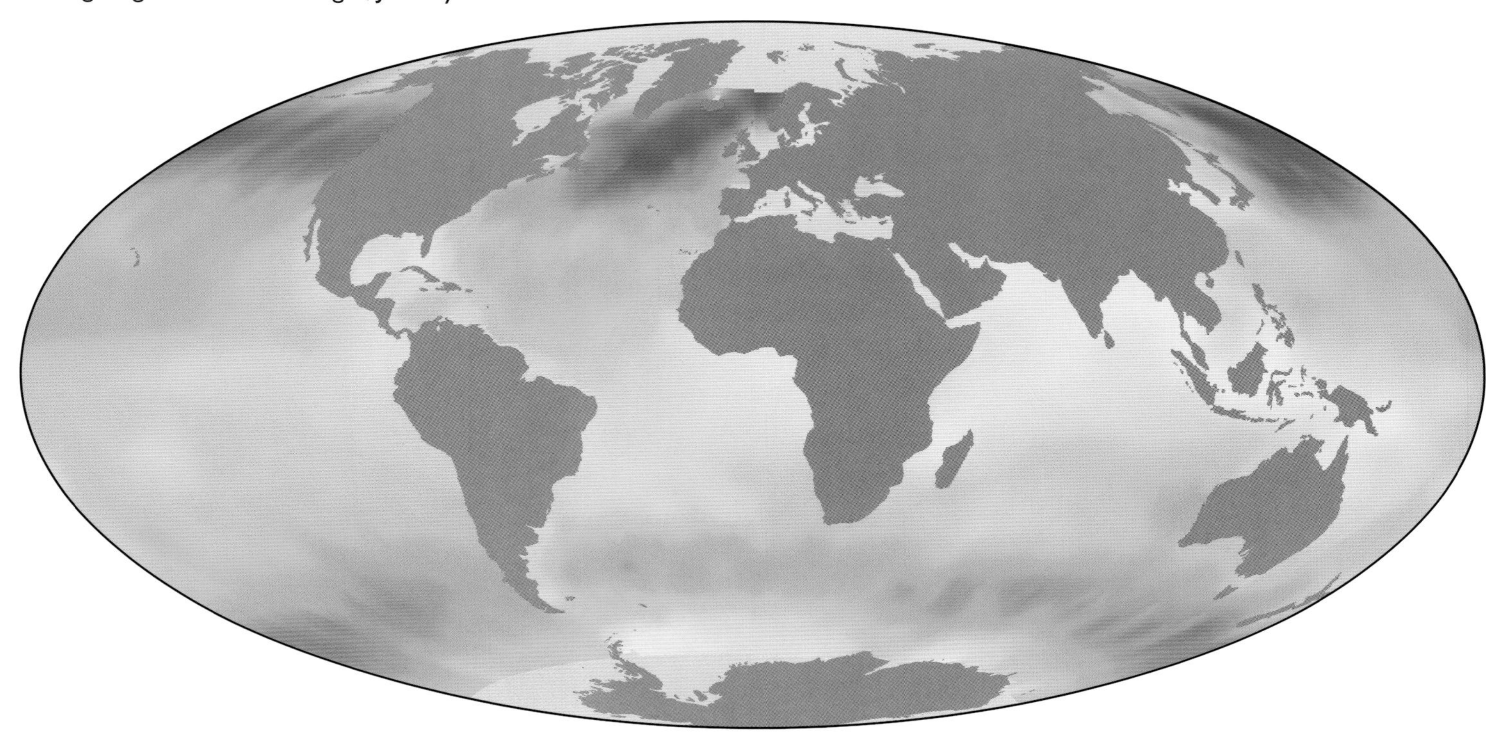

Average Significant Wave Height, July 2006

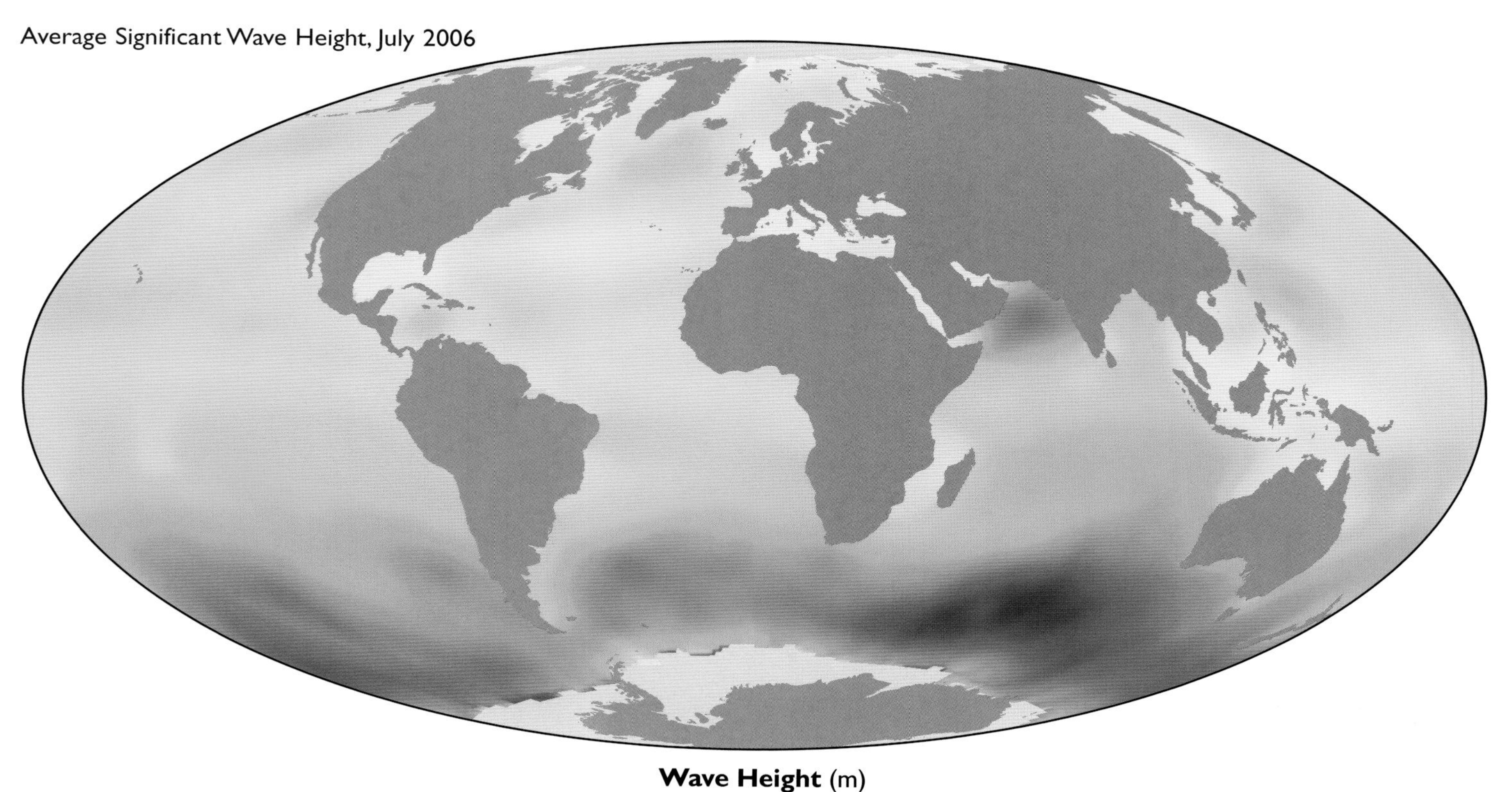


*Our Changing Planet*, ed. King, Parkinson, Partington and Williams
Published by Cambridge University Press 

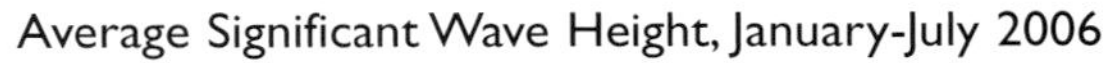
Average Significant Wave Height, January-July 2006

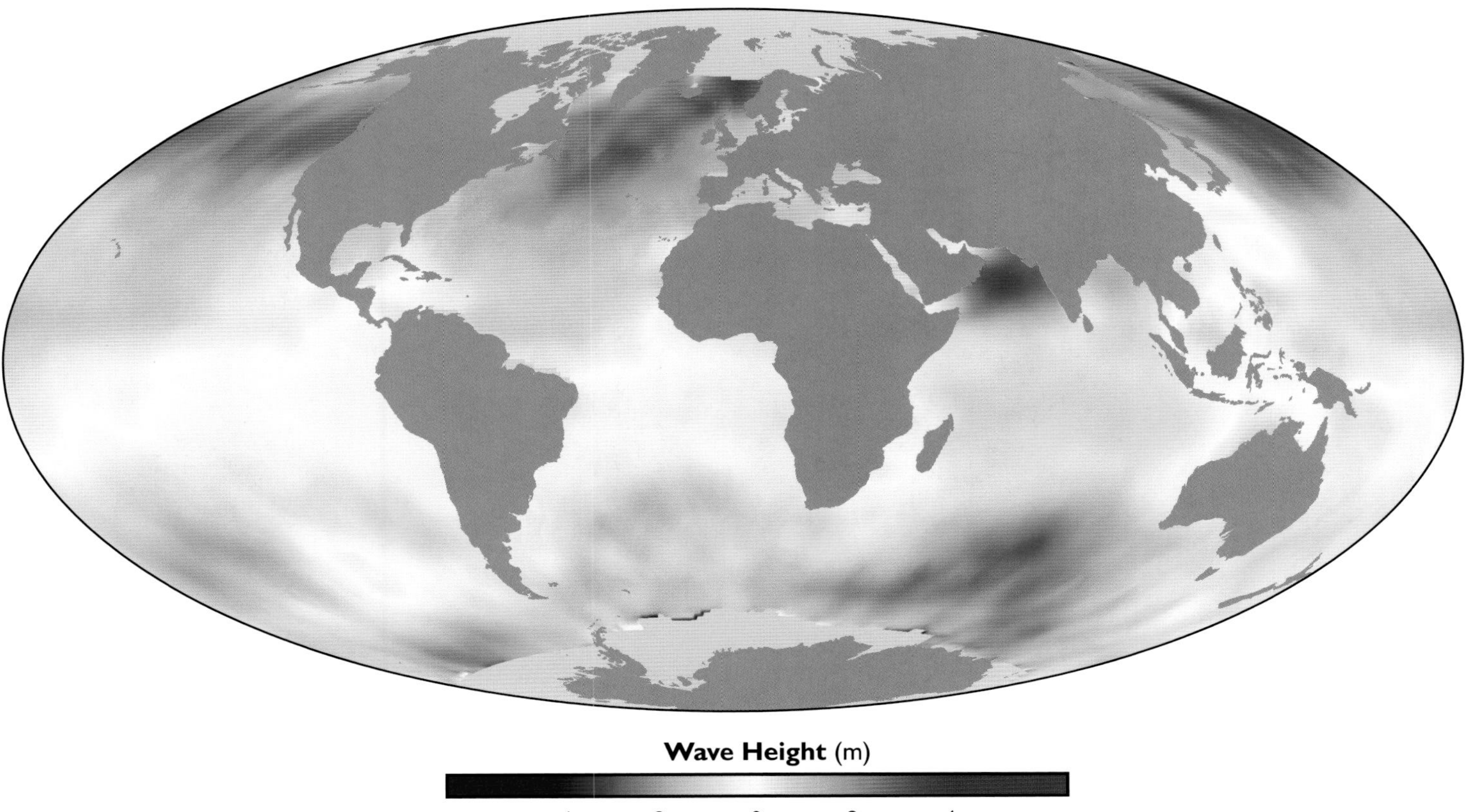


Images of strong winds and huge waves threatening the life of seafarers or battering coastal defenses are part of the winter scene. In common with the rest of our weather, it is impossible to predict individual storms more than a few days in advance. However, there is some prospect of being able to predict unusually stormy winters in advance, and to prepare for their impact. The ability to describe waves over the breadth of the oceans will help us to reach that goal.

Satellite altimetry provides global measurements of wave height. Previously, buoy and ship measurements were available only at a few scattered sites, mainly near land. Now we can observe from space the pattern of wave activity over the entire surface of the oceans. This enables us to detect seasonal changes and variations between years.

Some coastlines are noticeably more exposed to north Atlantic waves than other locations. For example, much of the coastline of Ireland and of Scotland is on the eastern border of the north Atlantic, where seas built by strong westerly winds on the open ocean come ashore. These beautiful coastlines can occasionally be ravaged by winter storms.

More than a decade ago it was discovered that winter wave heights had been generally increasing for the previous 30 years at isolated measurement sites in the northeast Atlantic. With the help of satellite altimeters, we have established that the variation in waves is just part

Left: Average significant wave height (m) in January and July. Above: difference between January and July average significant wave height. The observations confirm that waves are generally largest in the winter (January for the Northern Hemisphere and July for the Southern Hemisphere). In the central north Atlantic, winter waves usually exceed 5 m (16 ft) in height. (Data from the Poseidon-2 and other altimeters, and courtesy of the AVISO data service, copyright CNES-CLS 2006.)

Typical scenery on the islands of the Hebrides, on the west coast of Scotland. The characteristic dunes and 'machair' grasslands are the base of a unique ecology and attract many visitors. (Photograph by Susan Cooper.)

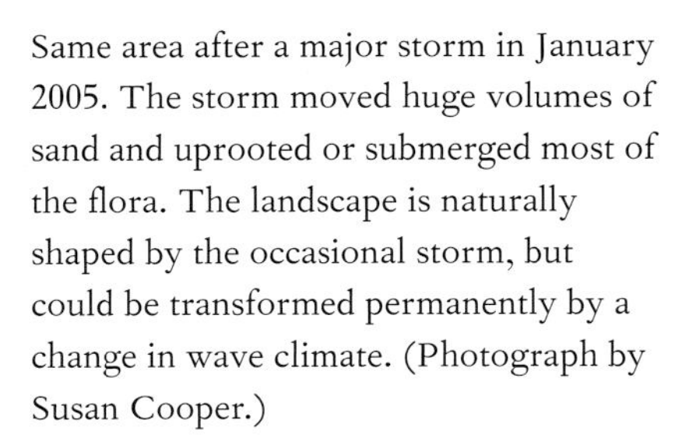

Same area after a major storm in January 2005. The storm moved huge volumes of sand and uprooted or submerged most of the flora. The landscape is naturally shaped by the occasional storm, but could be transformed permanently by a change in wave climate. (Photograph by Susan Cooper.)

of a large pattern of climate variability affecting rainfall, temperature, and other features of weather over a vast region centered on the north Atlantic. This pattern of variability extends back throughout the historical record, and is dominated by the 'North Atlantic Oscillation' (NAO). The NAO is one of many such features (the most famous being El Niño) that have emerged from studies of long-term variability in climate. The primary feature of the NAO is an episodic strengthening of the storm track over the north Atlantic. In northern Europe, an episode is characterized by a 'wet and windy' winter, with storms lashing the Atlantic shores. We can predict wave heights from the state (or index) of the NAO.

The recent increase in wave heights can thus be placed in context: the rise in wave heights resulted from a trend in the NAO that is unusually strong and sustained compared to its historical behavior, but does not clearly indicate a permanent shift in our climate. This trend ceased in the 1990s. The future behavior of the NAO, and how this relates to 'global warming' and other climate variability, is of huge interest.

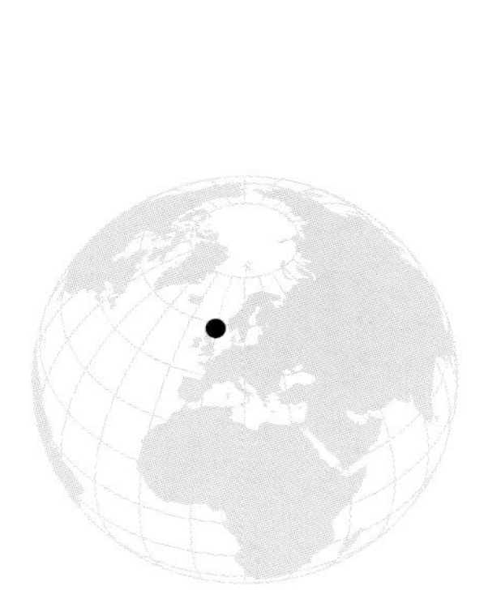

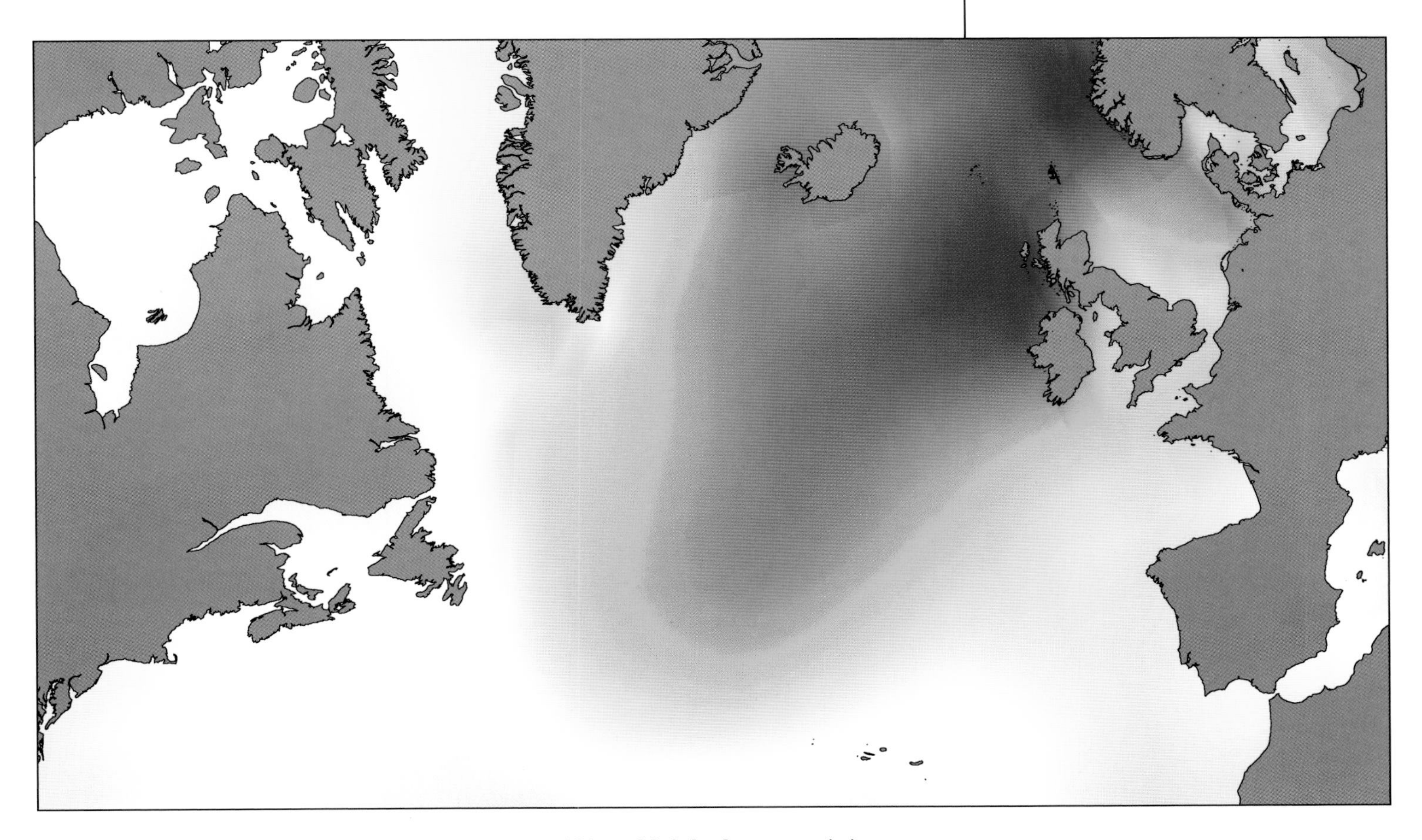


The future behavior of wave climate will have important impacts on us all. Waves are enormously powerful and can be extremely destructive, but they also are a potential source of energy if technology can be developed to harness their power. Plans for coastal defense management and for developing wave energy as a source of renewable energy, require predictions of future wave climate and will encourage further research.

Modeled change in wave height between the late 1960s and the early 1990s due to the North Atlantic Oscillation (NAO). Many historical observations can now be understood from the history of the NAO. (Image based on ECMWF ERA-40 data obtained from the ECMWF Data Server.)

The harbor at Wick in a storm. Large waves at the coast can destroy sea walls and lead to coastal flooding. Will this become more common in the future? (Photograph by William Gray.)

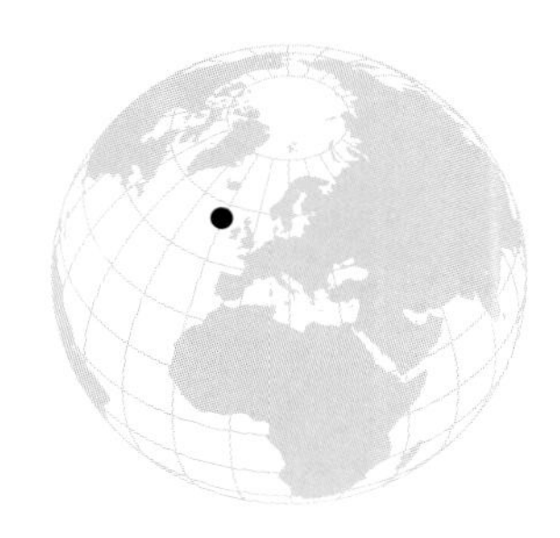

CHARLES R. McCLAIN
GENE C. FELDMAN

# The Ocean Biosphere

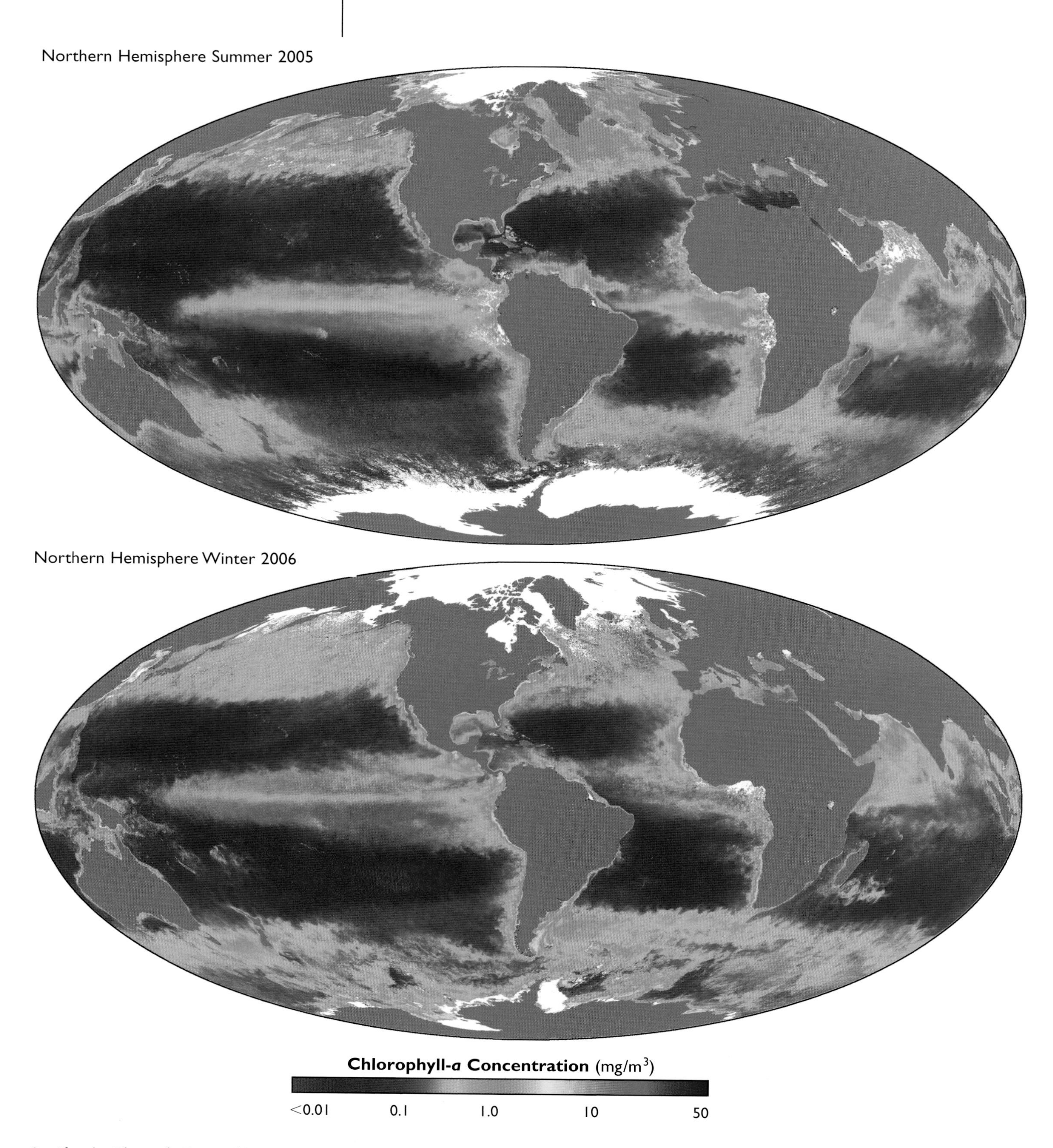


*Our Changing Planet*, ed. King, Parkinson, Partington and Williams
Published by Cambridge University Press 

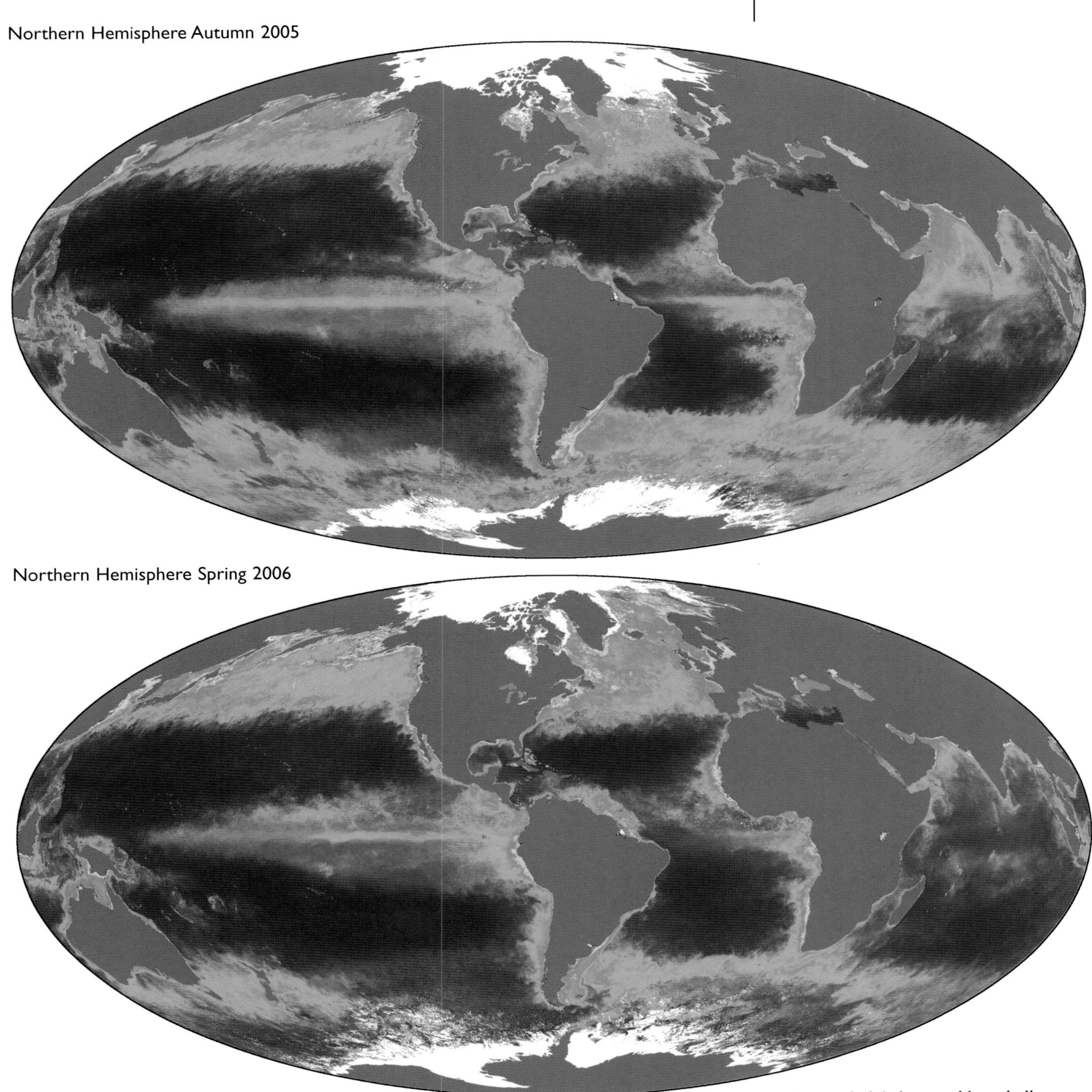


Seasonal global mean chlorophyll-*a* concentrations during 2005 and 2006. (Data from the SeaWiFS instrument on the OrbView-2 satellite.)

Prior to satellite observations of ocean color (the spectral reflectance of the ocean) oceanographers could only observe the biology of the ocean from ships, buoys, and the

January 1998

June 1998

**Chlorophyll-*a* Concentration** (mg/m$^3$)

$<$0.01 0.3 5 50

Chlorophyll-*a* concentrations during the 1997–1998 El Niño-La Niña event in the equatorial Pacific Ocean. The top panel is during the warm nutrient depleted phase (El Niño) and the lower panel is from the cool nutrient-rich phase (La Niña). (Data from the SeaWiFS instrument on the OrbView-2 satellite.)

shore. Sampling from ships allows only very limited coverage and very crude understanding of the distribution of marine plants and their variability over time and space. The first proof-of-concept ocean color mission, the Nimbus 7 Coastal Zone Color Scanner (CZCS), provided the first glimpse of what the real ocean looks like biologically and revolutionized the field of marine biology. The CZCS success led to a number of international ocean biology satellite sensors including the Sea-viewing Wide Field-of-view Sensor (SeaWiFS; a NASA data buy from Orbital Sciences Corporation) and two Moderate Resolution Imaging Spectroradiometers (MODIS; on the NASA Terra and Aqua Missions). The SeaWiFS images shown provide striking examples, and illustrate why conventional shipboard and buoy sampling strategies alone cannot adequately describe most ocean ecosystems.

The images are maps of ocean surface chlorophyll-*a*. Chlorophyll-*a* is the plant pigment central to photosynthesis, the transformation of carbon dioxide into organic compounds (termed 'primary production'). Phaeophytin also absorbs light, but is not photosynthetic. The marine plants responsible for photosynthesis are microscopic in size and are called phytoplankton. There are thousands of species of phytoplankton in the ocean. Overall, phytoplankton annual primary production is about equal to that of terrestrial plants, roughly 50

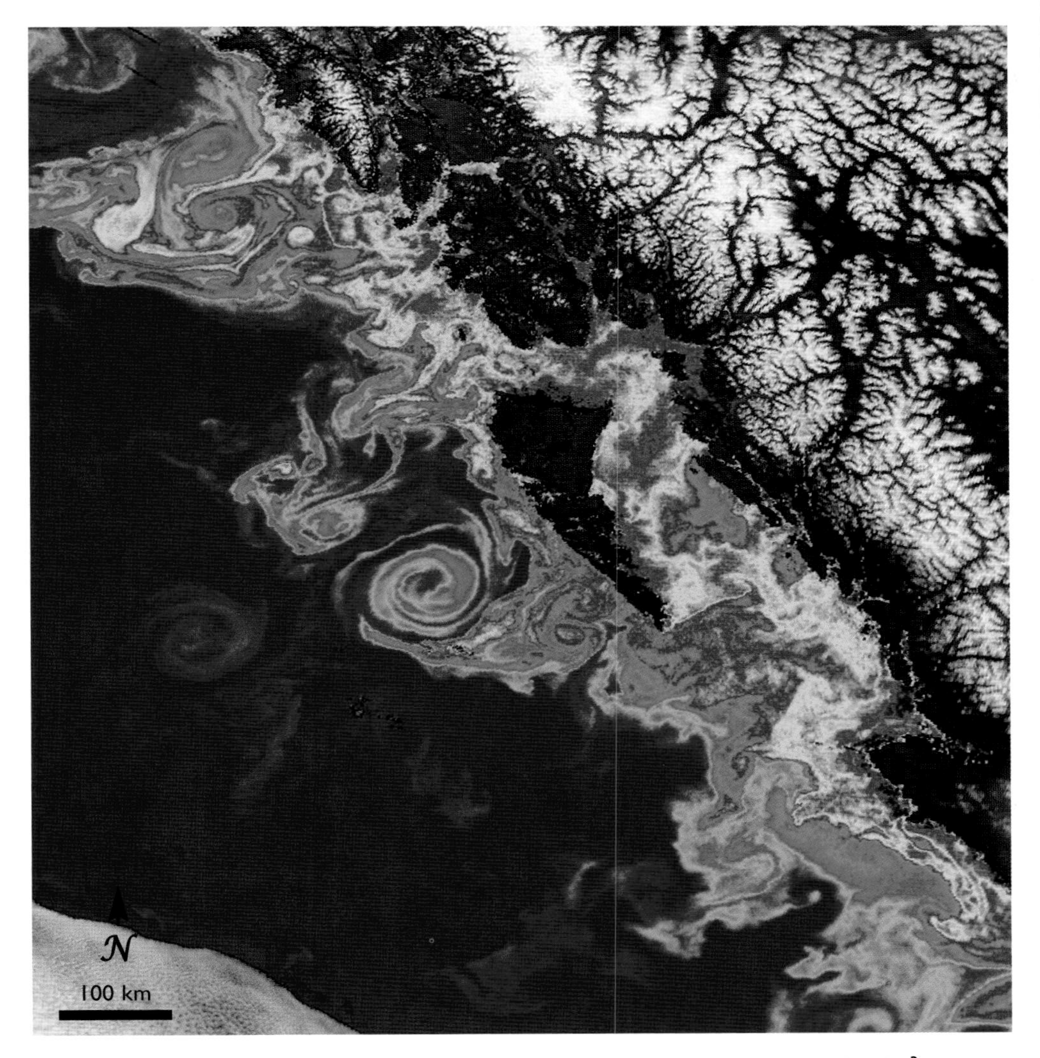


High concentrations of chlorophyll-*a* were observed to the west of British Columbia's Queen Charlotte Islands and Alaska's Alexander Archipelagoon on June 13, 2002. (Data from the SeaWiFS instrument on the OrbView-2 satellite.)

billion tons of carbon per year. This organic carbon fuels a vast food web ranging from microscopic animals called zooplankton) that graze on phytoplankton to large fish, like salmon and tuna, and the great whales. This process is important to regulating the amount of carbon dioxide in the atmosphere and research is currently underway to quantify the amount of anthropogenic carbon (i.e., the carbon released through fossil fuel combustion) being absorbed by the ocean in this way.

The spatial and temporal distribution of phytoplankton is controlled by the supply of nutrients (nitrate, phosphate, silicate, and dissolved iron), light, and the concentration of zooplankton. Nutrients and light regulate the rate of phytoplankton growth. In the open ocean, nutrients are supplied from the deep ocean by vertical mixing and upwelling processes. Upwelling is the result of atmospheric winds driving a divergent flow of surface water that causes subsurface water to rise to the surface. Upwelling typically occurs along coasts (e.g., off northwest Africa), and along the equator (e.g., the equatorial Pacific Ocean). In contrast, large expanses of the central ocean have downwelling circulations. As a consequence they have low nutrient and phytoplankton concentrations. These areas are called the sub-tropical gyres and are analogous to the land deserts.

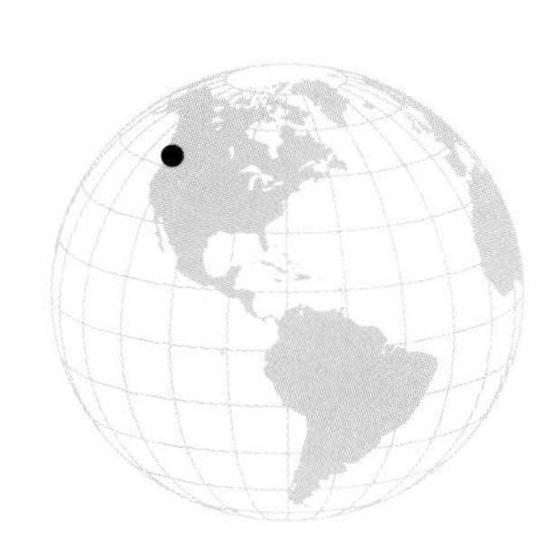

Light also regulates phytoplankton growth. As in mid and high latitudes on land, the ocean biology experiences seasons as well. The average chlorophyll-*a* (an important type of phytoplankton) concentrations for each season are shown. Some areas, such as in the north Atlantic, exhibit very pronounced seasonal cycles with a dramatic spring bloom. This is the result of deep winter mixing that replenishes the nutrient supply while light levels and photosynthetic rates are low.

Climate varies on longer timescales than seasons. The El Niño-La Niña cycle is one well-known example. During 1997–1998, there was an extreme El Niño event during which the usual upwelling of nutrient rich water along the Pacific equator was interrupted and warm, nutrient-poor water from the western Pacific migrated eastward causing the phytoplankton population to collapse. The concentrations of chlorophyll-*a*, as measured from ships, were the lowest on record.

During the spring and summer of 1998, the warm surface waters retreated back to the west and the upwelling of nutrient rich sub-surface water resumed. This led to a dramatic resurgence of the phytoplankton population with chlorophyll-*a* concentrations being the highest ever observed there.

As global warming and climate change continue, the ocean ecosystems will respond to the subsequent changes in ocean circulation and temperature. Continuous global observations of ocean color from satellites will provide the data required for understanding how marine ecosystems adjust.

# Coccolithophores and the 'Sea of Milk'

William M. Balch

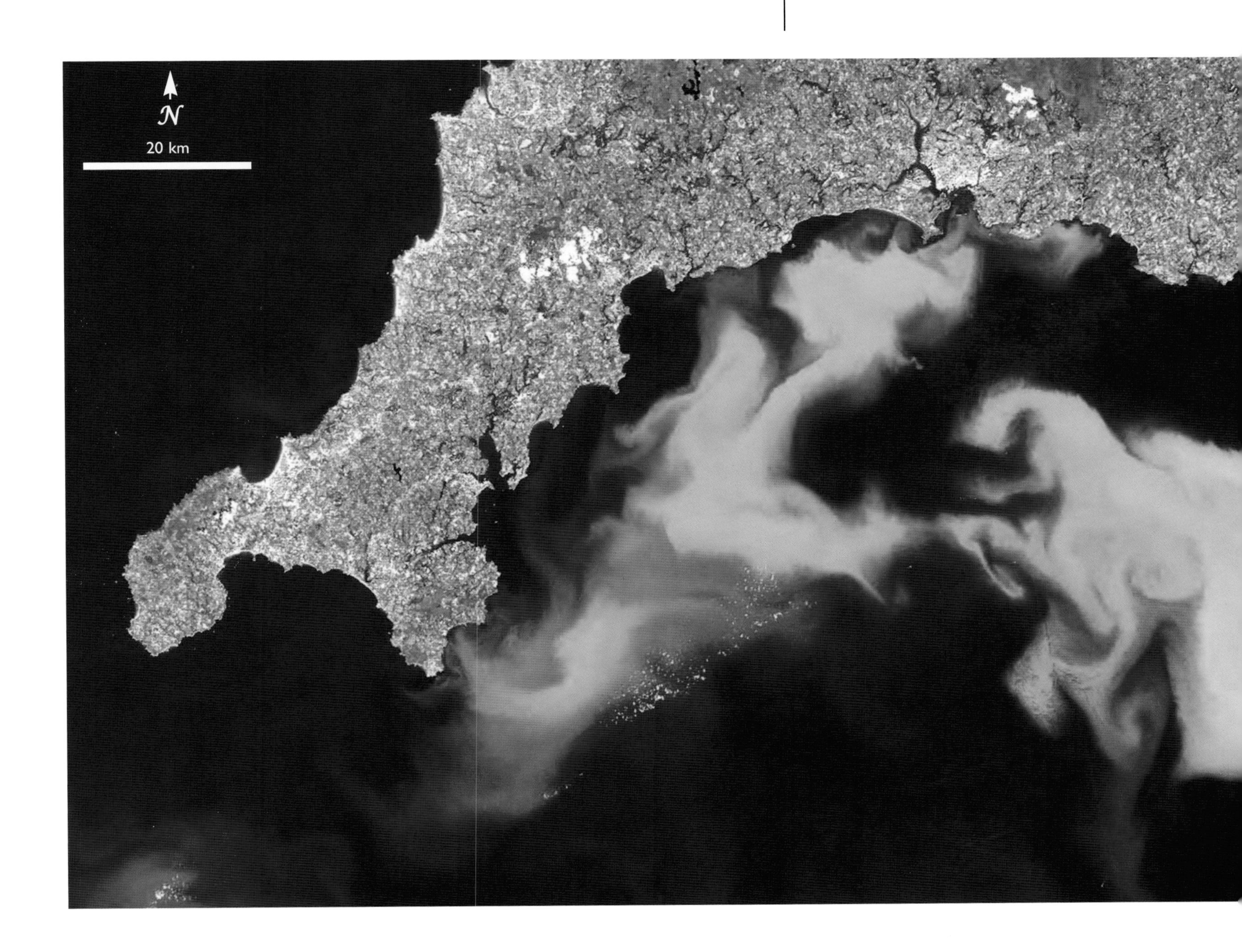


Coccolithophore bloom of *Emiliania huxley* in the western English Channel on July 24, 1999. [Data from Landsat 7 and courtesy of S. B. Groom (Plymouth Marine Laboratory, United Kingdom).]

Coccolithophores are microscopic marine plants (phytoplankton) that cover themselves with numerous small limestone scales, or 'coccoliths,' which they shed into the water. These coccoliths are ubiquitous in the world's oceans, and their mineral nature causes them to be highly reflective of visible light. About one quarter of all recent marine sediments are made of calcium carbonate, and a large fraction of that is contributed by coccolithophores. Typically, seawater contains about $10^5$ to $10^6$ detached coccoliths per liter, but when coccolithophores grow to high abundance, the coccolith concentration can exceed $10^9$ per liter.

*Our Changing Planet*, ed. King, Parkinson, Partington and Williams

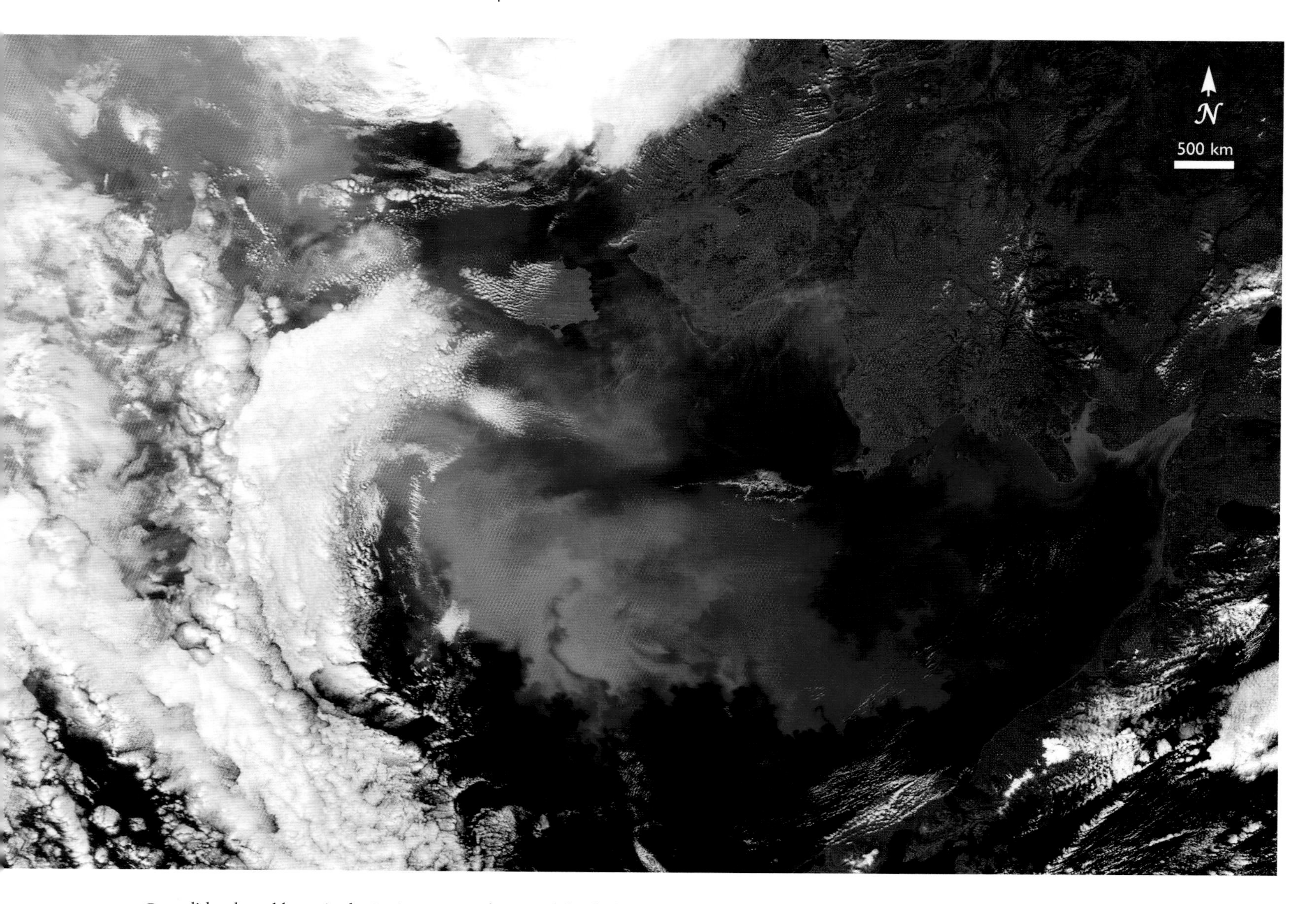


Coccolithophore bloom in the Bering Sea, September 15, 2000. (Data from the MODIS instrument on the Terra satellite.)

Such high densities of suspended calcium carbonate coccoliths can form spectacularly bright, milky, or turquoise-colored patches, spreading from horizon to horizon over hundreds of thousands of km$^2$. These blooms, forming over geological time, were responsible for the formation of the white cliffs of Dover, overlooking the English Channel.

Bright phytoplankton blooms, resembling coccolithophore patches, have been described since the 1870s. For example, in Jules Verne's book, *Twenty Thousand Leagues Under the Sea*, there is a passage describing patches of milky white waters, as big as 40 miles in width. Indeed, by its location in the Indian Ocean and large size, the feature described in the book may have been a coccolithophore bloom.

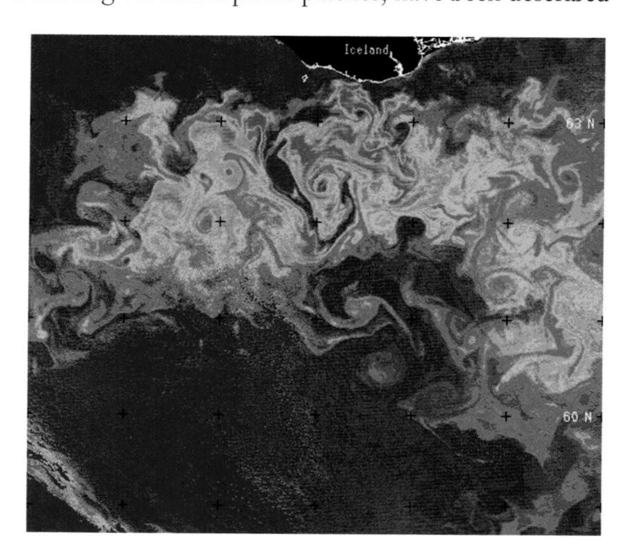


False-color image of visible reflectance from a coccolithophore bloom in the North Atlantic, south of Iceland, taken on June 18 1991. Color scale represents a continuum from purple areas representing regions of lowest visible reflectance to red areas representing regions of highest visible reflectance caused by dense concentrations of coccoliths. [Data are Rayleigh-corrected channel 1 results from the AVHRR instrument on the NOAA 11 satellite. Original data supplied by P. E. Baylis (Satellite Receiving Station, University of Dundee, United Kingdom) and processed by S. B. Groom (Plymouth Marine Laboratory, United Kingdom).]

In another quotation from the book, "Conseil could not believe his eyes, and questioned me as to

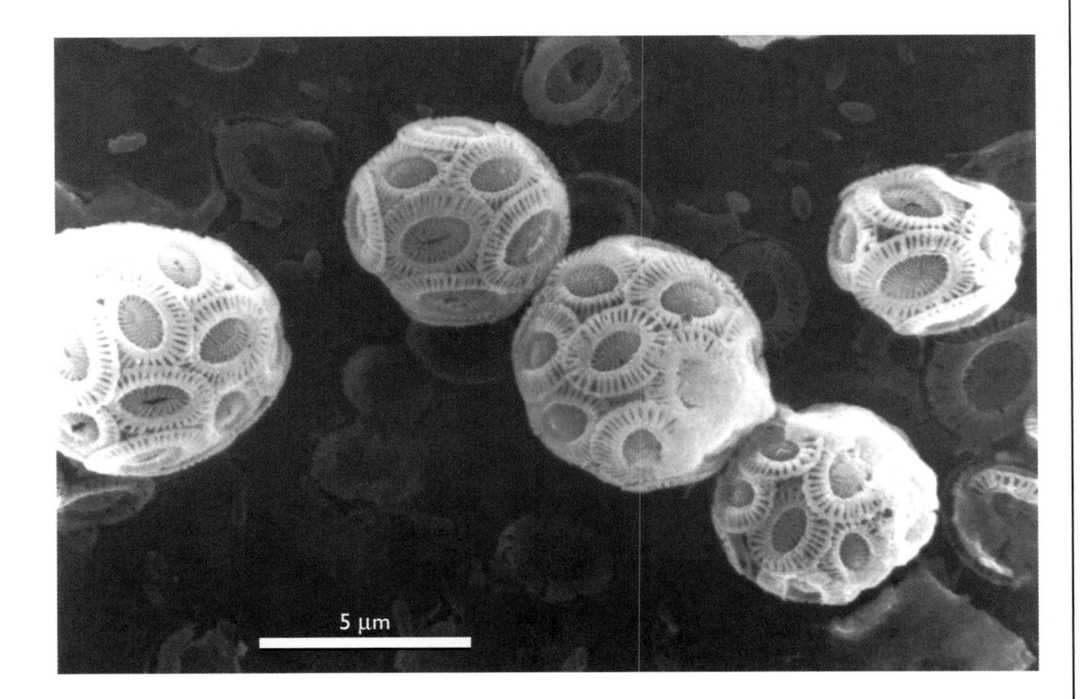

Left: Scanning electron micrograph of coccolithophores, showing various sized cells covered with calcium carbonate plates known as coccoliths. (Photograph by Jennifer Fritz, University of Miami.)

Right: Scanning electron micrograph of a glass fiber filter (0.7 µm pore size) through which was filtered seawater from the 1988 coccolithophore bloom from the Gulf of Maine. It can be seen that the water contained few particles other than detached coccoliths. (Photograph by Maureen Keller, Bigelow Laboratory for Ocean Sciences.)

the cause of this strange phenomenon. Happily I was able to answer him. 'It is called a milk sea,' I explained. 'A large extent of white wavelets often to be seen on the coasts of Amboyna, and in these parts of the sea.'" Anecdotally, this other region to which milky waters are attributed, Amboyna, is in eastern Indonesia, southeast of Sulawesi and north of Timor. The shores of Amboyna Bay consist of calcium carbonate chalk.

While the sediment record has shown that coccolithophore abundance varies strongly through geological time, satellites have provided new evidence of enormous coccolithophore blooms, varying over time scales of years to decades. The first such blooms over the western European Continental Shelf were described in 1983 by Patrick Holligan, using NASA's Coastal Zone Color Scanner (CZCS) and by Jacques Le Fevre using Landsat imagery. In the late 1980s more blooms were described in that region, as well as in the Gulf of Maine and the North Atlantic Ocean, just south of Iceland. The North Atlantic feature was one of the largest blooms of an algal species ever described, with an area exceeding a quarter million square kilometers. During the 1990s, blooms did not occur in the Gulf of Maine, but began appearing in the Bering Sea, and since 2000, they have begun appearing in the Gulf of Maine again. Coccolithophores prefer well-stratified ocean environments, often associated with warm surface waters. This has led some to consider whether increasing incidence of blooms indicates a gradual warming (and stratification) of the surface ocean.

With the advent of more sophisticated satellite sensors, we are now able to create global composite images of calcite concentration associated with coccolithophores. These images reveal remarkably calcite-rich regions in the Northern Hemisphere. They also reveal, for the first time, apparent high calcite concentrations in the polar convergence zone of the Southern Ocean, recently validated by shipboard measurements. These high concentrations of calcium carbonate represent a large fraction of the total suspended calcium carbonate in the global ocean, which itself is critical for understanding the planet's carbon cycle. With the new satellite sensors, we have the ability to examine long-term change in the distribution of the biogeochemically important coccolithophores.

# Coral Bleaching

Gang Liu
Alan E. Strong

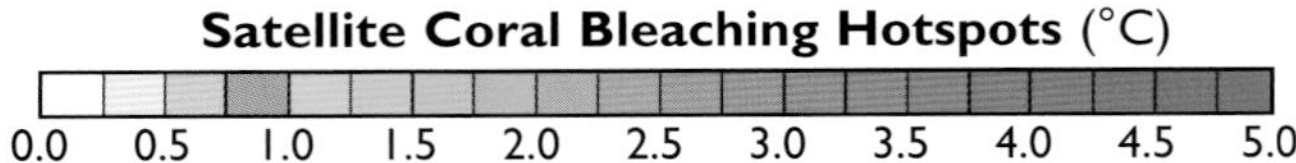


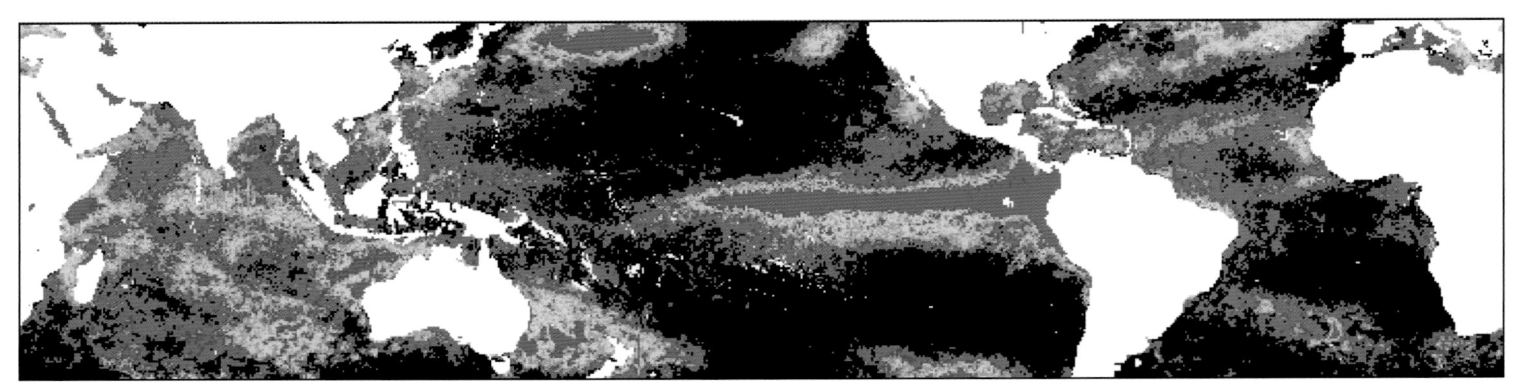

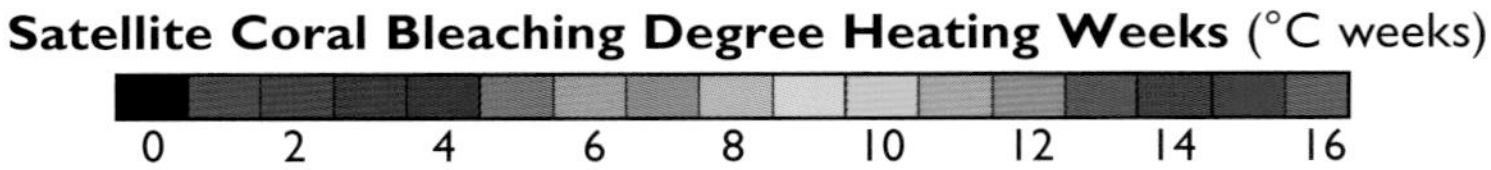


Top: Annual maximum satellite coral bleaching hotspots of 1998 showing areas of sea surface temperature anomalies above bleaching threshold. (Data from NOAA Coral Reef Watch.)

Bottom: Annual maximum satellite coral bleaching degree heating weeks of 1998 showing accumulated hotspots contributing to the coral bleaching in 1998. (Data from NOAA Coral Reef Watch.)

The breathtaking underwater coral reefs of the world's oceans constitute Nature's most diverse and complex marine ecosystem. Corals are ancient animals and well-developed reefs reflect thousands of years of history. However, coral reefs face numerous hazards and threats.

They are prone to various natural stresses, such as tropical storms, floods, diseases, and climate change. Today, these natural stresses are compounded by impacts from human activity, such as pollution, sedimentation, over-fishing, vessel groundings, and anchor damage. Many coral habitats worldwide have been declining rapidly. Current estimates are that 10% of all coral reefs are degraded beyond recovery and 30% are in critical condition. Experts predict that, if current pressures are allowed to continue unabated, 60% of the world's coral reefs may die completely by 2050.

Most coral species contain symbiotic microscopic algae. The host coral provides its symbiotic algae with a protective environment and the compounds necessary for photosynthesis.

*Our Changing Planet*, ed. King, Parkinson, Partington and Williams
Published by Cambridge University Press 

Coral reef ecosystems, made up of a vast assemblage of living organisms including various forms of plant, animal, and microbial life. (Photographs from Florida Keys National Marine Sanctuary.)

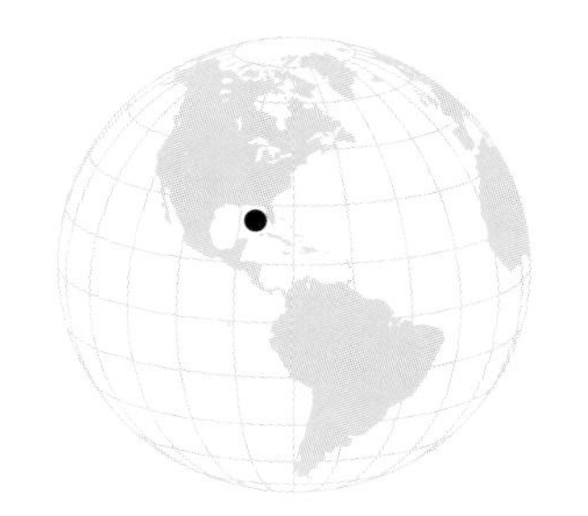

In return, the algae supply the coral with the oxygen and organic products produced by photosynthesis and help the coral to remove waste products. Healthy coral tissue displays colors ranging from yellow to brown. These are due to the photosynthetic pigments present in the symbiotic algae.

Under certain environmental stresses, the algae can be expelled by the host coral, and the coral colony can take on a stark white appearance or pale color, revealing its underlying white calcium carbonate skeleton through translucent coral tissue. This phenomenon is commonly known as 'coral bleaching.'

Localized bleaching has been observed since at least the beginning of the 20th century. The bleaching events reported prior to the 1980s were generally attributed to localized stresses such as major storm events, severe tidal exposures, high turbidity and sedimentation, pollution, abnormal salinity and temperature extremes, high levels of light and ultraviolet radiation, the presence of toxins, and bacterial infection. However, since 1980, anomalously high water temperatures, often linked to climate change and the El Niño/La Niña weather pattern, have contributed to the majority of the major bleaching events. Most corals live and thrive only in narrow temperature ranges between 18°C and 30°C. There has been an unprecedented increase in the number of coral

Bleached coral colonies found at Halfway Island, part of the Keppel Reef in the southern part of the Great Barrier Reef. (Photograph by Ray Berkelmans.)

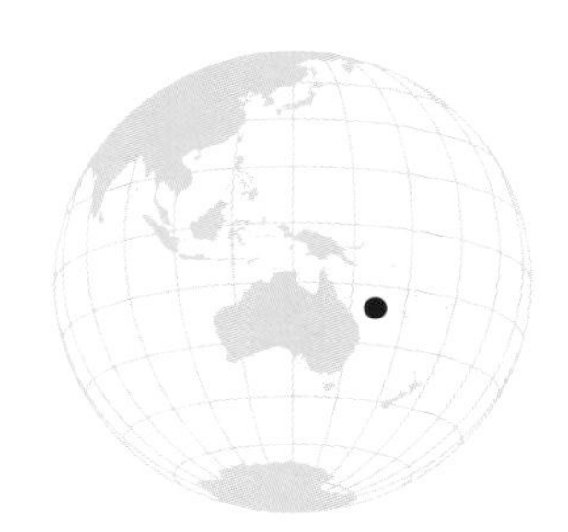

bleaching events during the past 2 decades (which have included some of the warmest years in recorded history).

During the very strong 1997–1998 El Niño, some of the most extensive and severe coral bleaching on record occurred with approximately 16% of the world's remaining coral reefs effectively destroyed. Most of the damaged reefs were in the central to northern Indian Ocean, in southeast and east Asia, and in the western Pacific.

A new wave of bleaching linked to changing climate conditions swept coral reefs worldwide in 2002, possibly tied to the most recent El Niño event. This makes 2002 the second worst year for the phenomenon after the major bleaching events of 1998. Australia's Great Barrier Reef, the largest coral reef system in the world, was among the reefs that was severely affected, with 2002 being the year of the worst bleaching event on record.

During previous bleaching events, once water conditions return to normal, most coral species were able to re-establish the symbiotic relationship with their symbiotic algae, and their color and health returned to normal within weeks to months. However, partially to totally affected coral colonies often die after prolonged bleaching events. Following the death of a coral colony, its skeleton is quickly colonized by algae and other encrusting organisms, and usually takes on a dirty brown appearance. It usually takes years, under favorable conditions, for healthy coral reefs to return. Severe bleaching events have dramatic long-term ecological consequences, with the loss of reef-building corals, changes in the assemblage of living organisms, degradation of reef structure, and in some cases, decrease in fish populations. These consequences are of major economic concern to the region, with possible negative impacts on fisheries and on coastal tourism.

Satellites, with the capability to continuously monitor the environmental conditions of global coral reef areas, provide us with a vital tool for understanding coral bleaching. The information provided by satellite remote sensing is critical for management decision making to preserve our precious coral reef ecosystems, the 'rainforests of the sea.'

Coral reef ecosystems benefit millions of people by providing income and food, offering material for medicine, protecting coastal communities, supporting tourism and recreation, and sustaining cultural traditions. (Photograph by Anthony R. Picciolo, NOAA.)

# Hunting Red Tides From Space

Kendall L. Carder
John J. Walsh
Jennifer P. Cannizzaro

Left: Water discoloration caused by a red tide bloom. (Photograph courtesy of Charlotte Sun Herald.)

Right: Fish killed by red tide wash ashore requiring cleaning up of Florida beaches. (Photograph courtesy of Florida Fish and Wildlife Conservation Commission, Fish and Wildlife Research Institute.)

Every year, thousands of tourists flock to Florida's pristine Gulf coast beaches, lured by the prospect of soaking up seemingly endless rays of sunshine. Instead of sun and fun, however, some unsuspecting guests and coastal residents alike have the misfortune of encountering one of Florida's most notorious visitors—red tides.

The term red tide originates from the reddish hue that sometimes emanates from oceanic waters when populations of certain types of microscopic plant-like cells called phytoplankton explode or bloom. Red tide blooms can also appear green, brown, or even black in color. The main culprit responsible for Florida's red tides is the toxic dinoflagellate *Karenia brevis*.

During red tide events, toxins produced by *Karenia brevis* accumulate in filter-feeding shellfish like oysters, clams, and mussels, and when ingested by humans can cause nausea, dizziness, and even paralysis. Human eyes and respiratory systems are sometimes irritated once the toxins become airborne in sea spray. Also, bird, fish, and marine-mammal mortalities have all been attributed to red tides. As a result, tourism and commercial shellfish industries have incurred millions of dollars in lost revenues due to red tide events.

Scientists have recently discovered that iron-deficient cyanobacterium, *Trichodesmium spp.*, found in Gulf of Mexico waters may provide a key nutrient source and fuel red tide blooms, when oceanic waters are fertilized by iron-rich African dust. Using satellite imagery provided by the Ozone Monitoring Instrument (OMI) on the Aura spacecraft and the Moderate Resolution Imaging Spectroradiometer (MODIS) on the Terra and Aqua spacecraft, forecasting of red tide blooms may be enhanced by monitoring iron-rich atmospheric dust that originates from Africa's Saharan desert.

As phytoplankton populations increase, concentrations of chlorophyll *a*, the pigment used by all phytoplankton to trap sunlight for photosynthesis, also increase. Increases in chlorophyll concentration cause a shift in water color from shades of blue to shades of green.

*Our Changing Planet*, ed. King, Parkinson, Partington and Williams
Published by Cambridge University Press 

Satellite ocean color sensors such as MODIS and the Sea-viewing Wide Field-of-view Sensor (SeaWiFS) on the OrbView-2 spacecraft can measure this shift allowing chlorophyll concentrations to be quantified from space. Ocean color imagery may be used to detect and monitor blooms from space, providing an early-warning system that allows state and local agencies to close shellfish beds and alert susceptible individuals with pulmonary weaknesses. Recent research results indicating that red tide blooms are less reflective than their non-toxic counterparts are promising and may further aid in the detection of red tides from space.

Today, many people believe that anthropogenic nutrient sources have caused red tide blooms to increase in frequency and duration in the Gulf of Mexico. Opponents argue that humans are simply more aware of red tides due to increased monitoring practices and the severe negative economic impacts associated with blooms. Using satellite imagery, scientists may now begin to address this debate.

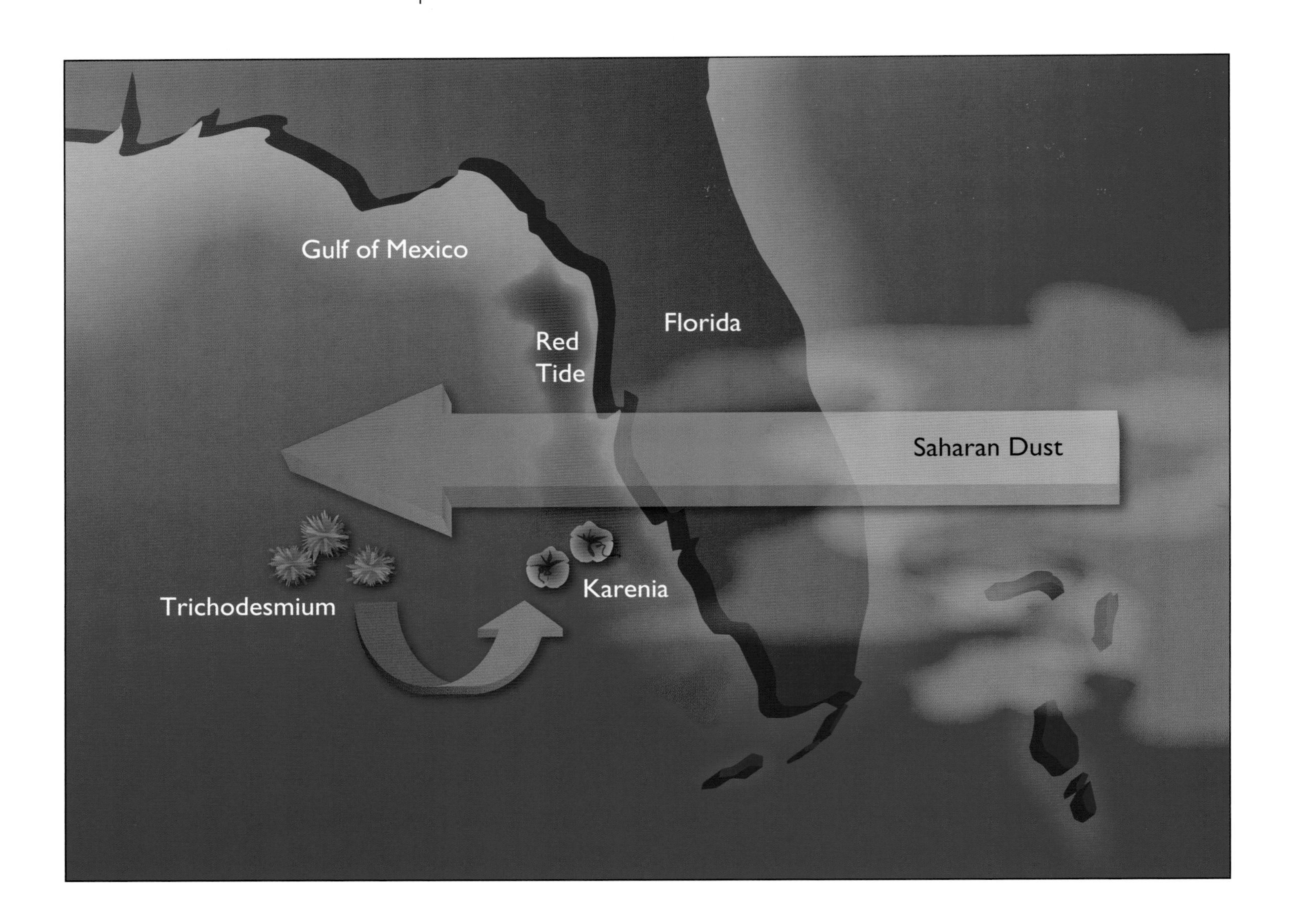


Arrival of iron-rich Saharan dust on the outer west Florida shelf each summer interrupts the normal seasonal succession of phytoplankton on temperate and subtropical shelves. Instead, iron-starved diazotrophs fix the ubiquitous nitrogen gas of seawater, passing this 'new nitrogen' in the form of dissolved organic matter of *Trichodesmium* origin to co-occurring toxic dinoflagellates for initial formation of red tides of *Karenia brevis*.

July 22, 2006

July 24, 2006

July 26, 2006

July 27, 2006

July 29, 2006

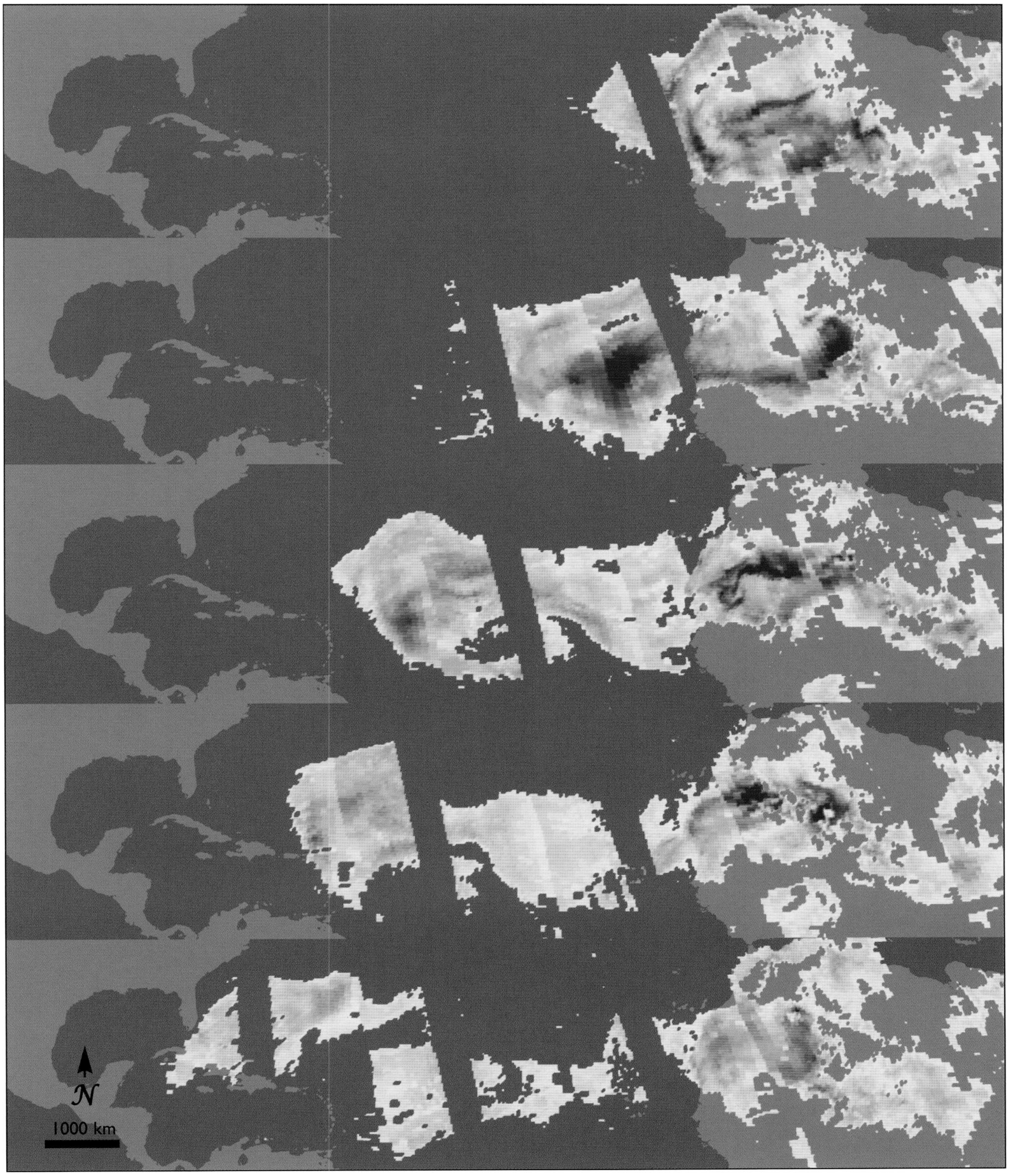

N

1000 km

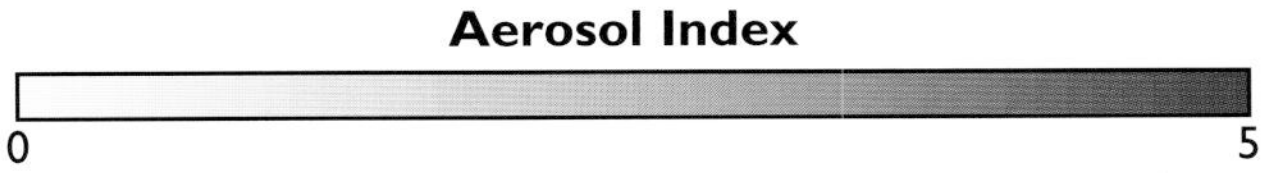

**Aerosol Index**

0 5

A time series of aerosol index values from July 22–29, 2006 dramatically showing the transport of dust from the Sahara Desert that moves across the Atlantic Ocean to the Gulf of Mexico. The dark red areas indicate high aerosol index values and correspond to the densest part of the dust cloud. (Data from the OMI instrument on the Aura satellite.)

Dust storm blowing from the northwest African desert westward across the Atlantic Ocean on July 24, 2006. (Data from the MODIS instruments on the Terra and Aqua satellites.)

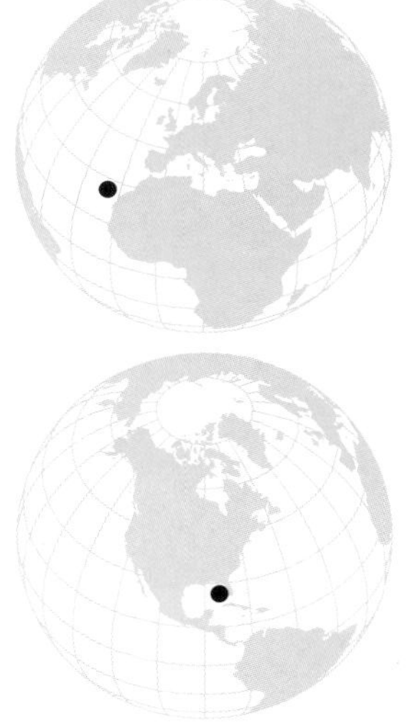

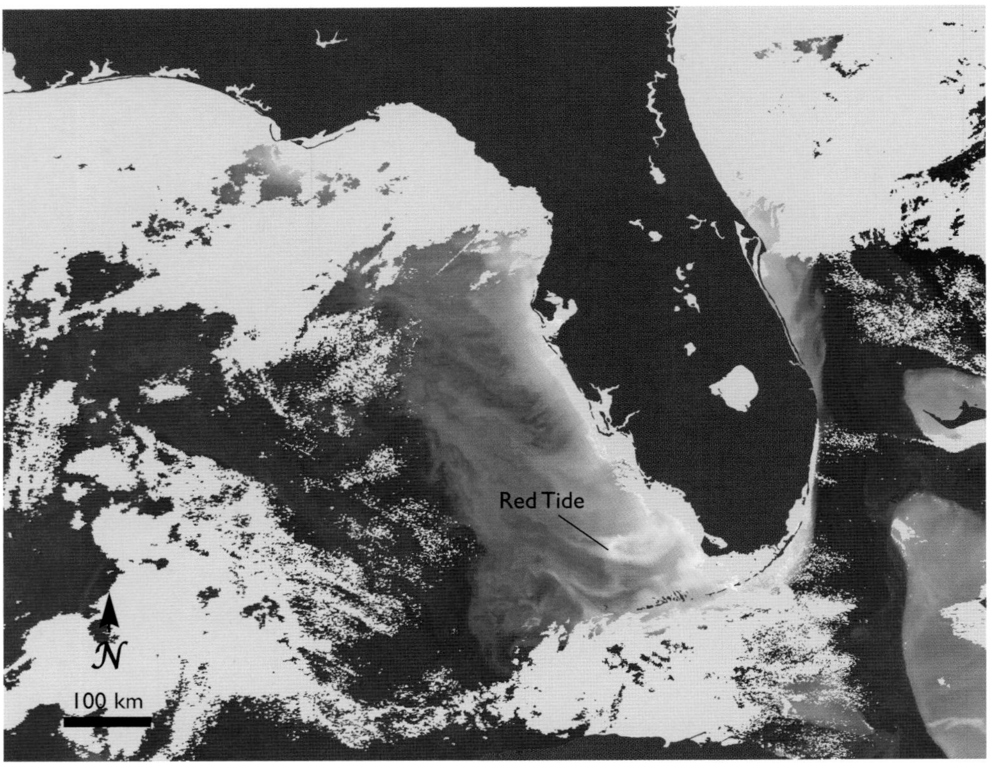


Chlorophyll-*a* concentration on November 20, 2004 in the eastern Gulf of Mexico. Red tides contain high concentrations of chlorophyll. (Data from the MODIS instrument on the Aqua satellite.)

# Marine Sediments

James G. Acker

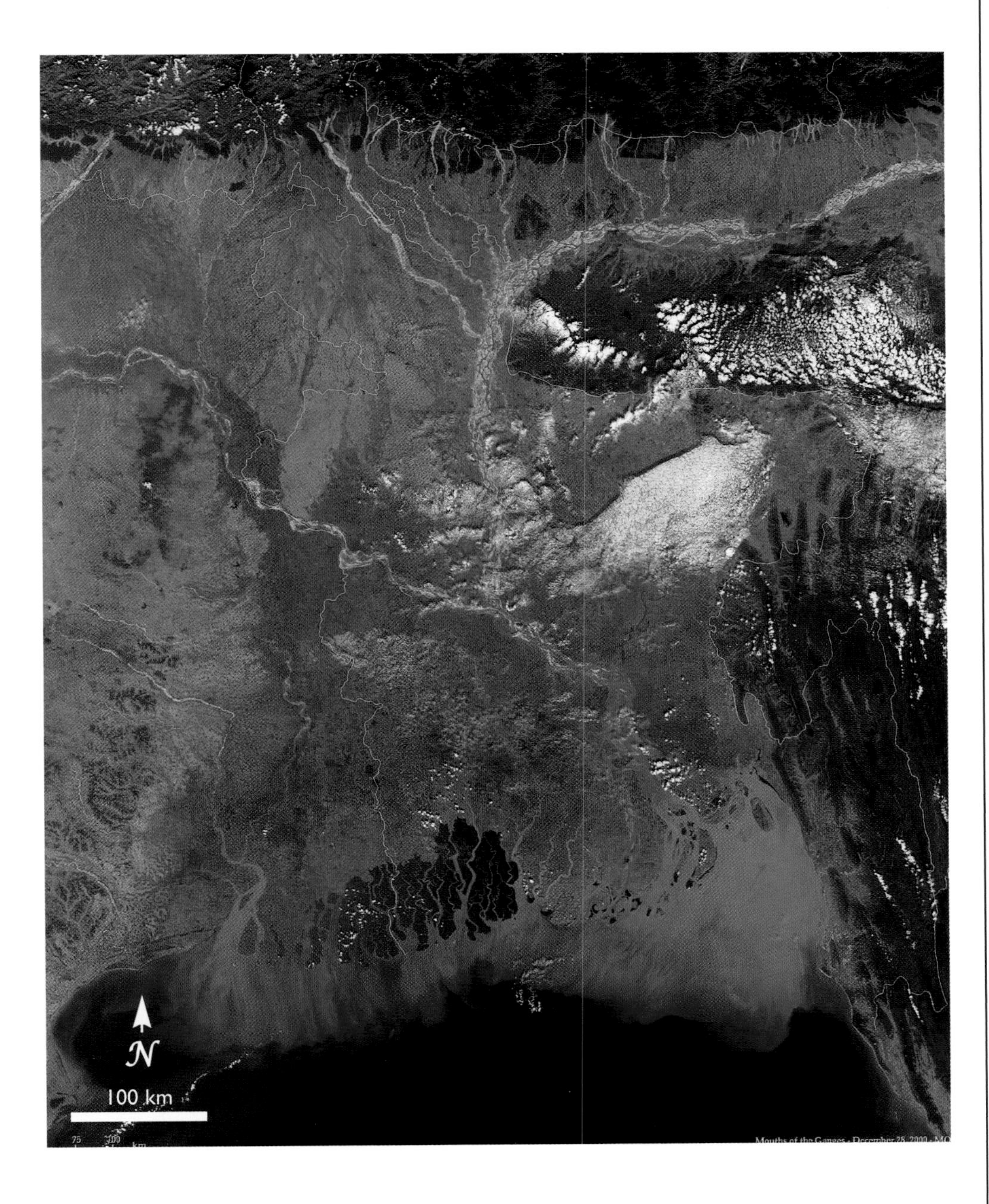


Ganges river/Brahmaputra river delta in Bangladesh whose source is in the Himalayan mountains. The immense amount of sediment carried by these rivers, derived from intense erosion of the Himalayas (in the north), is delivered into the Bay of Bengal. (Data from the MODIS instrument on the Terra satellite.)

Many of the rocks that are exposed at the Earth's surface are sedimentary rocks, which were initially deposited as marine sediments (e.g., sands, silts, and muds) on the seafloor. Over time, rock-forming processes convert these sediments to rock (e.g., sandstone, shale, and limestone), that are then raised above the sea surface by sea-level variability or tectonic movements.

Remote sensing from space allows observation of sedimentary processes near the sea surface. Rivers carry large amounts of material derived from continental erosion to the river mouth and the open ocean, and the plumes of sediment delivered by rivers or tidal currents

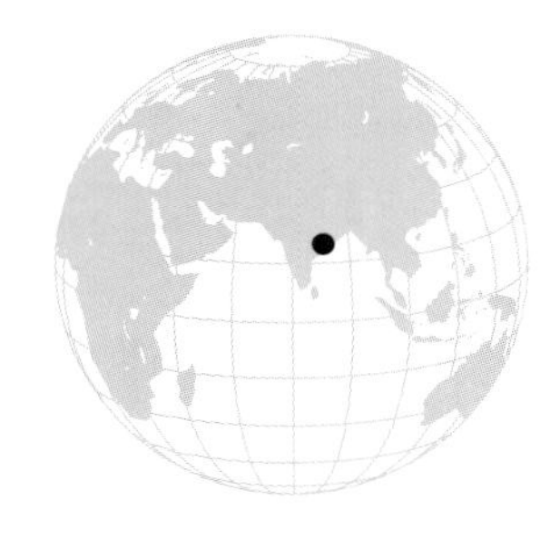

*Our Changing Planet*, ed. King, Parkinson, Partington and Williams

The Mississippi river delta, May 24, 2001. This is a classic example of a river delta created by the deposition of sediments carried by a large river. The low-lying land and numerous branching channels provide a fragile environment that is endangered by tropical storms, hurricanes, and sea level rise. (Data from the ASTER instrument on the Terra satellite.)

from estuaries are highly visible from space. The sediments formed by riverborne materials are termed clastic sediments, meaning that they are derived from existing rocks. The images of the Ganges and Brahmaputra river delta illustrate both the movement of these sediments and the massive amount of material that can be deposited by large rivers at river mouths or deltas.

The other primary types of marine sediment are chemical or biochemical sediments. These are formed on the sea floor by chemical action or by the action of marine organisms. Two familiar forms of rock that originated as chemical sediments are limestone, formed from the calcium carbonate shells of marine organisms, and salt, formed by the evaporation of seawater. Thick layers of salt found under the Mediterranean Sea indicate that this ocean basin repeatedly dried out and refilled in past geologic eras.

August 26, 2004

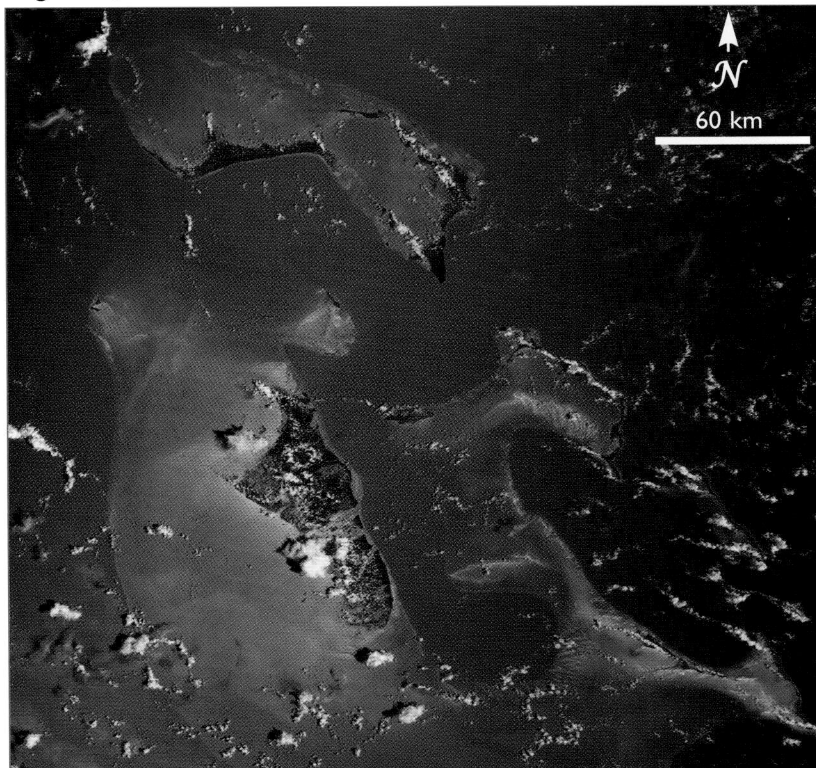


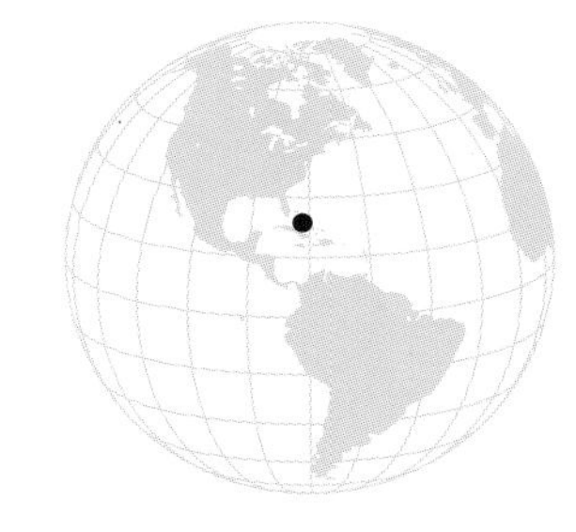

September 3, 2004

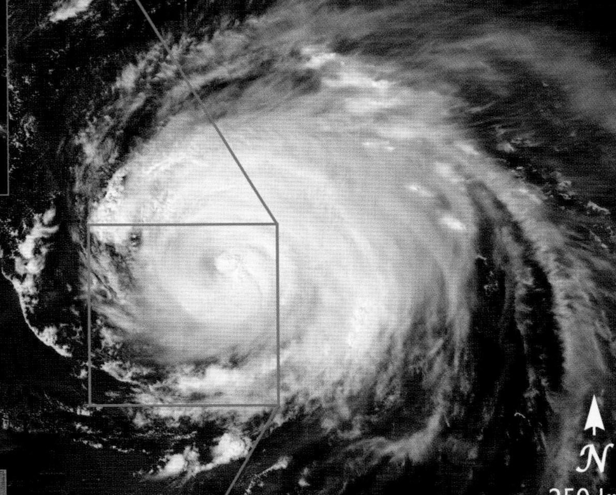


September 6, 2004

Carbonate platforms of the Bahamas Banks, before and after Hurricane Frances. Coral reefs and shallow carbonate platforms are the most productive regions for calcium carbonate in the ocean. Remote sensing can be used to observe the transport of these shallow water sediments to the deep ocean when the islands and banks are subjected to strong storm winds. The production of calcium carbonate and the dissolution of this mineral in the deep ocean are important processes in the marine carbon cycle. Increasing erosion rates and degradation of these tropical environments could be a consquence of sea level rise related to climate change. (Data for the Bahamas images from the MODIS instrument on the Terra satellite; data for the Hurricane Frances image from the MODIS instrument on the Aqua satellite.)

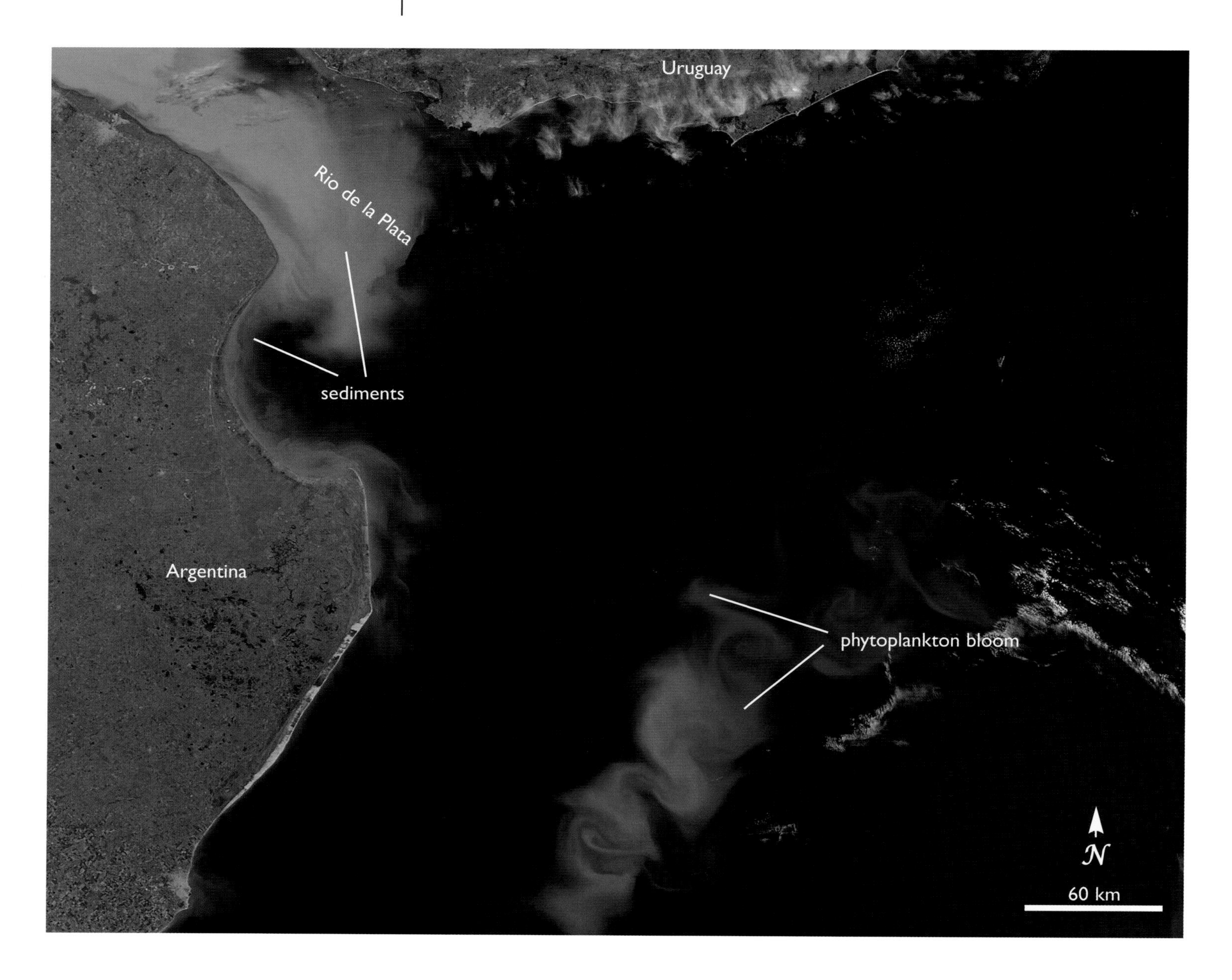


Phytoplankton bloom off the coast of Argentina on February 15, 2006. Light-colored sediments from the Rio de la Plata estuary and an offshore phytoplankton bloom (turquoise feature) are visible. Nutrients associated with river-borne sediments can encourage the growth of oceanic phytoplankton. (Data from the MODIS instrument on the Terra satellite.)

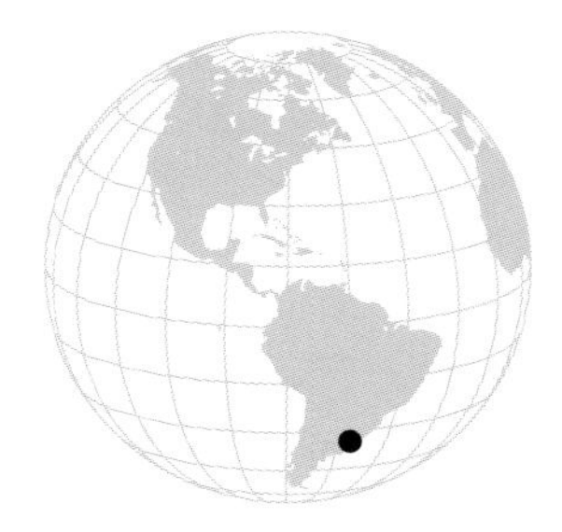

One of the main roles of remote sensing with regard to marine sediments is observation of sediment depositional patterns. Sediments can be carried long distances by ocean currents before eventually settling to the sea floor. Floods can provide additional energy that transports sediments into the open ocean. Hurricanes and storm winds can also move the biochemical sediments formed on coral reefs and shallow carbonate banks into deep ocean waters, where they will eventually dissolve. This process is an important facet of the marine carbon cycle.

Marine sediments are relatively easy to observe from space, but determining the mass of sediments is considerably more difficult. Sediments are highly variable in terms of particle size, particle reflectivity, and the absorption of light by other substances (such as dissolved organic matter) that may be associated with them. Quantifying suspended matter concentration in observable sediment features is the goal of current research.

Sediments are also important because they can reveal the pathways of movement of other substances in the marine environment. In particular, river sediments are commonly associated with nutrients that are utilized by phytoplankton in the ocean for growth. The delivery of nutrients by rivers can initiate enhanced phytoplankton growth in the coastal zone, which may be seen in the image off the east coast of South America. In this image, sediments from the Rio

A large muddy plume of sediment flows from the Mekong river delta into the South China Sea on October 24, 2006. The Monsoon season in southeast Asia causes annual flooding of the river, depositing fertile silt on the lower flood plain in Cambodia and Vietnam. (Photograph taken by astronauts aboard the International Space Station, photograph number ISS014-E-6242.)

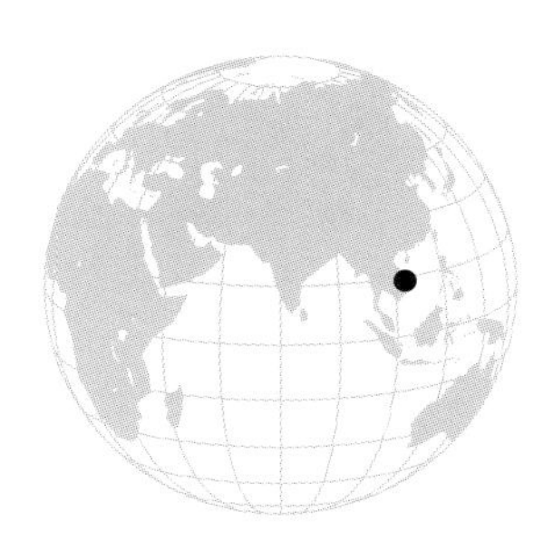

de la Plata estuary (which are rich in nutrients and organic matter) are flowing to the open ocean, and the offshore phytoplankton bloom may be related to increased nutrient availability.

River-borne sediments are vital for another reason, as the sediments of many rivers create fertile deltas (where they reach the sea) that are important regions both for agriculture and environmental diversity. Deltas in their natural state support diverse ecosystems due to the multitude of flora and fauna that utilize the fresh and brackish water environments of the delta. The rich soils of delta regions are very important for agricultural production, especially rice, in many countries. The fertile crescents of the Nile and Tigris-Euphrates deltas were important areas for the development of human civilization for this reason. However, deltas are highly changeable with regard to both natural and human impacts. Storms, particularly hurricanes and typhoons, can ravage delta regions with storm-driven surges. Human impacts include direct impacts such as the loss of sediments due to upstream reservoirs (a phenomenon occurring in the famous Nile River delta in Egypt), and erosion and subsidence due to sea level rise caused by global warming.

River sediments can also act as transport agents for substances that could be harmful to the oceanic environment. Both sediment particles and the organic matter that is produced in productive estuarine zones are very efficient at scavenging (adsorbing) metals that are dissolved in river waters. Many of these metals are highly toxic, in particular cadmium and mercury. By settling to the seafloor in deltas or at river mouths, the sediments can significantly reduce the metal concentration in the water and effectively bury it away from where it can be harmful to the environment. Natural erosion or dredging, however, can cause the metals in these sediments to be released into the ocean, with potentially harmful effects, especially in sensitive areas like coral reefs.

# Tsunamis

Remko Scharroo

This pair of images shows a stretch of the coastline of Thailand north of the island of Phuket. The Andaman Sea coast is a very popular tourist destination, studded with resorts and parks. The image on the left is a composite of 2 images taken on November 15, 2002, and February 28, 2003. The image on the right was taken on December 31, 2004, 5 days after the Indian Ocean Tsunami struck the area. Lush green low-lying areas with elevations less than 10 m have all been covered in a blanket of sand. The bowl-shape of the beach at the bottom of the image may have focused the energy of the tsunami wave and increased the run-up. The shallow beaches also contribute to larger waves. While rocky outcrop diffracts and breaks up the tsunami waves, a gentle sloping sea floor creates larger waves and longer run-out. (Data from the ASTER instrument on the Terra satellite.)

Composite Image from November 15, 2002 and February 28, 2003

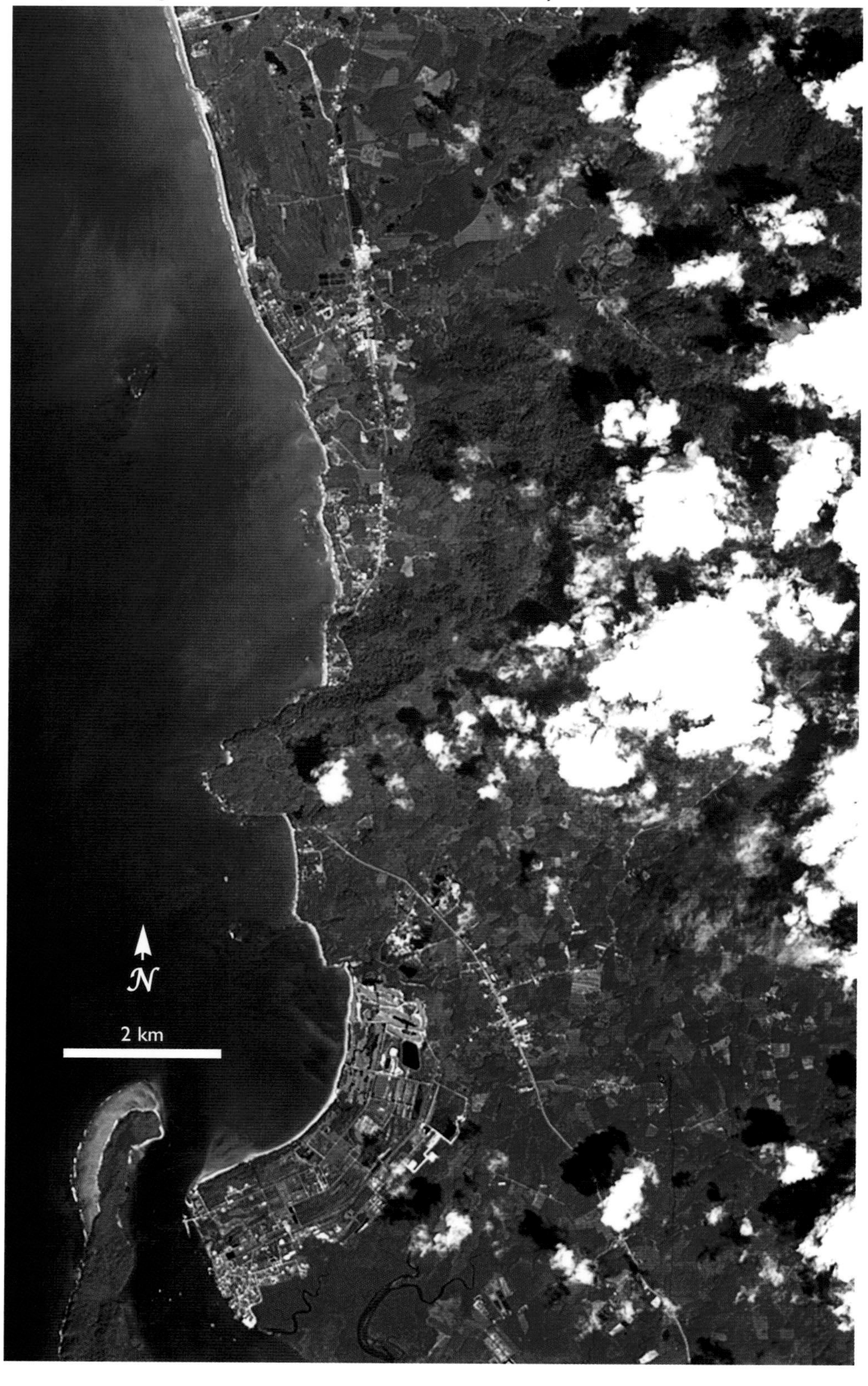


*Our Changing Planet*, ed. King, Parkinson, Partington and Williams
Published by Cambridge University Press 

December 31, 2004

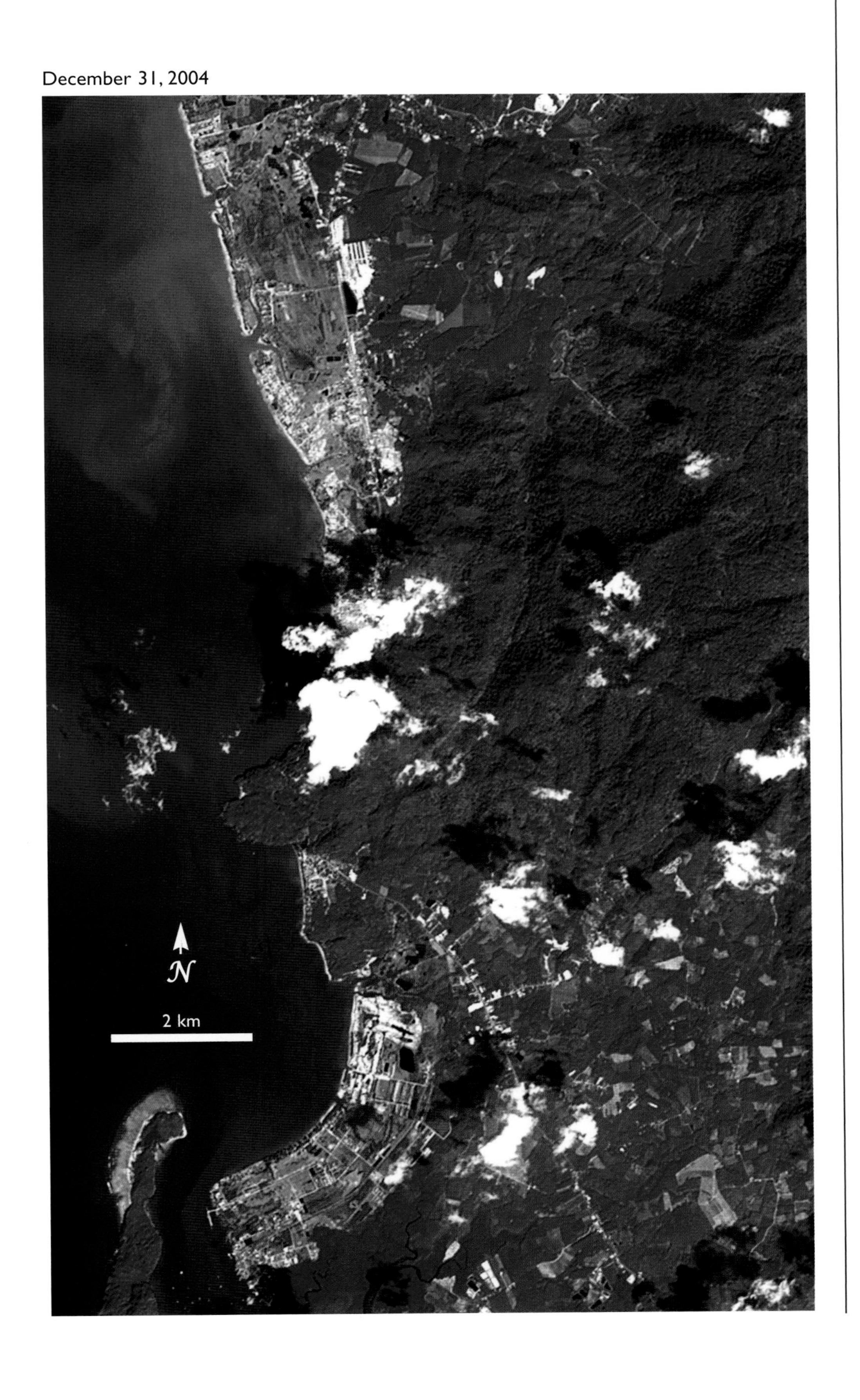


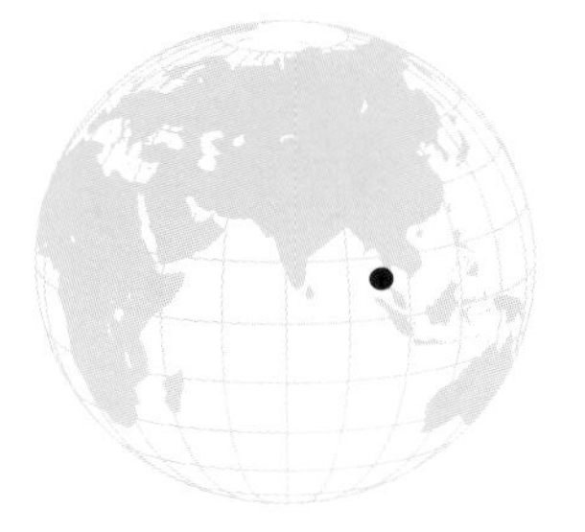

As we are changing our climate, and increasing the melting of glaciers and the massive ice shelves of Greenland and Antarctica, sea level keeps on rising. This jeopardizes the living condition of many of the estimated half billion people living within 10 km of the seacoast. Since global sea level rise is still relatively small—about 3 mm per year—coastal areas are not directly threatened by the rising sea level. However, any event on top of that will have a larger impact. Storms are more likely to destroy homes and coastal defenses and tsunamis will sweep further inland.

Of all the phenomena that have a devastating effect on coastal areas—sea level rise, hurricanes and tsunamis—tsunamis are not likely to increase in intensity as our climate warms up. Sea level will keep rising as the glaciers and the massive ice shelves of Greenland and Antarctica continue to melt faster than they accumulate snow. Hurricanes are expected to become more frequent and intense when the oceans contain more heat to transfer to tropical storms. Tsunamis, however, are not associated with climate change.

A tsunami can be generated when earthquakes, landslides, or volcanic eruptions abruptly deform the sea floor. The magnitude of the earthquake and the faulting mechanism responsible for the earthquake are important factors in determining the height of the tsunami. Along the continental margin, the oceanic tectonic plate drops below the continental plate. This slow process of subduction often takes the lip of the continental plate with it until stresses build up too large and the lip suddenly and violently snaps back, causing an earthquake and pushing up the water column. Undersea landslides and volcanic eruptions have a similar effect. Finally, a tsunami can occur when a meteor falls into the ocean, sending ripples across the globe.

What makes tsunamis particularly devastating is that they can barely be predicted. Research of plate tectonics has not yet developed a means to predict when earthquakes will

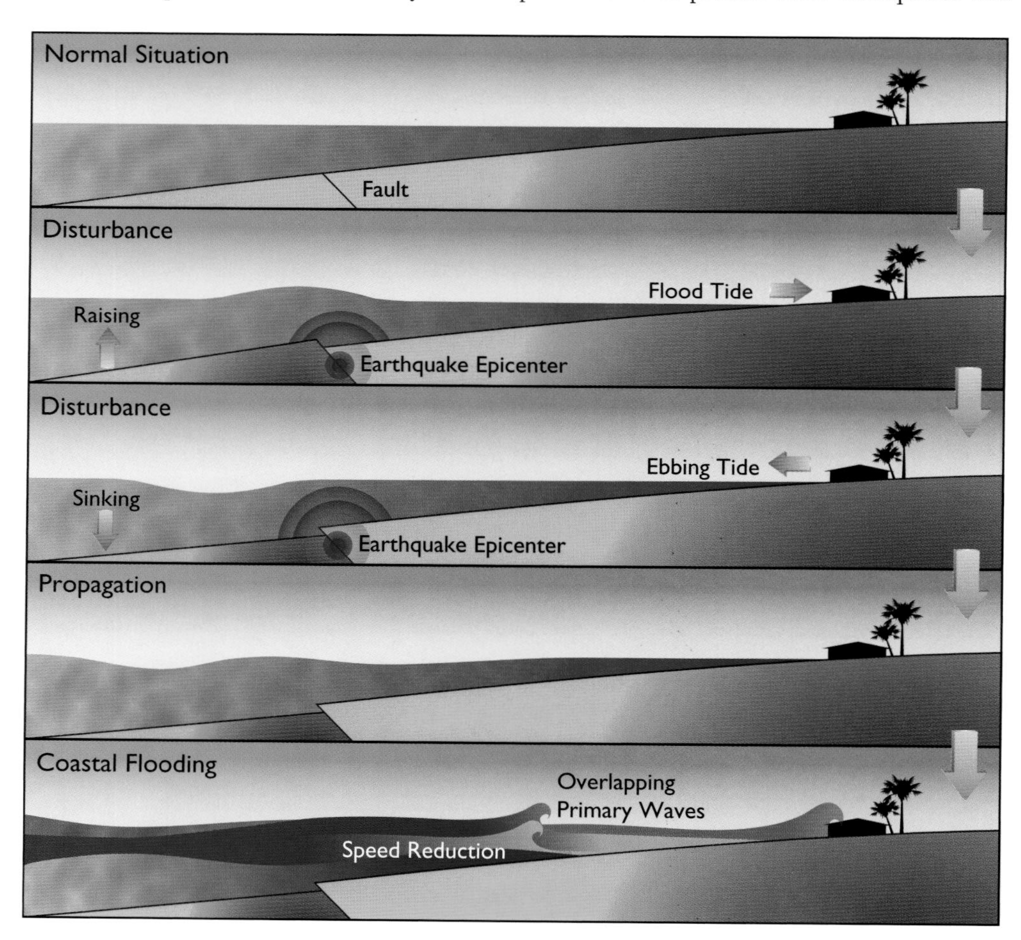


A tsunami can be generated by a sudden movement of the sea floor. In this illustration the oceanic plate rises and falls during a sudden release of tectonic motion (earthquake), creating a long wave train traveling outward from the epicenter. Ahead of the tsunami the coast will observe a small flood and ebbing tide. As the speed of propagation of the tsunami wave reduces as it nears land, the wave increases in height.

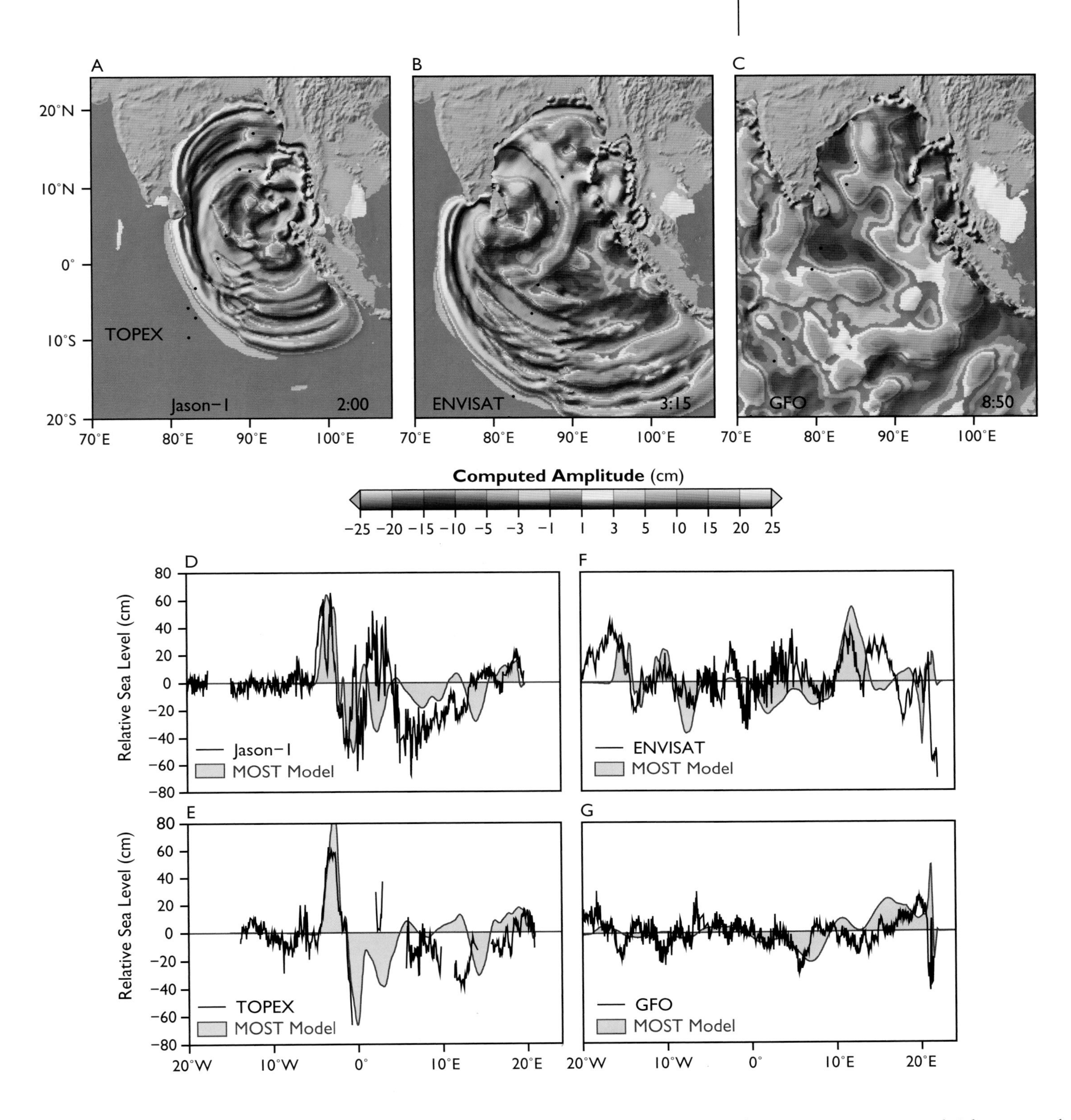


(A to C) Tsunami wave heights generated by the December 26, 2004 Sumatra earthquake, computed by the MOST model at (A) 2:00 hours, (B) 3:15 hours, and (C) 8:50 hours after the earthquake. These times coincide with the overflights of the satellites Jason-1 and TOPEX Poseidon, Envisat, and GFO, respectively. (D to G) Comparison between the sea height deviations as measured by the satellite altimeters and the modeled wave heights.

occur, or even how large they will be. The only warning time allotted to us is the travel time of the wave from the location of the earthquake to the coast. Since earthquakes generally occur at the coastal margins, this can be a very short time. The speed at which the tsunami wave propagates is equal to the square root of the product of the gravitational acceleration (9.8 m/s$^2$) and the water depth. In deep water (say 4,000 m depth or more) the wave travels 200 m/s or faster. At a depth of 40 m, the wave will have slowed down to only 20 m/s.

In the deep, ocean tsunami waves have a very long wavelength of tens of kilometers and a height of less than 1 m. This is completely different from the classical wind-driven waves or even tidal waves rolling in onto the seashore. The energy encapsulated in a tsunami wave is

N
154 m

immense and virtually unstoppable, wiping everything away in its path. As the wave reaches the coast and slows down, the kilometers-long wave gets bunched up, and—keeping the same energy—rises proportionally, so that a 1 m wave at open sea can become 10 m high when reaching the shore.

Geography also has a significant impact on the eventual height of the tsunami. Coasts sheltered by islands are generally well protected from tsunamis. Estuaries and harbors generally focus the waves so that even the smallest tsunami waves can create considerable damage. Hence the meaning of the Japanese word 'tsunami': 'harbor wave.' Although tsunamis cannot be prevented, measures can be taken to alleviate the risk to the population in these tsunami-prone areas, like building floodwalls or installing early warning systems.

On December 26, 2004 at 00:58:53 UTC a magnitude 9.3 earthquake with an epicenter about 160 km west of Sumatra, Indonesia, shook the neighboring islands and was felt as far as the Maldives and India. But much more damaging was the tsunami that ensued, eventually killing about 230,000 people and displacing many more.

Four satellites were able to observe the tsunami wave traveling across the Indian Ocean. These satellites were all equipped with a radar altimeter. Satellite radar altimeters are all-weather instruments intended to measure sea-surface height and wave heights along a narrow track about 2 km wide. In contrast to imaging sensors, they determine only profiles along this narrow track, but can do so very accurately (see illustration). After correction for ocean tides, ocean currents, and wind-driven waves, these altimeters showed for the first time a convincing cross-section of the tsunami wave trains. The altimeters on board TOPEX/Poseidon and Jason-1 spotted the highest wave front (60 cm), traveling in a southwesterly direction across the Indian Ocean. In only 2 hr the wave front had already traveled about 2,000 km. Even 6 hr later, the GFO satellite showed that the ocean was still riddled with waves of a few to hundreds of kilometers long. Some of those had by then reflected against the Indian, Bangladeshi, and Thailand coasts, making their way back across the ocean.

The direct observations of the tsunami wave by satellite altimeters were invaluable in assessing the accuracy of the MOST model run by the National Oceanic and Atmospheric Administration (NOAA). This model is meant to forecast the height of the tsunami wave anytime a submarine earthquake occurs. Bottom pressure gauges and buoys further aid the advance warning for tsunamis. It is key to determine where a tsunami will strike and how high the wave will be before the tsunami makes landfall so that the coastal population can seek higher ground. Accuracy is important: false warnings can be as damaging as missed warnings, since unnecessary evacuations come at significant cost and will cause the population to ignore future warnings.

The devastation after the Indian Ocean Tsunami was made clear in satellite images of the particularly hard-hit Banda-Aceh region of Sumatra (Indonesia) and the island of Phuket (Thailand). A grey coat of sand deposits replaced low-lying fields, roads and villages. Although it is nearly impossible to prevent tsunamis from damaging land and homes, researchers continue to study the information from satellites and other sources to improve tsunami forecasts. Better tsunami modeling and improved warning systems will be able to save numerous lives.

The town of Meulaboh in the Aceh Province in Indonesia is located only 150 km from the epicenter of the earthquake of December 26, 2004. This image made by the QuickBird satellite on January 7, 2005 shows the severe damage done to the west coast of Sumatra. The inset image was taken May 18, 2004. More than three quarters of the buildings are completely wiped away and others are heavily damaged. Bridges connecting the main island with the smaller islands are destroyed and the pier near the tip of the land jutting out into the ocean has collapsed. Beaches are eroded beyond recognition and most trees along the shore have been uprooted. (Image copyright DigitalGlobe.)

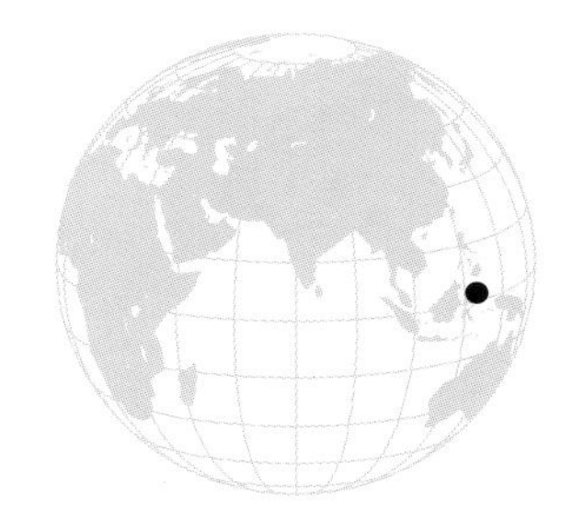

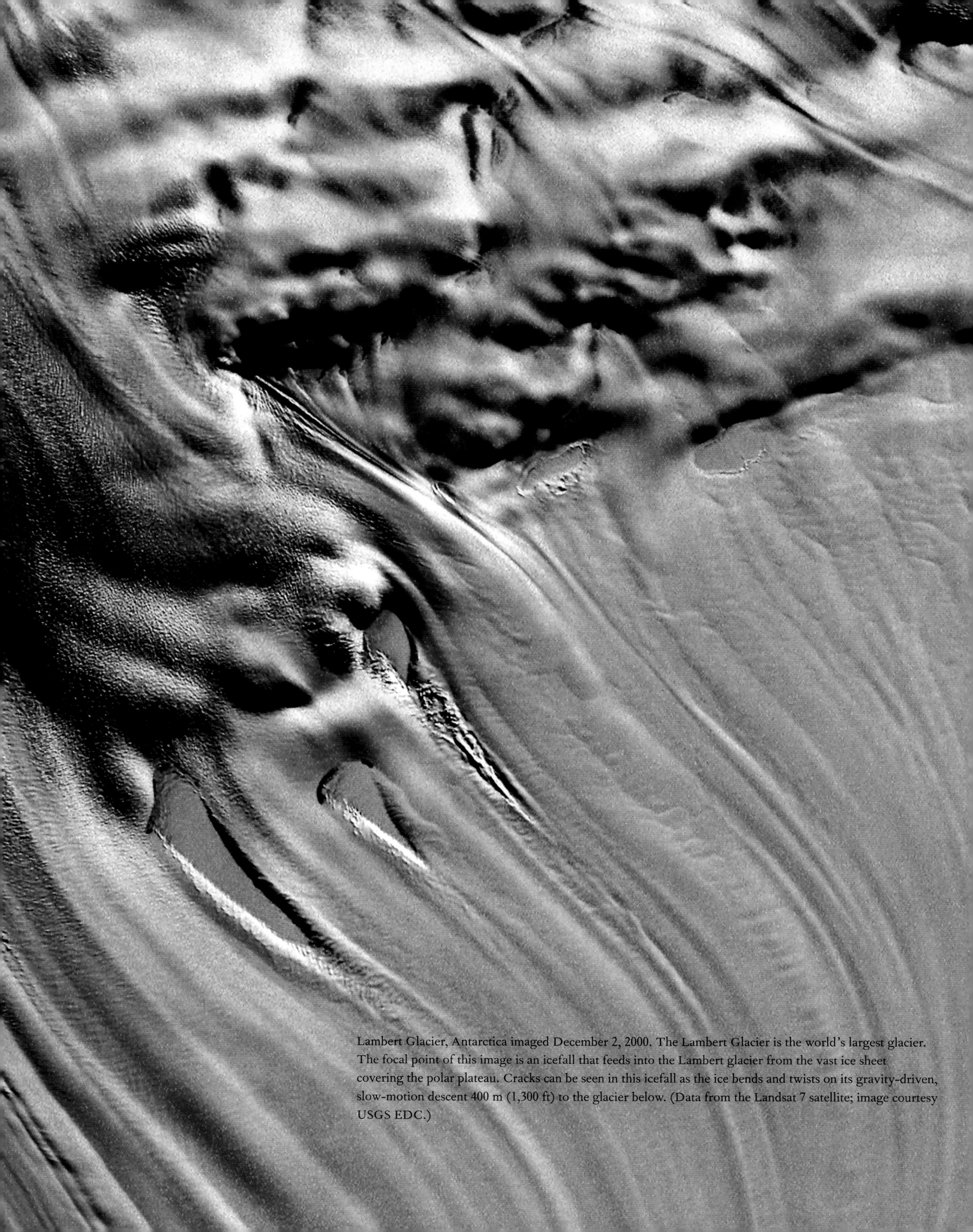

Lambert Glacier, Antarctica imaged December 2, 2000. The Lambert Glacier is the world's largest glacier. The focal point of this image is an icefall that feeds into the Lambert glacier from the vast ice sheet covering the polar plateau. Cracks can be seen in this icefall as the ice bends and twists on its gravity-driven, slow-motion descent 400 m (1,300 ft) to the glacier below. (Data from the Landsat 7 satellite; image courtesy USGS EDC.)

250 km

# The Frozen Caps: Introduction

Kim C. Partington

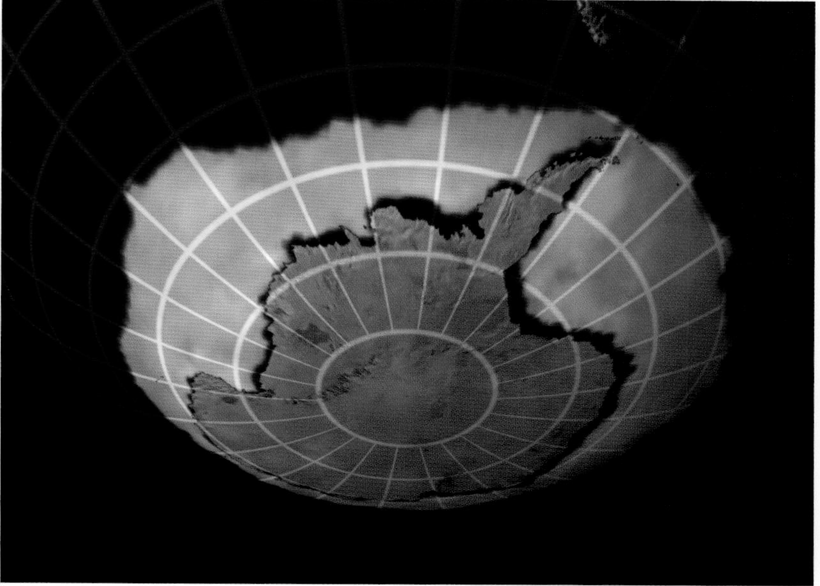

Less familiar views of the Earth from above the North Pole and South Pole respectively. Global sea ice coverage averages approximately 25 million km$^2$, the area of the North American continent, whereas ice sheets and glaciers cover approximately 15 million km$^2$, roughly 10% of the Earth's land surface area. (Data from the Special Sensor Microwave Imager (SSMI) instrument on a Defense Meteorological Satellite Program (DMSP) satellite.)

The cold extremities of our planet remain the last great frontier on the surface of the Earth and have, in comparatively modern times, provided the stage for truly heroic feats of exploration. Today, the interest in the Arctic and Antarctic extends to the scientific community, many of whom consider these regions to have a degree of influence on the Earth's population that belies their remoteness, notably through sea level and global ocean circulation.

To some extent, the two poles are mirror images of each other. The north is an ice-covered ocean (the Arctic) surrounded by bodies of grounded ice, with one massive ice sheet on Greenland, the World's largest island, and many smaller ice caps and glaciers. The south, on the other hand, is a frozen continent, Antarctica, surrounded by the ice-covered Southern Ocean.

Early interest in these cold regions was fostered by the search for new lands. Starting in the eighteenth century, national and commercial interests spurred more competitive incursions into the polar regions, notably to search for the fabled Northwest Passage and, later in the south, to develop a burgeoning whaling industry. The age of heroic exploration came to a head just after the turn of the twentieth century with the race for the poles. The North Pole was reached by Robert Peary in 1909 and the South Pole was reached overland by Roald Amundsen the following year. Later, the status of the two polar regions diverged when, in 1961, the Antarctic became subject to an international treaty protecting the continent from

Opposite: Greenland ice sheet, July 7, 2002. (Data from the MODIS instrument on the Terra satellite.)

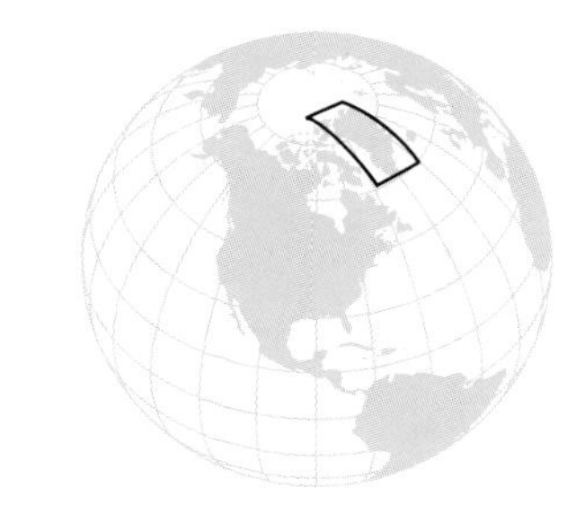

*Our Changing Planet*, ed. King, Parkinson, Partington and Williams
Published by Cambridge University Press 

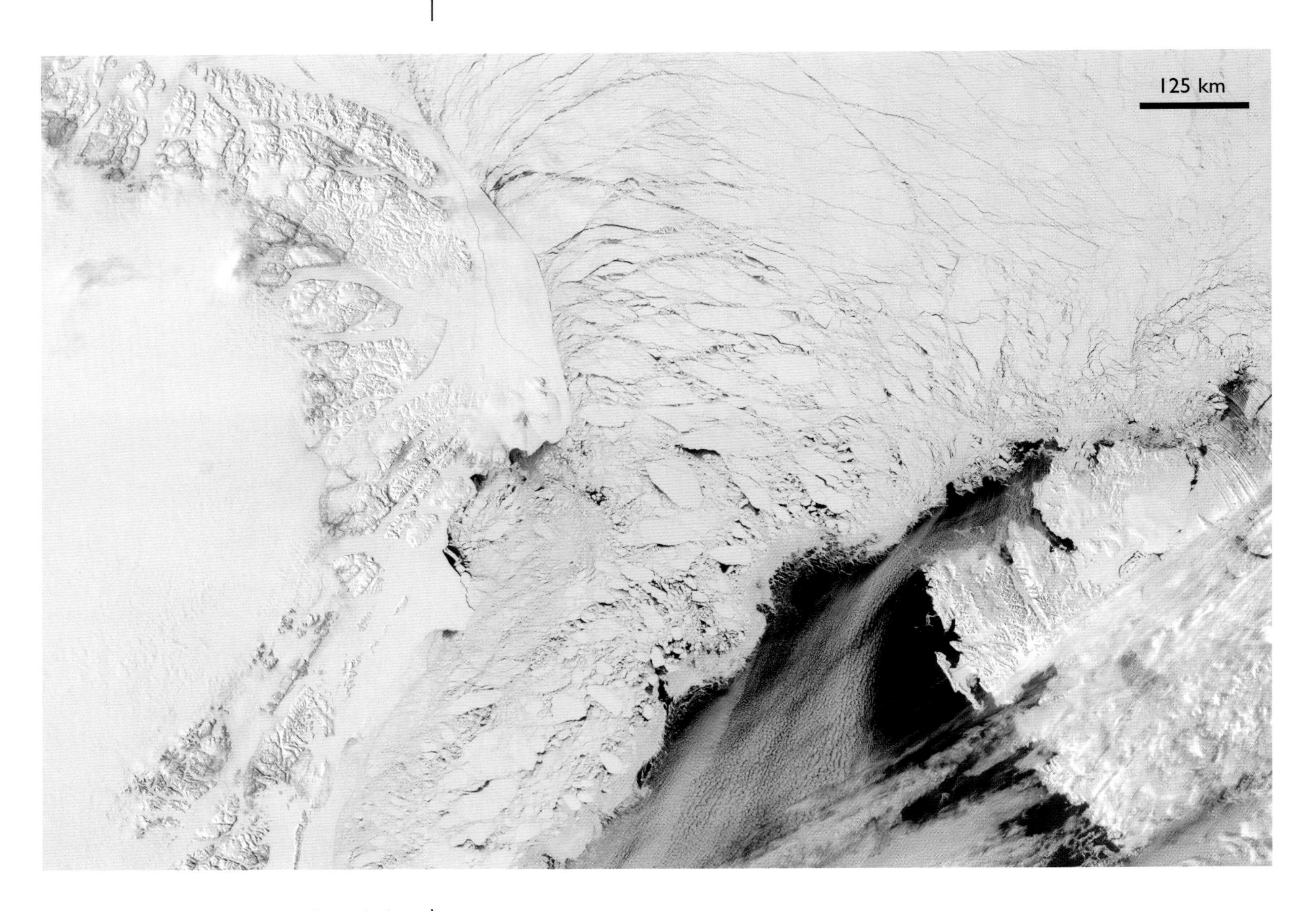


Sea ice exiting the Arctic Ocean through the Fram Strait, between Greenland (left) and Svalbard, May 3, 2002. This export of ice into the Atlantic Ocean is equivalent in magnitude of fresh water to that of a river larger in size than any on Earth with the exception of the Amazon. Ice export through the Fram Strait has an important impact on oceanography in the north Atlantic. (Data from the MODIS instrument on the Terra satellite.)

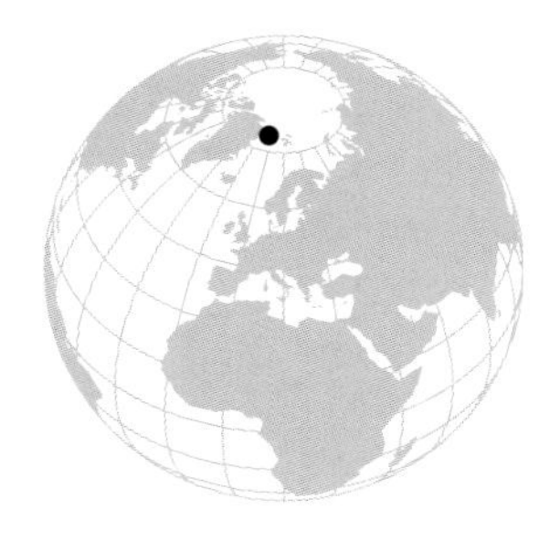

exploitation, while in the north, the Arctic became a frontier of the cold war. Early warning stations were established along the margins of the Arctic and nuclear submarines patrolled under the ice. These activities in the north galvanized the first systematic observations of the ice, providing a means to ensure that submarines could surface between the ice when needed, and supporting resupply of remote stations which served to exercise sovereignty. In the Antarctic, the scientifically driven International Polar Year of 1957–1958 spurred a number of research programs that involved the establishment of nationally supported research stations around the Antarctic and extensive aircraft surveys.

Although these early scientific expeditions were invaluable in providing such important information as the depth of ice in Antarctica and details of the coastline, airborne and ground-based expeditions are extremely expensive as well as limited in coverage. Truly extensive, and comprehensive mapping awaited the arrival of satellites in the final decades of the twentieth century, and some of the achievements of satellites are described in the following chapters.

Microwave sensors combine the ability of radar to view the ice through the long polar night and through thick cloud and these features have enabled the first continent-wide, detailed image of Antarctica to be produced using the Canadian RADARSAT-1 satellite, as described in "Antarctica: A Continent Revealed." This mapping revealed many features including megadunes of snow and lakes isolated as much as 3 km beneath the surface

View along the edge of the B-15A Iceberg. Icebergs transport ice from the ice sheets into the ocean and can circulate for years in the Southern Ocean. B-15 formed from the Ross Ice Shelf in March 2000 with dimensions of nearly 300 km in length and 40 km in width. Northern Hemisphere icebergs are significantly smaller, but can be deadly, one sinking the Titanic in the western Atlantic at 43°N, well into mid latitudes. Satellites are used to track the larger icebergs. (Photograph courtesy Josh Landis, National Science Foundation.)

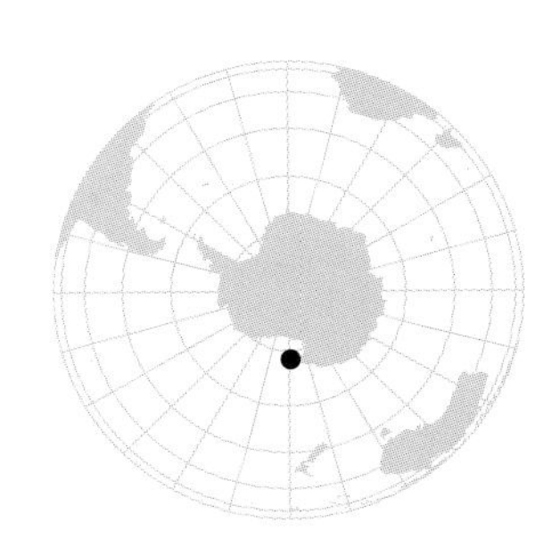

of the ice sheet. Ice sheets are not static features, but flow slowly under the influence of gravity. It is this slow motion of the ice that enables scientists to use ice cores to reveal information about the composition of the atmosphere hundreds of thousands of years ago, when the ice originally formed near the surface of the ice sheet, trapping minute quantities of air. The slow progression of the ice from the interior of the ice sheet to the ocean was revealed by the same radar that was used to map the continent some 3 years previously, as described in "Ice Sheets on the Move." This mapping of the motion of the ice reveals tentacles of fast moving ice streams extending hundreds of kilometers into the interior of the ice sheets, acting much like rivers on land to collect and transmit the ice to the ocean where it is delivered in the form of icebergs. The ultimate question for many scientists relates not to the current behavior of the ice sheets, but to whether they can respond catastrophically to changes in climate, and in particular to changes in climate related to current projected increases in greenhouse gases. The chapter on "Ice Sheets and the Threat to Global Sea Level" describes how terrestrial ice sheets have responded abruptly to changes in climate in the past, and suggests that Greenland and the West Antarctic ice sheets are, for different reasons, each likely to be sensitive to warming temperatures. The impact on global sea level of any changes could be very serious given that at least 39% of the world population lives within 100 km of the coast.

The high-latitude oceans are also covered by ice, which is very different to the freshwater ice anchored on land or, as ice shelves, to the coast. Because it floats, it does not impact on sea level when it melts, but because it is floating on a relatively warm ocean underneath a relatively cold atmosphere, in some senses it acts rather like a lid on a pan of boiling water. When some of the ice melts, huge quantities of moisture and heat escape into the atmosphere. Furthermore, because the ice is highly reflective to the Sun's incoming radiation, the Earth is a more efficient absorber of the Sun's energy when there is less sea ice. For these reasons, the coverage of sea ice is considered to be important as a stabilizing influence on the Earth's climate. The chapter entitled "The Great White Ocean: The Arctic's Changing Sea Ice Cover," shows that Arctic sea ice has been declining in coverage for the last quarter century with implications not only for wildlife and hunter-gatherer populations fringing the Arctic, but also for the rest of the Earth's population as a possible sign of instability in the climate system. This wider significance of sea ice to the ocean and atmosphere is explored further both for the case of the Northern Hemisphere, in "Bound Together: Arctic Sea Ice, Climate and Atmosphere," and, at the other end of the planet, for the case of the Southern Hemisphere in "Sea Ice: The Shifting Crust of the Southern Ocean." Finally, in "Antarctic Polynyas: Ventilation, Bottom Water, and High Productivity for the World's Oceans," focuses on some of the most isolated, yet biologically most productive environments found on Earth, often referred to as icy equivalents to desert oases. The extent of polynyas was revealed dramatically in some of the earliest satellite images, and they remain of interest today as scientists assess the impacts of changes taking place in the polar regions.

These chapters show how satellites have been able to demonstrate that the frozen caps are far from being static, remote, and inconsequential regions at the ends of the Earth. The brief history of satellite observations of the polar regions has in many respects been a highly accelerated program of basic mapping, discovering how the ice typically grows, melts, and moves, and establishing whether the frozen caps are changing systematically, with implications for mankind, all within the space of 30 years. New insights will be gained from data collected during the 2007–2009 International Polar Year of which satellite observations form a key component.

# Antarctica: A Continent Revealed

Kenneth C. Jezek

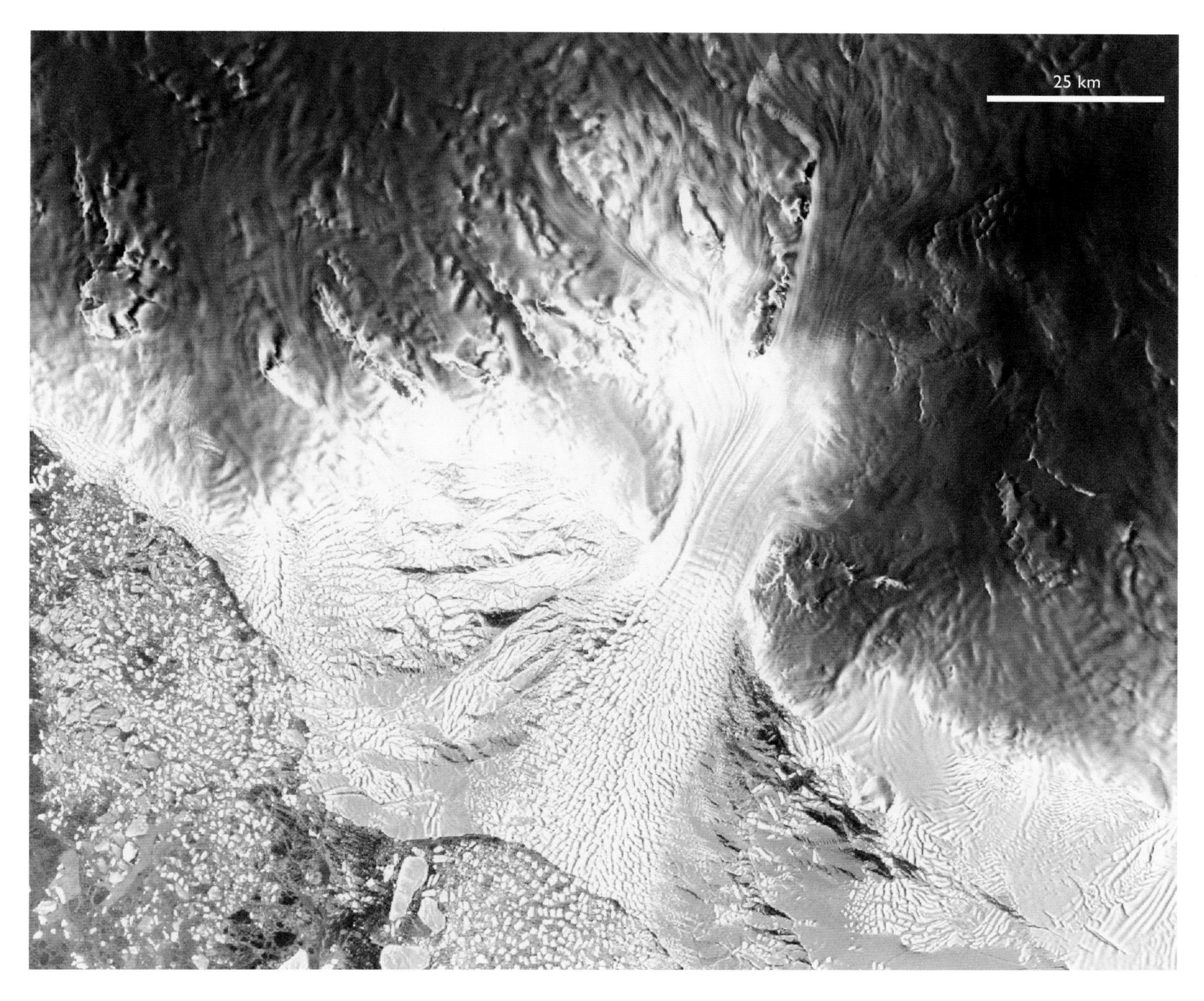


Land Glacier (center) and the complex of inland ice (top), ice tongues (seaward extent of Land Glacier), icebergs, sea ice, and ocean taken in 1997. (Data from the SAR instrument on the RADARSAT-1 satellite.)

The existence of a great southern continent was a matter of conjecture until the latter part of the eighteenth century when Captain James Cook set out in 1772 on the first circumnavigation of the icy waters of the south. Cook concluded that there was no hospitable, southern landmass, even going so far as to speculate that the impenetrable sea ice he encountered reached to the pole. Another 49 years would pass before the Russian explorer, Thaddeus Bellingshausen on the ship Vostok, would sight mountain peaks in the Antarctic Peninsula, finally proving that there was land within the Antarctic Circle. The news of a southern land mass kindled the interests of nineteenth and twentieth century explorers and scientists who asked how big is

*Our Changing Planet*, ed. King, Parkinson, Partington and Williams
Published by Cambridge University Press 

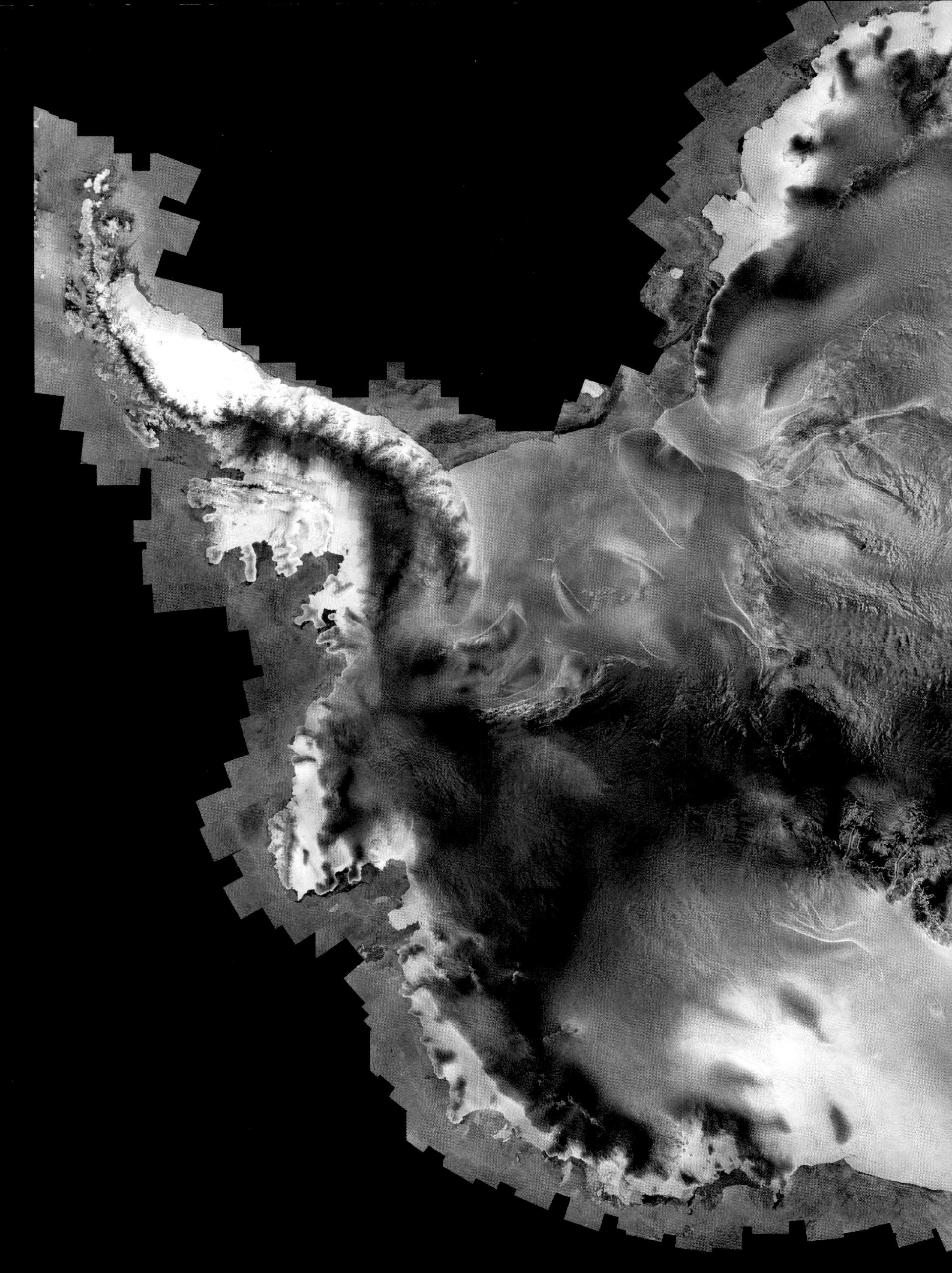

1997 RADARSAT-1 mosaic of Antarctica. The bright regions ringing the continent indicate locations where seasonal summer melting occurs on the ice sheet. (Courtesy National Snow and Ice Data Center.)

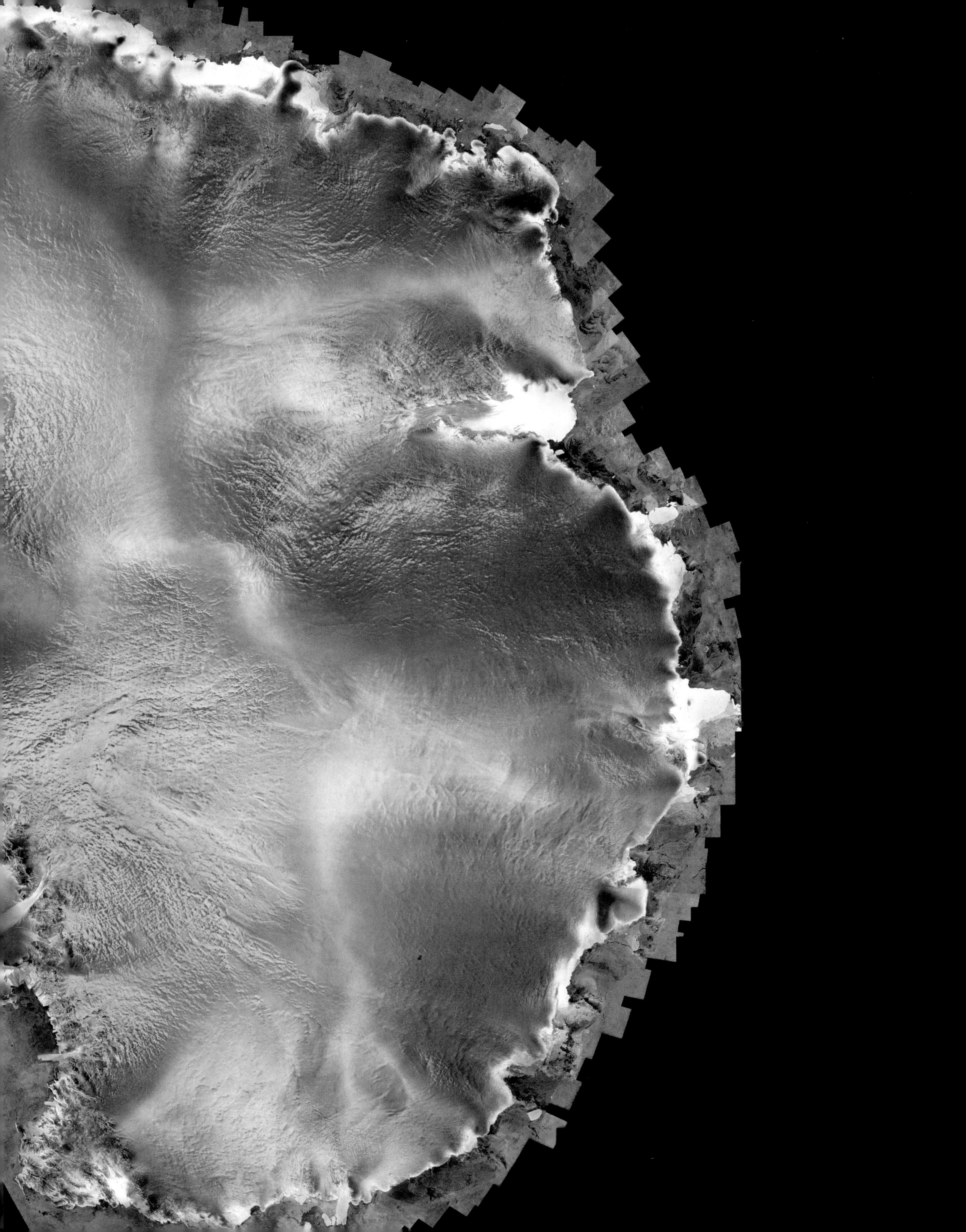

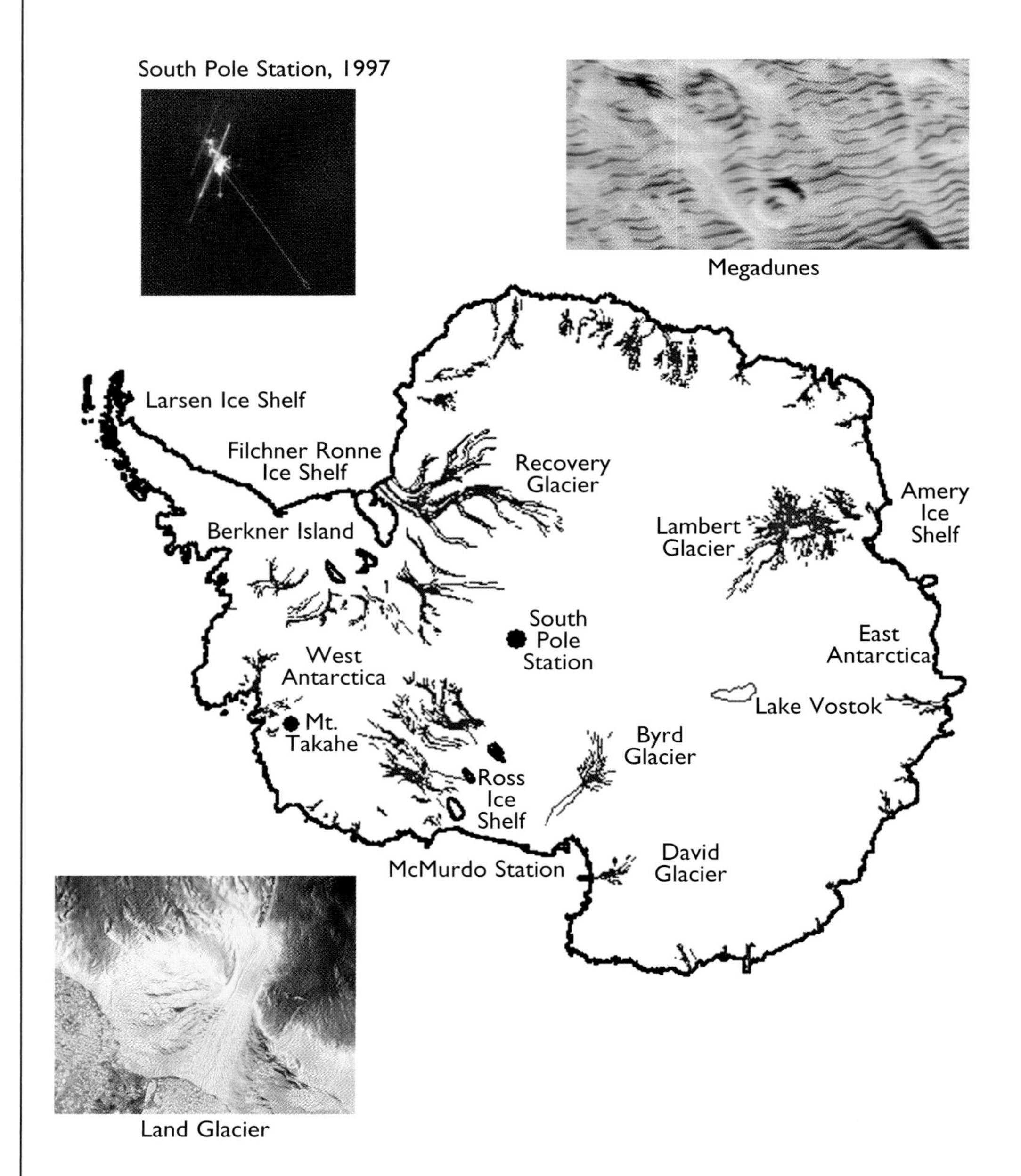


Map of Antarctica showing the locations of major geographic features, for comparison to the RADARSAT-1 image mosaic.

Insets: (upper left) An aircraft runway is revealed by RADARSAT-1 in this image segment from the South Pole, where the United States maintains a research station. (Data from the SAR sensor on the RADARSAT-1 satellite.)

(upper right) Snow 'megadunes.' These features, which remained unobserved until the RADARSAT-1 mapping in 1997, are analogous to dunes found in sandy deserts and are formed by wind. Here, the dunes are some 2–5 km apart and approximately 2–4 m in height. (Data from the SAR sensor on the RADARSAT-1 satellite.)

Antarctica? How much of the continent is ice-covered? What is the location of the coastline and the shape of the continent? These simply stated questions spurred the earliest explorers and provide similar motivations for the scientists of today. But finding answers has proved a daunting task that in large measure has had to await the era of satellites.

An important milestone in the spaceborne exploration of Antarctica was the dedicated mapping of the continent carried out by an international team of scientists and engineers led by NASA and the Canadian Space Agency, in 1997. Investigators used the Synthetic Aperture Radar (SAR) which is capable of operating day and night and 'sees' through cloud cover. The mission was risky and involved turning the satellite around 180° so that it could see all the way to the South Pole. The mission extended over 18 intensive days of November 1997, and amounted to a major feat of exploration using modern technology. The project created the first, detailed map of Antarctica and formed the final piece of the jigsaw in radar mapping of Earth by satellites.

On the RADARSAT-1 image (previous page), the bright ring around the edge of the continent shows where the ice melted during the previous austral summer (November being late spring in the Southern Hemisphere). Elsewhere in the image can be seen all sorts

Photograph of an Antarctic megadune field as viewed from an airplane. The white areas are areas with high surface roughness due to snow redistribution, and the dark areas are areas of extremely low or no accumulation which have a glazed surface much of the time. Photo taken December 2002. (Photograph courtesy Mary R. Albert, Cold Regions Research and Engineering Laboratory.)

of strange and intriguing shapes. Buried lakes, some 3 km beneath the surface of the ice, are indicated by extraordinary flat regions of ice. The South Pole is identified by the aircraft runway abutting the manned station. Megadunes—snow equivalents to the dunes in the Sahara—extend hundreds of kilometers across the interior of the ice sheet and are revealed by their crinkly pattern. Pinnacles of rock extend just above the ice in the largely buried Trans-Antarctic Mountains while just inland from many of these peaks, areas of snow-free or 'blue' ice can be seen, a source of ancient meteorites brought to the surface by the movement of the ice and scouring by strong winds. Scientists are still exploring the implications of these findings from 1997, and the map will provide a reference to them for many years to come.

Captain Ashley McKinley holding the first aerial surveying camera used in Antarctica. It was mounted in the aircraft 'Floyd Bennett' during Byrd's historic flight to the pole in 1929. (Photograph from The Ohio State University Archives, Papers of Admiral Richard E. Byrd, Box 232, Byrd Antarctic Expedition 1928–1930 photo album.)

Eric Rignot
Ian Joughin

# Ice Sheets on the Move

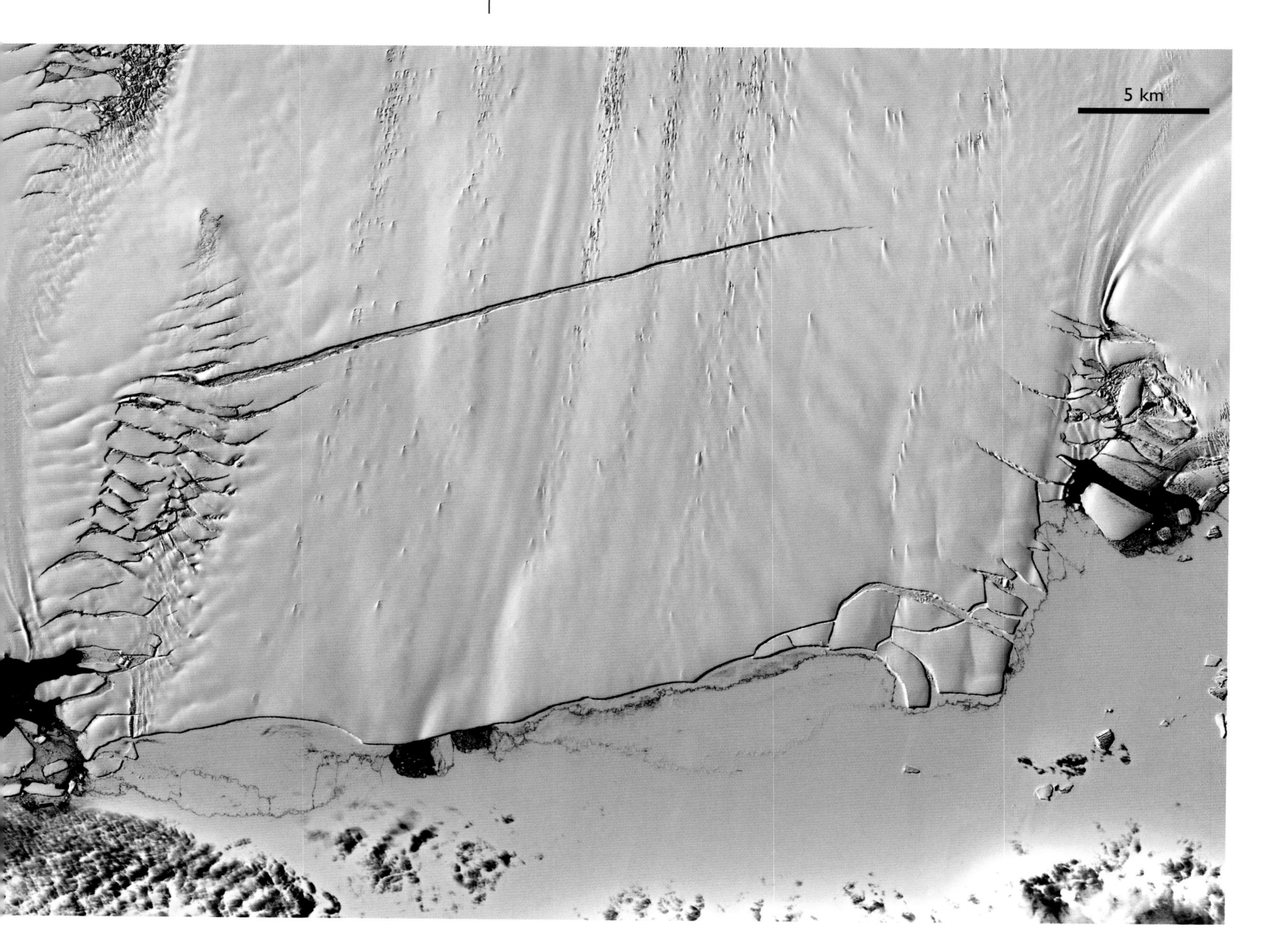


The large Pine Island Glacier in the process of shedding itself of an iceberg over 40 km long, December 12, 2000. This glacier produces an iceberg of this size about every 3 years. The truly massive icebergs emanate from Antarctica and in extreme cases can cover areas as large as Belgium or New Jersey. (Data from ASTER on the Terra satellite courtesy of NASA/GSFC/METI/ERSDAC/JAROS and US/Japan ASTER Science Team.)

It was realized a long time ago that the ice sheets are not static masses of ice, but in fact move under their own weight, which means that the ice must periodically break off into the ocean. This was indicated to many seafarers including the ill-fated passengers on the Titanic, which was hit by an iceberg that probably originated from the Greenland ice sheet thousands of kilometers to the north. In fact the whole ice sheet is in a constant state of movement. But how does the ice move? Does it slide like a conveyor, or does it creep like treacle? The answer to this question is important because it can indicate how quickly the ice sheet can change. A rapid reduction in the mass of the ice can mean a major increase in sea level.

*Our Changing Planet*, ed. King, Parkinson, Partington and Williams
Published by Cambridge University Press 

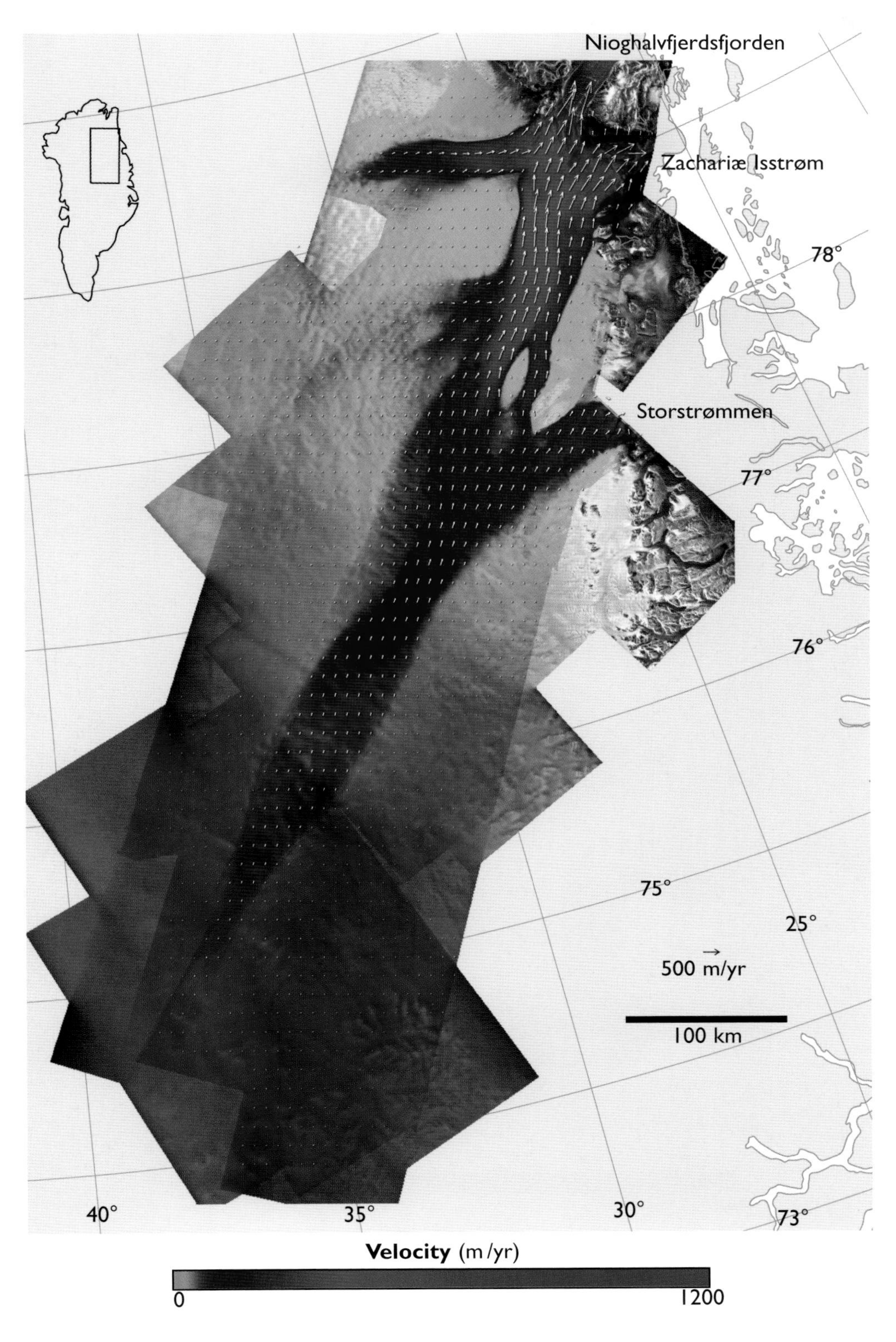


Northeast Greenland ice stream showing ice velocities collected over the period 1991–1996. Although this ice stream was only recently discovered using ERS imagery, it is now one of the most completely mapped ice streams in terms of its velocity field. Vector velocity measurements were made using crossing orbits from ascending and descending passes. Speed is displayed as color over SAR imagery. Sub-sampled velocity vectors are shown with white arrows. (Data from the SAR instrument on the European Space Agency's ERS-1 and ERS-2 satellites.)

An early discovery was that the ice moves both by creep and by sliding, creeping under its own weight over most of the ice sheet, but in some areas, known as ice streams, sliding over its bed. One of the most invaluable roles of satellites has been to reveal the extent of these 'rivers of ice.' These ice streams reach hundreds of kilometers inland and in many cases create sufficient momentum to form huge floating shelves of ice, up to 1 km thick, attached to the continent but floating hundreds of kilometers across the ocean. The Ross ice shelf, for example, covers an area the size of half a million square kilometers, approximately the size of Spain. These ice shelves form the breeding grounds for most of the truly massive icebergs.

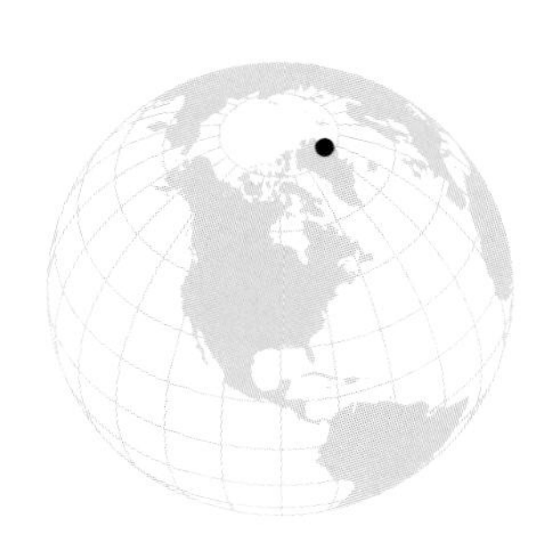

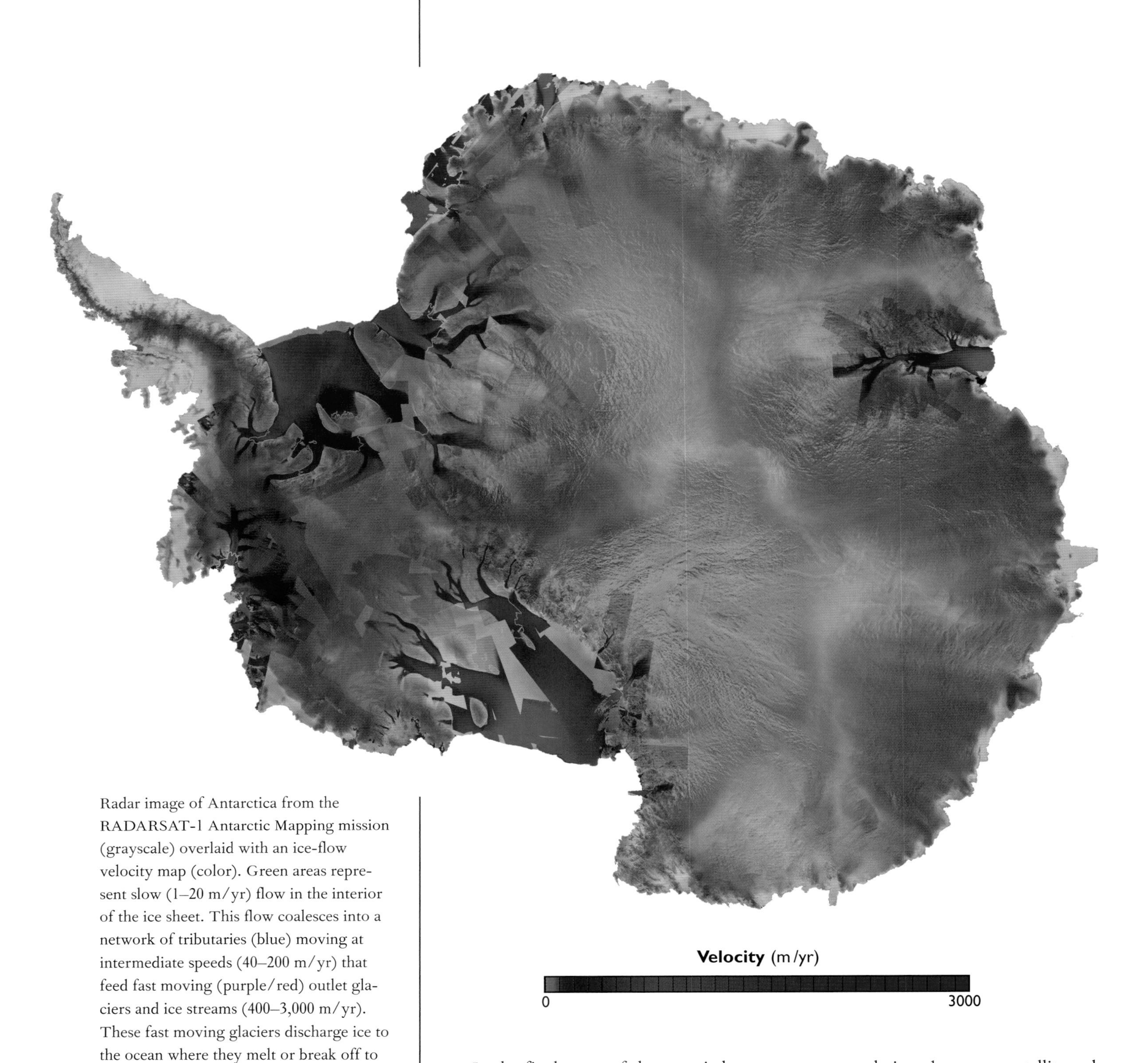


Radar image of Antarctica from the RADARSAT-1 Antarctic Mapping mission (grayscale) overlaid with an ice-flow velocity map (color). Green areas represent slow (1–20 m/yr) flow in the interior of the ice sheet. This flow coalesces into a network of tributaries (blue) moving at intermediate speeds (40–200 m/yr) that feed fast moving (purple/red) outlet glaciers and ice streams (400–3,000 m/yr). These fast moving glaciers discharge ice to the ocean where they melt or break off to form icebergs. (Data from the SAR instrument on the RADARSAT and ERS-1 and ERS-2 satellites.)

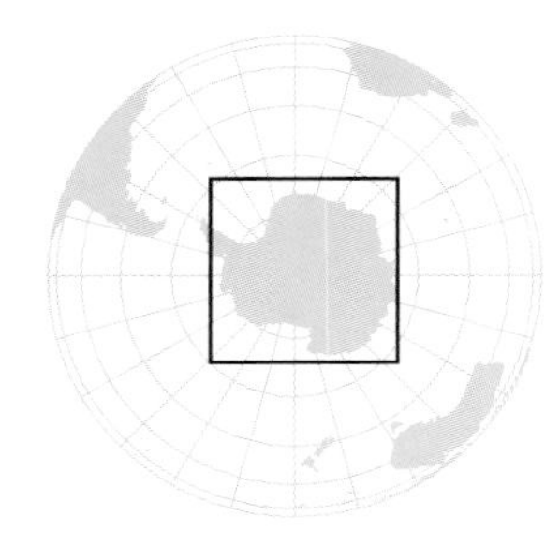

In the final years of the twentieth century a new technique known as satellite radar interferometry became available with which to map the motion of the ice and so, in 2000, NASA and the Canadian Space Agency combined forces to use radar to map the continent for a second time, but this time to map the extent of these 'rivers of ice.' SAR interferometry allowed the movement of the ice to be mapped, even though the ice is apparently featureless over large areas. In the result, the white continent is transformed into artificial color, with red indicating movement of 1 km or more per year and other colors indicating slower movement, mainly within the interior of the ice sheet. Ice streams are revealed as sinuous, river-like features discharging ice into the ocean. The massive extent of these ice streams was surprising to many scientists, suggesting that the ice sheets may be more

Crevasses photographed from the air. (Photograph courtesy Ian Joughin.)

An Antarctic crevasse, often partially hidden evidence of the movement of the ice which scientists need to be ever vigilant for. This crevasse was 9.8 m wide and 25 m deep. Though thick enough to support a person, this snow bridge could not support a heavy vehicle and a traverse team had to fill the crevasse with snow so that heavy vehicles could cross it safely. (Photograph courtesy Russ Alger, Cold Regions Research and Environmental Laboratory, National Science Foundation.)

responsive to climate change than has previously been assumed. However, observations of how their dynamics change over time will be required to confirm whether this is the case.

Robert H. Thomas
Robert A. Bindschadler

# Ice Sheets and the Threat to Global Sea Level

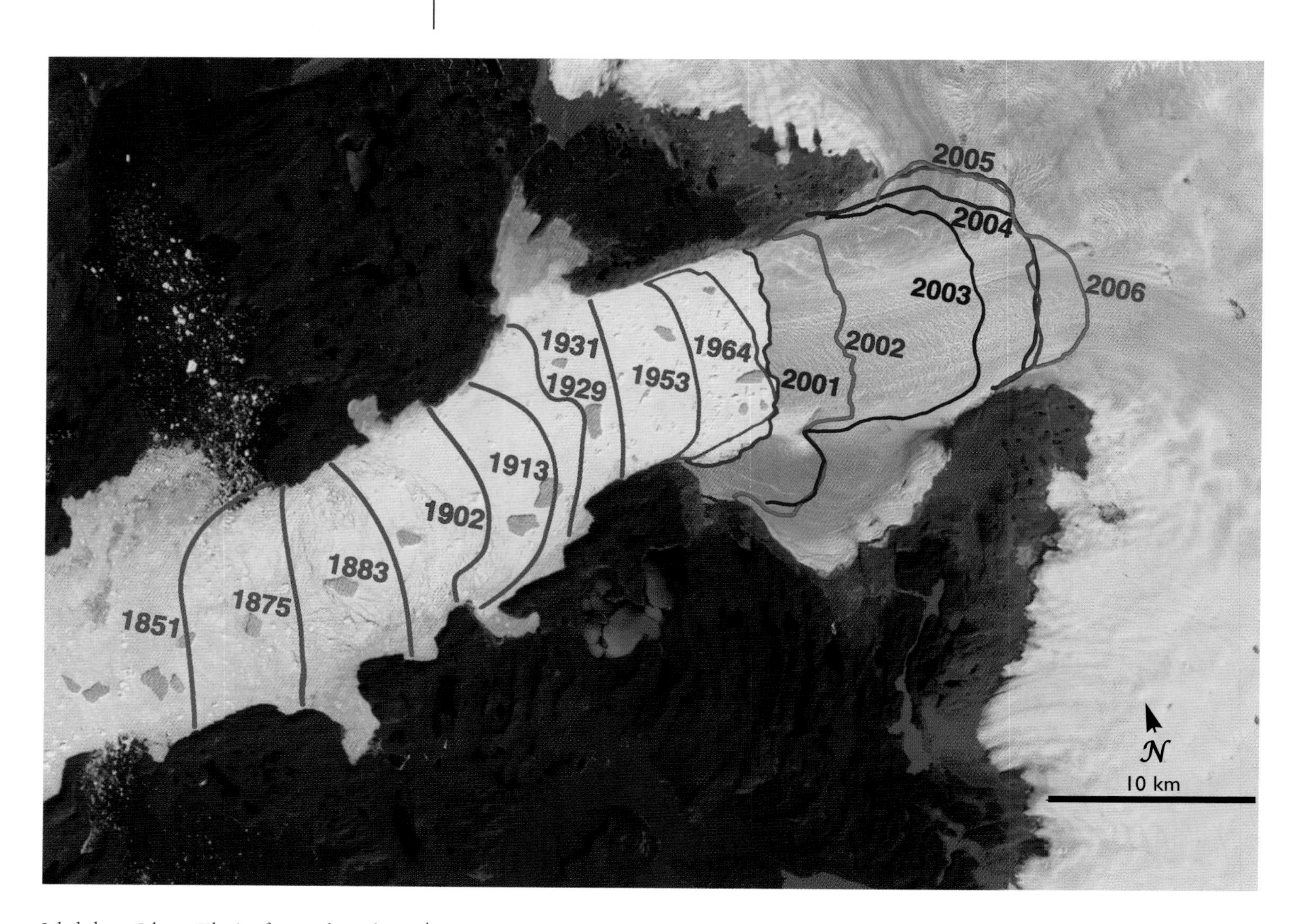


Jakobshavn Isbrae. The ice front, where the glacier calves into the sea, receded more than 40 km between 1850 and 2006. Between 1850 and 1964 the ice front retreated at a steady rate of about 0.3 km/yr, after which it occupied approximately the same location until 2001, when the ice front began to recede again, but far more rapidly at about 3 km/yr. After 2004, the glacier began retreating up its 2 main tributaries: one to the north, and a more rapid one to the southeast. (Historic calving front locations (gray) are from Anker Weidick and Ole Bennike; recent calving fronts (colors) are from the ETM+ instrument on the Landsat 7 satellite and the ASTER instrument on the Terra satellite.)

Satellites are providing a remarkable suite of tools to provide scientists with a remote set of eyes, giving them extraordinarily detailed views of how the ice sheets are thickening or thinning, and revealing patterns of movement and how these patterns are changing. The potential for ice sheets to raise sea level with catastrophic impact, even from a relatively modest, partial collapse, has provided the motivation for a small but committed body of scientists to monitor their 'health' and to assess their inherent stability. The key to significant changes may be found in looking at the ice streams, vast conveyors transporting ice to the ocean from the ice sheet interior. These icy rivers can change their speed surprisingly fast, switching the ice sheet from a state of net growth to net reduction, or even collapse. As huge icebergs calve off periodically, tearing away more frozen water in a single event than the population of the world uses in many years, the challenge for scientists is to look beyond this natural ebb and flow of the ice sheet to determine whether any of this is indicative of persistent and unusual change. Although the current rate of ice loss from the ice sheets

*Our Changing Planet*, ed. King, Parkinson, Partington and Williams
Published by Cambridge University Press 

explains only about 14% of the current rate of sea level rise of about 3.1 mm/yr, the West Antarctic and Greenland ice sheets may be particularly sensitive to climate change, with the potential of each to release sufficient ice to raise sea level by about 6 m.

Such a calamity is not fantasy. During the last major cold period in the Earth's history, culminating some 20,000 years ago, there were many ice sheets filling deep marine basins and consequently called 'marine ice sheets.' The largest were in Canada's Hudson Bay and in Europe's Baltic Sea. These disappeared thousands of years ago, but did not melt quietly away. Rather, they collapsed quickly into the oceans, setting free a vast armada of icebergs strewn across the north Atlantic Ocean and leaving a record of sea level that rose 35 times faster than currently observed rates.

As the only remaining marine ice sheet, West Antarctica bears watching to see if a similar climatic drama will unfold. Although the West Antarctic Ice Sheet experiences very little summer melting, it is based on ground located in large part (but not entirely) below sea level. Most of its outlet glaciers flow into vast floating ice shelves, two of which are about as large in area as Texas with ice thicknesses of up to 1 km or more. The question that scientists have posed for the West Antarctic Ice Sheet is: could these ice shelves be acting like corks in a bottle to make the ice sheet particularly sensitive to ocean temperature changes as

The calving front of Jakobshavn in 2002. The image was taken while flying low over the ice fjord toward the floating terminus of the Jakobshavn Glacier in Greenland. The vertical ice wall represents the terminus of the floating ice tongue. The bluish color of the ice wall indicates recent calving of icebergs. (Photograph courtesy Konrad Steffen, Cooperative Institute for Research in Environmental Sciences (CIRES), University of Colorado.)

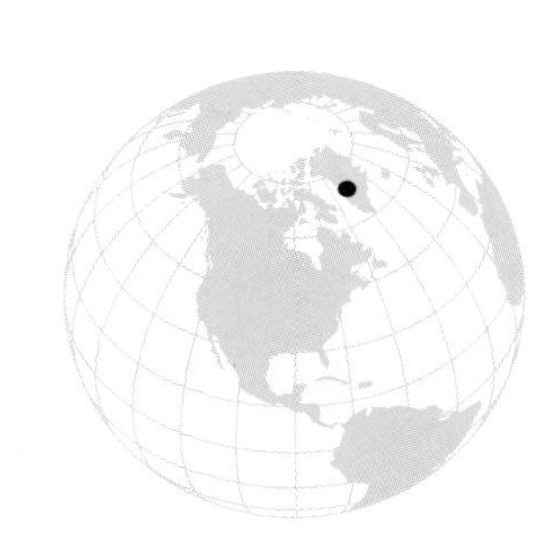

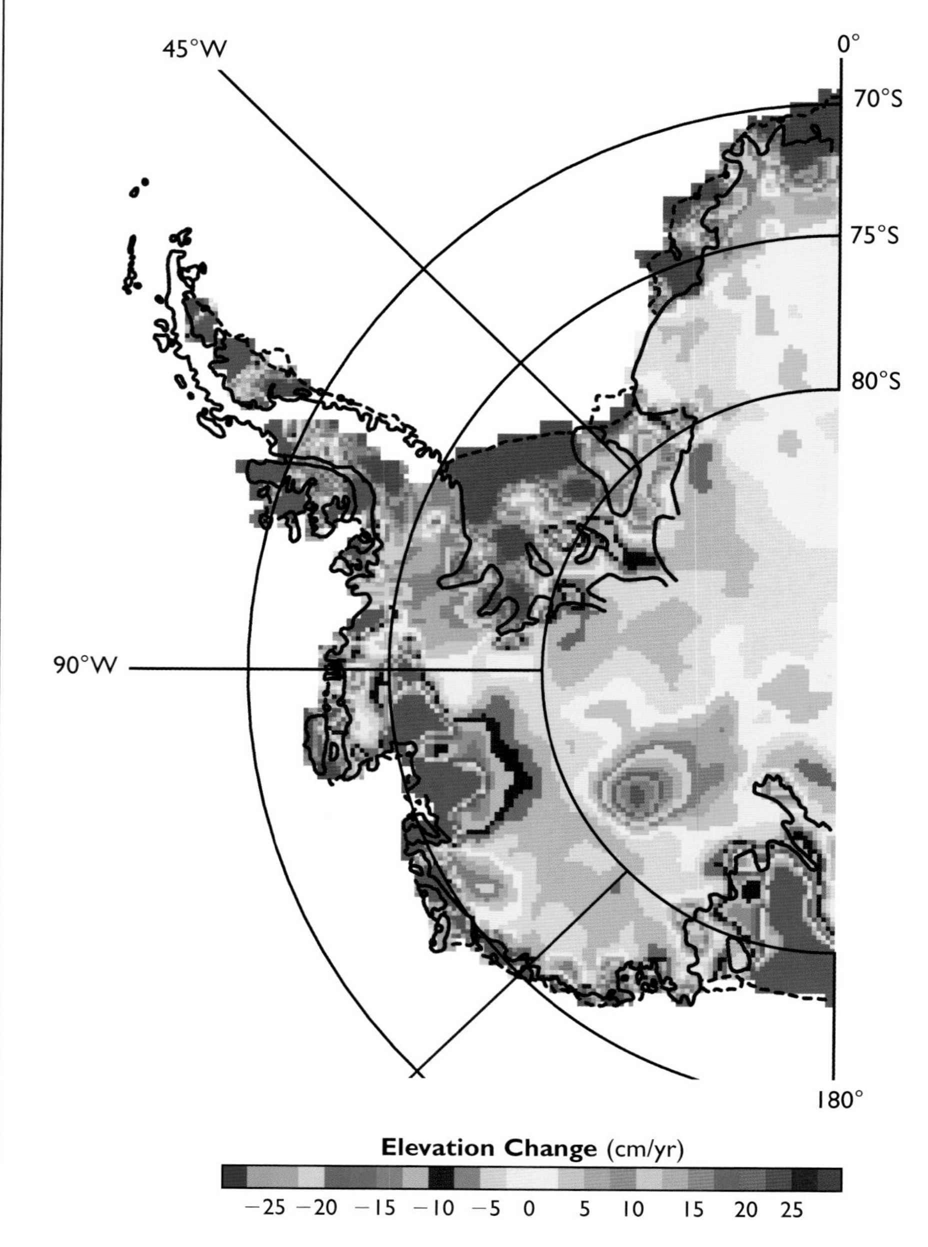


Surface elevation change across the West Antarctic ice sheet. Orbiting satellites with altimeters send electromagnetic pulses earthward that reflect off the surface. Their time of flight is used to measure the elevation of the ice sheet and repeated measurements can detect where the ice sheet is growing or thinning. Here color is used to depict this growth (yellow, brown, red and purple) or shrinkage (gray, green and blue) over the period from April 1992 to April 2001. A complex pattern illustrates the variety of ways in which the ice sheet responds to climate. (Data from radar altimeters on the European Space Agency ERS-1 and ERS-2 satellites. Figure courtesy H. J. Zwally.)

well as atmospheric warming? If the ice shelves were to collapse, would the grounded ice behind the ice shelves surge into the sea to cause a massive sea level rise?

The prognosis is not encouraging. While some areas of the ice sheet are growing, large portions can be seen to be thinning rapidly, resulting in net ice losses sufficient to suggest that a collapse may be underway. In some areas, notably those areas in the north of Antarctica most susceptible to change, rapid disintegration of ice shelves to form many small icebergs is evidence that climate is changing. Catastrophic retreat of the Larsen Ice Shelf, on the east side of the Antarctic Peninsula, is almost certainly a consequence of regional warming over the past several decades, and was followed by as much as an 8-fold increase in the speed of glaciers that had flowed into it. By Antarctic standards, these glaciers are quite small, but Pine Island Glacier, which drains more than 80 $km^3$/yr of ice from the West Antarctic ice sheet has also accelerated as its ice shelf has thinned, resulting in an associated sinking of its surface far inland from where it flows into the sea.

Attention has also been focused on the Greenland Ice Sheet which, although not a marine ice sheet, reaches as far south as 60°N and so is sensitive to warming temperatures.

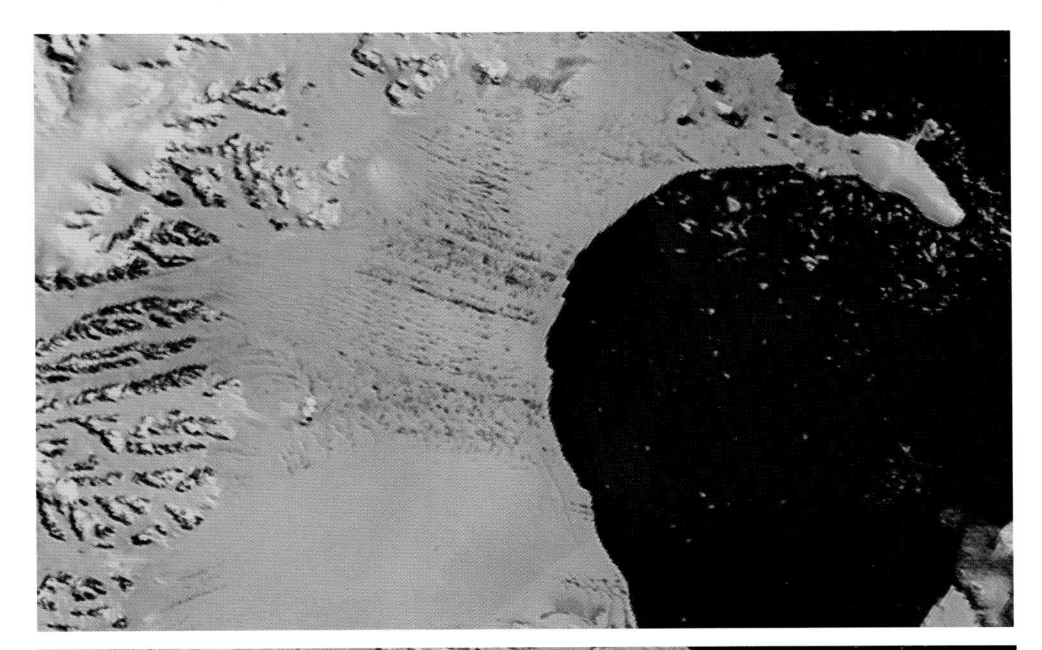

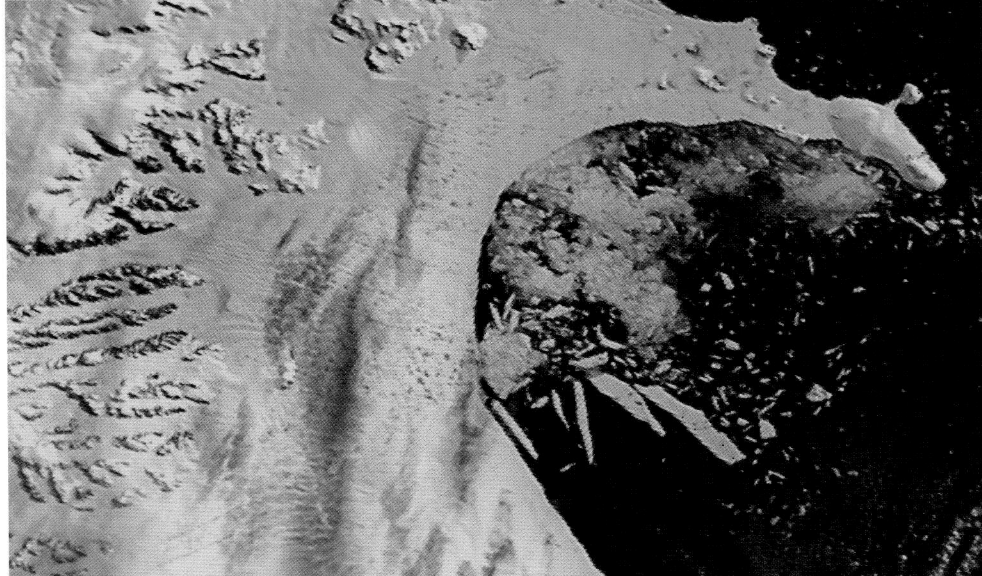

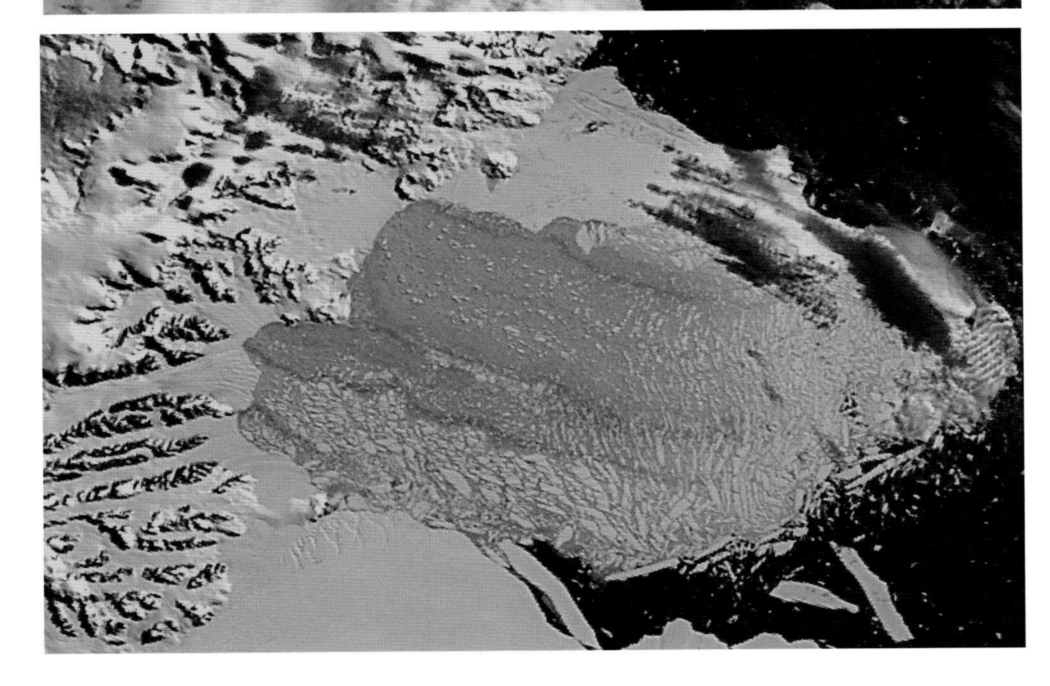

Three true color images showing the collapse of the Larsen B ice shelf in 2002.

Top: January 31, 2002. This image shows the shelf in late austral summer with dark bluish melt ponds dotting its surface.

Center: February 23, 2002. This shows minor retreat amounting to about 800 km$^2$, during which time several of the dark bluish melt ponds well away from the ice front drain through new cracks within the shelf.

Bottom: March 7, 2002. This image shows thousands of sliver icebergs and a large light blue area of very finely divided bergy bits where the shelf formerly lay. Brownish streaks within the floating chunks mark areas where rocks and morainal debris are exposed from the former underside and interior of the shelf. The last phases of the retreat totaled approximately 2,600 km$^2$.

(Data from the MODIS instrument on the Terra satellite.)

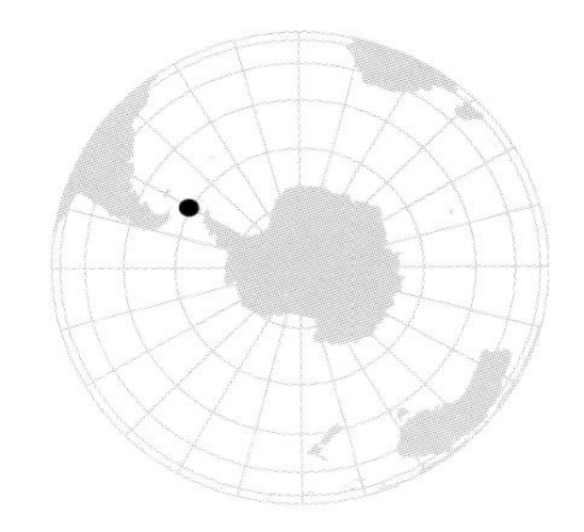

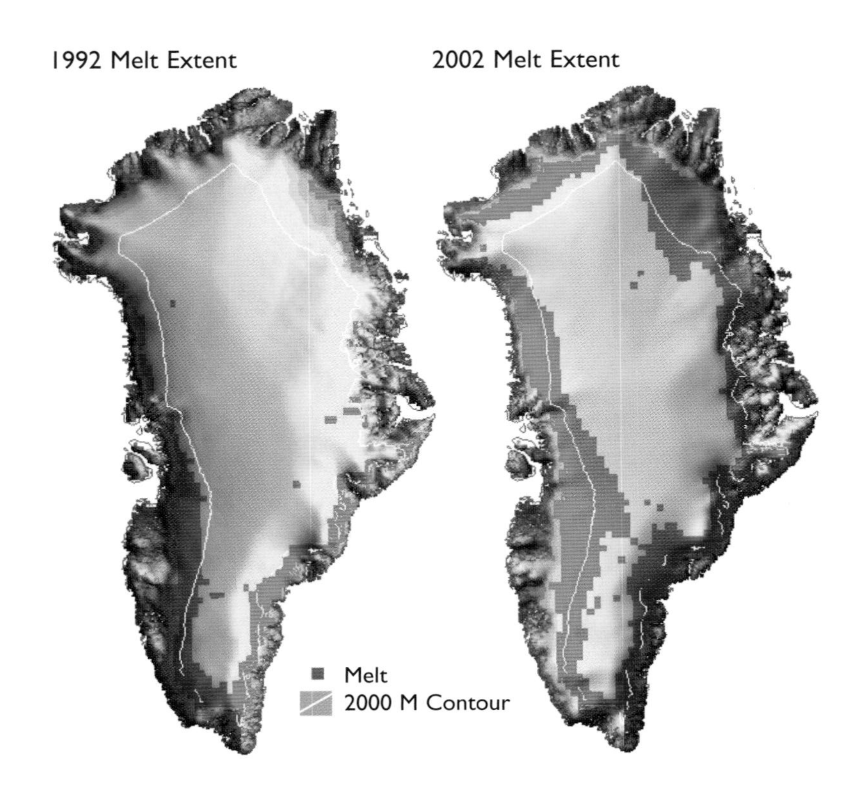


Area of summer melting on the Greenland ice sheet, derived from satellite passive-microwave measurements. Cool (1992) and warm (2002) summers are shown. (Images courtesy Konrad Steffen, University of Colorado.)

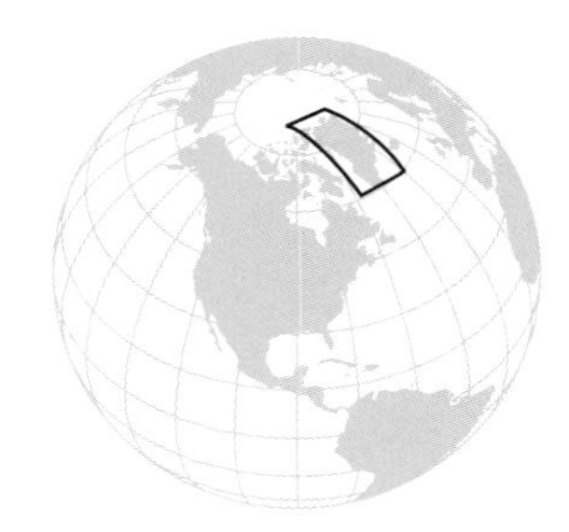

As much as half of its surface experiences summer melting. Summer temperatures have been rising around Greenland since the 1980s, and associated ice loss by melting has increased substantially. Although it is expected that warmer temperatures will also increase the amount of snow falling on the ice sheet, this has not balanced the increased melting, and net ice loss from Greenland is currently sufficient to raise sea level by more than 0.2 mm/yr. Moreover, only part of this loss is from increased melting; the remainder results from large increases in the speeds of some glaciers, with one—Jakobshavn Isbrae—now moving at about 13 km/yr, doubling its speed in only 5 years. This remarkable acceleration appears to have been initiated by very rapid thinning and breakup of its floating ice tongue. More significantly, satellites have shown that many of the coastal glaciers are accelerating in

Plot showing the overall increase in the summer melt area of Greenland since 1978 as summer temperatures progressively increased. (Courtesy Konrad Steffen, University of Colorado.)

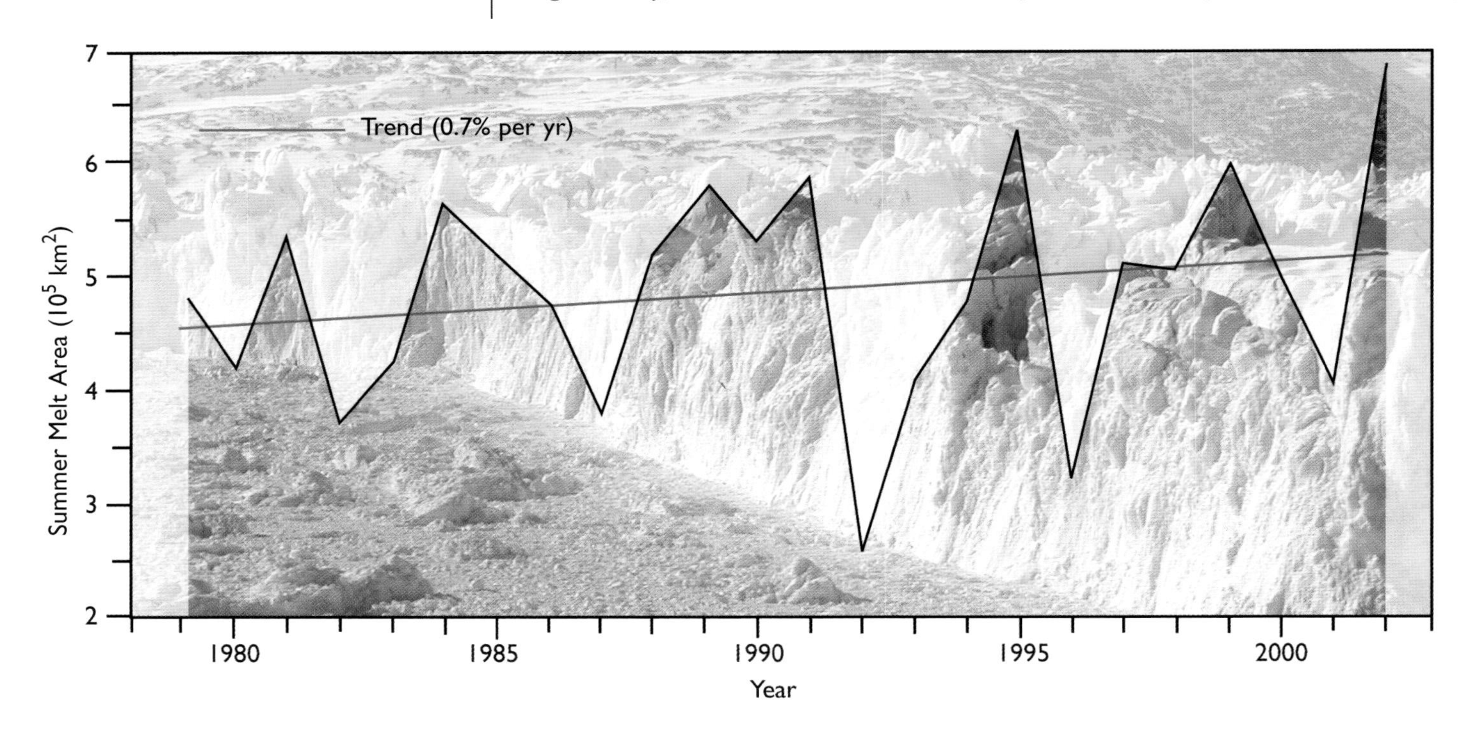

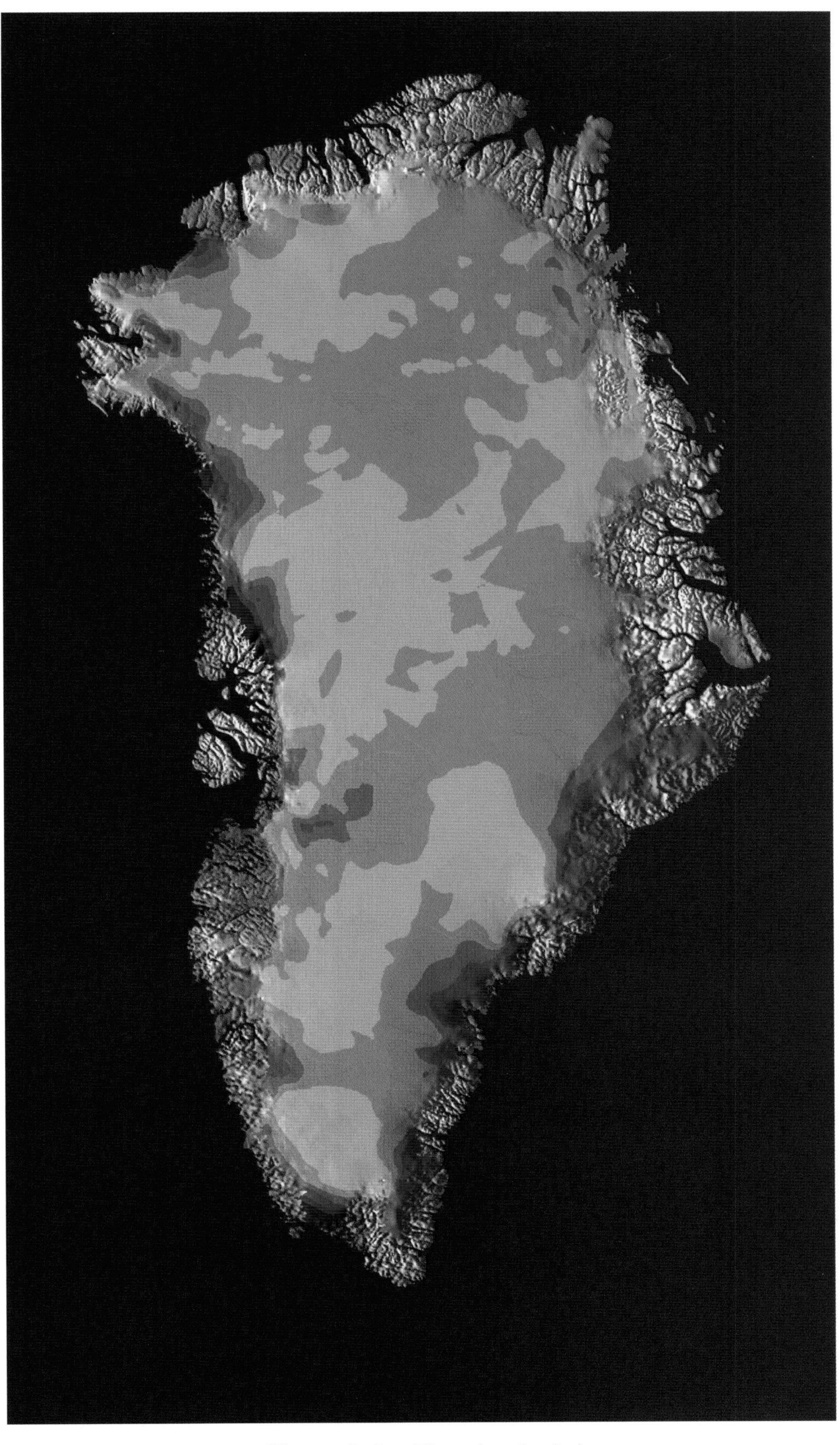


Greenland, showing ice thickening/thinning rates, derived from airborne laser-altimeter surveys. (Courtesy W. Krabill, NASA Wallops Flight Facility.)

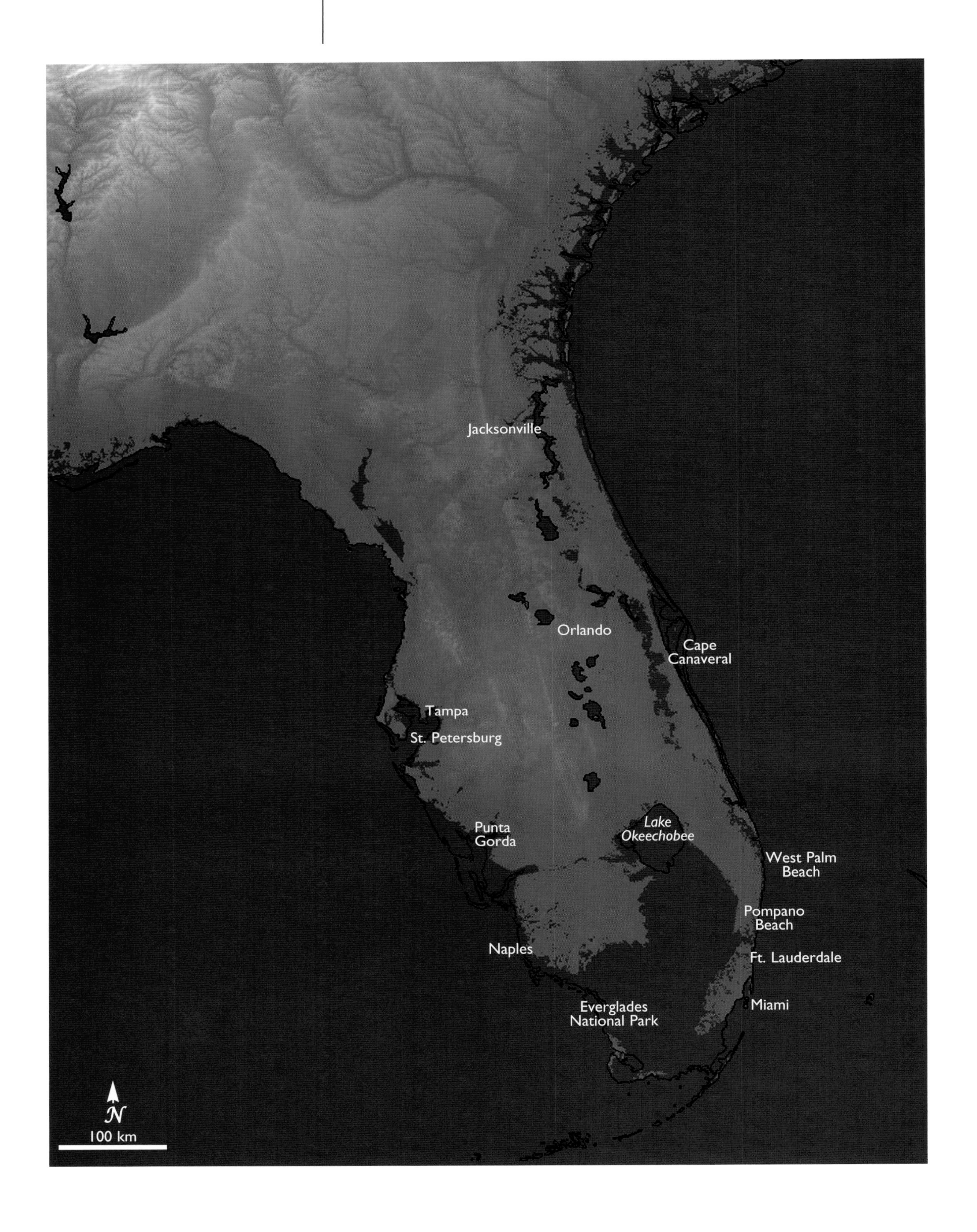

Jacksonville
Orlando
Cape Canaveral
Tampa
St. Petersburg
Punta Gorda
Lake Okeechobee
West Palm Beach
Pompano Beach
Naples
Ft. Lauderdale
Miami
Everglades National Park
N
100 km

unison, suggesting a common cause. Not all of these glaciers have floating ice tongues, and it is possible that increased surface meltwater is draining to the glacier beds and acting as a lubricant for the ice to slide more effectively over underlying rock. If this is so, continued warmer summers are likely to result in a substantial increase in ice discharge from Greenland and an added contribution to sea level rise.

A major concern is that the acceleration of Jakobshavn Isbrae in Greenland may represent a precursor of future conditions in parts of Antarctica where ice shelves are under threat. The big question for glaciologists is: are similar changes occurring elsewhere within the ice sheets, and will they continue and extend inland towards the heart of the ice sheets? If so, then rising sea levels will become a key concern for our children and grandchildren.

Opposite: The Florida coastline. Were all of the West Antarctic ice sheet to flow into the ocean, the coastline of the southern United States would be changed dramatically, as seen in this illustration of the impact of a 5 m rise in sea level (Image courtesy William Haxby, Lamont Doherty Geophysical Observatory).

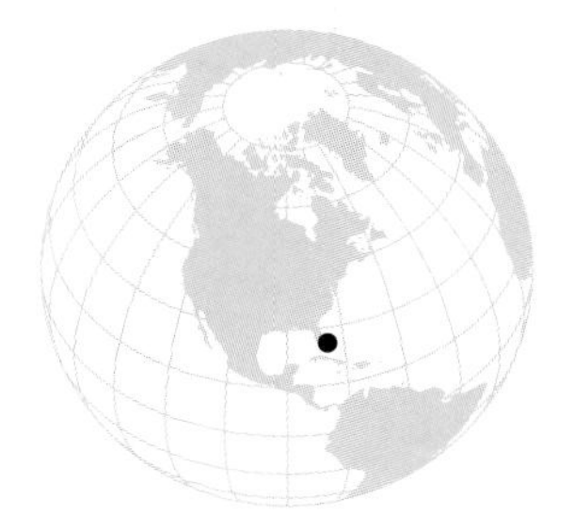

Claire L. Parkinson

# The Great White Ocean: The Arctic's Changing Sea Ice Cover

Dogsledding over the sea ice of the Great White Ocean, in the near vicinity of the North Pole, April 27, 1999. Despite the water underneath, sea ice floes in the Arctic can be substantial enough for considerable human and other animal activity, and for airplanes and helicopters to land and take off. Still, the danger is ever present, as cracks can quickly form, splitting ice floes into smaller pieces, and winds and waves can cause the floes to crash into each other or spread apart significantly. (Photograph by Claire Parkinson.)

In the context of human experience, the Arctic Ocean began as the great frozen north, known intimately only to Inuit peoples hunting along its margins. Later, it was an exciting but treacherous environment for explorers seeking the fabled Northwest and Northeast Passages and the geographic and magnetic North Poles. In the twentieth century, it became a frontier across which the Soviet Union and the United States faced each other, as well as the scene of numerous scientific studies. Today, the scientific studies continue and concern centers on climate changes, including possibly a shrinking sea ice cover, and the impacts of these changes on people, wildlife, and ecosystems.

The sea ice that transforms the Arctic Ocean into a predominantly white expanse forms through the freezing of seawater, resulting from the below-freezing temperatures typical of the high latitudes. Its annual expansion and retreat follow a predictable seasonal cycle, with expansion occurring during fall and winter and retreat during spring and summer. At its annual maximum extent, at the end of winter, the sea ice cover spreads over 15 million $km^2$, one and a half times the area of Canada, and even at its annual minimum, at the end of summer, it typically covers at least 6 million $km^2$, twice the combined area of Argentina and Uruguay.

*Our Changing Planet*, ed. King, Parkinson, Partington and Williams
Published by Cambridge University Press 

248 THE ILLUSTRATED LONDON NEWS. [Oct. 13, 1849.

PICTURES OF THE POLAR REGIONS.

Engraving of life in the polar regions that appeared in *The Illustrated London News* on October 13, 1849, as part of the newspaper's extensive coverage of the missing Arctic expedition of Sir John Franklin. This and other sketches in the series were taken from journals of other Arctic expeditions, as nothing was available from the missing Franklin expedition itself. (Illustration digitally scanned by and provided courtesy Russell A. Potter.)

In view of the harsh polar conditions and the vast area involved, the Arctic ice cover would be particularly difficult to monitor in full from ground level. Fortunately, the ice is quite visible from space, against the very different background of the liquid ocean, and consequently the vast ice cover of the Arctic can now be routinely monitored with satellite instruments. This has been true since the late 1970s, and the satellite data have revealed a great deal about the Arctic sea ice cover. This includes not only many details about the seasonal advance of the ice every fall and winter and the retreat of the ice every spring and summer but also the fact that the ice cover varies considerably from year to year and from region to region. Furthermore, of most interest for studies of climate change, these satellite data have shown that the ice cover has, overall, decreased in extent from the late 1970s to the early twenty-first century. In fact, analysis of the satellite data indicates that over this period the ice cover shrank in extent by an area exceeding the size of Texas, which is slightly larger than the size of Afghanistan. Ice decreases occurred in all seasons and all major regions, although the decreases were not uniform, and the timing and magnitude of the decreases differed from one region to another and from one time period to another.

Sir John Franklin, the leader of an 1845–1847 expedition that disappeared while searching for the Northwest Passage, precipitating at least 32 search and rescue expeditions over the next 15 years. Franklin was one of many explorers seeking a shortened route to the riches of the Orient. He and his crew, all of whom died during the expedition, would have had an easier time in many recent summers, in the 1990s and early 2000s, when the ice conditions in the Canadian Archipelago were considerably less severe. The portrait engraving appeared in the *Illustrated Family Paper* in the mid-1800s after all hope for a successful rescue had been abandoned. (Illustration digitally scanned by and provided courtesy Russell A. Potter.)

Although monitoring the ice distribution and extent from satellites is now routine, that is not yet the case for monitoring ice thickness, which is a more complicated measurement from the remote reaches of the satellite orbits, hundreds of

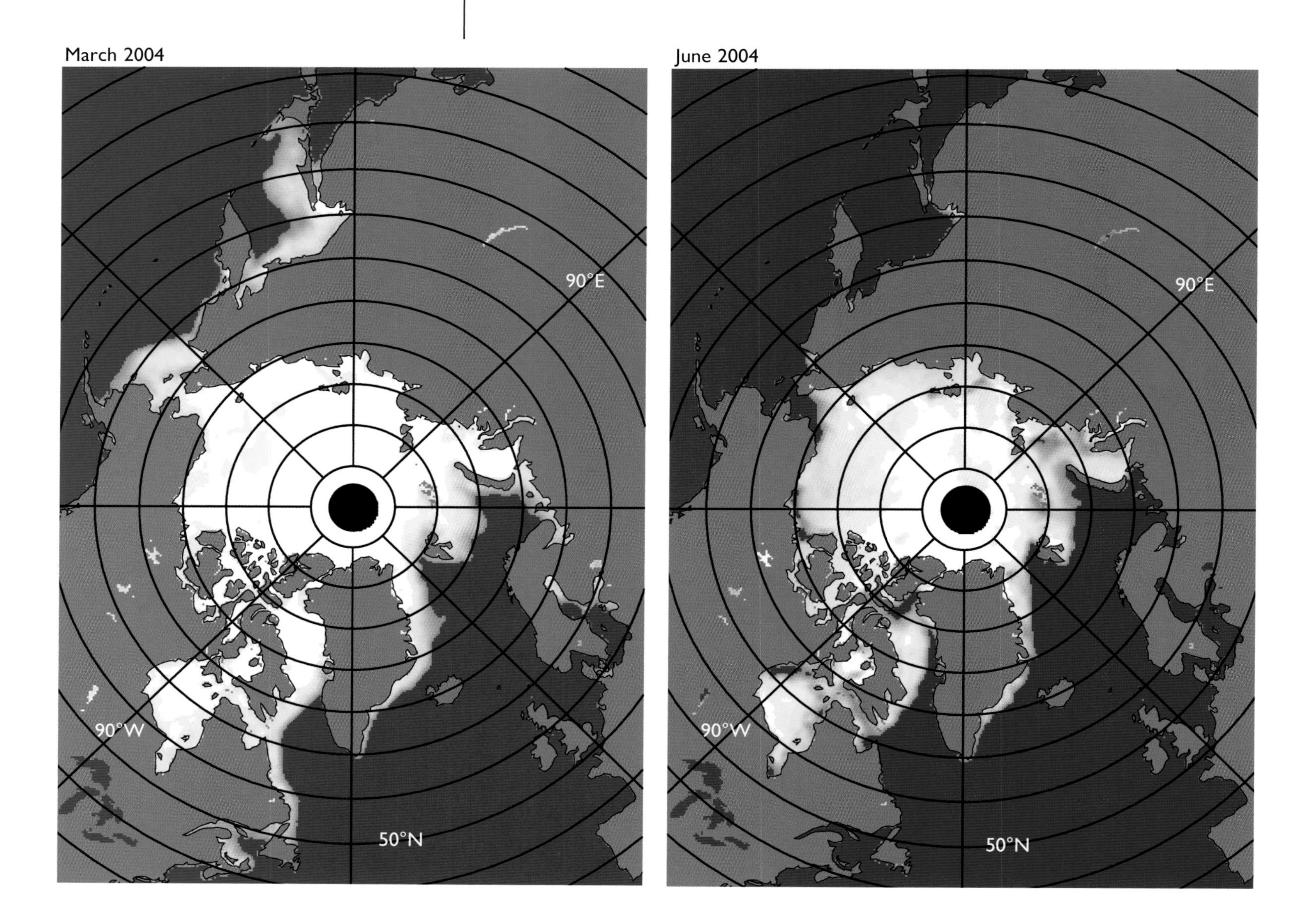

Seasonal cycle of Arctic sea ice coverage in 2004, as determined from satellite data. The maps show sea ice concentrations (percent areal coverages) in March, June, September, and December, with white indicating total ice coverage and deep blue no ice coverage. March is typically the month with the greatest Arctic ice coverage and September the month with the least. No data were available from north of 87.6°N, because of the particular orbit of the satellite, and so that region has been colored black to indicate missing data. (Data from the SSMI instrument on the DMSP F13 satellite.)

kilometers or miles above the Earth's surface. Still, there are many ice thickness measurements, both from people on the ice itself, drilling holes and measuring the ice thickness directly, and from people in submarines measuring the ice remotely as the submarine passes underneath. These *in situ* and submarine measurements, even in total, are quite limited, neither being available from the entire Arctic nor being available throughout the year; but despite their limitations, the ice thickness measurements show a thinning of the ice cover that matches well with the decreases in the extent of the ice found from the spatially and temporally much more complete satellite data. Moreover, satellite laser altimetry measurements are showing promise that the ice thicknesses throughout the Great White Ocean will, in the future, be able to be routinely determined, at least approximately, from space.

No one knows for sure whether the ice decreases from the late 1970s to the early twenty-first century will continue, as they might be connected with large-scale oscillations or other variabilities within the climate system rather than being part of a continuing long-term trend. Nonetheless, considerable concern exists that the decreases might continue,

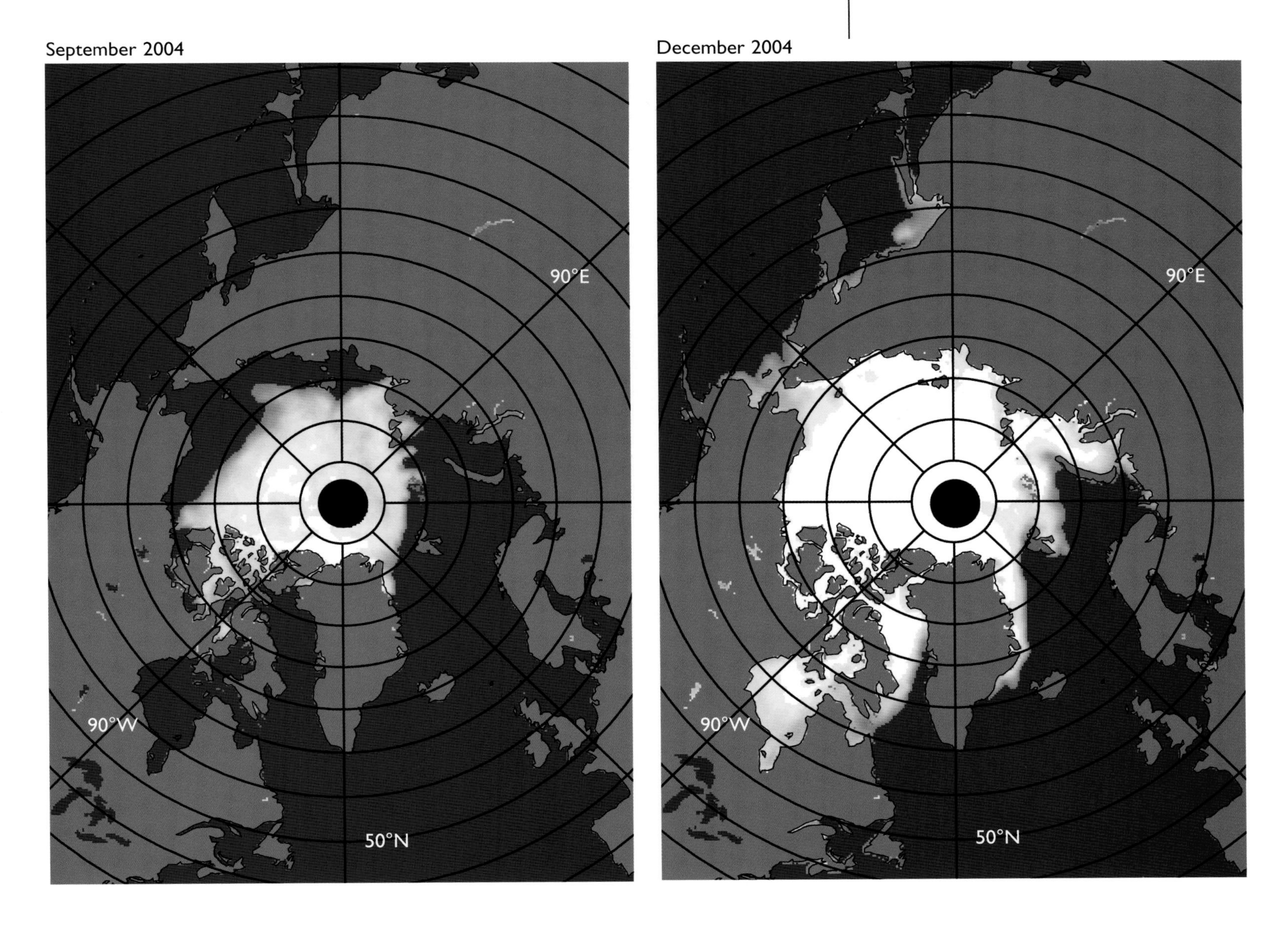

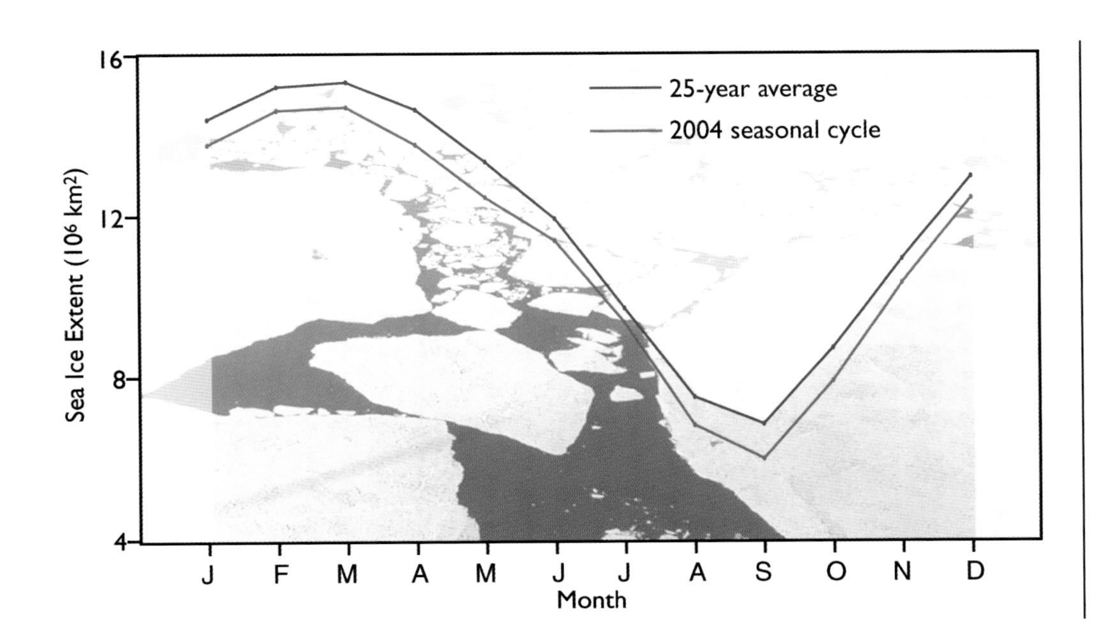


Seasonal cycle of sea ice extents in the Arctic for the year 2004 (in red) and for the average of the 25 years 1979–2003 (in blue). 'Sea ice extent' is the total area of the ocean having an ice concentration of at least 15%. In line with the general decrease of ice in the Arctic, the ice extents in 2004 were lower than the 25-year average in each month of the year. Still, as is typical, the ice in 2004 reached its maximum extent in March and its minimum extent in September. (Data from the SSMI instrument on the DMSP F13 satellite, plotted by Nick DiGirolamo; background photograph by Claire Parkinson.)

Jumbled mixture of ice pieces and floes at the ice edge in the Bering Sea, between Alaska and Siberia, March 8, 1981. (Photograph by Claire Parkinson.)

A polar bear roaming the Canadian Arctic, date unknown. Polar bears are among the many animals greatly influenced by the Arctic sea ice cover, using the ice as a platform from which they feed on the life within the surrounding waters. They will be forced to undergo significant changes in their lifestyles, and will be seriously threatened, if the Arctic ice cover lessens dramatically. (Photograph © Robert Taylor.)

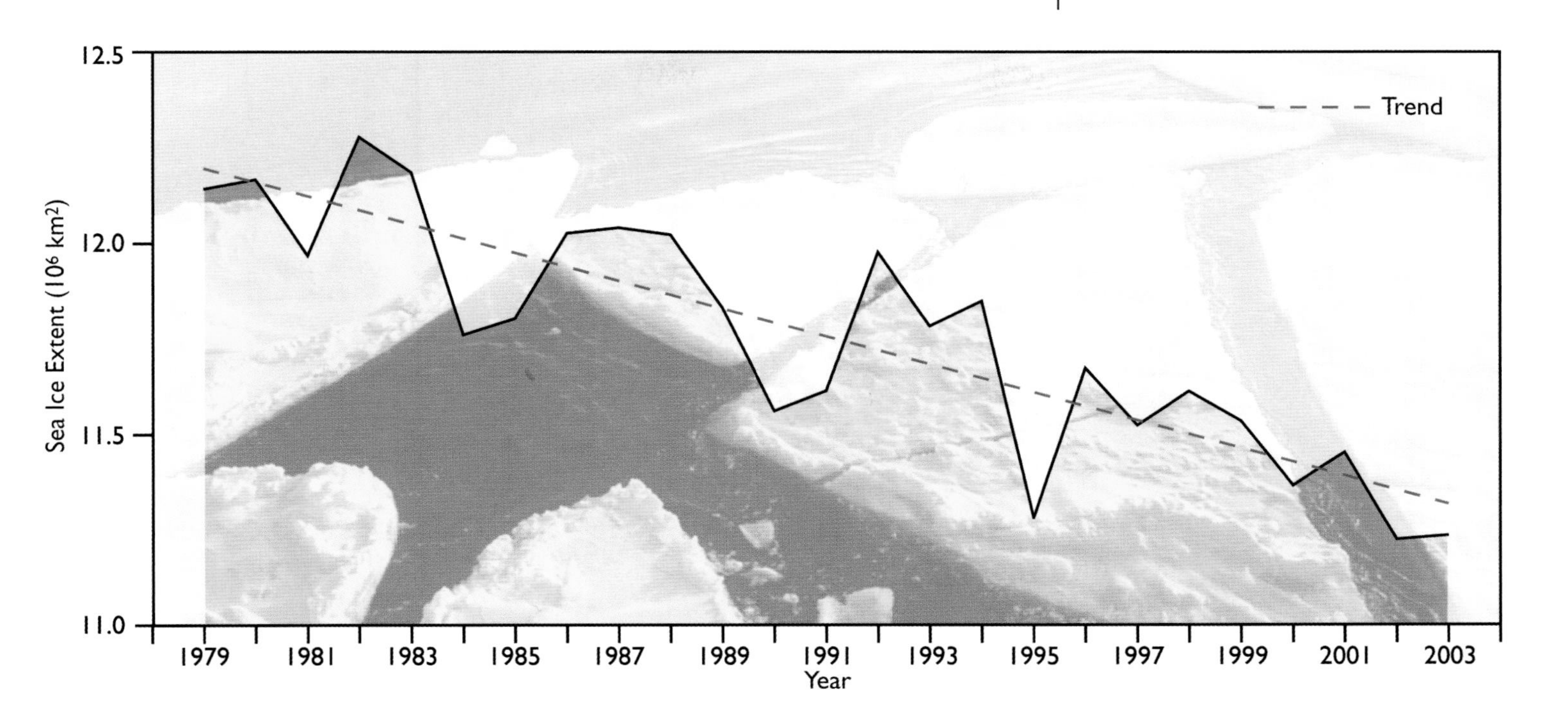


Retreating Arctic sea ice cover, 1979–2003. The points plotted are annual average sea ice extents. (Data for 1979–1987 from the SMMR instrument on the Nimbus 7 satellite and for 1987–2003 from the SSMI instruments on the DMSP F8, F11, and F13 satellites; figure extended from a graph in C. L. Parkinson, D. J. Cavalieri, P. Gloersen, H. J. Zwally, and J. C. Comiso, *Journal of Geophysical Research*, vol. 104, no. C9, pp. 20,837–20,856, 1999; background photograph by Claire Parkinson.)

especially in light of the expected continued warming of the Arctic from increases in greenhouse gases due to human activities. Some scientists even project that the late summer ice cover might disappear altogether within the next few decades. If this were to happen, it could have far-ranging consequences, including positive ones, such as simplifying ship passage through Arctic waters, and negative ones, such as eliminating both the habitat of the microorganisms living within the ice and the platform the ice provides for polar bears and other Arctic animals. It would also increase the amount of the Sun's radiation absorbed by the Arctic Ocean, by taking away the highly reflective white covering, and could thereby contribute to further Arctic warming. Furthermore, its consequences could extend globally, as it could affect the entire global ocean's 'conveyor belt' circulation by impacting the amount of deep water formed in the northern Atlantic. Such interconnections among the ice and other aspects of the climate system are the central topics of the next chapter.

CLAIRE L. PARKINSON

# Bound Together: Arctic Sea Ice, Ocean, and Atmosphere

Ice flow through the Bering Strait, May 7, 2000. Winds helped drive some of the ice southward from the Arctic Ocean through the Strait and into the Bering Sea. The Bering Sea is also cold enough each winter that considerable ice forms each year by water freezing within the sea itself. (Data from the MODIS instrument on the Terra satellite.)

The sea ice of the Great White Ocean (previous chapter) is intricately connected to the rest of the north polar climate system and has ties also extending globally. The ice forms because the ocean temperatures decrease to or below the freezing point; it moves around forced by winds, waves, and currents; and it melts because of energy received from the ocean, the atmosphere, and the Sun. The tie-in between the ice and the rest of the climate system is much apparent in the asymmetrical distribution of the ice: warm currents from the south (especially the Gulf Stream, North Atlantic Current, and Norwegian Current) delay and sometimes even prevent the formation of ice in some quite high-latitude regions, while cold currents carry ice sometimes far south of where it would typically form.

*Our Changing Planet*, ed. King, Parkinson, Partington and Williams

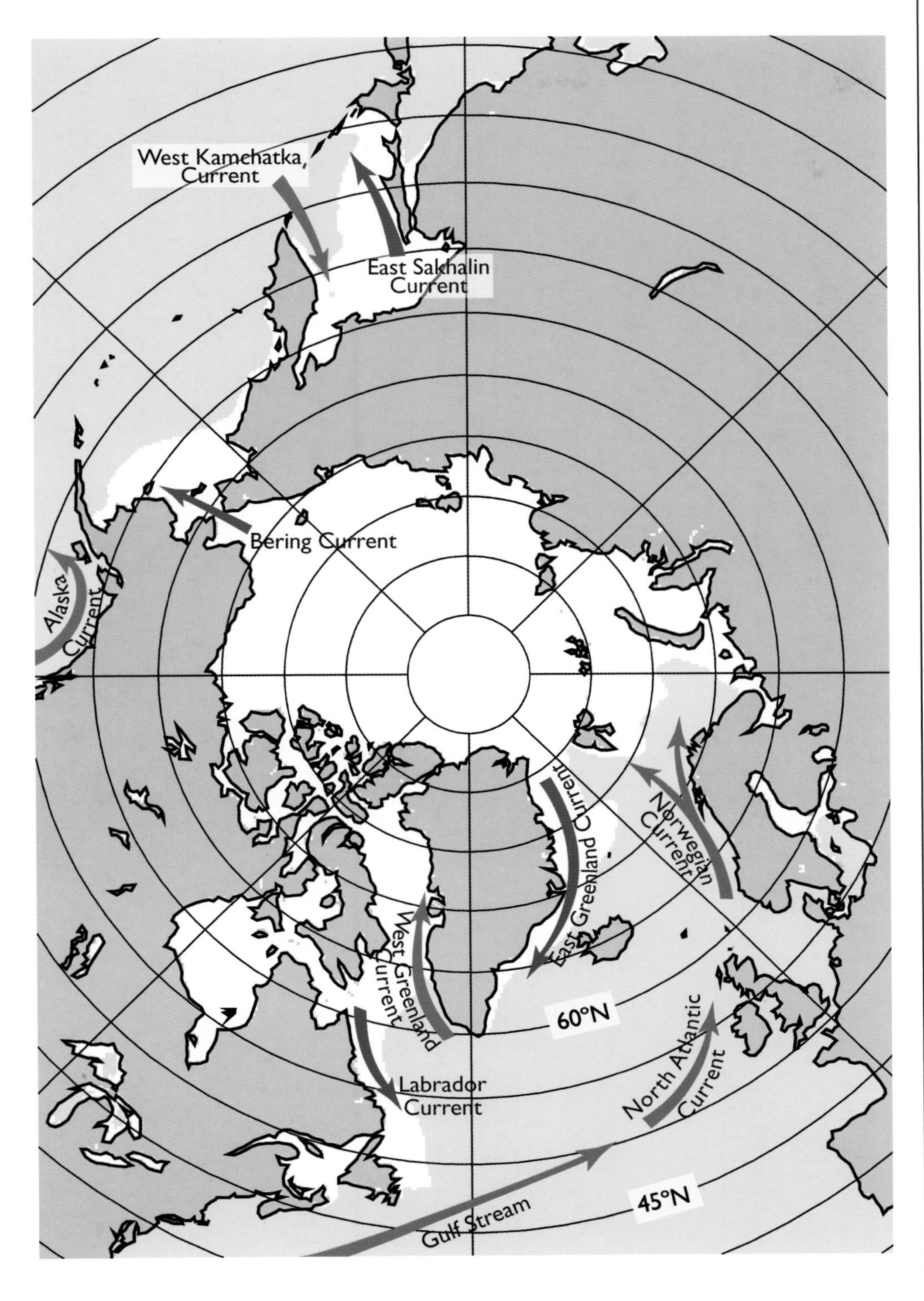


Wintertime sea ice distribution (in white) in the north polar region, juxtaposed with arrows indicating prominent warm (red) and cold (blue) currents. The ice distribution is the average February distribution for 1979–2005, from satellite data. Warm currents limit ice coverage, by keeping temperatures above the freezing point, while cold currents often transport ice far to the south of where it initially formed. Note in particular the influence of the Gulf Stream and North Atlantic Current, which transport warm water across the Atlantic and northward to the west and north of the British Isles, restricting the southward expansion of the ice cover in the Barents Sea (latitudes 70–80°N). (Ice data for 1979–1987 from the SMMR instrument on the Nimbus 7 satellite, and for 1988–2005 from the SSMI instruments on the DMSP F8, F11, and F13 satellites.)

Sea ice in turn has major impacts throughout the polar climate system. It is a strong insulator between the atmosphere and the ocean, helping the ocean to retain heat during the bitterly cold Arctic winters. It also limits chemical exchanges, including exchanges of carbon dioxide and water, between the ocean and the atmosphere, so that, for instance, the Arctic atmosphere generally has much less water vapor than it would in the absence of an ice cover. Similarly, the ice limits the transfer of energy from overlying winds to the water, often making the waters within an ice-covered region much calmer than the adjacent ice-free waters. (This latter impact can be a considerable benefit for ship passengers subject to seasickness.)

Another major effect of the ice is its strong tendency to reflect the Sun's radiation that reaches it. Because of the high reflectivity of the ice, increased even further when the ice is

Schematic illustrating the insulation effect of sea ice. Not much of the ocean's heat makes its way through the ice to the atmosphere, or vice versa; but where the ice floes separate, considerable ocean/atmosphere exchanges can take place.

overlain by a fresh snow cover, most of the Sun's radiation incident on the ice gets reflected away, much of it returning to outer space. In great contrast, in the absence of an ice cover, the overwhelming majority of the Sun's radiation incident on the ocean gets absorbed into the ocean, helping to heat the ocean/atmosphere system. This contrast is one of the prime reasons why climate changes are expected to be amplified in the polar regions. Specifically, polar warming encourages more ice melt and less ice formation, leading to reduced ice cover and therefore, through the reflectivity contrast between the ice and ocean, to more absorption of the Sun's radiation and hence a further warming. Similarly, cooling leads to more ice and therefore less absorption of the Sun's radiation and consequently further cooling. This feedback, brought about by the high reflectivity (or 'albedo') of the ice, is called the ice–albedo feedback. Because it tends to amplify the initial change—warming leading to more warming and cooling to more cooling—it is termed a 'positive feedback,' even though sometimes the outcome is not desirable. The climate system has negative (non-reinforcing) as well as positive feedbacks, but this positive ice–albedo feedback is particularly important for ice-covered regions.

The insulation effect and the high reflection of solar radiation are two of the most important impacts that sea ice has on polar climates, but there are other impacts as well. For instance, it takes energy to melt ice, and so during the melt season temperatures rise less than they would in the absence of ice because some of the energy that otherwise would be available to raise temperatures is instead melting the ice. Conversely, as ice freezes, energy is released, helping to keep temperatures from plummeting as far as they otherwise would. So the melting of the ice restricts the temperature increases in summer and the freezing of the ice restricts the temperature decreases in fall and winter.

The many impacts among the Arctic ice, ocean, and atmosphere, and the variability of all of these components from year to year, contribute to the complications of sorting out issues regarding climate change. Certainly the overall decreases in Arctic ice coverage found from the satellite record since the late 1970s (previous chapter) can logically be tied to the warming of the Arctic region reported in temperature records and by local inhabitants. However, neither the ice decreases nor the temperature increases have been uniform,

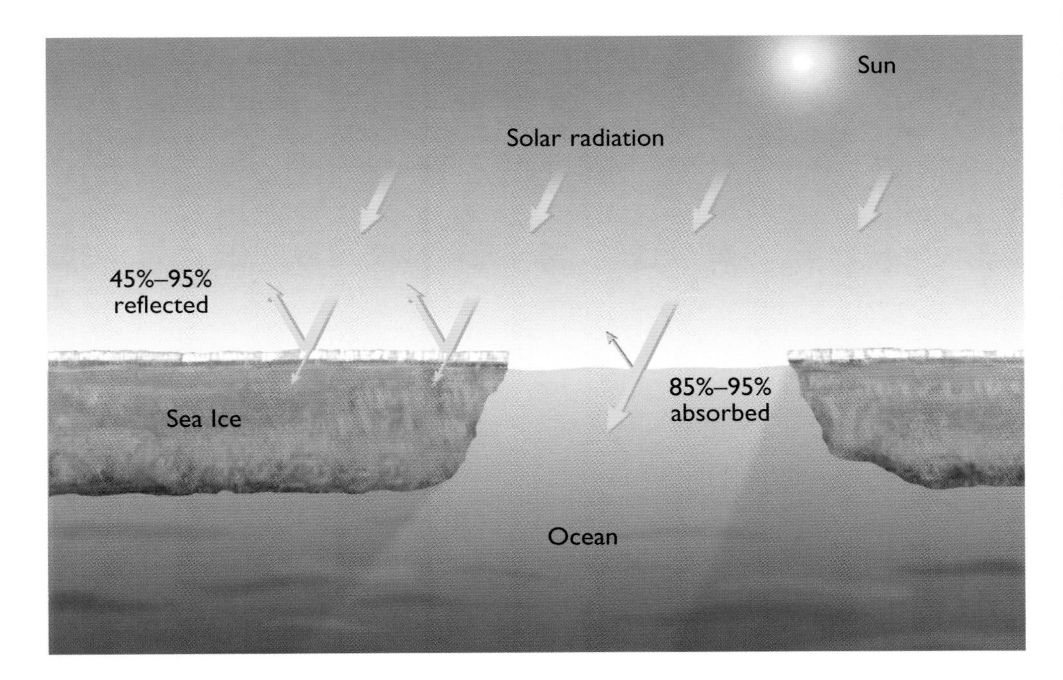


Schematic illustrating the impact of sea ice on the retention (absorption) of radiation from the Sun in the polar climate system. Because of the high reflectivity of the ice (and the snow if the ice is snow-covered), most of the solar radiation incident on it gets reflected away. In contrast, most of the solar radiation incident on the liquid ocean gets absorbed into the ocean, warming the system.

and some of the patterns of change in the ice cover suggest the possibility of connections with large-scale oscillations in the climate system, although these connections have appeared weaker as the data record has lengthened into the twenty-first century and the warming and ice decreases have continued.

One of the most studied oscillations having an impact on the Arctic sea ice is a large-scale atmospheric pressure oscillation termed the North Atlantic Oscillation, abbreviated NAO. At one extreme of the NAO, with a high positive NAO index, a deep low pressure system exists in the vicinity of Iceland and a strong high-pressure system exists further south over the Atlantic. At the other extreme of the NAO, with a negative NAO index, the low pressure system in the vicinity of Iceland and the high pressure system further south are both much weaker. The reason this pressure oscillation affects the Arctic sea ice is that when the low-pressure system near Iceland is strong, more warm air is brought north into the Arctic to the east of the low pressure, inhibiting ice coverage in the northern Atlantic and the central Arctic, and more cold air is brought south from the Arctic to the west of the low pressure, often advancing the ice cover to the west of Greenland. As the oscillation fluctuates, the ice cover can fluctuate as well, so that, for instance, a long-term trend could be temporarily masked by the oscillation.

Looking at even longer timescales, one of the most important ways that sea ice might have impacts extending well beyond the Arctic region is through its influence on the ocean's thermohaline circulation. The thermohaline circulation is a major circulation pattern throughout the Earth's oceans and is driven by differences in sea water density. These density differences are controlled by temperature ('thermo') and salinity ('haline'). The thermohaline circulation is also known as the global conveyor belt, because it extends globally and transports heat and salt around the Earth's oceans. The large-scale flow pattern includes the near-surface transport of warm salty water from southwest to northeast across the northern Atlantic in the Gulf Stream, sinking of the water at high latitudes in the Atlantic, and return of the water to lower latitudes in the deep ocean. The water that sank in the northern Atlantic can remain in the deep ocean for hundreds or even a thousand years, as it slowly moves southward through the north and south Atlantic and then onward into

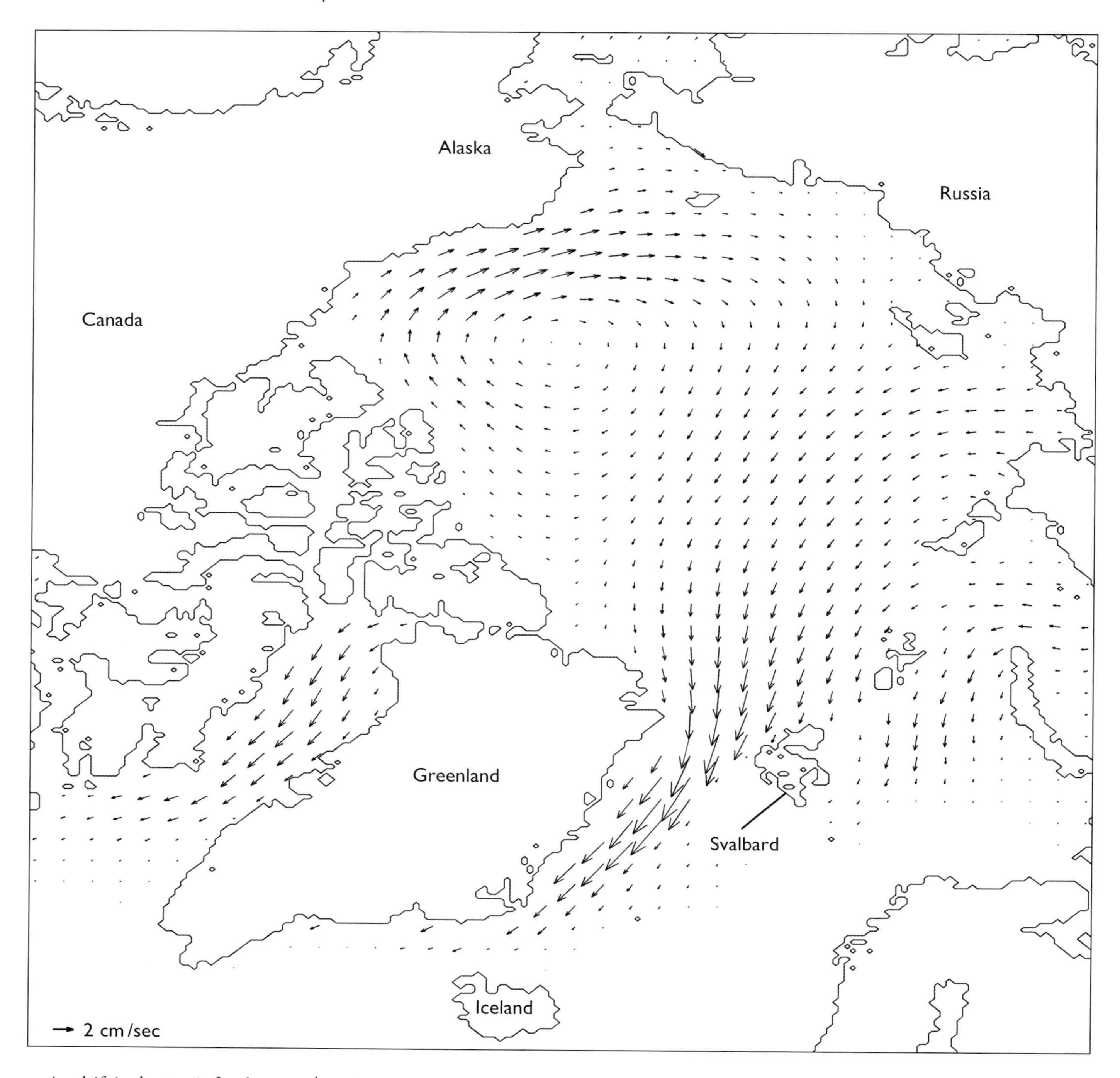


Average sea ice drift in the Arctic for the period January 1979–December 2002. Note in particular the clockwise circulation in the Beaufort Sea, north of Alaska, and the flow out of the Arctic Basin through the Fram Strait between Greenland and Svalbard. (Data from the AVHRR instrument on NOAA satellites, the SMMR instrument on the Nimbus 7 satellite, the SSMI instruments on DMSP satellites, and the International Arctic Buoy data set. Figure provided by David Simonin, using data provided by Charles Fowler and the National Snow and Ice Data Center, Boulder, Colorado.)

the Indian and Pacific Oceans. This global thermohaline circulation plays a major role in the transport of heat between the equatorial and polar regions, with one notable consequence being the mild winters in the British Isles versus in other regions of comparable latitudes. Another gigantic impact of this global conveyor belt is the promotion of biological productivity within the oceans, through maintaining a continual supply of nutrients from the ocean depths.

In view of the importance of the thermohaline circulation, major changes in it could have consequences worldwide; and that's where an additional potentially significant impact of the Arctic sea ice enters. There is evidence that the global conveyor belt has slowed or even come to a complete stop during periods in the distant past, notably near the end of the last ice age. A concern is that this might happen again and that this could be precipitated in part by changes in the flow of Arctic sea ice into the North Atlantic. Specifically, if the

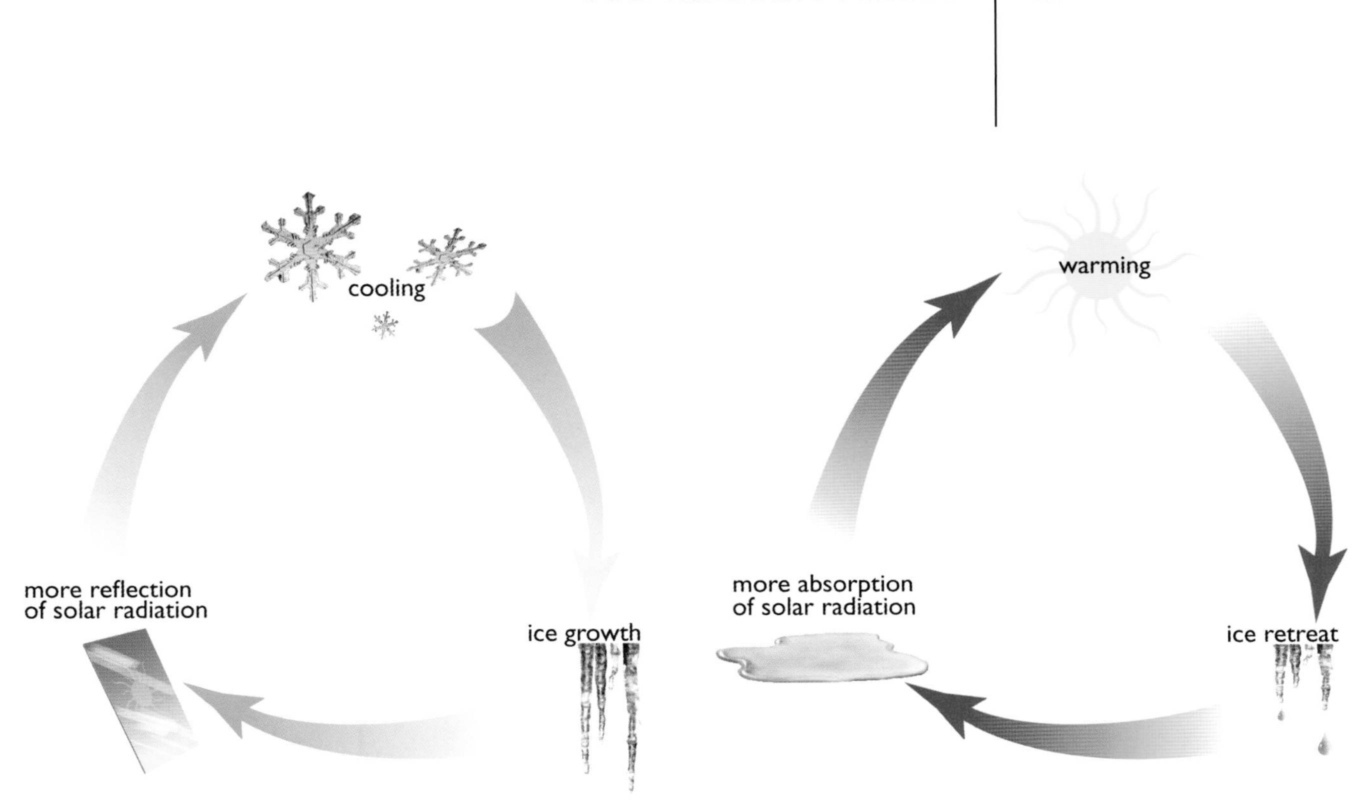


Schematic of the ice–albedo feedback. Because of the high albedo of ice, ice growth increases the reflection of solar radiation at the surface and ice retreat decreases the reflection of solar radiation (the two bottom arrows), with the result that, in the absence of other factors, an initial cooling gets amplified to further cooling and an initial warming gets amplified to further warming.

export of sea ice from the Arctic into the Atlantic were to increase, then the increased release of fresh water upon melting, along with increased melt from the Greenland ice sheet, could act as a brake on the global conveyor, through the buoyant properties of freshwater. Although the thermohaline circulation is unlikely to completely shut down, any major disruption to it could have serious consequences, providing an additional reason for keeping a watchful eye on the Arctic sea ice cover and any changes in the amount of sea ice flowing out of the Arctic into the Atlantic.

Mark R. Drinkwater
Josefino C. Comiso

# Sea Ice: The Shifting Crust of the Southern Ocean

Ross Sea ice jams the channel where the mighty B-15A iceberg broke apart in October 2003. The iceberg on the right retains the name B-15A, while the iceberg on the left is now B-15J. Barely visible is iceberg C-16 and Beaufort Island. (Photo courtesy Brien Barnett, National Science Foundation.)

To most people the frozen canopy of ice in the polar regions represents a static remnant or artifact, moulded by the forces of nature in Earth's icy past. To the eye, the endless sea of ice gives the impression of a silent, motionless landscape. However, the thin veil of sea ice that encircles the Antarctic continent is one of the most dynamic components of the Earth's climate system.

For around 9 wintery months of a year, pack ice grows upon the ocean surface surrounding Antarctica. At its winter peak (typically in September) this frozen crust extends north to around 55° latitude, covering an area of up to approximately 19 million km$^2$, equivalent to an area 36% larger than that of the Antarctic continent itself. In summer, most of the ice breaks up and drifts equatorward to encounter warmer air and water temperatures where it melts rapidly. The southernmost thick ice that survives the

*Our Changing Planet*, ed. King, Parkinson, Partington and Williams
Published by Cambridge University Press 

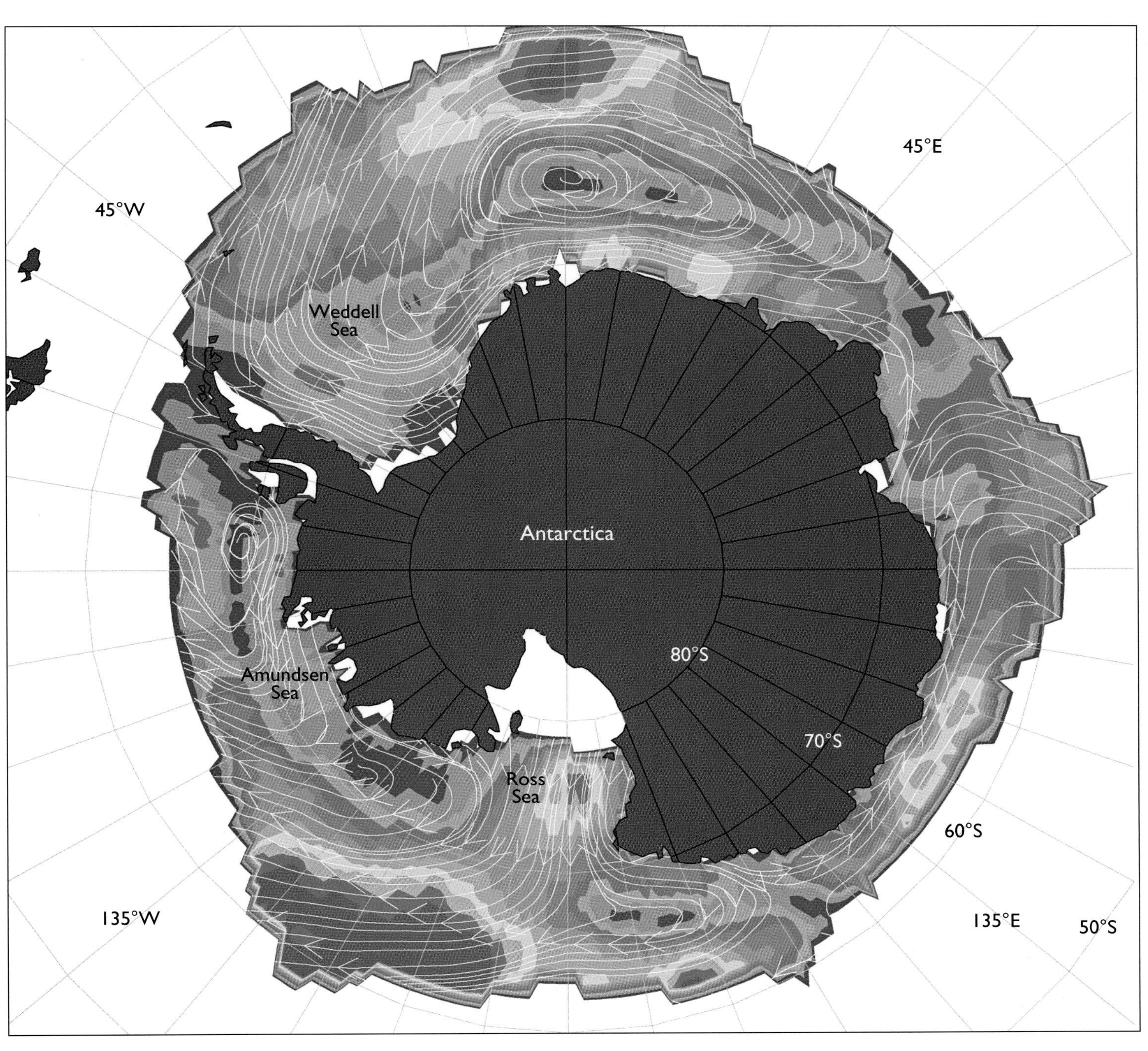


Average daily sea ice drift recorded during the period July–September 1998. The white lines with arrows indicate the direction of ice drift and the color indicates the speed. (Data from the SSMI instrument on a DMSP satellite.)

summer experiences regrowth the following winter, and is known as 'multi-year' or 'perennial' ice. Up to 15% of the total ice pack survives the summer, with the largest concentrations found typically in the Weddell and Ross/Amundsen Seas. The coverage of the multiyear ice varies from year to year depending on the oceanic and atmospheric conditions and circulation patterns.

It was the early explorers who first discovered that the sea ice was neither motionless, nor lifeless. However, to them the ice formed a randomly shifting barrier that blocked their passage southward from the roaring forties to the coast of Antarctica. Equipped with wooden

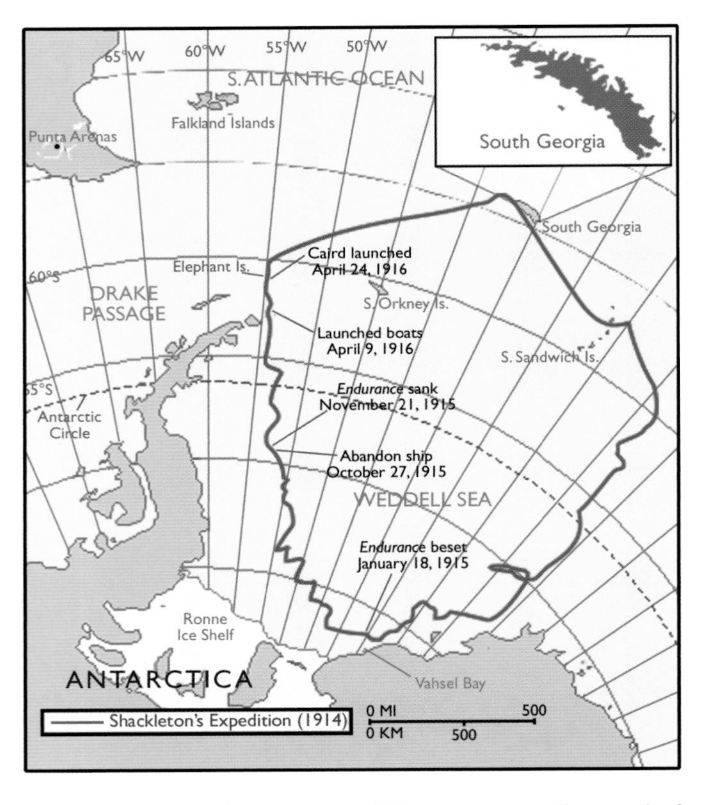


Map of the Shackleton expedition, indicating the locations where Endurance became trapped, the ship's icebound drift track, and the location where the ship was abandoned before sinking. (Adapted from Nova Online—*Shackleton's Voyage of Endurance*, PBS.)

vessels the ice pack threatened a watery grave. Yet, in spite of these dangers, heroic and sometimes reckless quests led them to new scientific discoveries. Shackleton's 2-year-long adventure on the Endurance provides the most astonishing case in point.

This veteran explorer, single-minded and ambitious, sailed on board the Endurance with the aim of reaching Antarctica to launch the first attempt at a coast-to-coast ice sheet traverse. His attempt was thwarted when Endurance became stuck in sea ice close to the continent in January 1915. Subsequently, his ship was swept west then north for hundreds of kilometers, gradually being crushed by the ice, before being abandoned and then sinking in November 1915. Shackleton's men were forced to carry their lifeboats to where the ice thinned north of the Antarctic circle before they could be launched. In an attempt to save his crew, Shackleton succeeded in sailing in a lifeboat from Elephant Island to South Georgia in late 1916, to raise a rescue party and thus complete the final leg of their extraordinary expedition. In carrying out one of the greatest feats of adventure ever recorded, by completing a huge circle Shackleton's group had unwittingly traced the existence of the western limb of a massive ocean vortex. This so-called eddy circulation traced out by the drifting ice, is now known as the Weddell Gyre. This circulation had swept them 15 degrees of latitude northwards before finally releasing them into the open ocean. Today, scientists recognize these gyres as important features of the Earth's climate,

HMS Endurance trapped in the sea ice pack of the Weddell Sea in early 1915. Eventually, the shifting motions of the ice pack ruptured the hull, sinking the ship. This ship carried the exploration party led by Sir Ernest Shackleton on an ill-fated attempt to be the first humans to cross Antarctica. (Photograph by Frank Hurley, © Royal Geographical Society.)

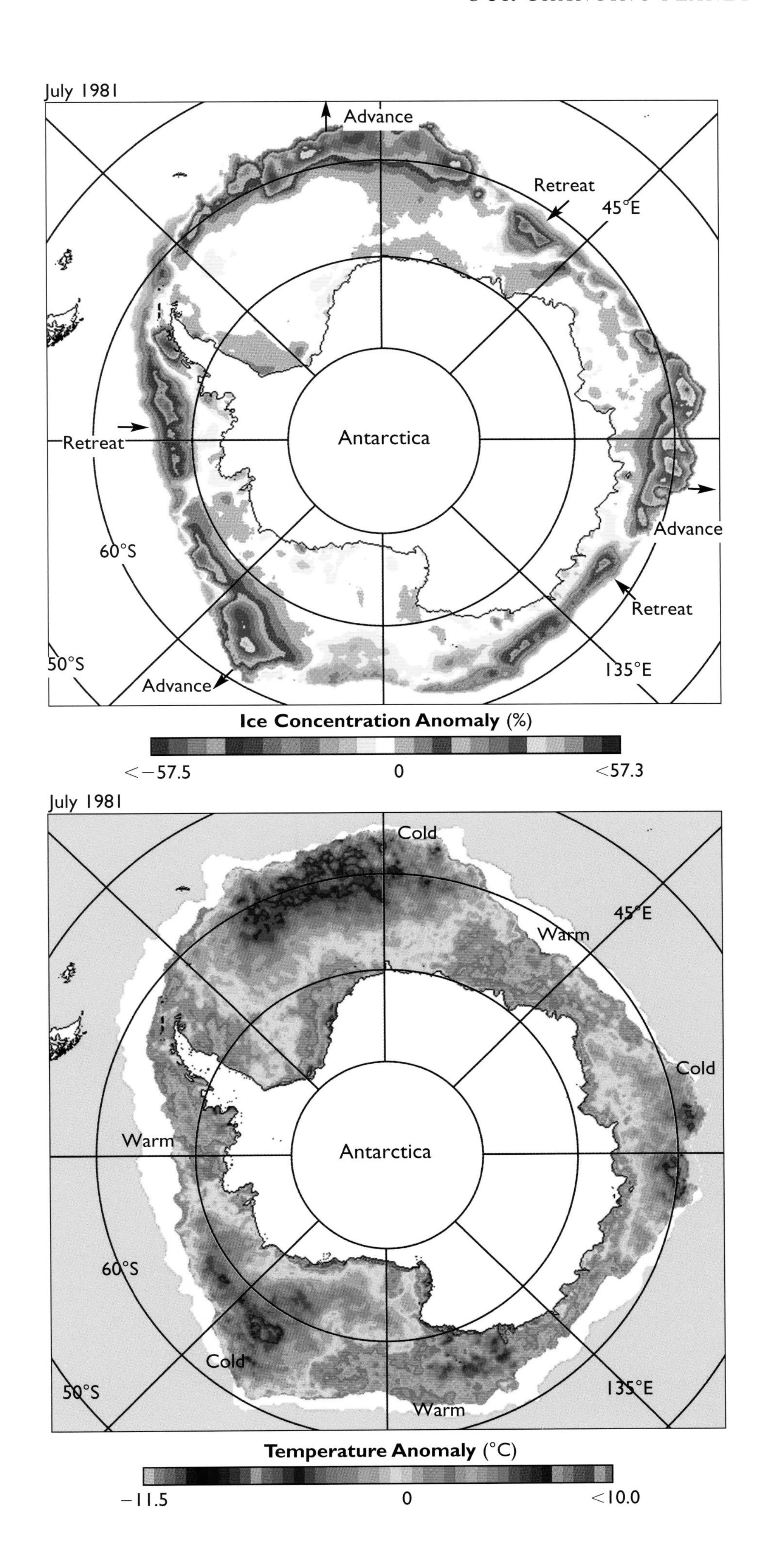


The effect of the Antarctic Circumpolar Wave on sea ice and surface temperatures, demonstrated using passive microwave and thermal infrared data respectively.

Top: Sea ice concentration anomaly map for July 1981 showing a regular pattern of sectors of ice edge advance and retreat around the continent associated with the Antarctic Circumpolar Wave.

Bottom: Corresponding surface temperatures derived from satellite thermal infrared data. This image shows a similar surface temperature anomaly map for July 1981 that indicates the correspondence of relatively cold surface temperatures with ice edge advance, and relatively warm air temperatures in locations of ice edge retreat. The white area shows the maximum ice extent for the period 1979 to 1999.

(Data from AVHRR instruments on NOAA satellites and SMMR and SSMI instruments on the Nimbus 7 and DMSP satellites, respectively, augmented by recent data from the AMSR-E instrument on the Aqua satellite.)

providing a mechanism for transmitting energy and momentum around the globe.

Sea ice drifts at a rate largely governed by the effect of the winds and currents and their direction with respect to the shape of the coastline. In the Southern Hemisphere, low pressure conditions result in winds that drive the ice in large-scale clockwise circulation patterns. Loose ice pack can reach speeds that transport the ice several tens of kilometers each day, as confirmed by the ice-bound drift of the Endurance.

Gyres and other features of ice drift can be mapped using satellite data, by tracking features in the ice as they move. An example is shown here for the period July to September 1998. The pattern reveals the large clockwise, cyclonic gyre circulations known now to be characteristic of both the Weddell and Ross Seas. Each gyre carries ice floes towards the open seas at speeds of up to 65 km a day, with average daily values over the ice pack as a whole closer to 5 km a day.

Using satellite data, year-to-year variations in sea ice extent and surface temperature have been shown to be large and significant in the context of climate. In particular, these variations are now known to be connected with the Antarctic Circumpolar Wave, which has links to the El Niño Southern Oscillation (ENSO) ocean-climate phenomenon. The Antarctic Circumpolar Wave comprises an interconnected pattern of atmosphere, ocean, and ice behavior that roams in a clockwise direction around the entire continent over a period of about 8 years. A visual 'snapshot' representation of the impact of the Antarctic Circumpolar Wave is shown here in the form of anomaly maps for July 1981 (austral winter) for both ice concentration (percentage ice cover) and surface ice temperature. These anomaly maps show the difference between actual ice concentration and surface air temperature and average ice concentration and surface ice temperatures, respectively. Sea ice around the northern margin of the winter ice pack displays alternate patterns of advance and retreat that are linked to patterns of warming and cooling, respectively.

The imprint of the Antarctic Circumpolar Wave is most readily visible in the ice pack and is not as apparent in the open ocean or on the continent. Ice edge advance occurs when cold southerly winds blow off the continent. The flow of cold air drives the ice pack to the north and allows conditions favorable for ice growth to persist. Conversely, in areas of ice retreat surface temperatures are warm because of northerly winds that carry warmer air masses from lower latitudes. Northerly winds cause ice edge melting and poleward retreat of the pack. Atmospheric pressure and wind are thus also essential components of the Antarctic Circumpolar Wave.

The origin of the physical forces that drive the Antarctic Circumpolar Wave is still a source of debate, but recent studies indicate a strong correlation and potential relationship with the more well-known ENSO phenomenon in the South Pacific, which influences weather as far away as North America. Examples shown here highlight the mutual interactions between the sea ice cover and the atmosphere and ocean. Since sea ice acts as an insulator, it controls the heat lost from the ocean to the atmosphere. It is unique in that, in this way, it helps regulate the temperatures of the waters carried south by currents from the equatorial Pacific, Indian, and Atlantic Oceans. The examples shown here illustrate dramatically the role that sea ice plays in transmitting patterns of climate behavior around the entire Southern Hemisphere circumpolar belt. Moreover, since the oceans drive the world's weather patterns, the sea ice around Antarctica is thus a major factor influencing the Earth's wider climate system.

# Antarctic Polynyas: Ventilation, Bottom Water, and High Productivity for the World's Oceans

JOSEFINO C. COMISO
MARK R. DRINKWATER

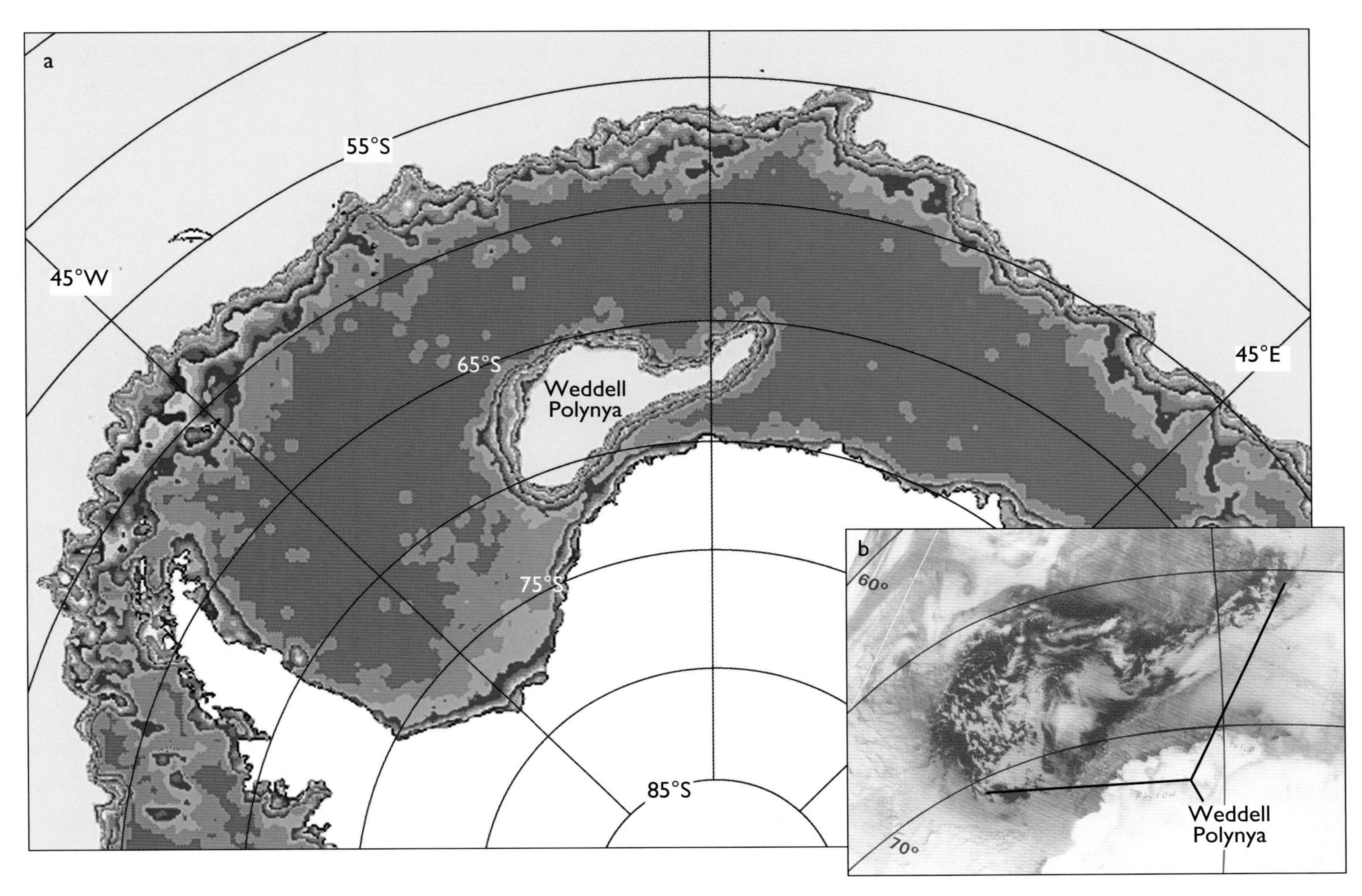


The Weddell Polynya, September 1975. In (a), purple represents highly consolidated ice while light blue represents open water while in (b), black represents open water while white represents ice covered areas or clouds. (Data from the (a) ESMR instrument on the Nimbus 5 satellite and (b) AVHRR instrument on the NOAA-4 satellite.)

The voyages of James Cook and others into the Southern Ocean in the eighteenth century led to the realization that the Antarctic sea ice cover is very expansive yet changes dramatically, not only with the seasons, but also with winds and currents. Ever since the Belgica was beset in the ice in 1898, many vessels have been forced to 'winter over' in the unforgiving ice, including the Magdalena Oldendorff as recently as 2002.

The Antarctic ice pack was generally regarded as a continuous blanket of ice until the advent of images from space in the 1960s, starting with those from TIROS satellites. Early images revealed surprisingly that in addition to innumerable leads (or cracks), there exist large areas of open water within the pack ice, called 'polynyas.' The most spectacular of these polynyas remains the large Weddell Polynya which occurred in three consecutive winters in 1974, 1975, and 1976 and was first observed and monitored continuously by the Electrically Scanning Microwave Radiometer (ESMR) on board the Nimbus 5 satellite.

*Our Changing Planet*, ed. King, Parkinson, Partington and Williams
Published by Cambridge University Press 

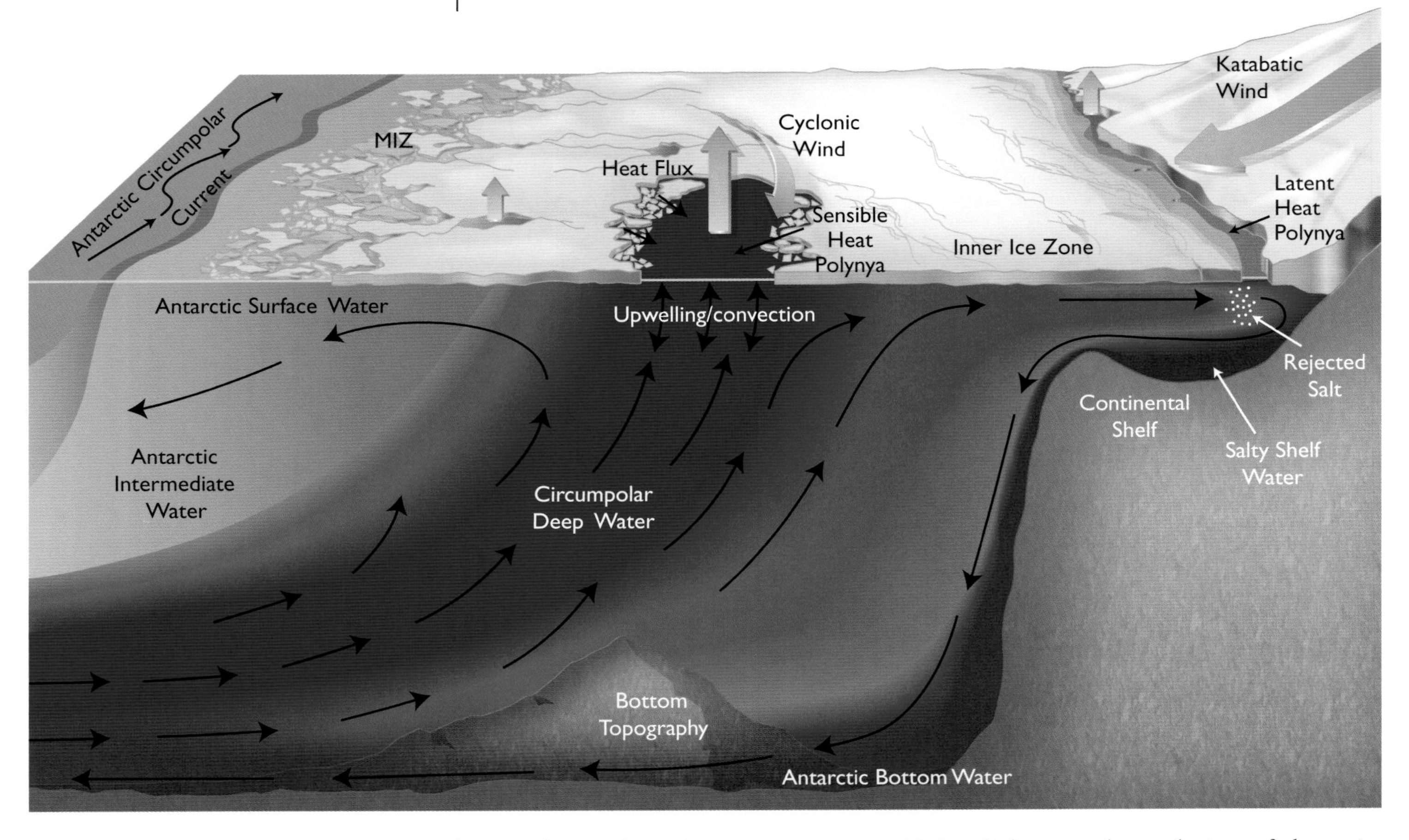


Schematic diagram illustrating the formation of latent and sensible heat polynyas. MIZ refers to the Marginal Ice Zone. (Diagram from Arnold Gordon and Josefino Comiso, modified by Josefino Comiso.)

The passive microwave sensor provided a 24-hour unobscured view of the entire polynya since microwave radiation is insensitive to clouds. This polynya was also observed using the visible channel of the NOAA Advanced Very High Resolution Radiometer (AVHRR) sensor. The Weddell Polynya was unusually large and nearly equivalent to the size of the state of California. In the AVHRR image, the polynya can be seen to be partly covered by clouds which were likely formed due to the unusually high heat and humidity fluxes from the exposed surface of the ocean. Simulation studies using combined sea ice and ocean physical models have indicated the importance of cyclonic winds in at least the initial stages of the formation process for the Weddell Polynya, but wind by itself is not able to explain its sustained duration or reappearance. Oceanographic conditions are therefore believed to play a significant role, through upwelling of warm water over bottom topographic features, like the Maud Rise, maintained in this case by the motion of the Weddell Gyre. As this type of sustained polynya has not reappeared since the 1970s, a debate over the precise formation mechanism continues.

Emperor penguins in front of a polynya. (Photograph by Josefino Comiso.)

The impact on the ocean of such a large polynya event can be dramatic. Hydrographic measurements in 1977 in this region of the waters of the Weddell Sea revealed that the ocean temperature

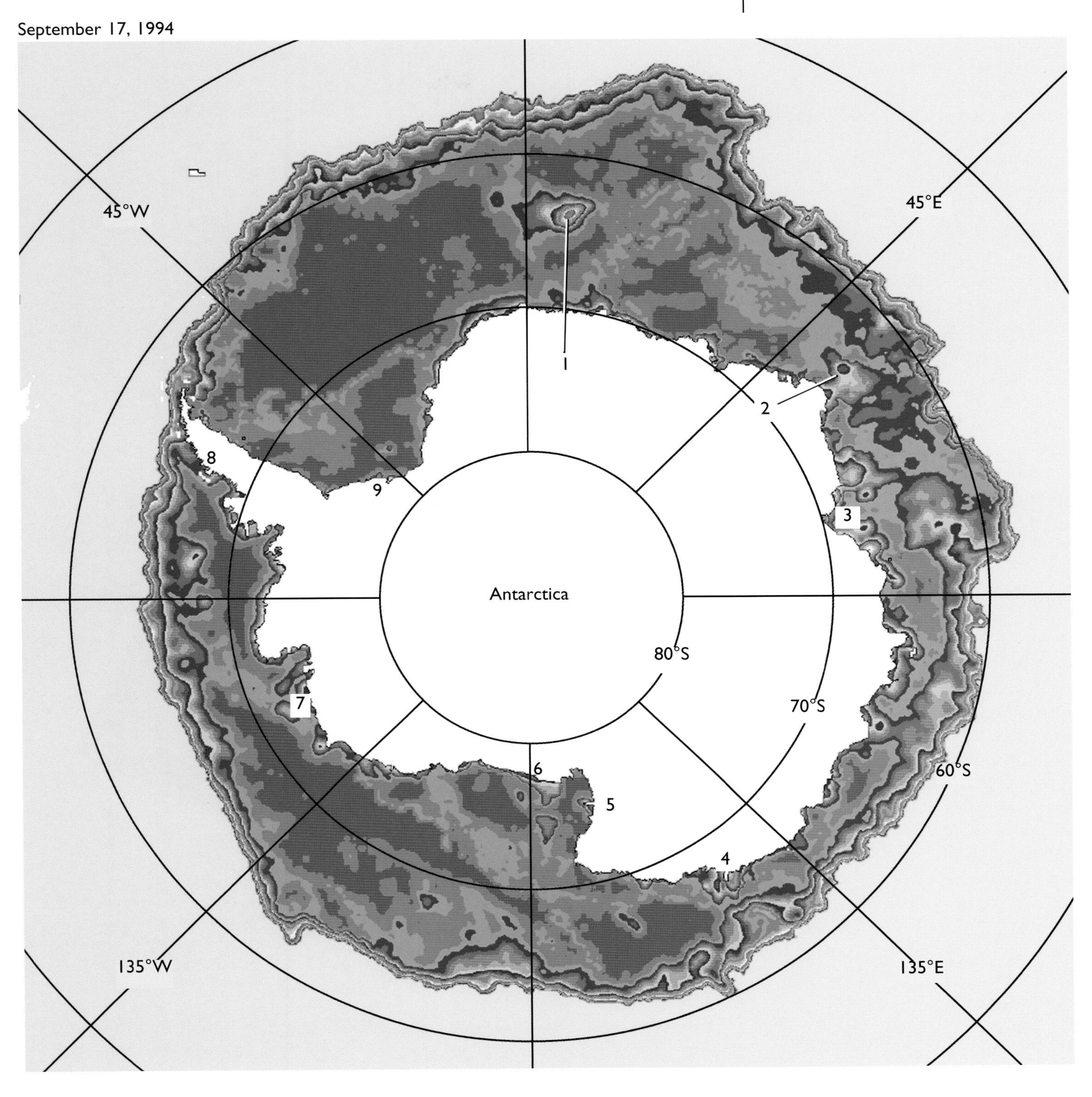


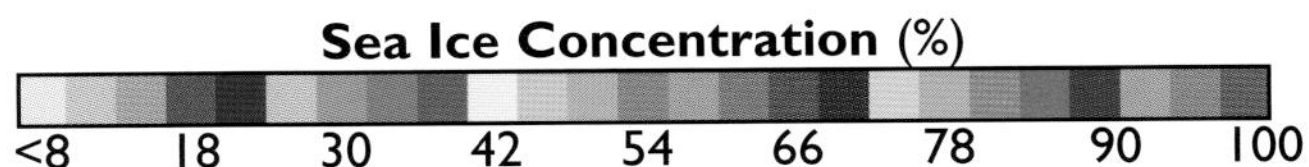


Examples of sensible heat (SH) and latent heat (LH) polynya developments. Among the ones studied are: 1. Maud Rise polynya (SH), 2. Cosmonaut Polynya (SH), 3. Prydz Bay (LH), 4. Mertz Glacier (LH), 5. Terra Nova Bay (LH), 6. Ross Sea (LH), 7. Pine Island (LH), 8. Marguerite Bay (LH), and 9. Ronne Ice Shelf (LH). The polynyas are highlighted by areas of lower ice concentrations adjacent to the numbers. (Image from the SSMI instrument on a DMSP satellite.)

down to 3,000 m depth had cooled by 0.8°C after the polynya event. This is a sure sign that significant convection or 'overturning' took place, allowing the heat to be drawn from the water column by the cold atmosphere. Through the convective process the deep ocean gets 'ventilated,' thereby restoring oxygen levels that are depleted by the oxidation of organic material, as well as introducing elements of anthropogenic origin when the water is exposed to the atmosphere. This type of polynya is sometimes called a 'sensible heat polynya' because

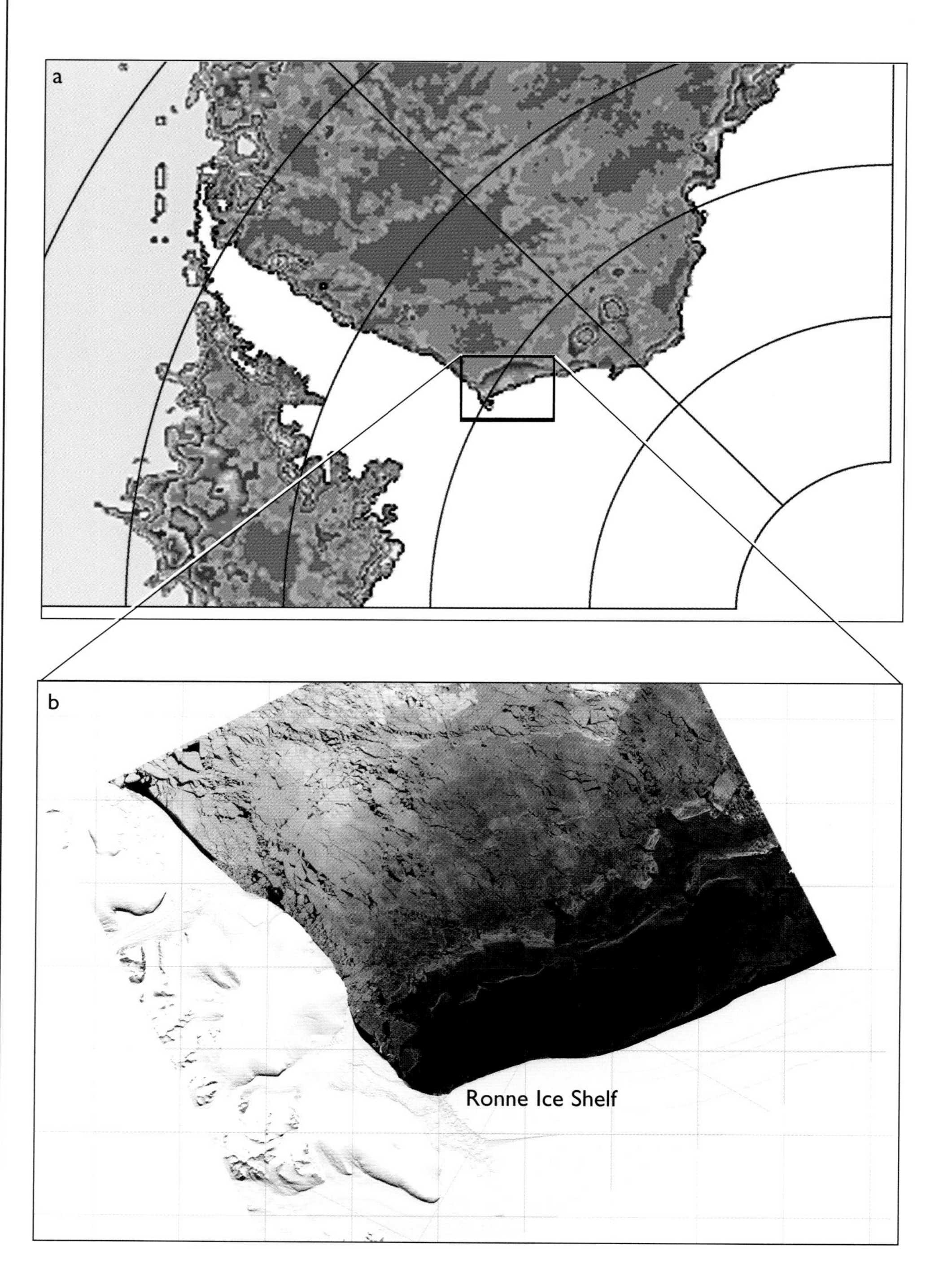


A Weddell Sea polynya located adjacent to the Ronne Ice Shelf, shown in a map of ice concentration derived from passive microwave data at 12.5 km resolution (a) and a visible Landsat image at 15 m resolution (b), on November 26, 2002. Weddell coastal polynyas are believed to be among the main sources of high density and high salinity bottom water that circulates around the globe through the thermohaline circulation. These coastal polynyas are created through the advection of sea ice to the north due to wind, current and other processes causing the creation of ice-free water adjacent to the coastline that gets frozen almost immediately in an extremely cold environment. Persistent wind keeps the ice open for long periods with the result that there is almost continuous growth of sea ice, making the region, in effect, an ice factory. The different shades of gray in the Landsat image represent different stages of growth of the sea ice cover. For example, the darkest gray area corresponds to nilas, a very young type of thin ice which is shown to exhibit little cracks (called leads). Adjacent to this region is a gray area corresponding to a slightly older ice type which is separated from the dark gray area by a narrow light gray area which is likely an area of wind-forced rafting of ice. (Data from the AMSR-E instrument on Aqua and from the ETM+ on Landsat 7.)

the initial warming of surface temperature of the ocean causes melting of sea ice, which in turn leads to the formation of the polynya. Cyclonic winds and the upwelling of relatively warm water may be needed to initiate its formation, but it takes ocean convection to keep the surface water warm and sustain the large ice-free region for a relatively long period.

Elsewhere around the Antarctic, along the coasts, an additional type of polynya is formed in response to strong winds blowing off the ice shelves and glaciers forcing the ice to drift away from the coast creating openings at the land/ice boundary. These are usually called coastal polynyas. Such openings are also known as 'latent heat polynyas' because the surface releases 'latent heat' while in the process of forming ice. Due to extremely cold air temperatures experienced when winds blow off the Antarctic continent, exposed open water areas do not survive ice-free for very long. Depending on the combination of wind, air temperature and wave conditions frazil ice (a 'soup' of ice crystals) accumulates rapidly at

December 1997

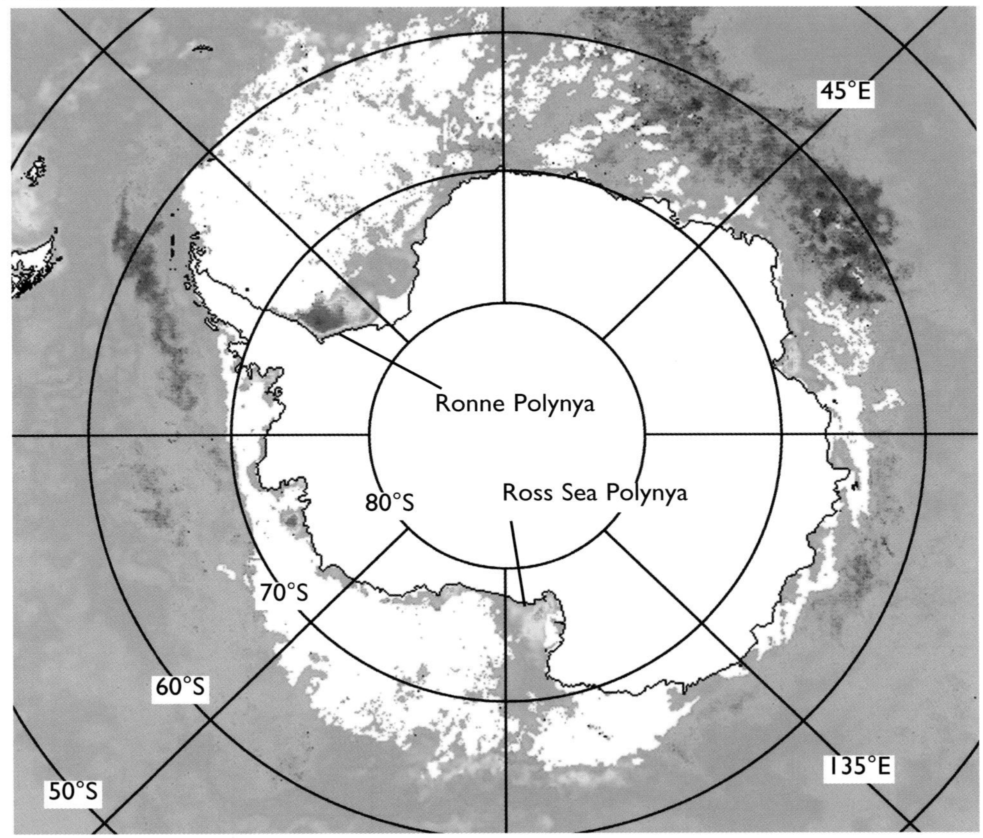


February 1998

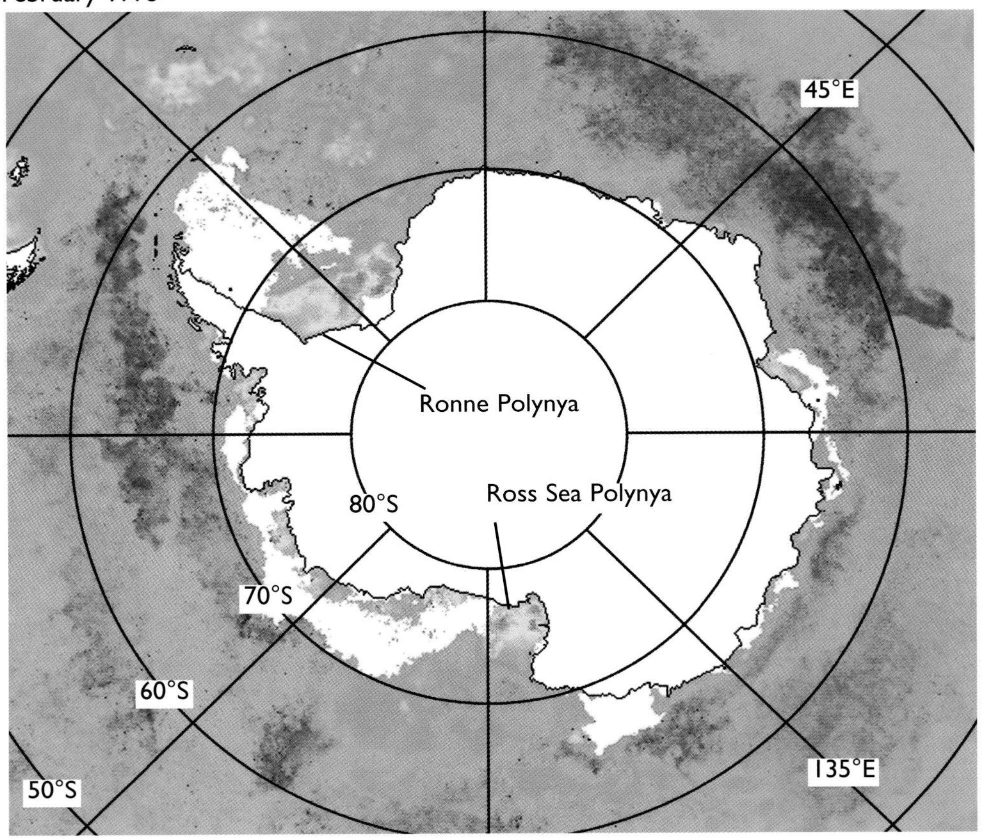


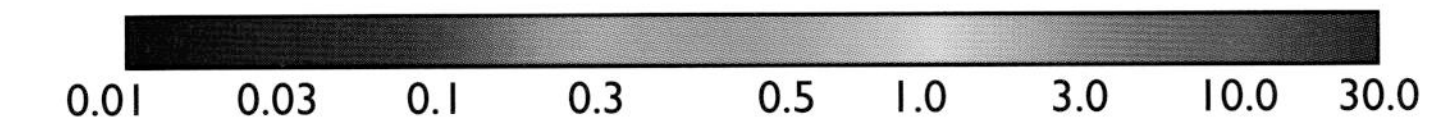


Examples of highest chlorophyll *a* pigment concentrations (in yellow, orange and red) are located in the Weddell Sea and Ross Sea polynya regions during December 1997 (top) and February 1998 (bottom). White represents sea ice or continental ice. (Images from the SeaWiFS instrument on the OrbView-2 satellite.)

The B15 iceberg in the Ross Sea, adjacent to Antarctica, on October 26, 2003. Such calving events affect the production of ice from coastal polynyas and the primary productivity of the region. The iceberg had a length of 295 km, width of 37 km, and an estimated average thickness of 200 m, for a total volume of 2,200 km$^3$. The calving commenced on approximately March 17, 2000, and the iceberg was subsequently grounded. This affected the circulation of water in the ocean shelf region and coastal polynya formation along the edge of the Ross Ice Shelf, thereby impacting bottom water formation. (Data from the MODIS instrument on the Terra satellite.)

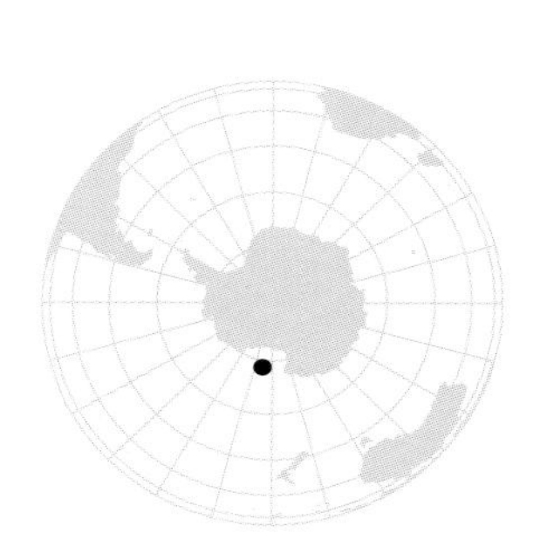

the ocean surface, or the surface water quickly freezes to form a sheet of thin ice called nilas. These winter coastal polynyas have been regarded as ice factories since ice continues to form as long as the polynya is maintained by persistent off-shore wind.

Beneath the surface of the ice forming rapidly in the coastal polynya an interesting process is concurrently taking place. Formation of sea ice is accompanied by the rejection of brine (salty liquid) that in turn increases the saltiness of the underlying water. When polynyas are formed in shelf regions where the water is relatively shallow, as in the western Weddell Sea, the saltier waters collect on the shelf. Eventually the density of the shallow shelf water matches or exceeds that of the surrounding deeper waters and so the shelf water sinks off the shelf to become part of what is known as bottom water. Antarctic coastal polynyas have been postulated as one of the key sources of the world's bottom water and it is this process which drives the circulation of the world's oceans in a process known as thermohaline circulation. If the rate of coastal polynya formation changes, the rate of formation of the bottom water may change as well and this may have a significant impact on the thermohaline circulation and the characteristics of ocean and hence, the global climate.

Passive microwave data have so far provided the longest and most comprehensive characterization of the sea ice cover. From the time of ESMR, we now have more than 30 years of spatially detailed and consistent sea ice data with which to evaluate the size and frequencies of polynya formation. While the large Weddell Polynya has not reappeared, several brief but deep ocean polynya events have occurred. By contrast, coastal polynyas form almost continuously where strong, gravity-driven katabatic winds flow offshore from the continent. Examples of these transient and coastal polynyas are depicted here using passive microwave data with the well-known polynyas labeled numerically. The recurrence of polynyas near Maud Rise (1, top middle, near 0° longitude) is indicative that topography can have a significant influence. Another area of deep ocean polynya formation is located in the Cosmonaut Sea where interactions between the Antarctic circumpolar current and local currents can cause an upwelling of warm water. Coastal polynyas can be monitored continuously by passive microwave sensors during day and night and through almost all weather conditions, providing information on ice and salt production. However, full interpretation of passive microwave data requires the use of high-resolution sensors. A set of passive microwave and high-resolution visible images show more detailed surface conditions during the formation stages of a polynya.

Antarctic polynyas provide valuable regions of high ocean productivity during the spring and summer months. Satellite images of ocean color show elevated chlorophyll-*a* concentrations in the Southern Ocean. Importantly, they also reveal that the highest values are located in polynya areas, especially those near the coastal regions in the western Weddell Sea (Ronne Polynya) and the Ross Sea (Ross Sea Polynya). Melting of ice in spring creates an extensive vertically stable ocean surface with abundant light for photosynthesis and therefore enhanced phytoplankton growth. It is not surprising that penguin colonies are usually located near coastal polynya areas, since they feed on the krill and zooplankton that require the phytoplankton for their own grazing and nourishment. More recent studies also show that whales have spring migration patterns linked to the polynyas of the Ross Sea providing an indication of the extensive nature of the food chain ultimately supported by these phytoplankton blooms. Having been used to discover these remote and productive environments, satellites will continue to form an important tool for their study.

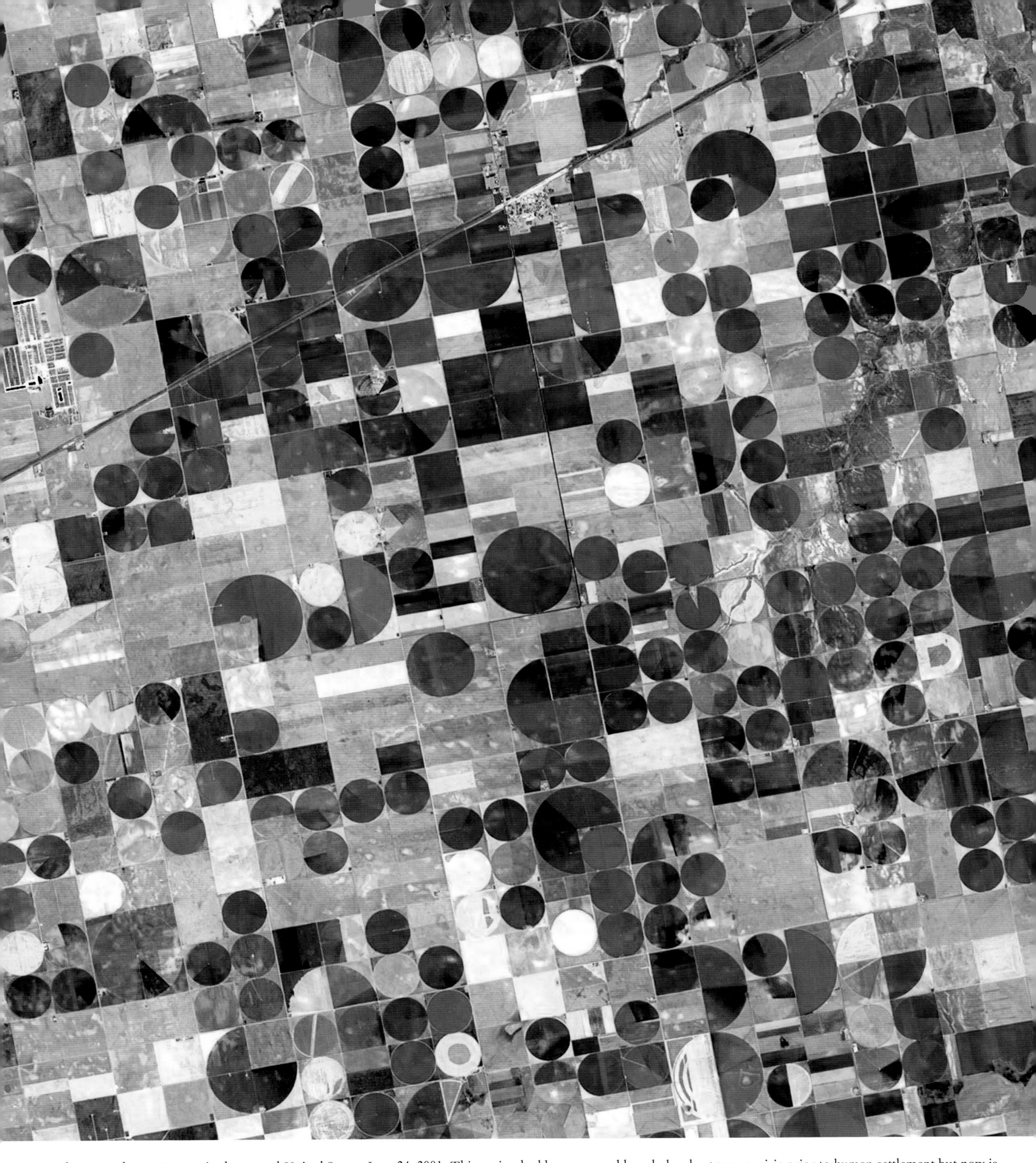

Crop circles in southwest Kansas, in the central United States, June 24, 2001. This region had been covered largely by shortgrass prairie prior to human settlement but now is dominated by agricultural plots, particularly corn, wheat, and sorghum. The prominent circular shapes are due to central pivot irrigation, with water drawn out of a well in the center of the field and long pipes on wheels rotating around the pivot and spraying the water on the crops. The small circular fields in this image are 0.8 km (0.5 mi) in diameter, and the large fields are 1.6 km (1 mi) in diameter. In general, the mature growing crops are dark green, the less mature growing crops are paler green, the recently harvested fields are gold, and the fields that have been plowed under or lie fallow are brown. (Data from the ASTER instrument on the Terra satellite; figure courtesy United States/Japan ASTER Science Team and the NASA Earth Observatory.)

# Evidence of Our Tenure: Introduction

Claire L. Parkinson

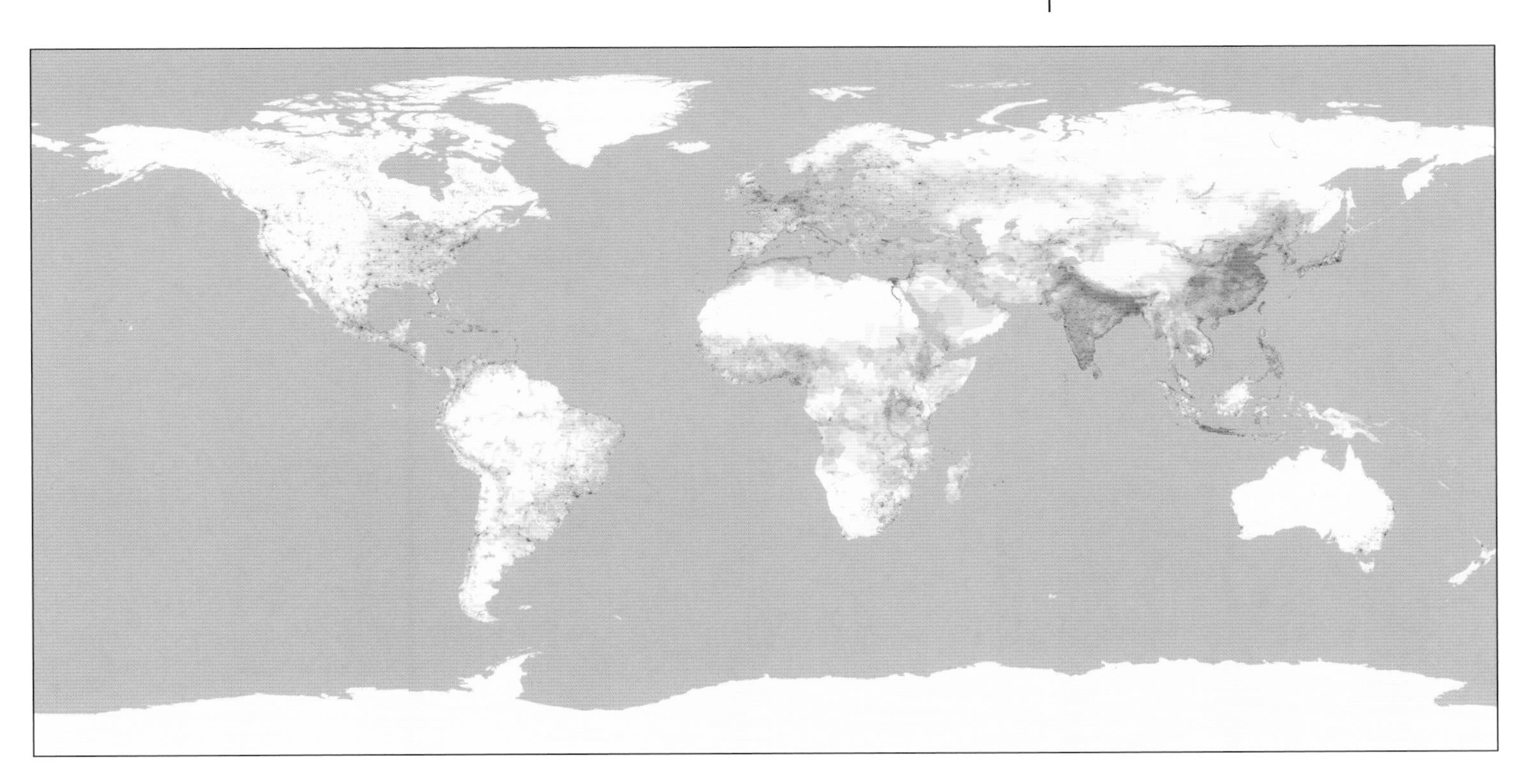


Above: World population density. Beige areas have less than 10 people per km$^2$, while dark red areas represent population densities of 10,000 per km$^2$ or greater. (Map based on Gridded Population of the World, version 3, Socioeconomic Data and Applications Center, Columbia University; image, NASA/SVS.)

Humans have impacted local environments ever since the first humans arrived, irrespective of the varied understandings of when, where, and how that occurred. From the beginning, walking, crawling, climbing, lying down, eating, breathing, and depositing waste have all had impacts on the immediate environment. These impacts accelerated as human capabilities increased, most notably with the development of fire, the hunting of game, the construction of dwelling places, and, among the most crucial of all, about 10,000 years ago, the development of agriculture. In addition to involuntary impacts, such as from breathing, humans have an unsettling history of using the atmosphere, oceans, and land as dumping grounds for a vast array of waste products, sometimes then moving elsewhere as the local region became too wasted or otherwise unpleasant. As the human population has increased, from 1–10 million 12,000 years ago, to about 300 million 1,000 years ago, to about 1 billion in 1800, and to over 6.5 billion in 2007, the impacts have expanded from local impacts to regional impacts and, by now, to global impacts.

The fact that humans are having massive impacts is undeniable. Cities transform generally vegetated land surfaces into mixtures of buildings, pavement, and other artificial surfaces, along with some vegetation, breaking up the habitats of other species of life and affecting air flow and air quality. We have spewed vast mixtures of chemicals into the atmosphere,

Opposite: San Francisco, California, August 28, 2004. The water in the lower right is San Francisco Bay, and the prominent bridge is the Bay Bridge. (Data from the IKONOS 2 satellite, courtesy Space Imaging.)

*Our Changing Planet*, ed. King, Parkinson, Partington and Williams
Published by Cambridge University Press 

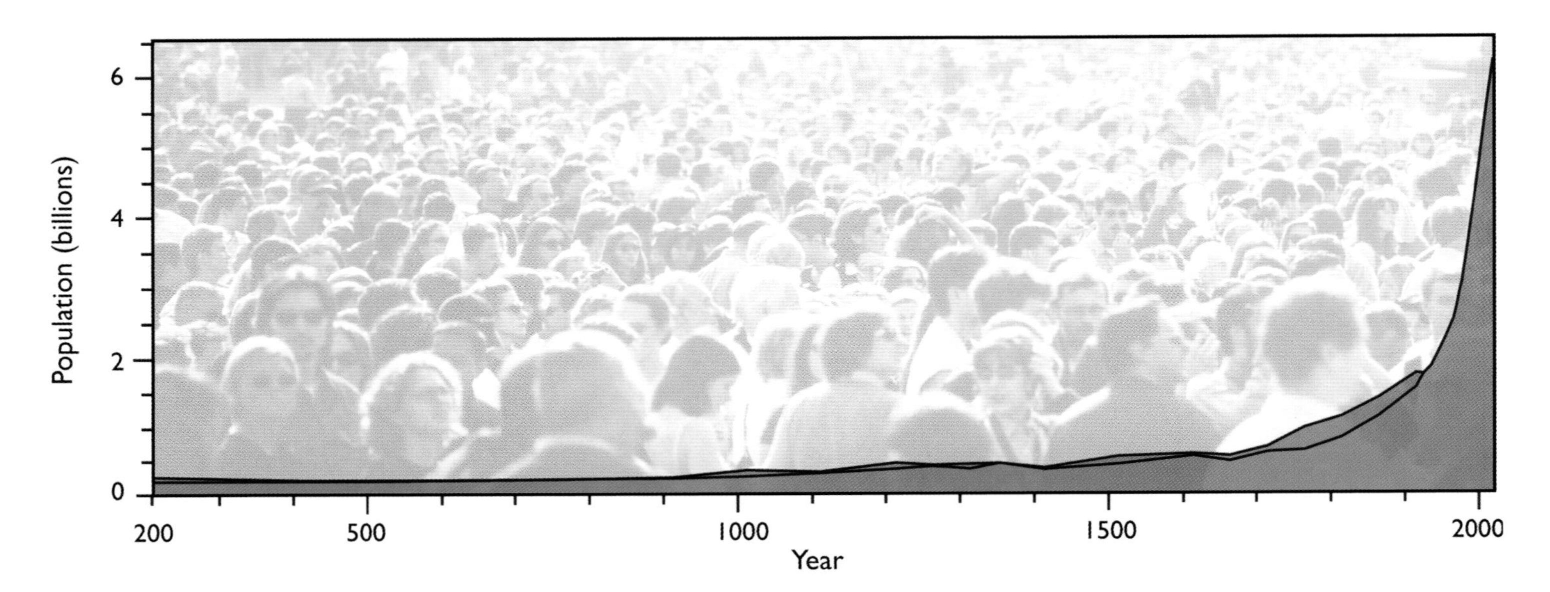


Global human population, AD 200–2004, with both high-end and low-end estimates given for years prior to 1910. (Data from the United States Census Bureau, obtained from http://www.census.gov/ipc/www/worldhis.html on July 14, 2005, and plotted by Nick DiGirolamo; photograph © 2006 Fotosearch Stock Photography.)

with undeniable impacts on atmospheric composition, as registered in worldwide increases in carbon dioxide and other gases, and have tossed other wastes into the oceans, lakes, and other waterways of the world. Fortunately, concern for the environment exists also, and as a result of both conscientious citizens and government regulations, many individual lakes and rivers have been notably cleaned up and there is in many locales increased recycling of materials and much less unrestricted dumping of waste into the environment. Furthermore, in many areas purposeful reforestation is occurring, helping to reverse the vast human-caused deforestation that occurred historically first across much of Europe and Asia, then across much of North America, and more recently across significant portions of the tropical rain forests.

Humans have had both positive and negative impacts on the environment, and many of these, both good and bad, are sufficiently consequential to be observable from satellite observations. Several categories of human impacts on the atmosphere are illustrated under "The Dynamic Atmosphere" in chapters on ship tracks, airplane contrails, atmospheric pollution (both carbon monoxide and nitrogen dioxide), stratospheric chlorine monoxide (ClO) increases, and, related to the ClO increases, stratospheric ozone depletion. In this section, 15 chapters illustrate a sampling of the many ways that satellite data are being used to reveal evidence of human impacts on the Earth's land and ocean surfaces.

Illustrating the application of satellite imagery to the study of past human impacts, the chapter on "Mapping the Ancient Maya Landscape from Space" describes satellite revelations about ancient Mayan features, such as roads, water reservoirs, and cities, in what is now northern Guatemala. It also depicts damaging modern deforestation that is contributing to the destruction of Mayan archeological sites.

The chapter entitled "Global Land Use Changes" provides an overview of land use changes over time, illustrating the recent situation with satellite-derived global maps of percent croplands and pastures. This chapter is followed by a chapter on "The Changing Role of Fire on Earth," describing the human use of fire around the globe for agricultural practices and land management. "The Tropical Rain Forest: Threatened Powerhouse of the Biosphere" highlights deforestation specifically in the tropical rain forests; and "The

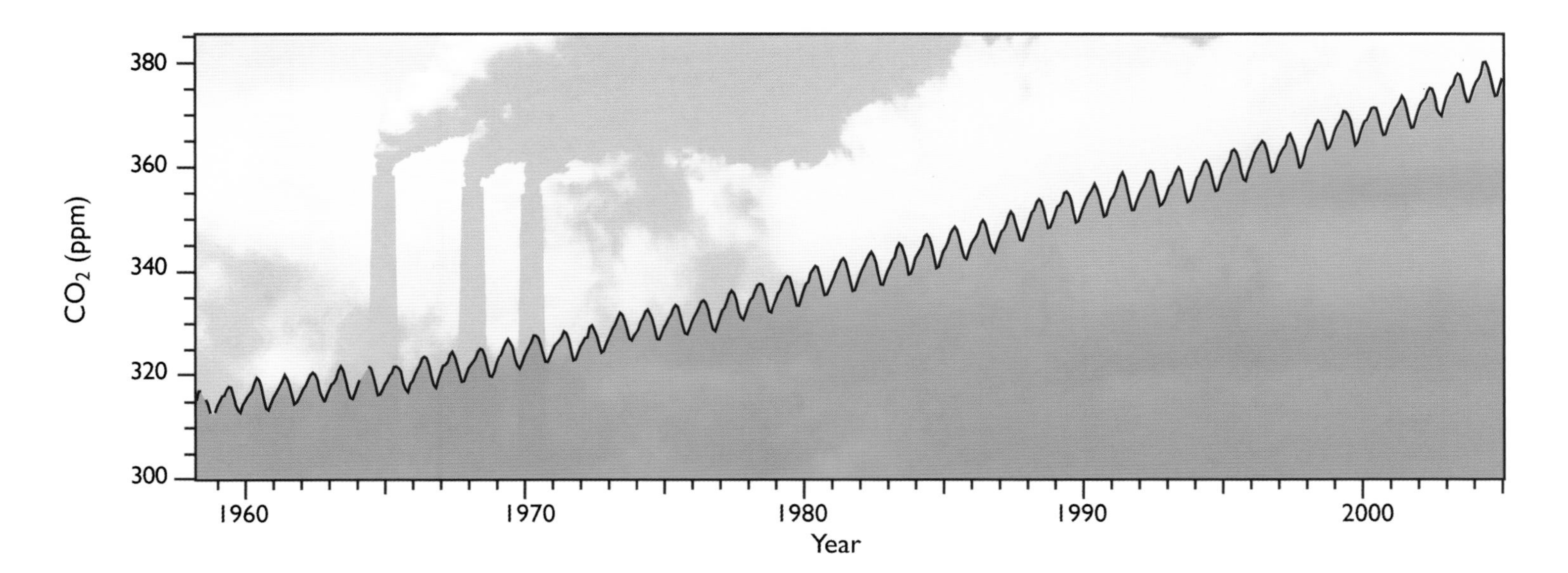


Atmospheric carbon dioxide concentrations in parts per million (ppm), plotted by month for the period March 1958–December 2004, as measured at the Mauna Loa Observatory in Hawaii. No data are available for June 1958, October 1958, or February–April 1964. The monthly data reveal both the natural cycle of seasonal increases and decreases in $CO_2$ and the prominent overall increasing trend, largely caused by human activities. (Data from C. D. Keeling, T. P. Whorf, and the Carbon Dioxide Research Group, Scripps Institution of Oceanography, University of California, July 14, 2005; photograph © 2006 Rinderart, Arkamind Technologies Limited.)

Green Wave" chapter demonstrates that, in contrast to purposeful deforestation, some changes occurring in the Earth system are favorable for vegetation growth, showing that in both Eurasia and North America the highly publicized warming of the late twentieth century has been accompanied by a 'greening' of the surface and a lengthening of the growing season.

Several chapters center on urbanization, which is among the most prominent intentional transformations on the planet. This group of chapters begins with a brief historical overview of urbanization and illustrations of the satellite monitoring of urban areas, in "Monitoring Urban Areas Globally and Locally," followed by a discussion of the many impacts of urbanization, in "Gray Wave of the Great Transformation," with results shown in particular for the mid-Atlantic region of the United States, along with a global composite image of nighttime lights, predominantly from cities. The chapter "Urban Heat Islands" discusses the 'urban heat island' effect, wherein cities are typically measurably warmer than the surrounding communities, as a result both of their generally less reflective surfaces and the heat generated within them, illustrating with results from Atlanta, Georgia. Attention then shifts to China, in "Urbanization in China: The Pearl River Delta Example," illustrating the extensive recent urbanization taking place in the world's most populous country.

The human need for water has caused a variety of intentional and unintentional changes in the environment, and several chapters examine these. The use of irrigation around the globe, beginning some 6,000 years ago in the region between the Tigris and Euphrates rivers, is summarized in "Water Issues in the Fertile Crescent: Irrigation in Southeastern Turkey," which also illustrates the landscape impacts with satellite imagery before and after the damming of a portion of the Euphrates in southeast Turkey. This chapter is followed by 2 chapters highlighting hugely detrimental consequences that have accompanied some purposeful water diversions. The first, "Chronicling the Destruction of Eden: The Draining of the Iraqi Marshes," discusses and illustrates the damage to Iraqi marshlands brought about by siphoning off water both for irrigation and for brutal political purposes. The second, "Destruction of the Aral Sea," describes the horrendous and widely consequential decay of the Aral Sea, caused in large part by humans siphoning off, for irrigation of fields of cotton and rice, water from the Amu Dar'ya and Syr Dar'ya rivers flowing into the sea. This is

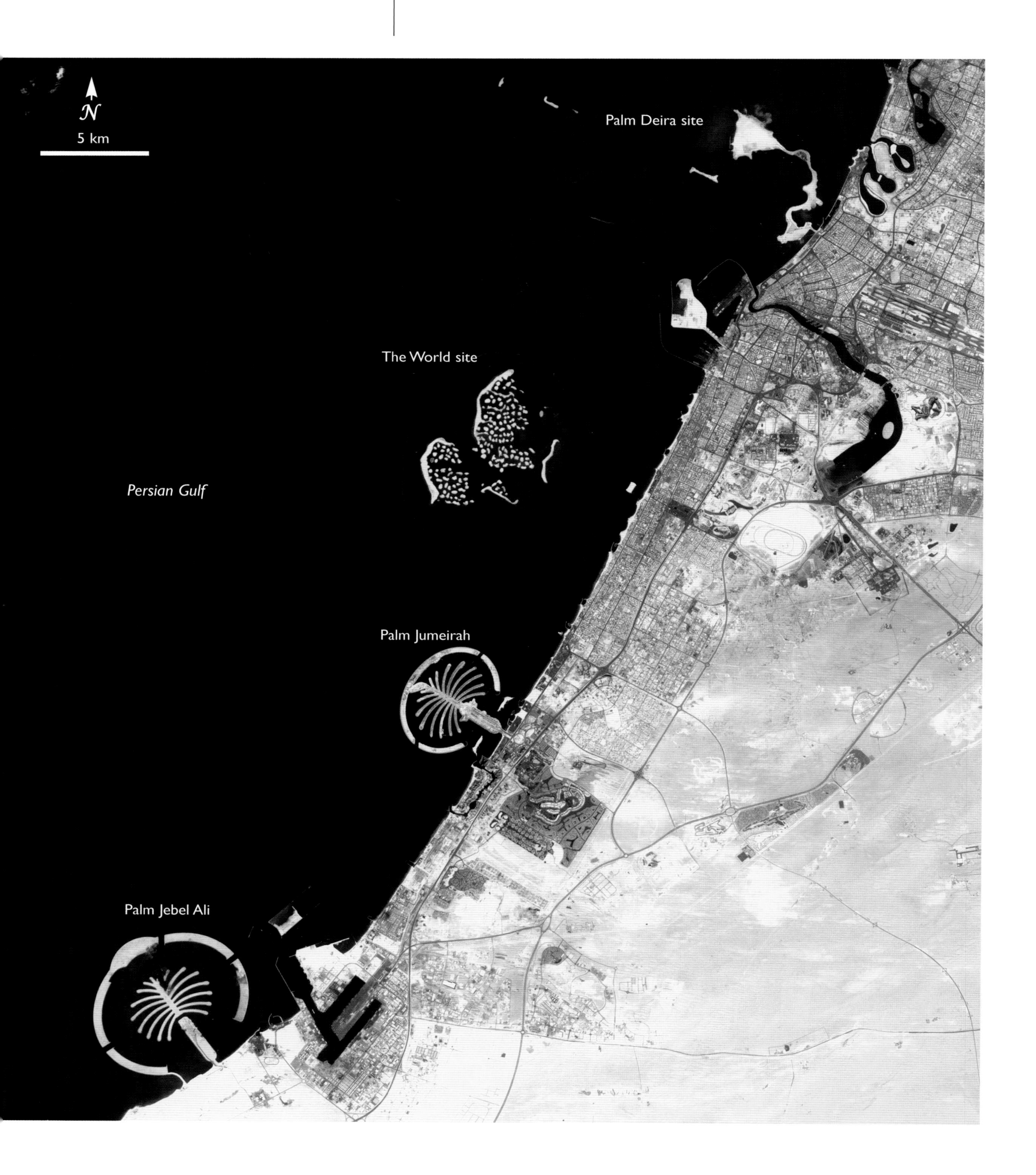
N
5 km
Palm Deira site
The World site
Persian Gulf
Palm Jumeirah
Palm Jebel Ali

followed by a chapter on "The Desiccation of Lake Chad," describing a case where the consequences have also been severe, but it is less certain to what extent humans have contributed. The central reason for the decay of Lake Chad is an extended period of drought, but an additional reason is the use of lake water for irrigation projects in 4 neighboring countries: Chad, Niger, Nigeria, and Cameroon.

The section concludes with 2 chapters highlighting largely unintended consequences of industrial activity. "Industrial Pollution in the Russian Arctic" centers on a region where industrial activity releases massive amounts of pollutants to the atmosphere each year, including hundreds of thousands of tons of sulfur, causing widespread damage to the landscape as well as to the atmosphere. "Oil Spills at Sea" describes accidental and intentional oil spills and the consequent pollution of oceans and waterways, illustrating in particular the accidental case of the oil tanker Prestige in 2002 and the intentional (and illegal) release of oil into the Mediterranean Sea in 2004.

The chapters in this section illustrate a sampling of the many impacts that humans have had on the environment and a sampling of the many ways that satellite data are helping us to see and understand the resulting changes. In many instances, satellite imagery can show more readily than through other means the large-scale environmental changes. A hope is that through this ready depiction of changes, satellite imagery is contributing to increased consciousness of the environmental consequences of human activities and will eventually help lead us to improved stewardship of our home planet.

Opposite: Artificial islands along the coast of Dubai, United Arab Emirates, September 18, 2006. These islands, still under construction, are being built by dredging sand from the bottom of the Persian Gulf and spraying it from ships to create the desired shapes, which in the case of Palm Jebel Ali, Palm Jumeirah, and, eventually, Palm Deira are palm trees enclosed by arcs. The World site already resembles a world map but will do so even more accurately once the construction is completed. On the mainland, other quite visible human-created features include the network of roads and buildings along the coast and the green color of the well-irrigated areas, contrasting with the tan color of the sand dominating much of the lower right of the image. (Data from the ASTER instrument on the Terra satellite.)

Below: Aerial photograph of a portion of Palm Jumeirah, along the coast of Dubai, United Arab Emirates, February 22, 2006, showing extensive construction on this artificial island (see facing ASTER image). (Photograph © 2006 Brian McMorrow.)

Thomas L. Sever
Daniel E. Irwin

# Mapping the Ancient Maya Landscape from Space

November 5, 1988

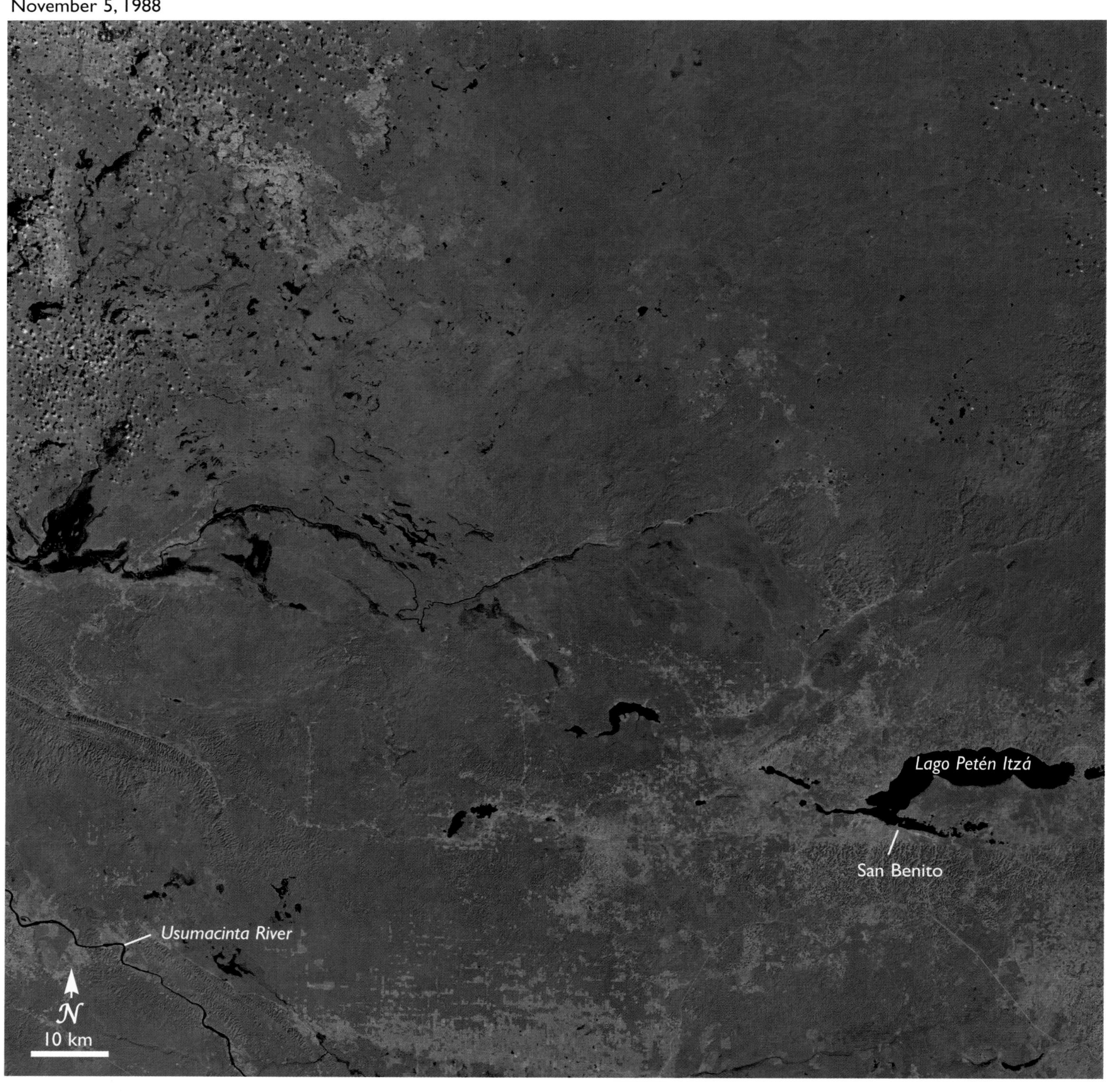


*Our Changing Planet*, ed. King, Parkinson, Partington and Williams

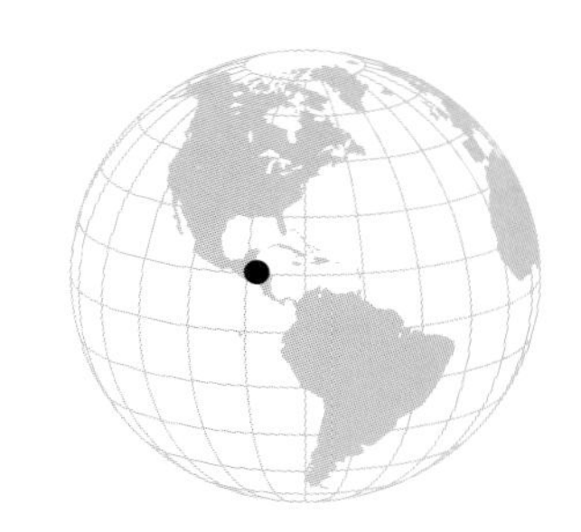

March 27, 2000

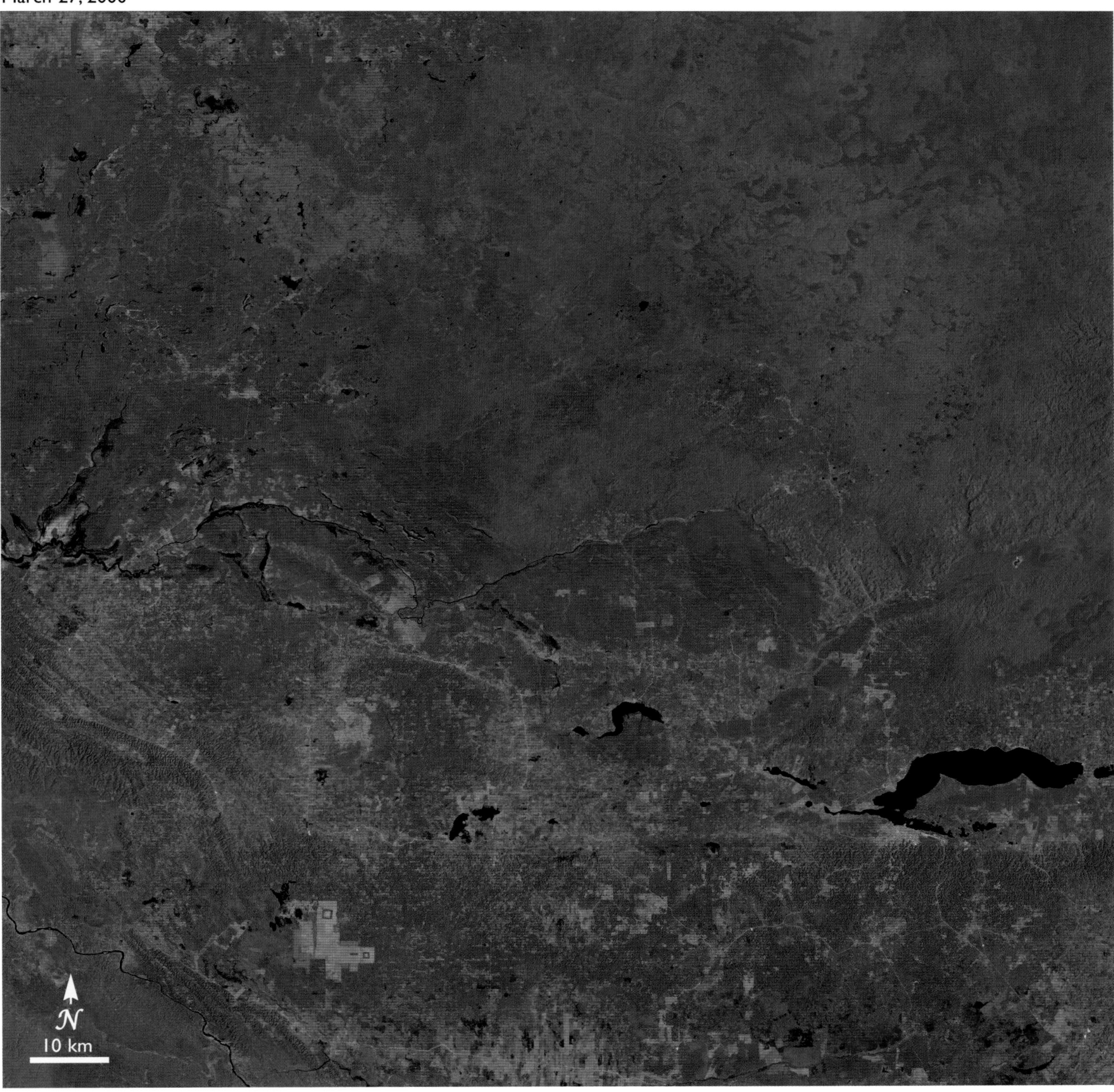


Comparison of satellite imagery over the Petén region of northern Guatemala showing the effects of deforestation over the 11-year period from November 5, 1988 to March 27, 2000. The green color indicates healthy forest, while the red color represents deforested areas. Deforestation is destroying the Maya archeological sites in the region. (Data from the TM instrument on the Landsat 4 satellite [opposite] and from the ETM+ instrument on the Landsat 7 satellite [above].)

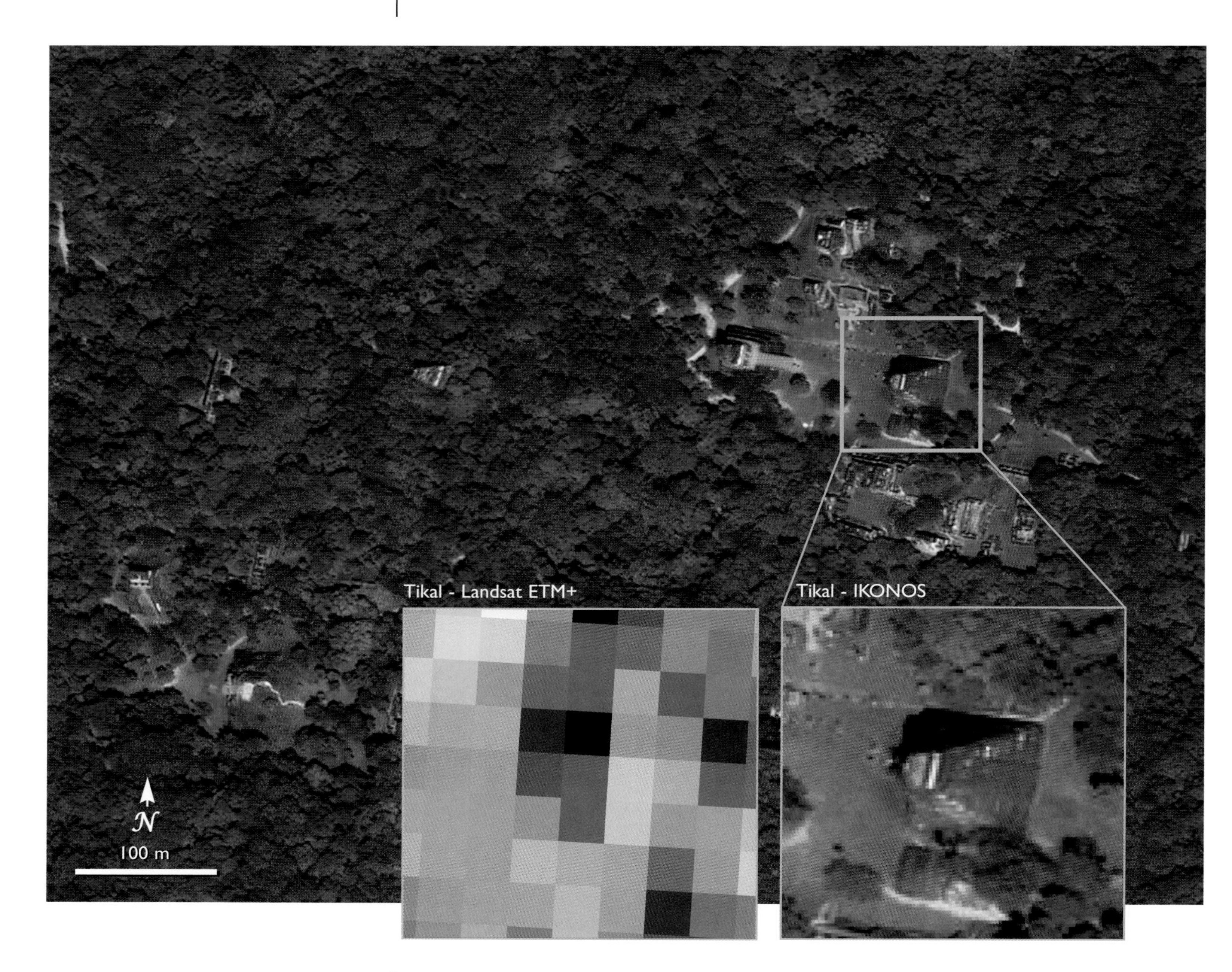


Tikal National Park in Petén, Guatemala, March 13, 2001. (Data from the IKONOS satellite, courtesy Space Imaging.)

Insets: Comparison of high-resolution Landsat imagery (left) and higher-resolution IKONOS imagery (right) over Tikal National Park in Petén, Guatemala. The high-resolution capability of IKONOS data reveals detailed temples and plaza areas. (Data from the ETM+ instrument on the Landsat satellite and from the IKONOS satellite.)

The Petén region of northern Guatemala is one of the key places on Earth where it is believed that major archeological sites remain to be discovered. It was in this region that the Maya civilization began, flourished, and abruptly disappeared. Remote sensing technology is helping to locate and map ancient Maya sites that are threatened today by accelerating deforestation and looting. Thematic Mapper and IKONOS satellite and airborne Star3i radar data, combined with Global Positioning System (GPS) technology, are successfully detecting ancient Maya features such as cities, roadways, canals, and water reservoirs. Satellite imagery is also being used to map the bajos, which are seasonally flooded swamps that cover over 40% of the land surface. The use of bajos for farming has been a source of debate within the professional community for many years. However, the recent detection and verification of cultural features within the bajo system are providing conclusive evidence that the ancient Maya had adapted well to wetland environments from the earliest times and utilized them until the time of the Maya collapse. The use of the bajos for farming is also an important resource for the future of the current inhabitants and the region, which is experiencing rapid population growth.

Mayan pyramids in northern Guatemala, March 1991. (Photographs by Thomas Sever.)

Remote sensing imagery is also demonstrating that in the Preclassic period (600 BC–AD 250), the Maya had already achieved a high organizational level as evidenced by the construction of massive temples and an elaborate interconnecting roadway system. Although they experienced several setbacks such as droughts and hurricanes, the Maya nevertheless managed the delicate forest ecosystem successfully for several centuries. However, around AD 800, something happened to the Maya to cause their rapid decline and eventual disappearance from the region. The evidence indicates that at the time there was increased climatic dryness, extensive deforestation, overpopulation, and widespread warfare. This raises a question that is relevant to the contemporary world—namely, how severe do internal stresses in a civilization have to become before relatively minor climate shifts can trigger a widespread cultural collapse?

Jonathan A. Foley
Navin Ramankutty
Billiana Leff
Holly K. Gibbs

# Global Land Use Changes

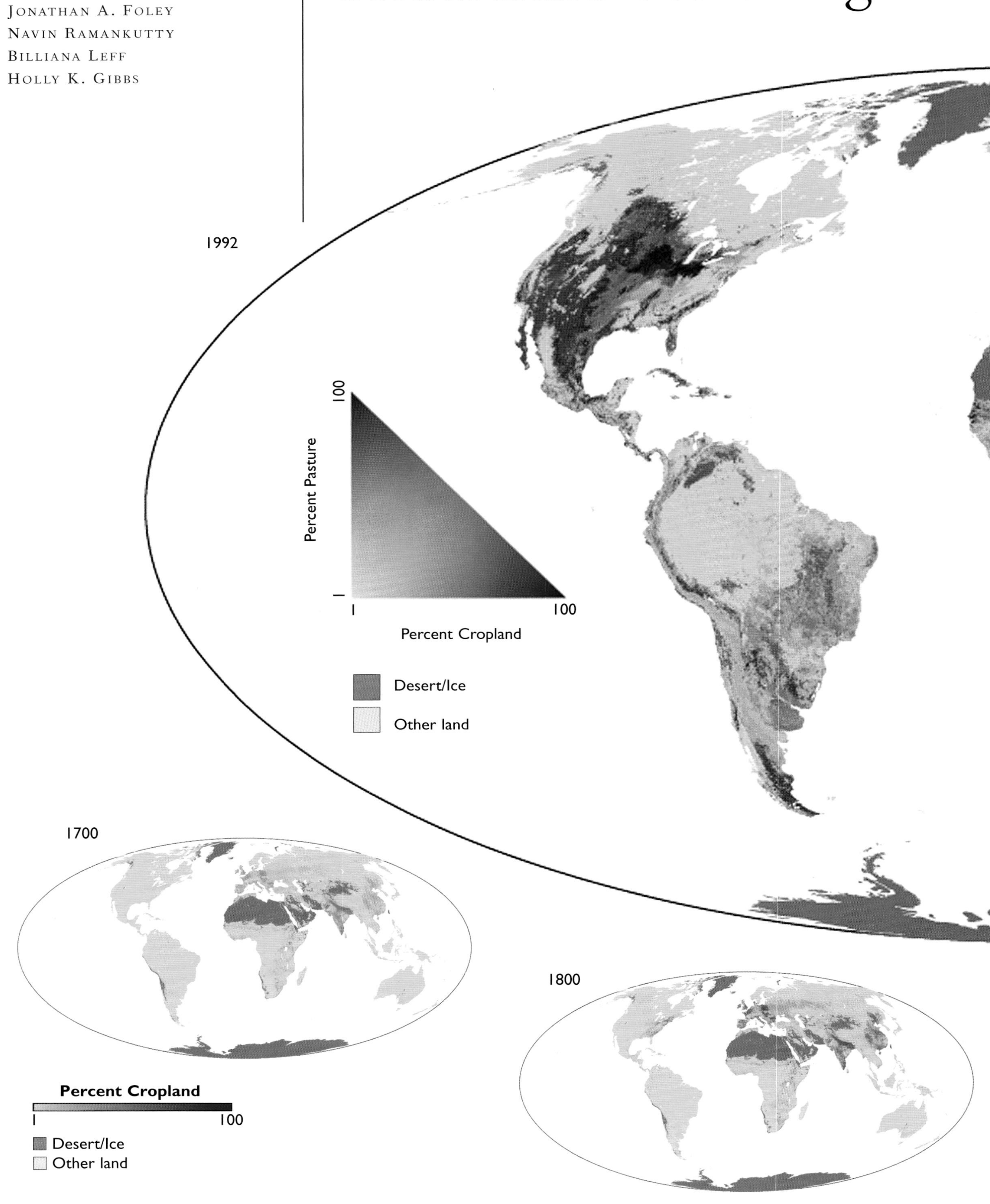


*Our Changing Planet*, ed. King, Parkinson, Partington and Williams
Published by Cambridge University Press 

Extent of worldwide use of land for pasture and crops as of 1992 (large map). The world's productive agricultural lands are found in a few places—largely where climatic conditions and soils allow for productive landscapes. The four small maps across the bottom show the progressive use of land for the production of crops over the last 300 years. (Data from ground-based census information and from the AVHRR instrument on the NOAA 11 satellite; Foley *et al.*, Global consequence of land use, *Science*, vol. 301, pp. 570–574, 2005.)

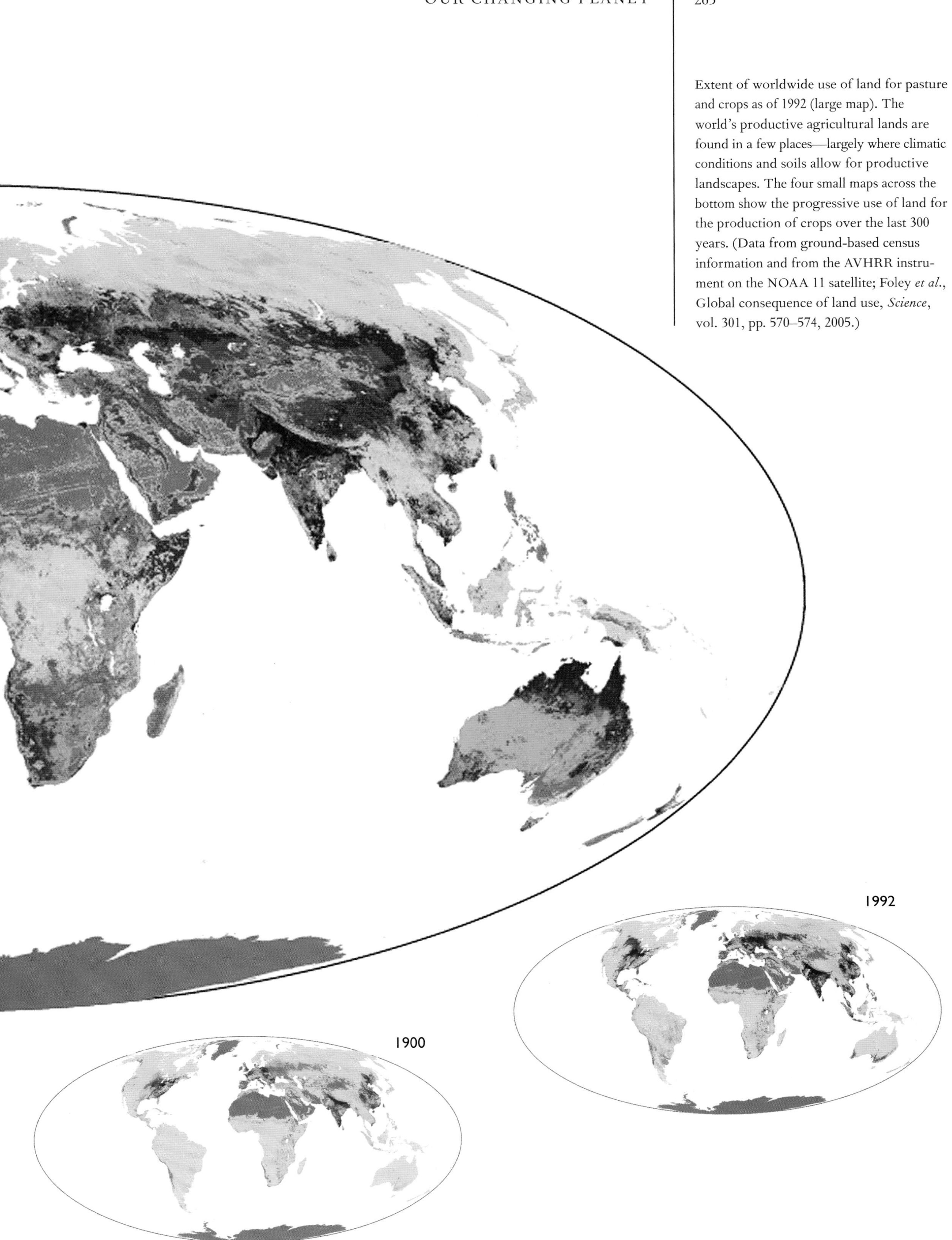

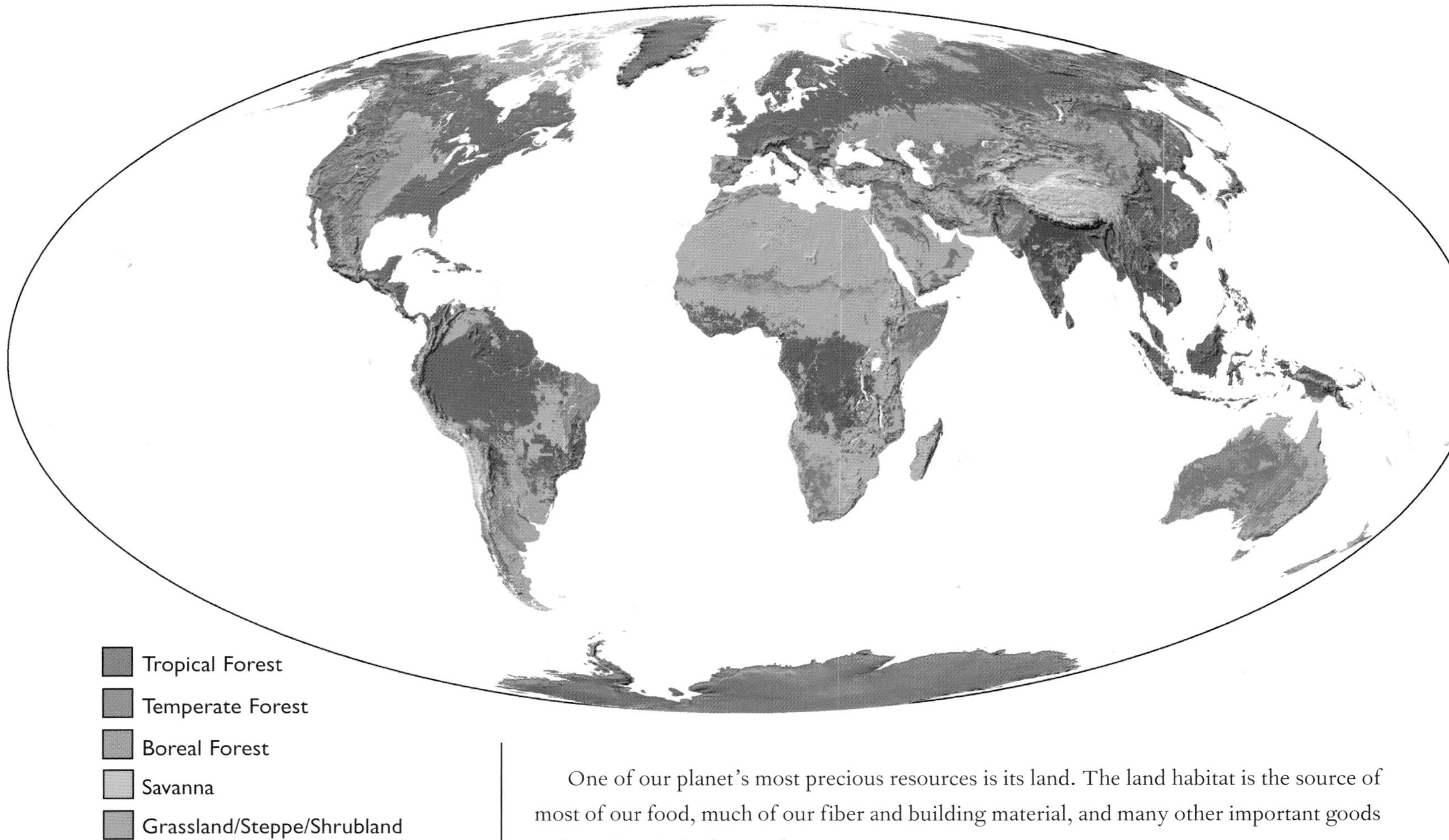


Map of natural vegetation types that likely would exist if there were no human agricultural utilization. The expansion of croplands and pastures came at the expense of natural ecosystems, including forests, grasslands, and savannas.

One of our planet's most precious resources is its land. The land habitat is the source of most of our food, much of our fiber and building material, and many other important goods and services. It is also our home.

But our relationship to the land has been changing. For thousands of years, people have used the land for agricultural practices and settlements. But in the last 300 years, we have seen an explosion of human land use, and this has come mainly in the expansion of croplands and pastures.

Now, at the beginning of the twenty-first century, human activities clearly dominate over one third of the Earth's land surface—and we have influence over a great deal more. For example, we use nearly 18 million $km^2$ (roughly the size of South America) for growing our crops; and we use another 32 million $km^2$ for our cattle and pastures. Our cities, villages, and towns are growing too: we now use roughly 0.5 million $km^2$ for those. The lands that are left are rarely untouched: they often feel the presence of *Homo sapiens* through our roads, mining efforts, and logging practices.

Our growing demand for land comes at the expense of the natural biosphere. Croplands and pastures have been established in lands that used to be covered with forests, savannas, and grasslands. Our cities and towns have been built on what were natural areas as well—ranging from tropical rain forests (like the city of Manaus in the middle of the Amazon) to desert (like Las Vegas, in the arid lands of Nevada). Since we now dominate nearly a third of the land surface with our agriculture and settlements, we only leave the remaining two-thirds for all other species to use without major human disruption.

The effects of our land use practices go far beyond their immediate surroundings. Clearing forests and replacing them with pastures, for example, transfers the carbon in

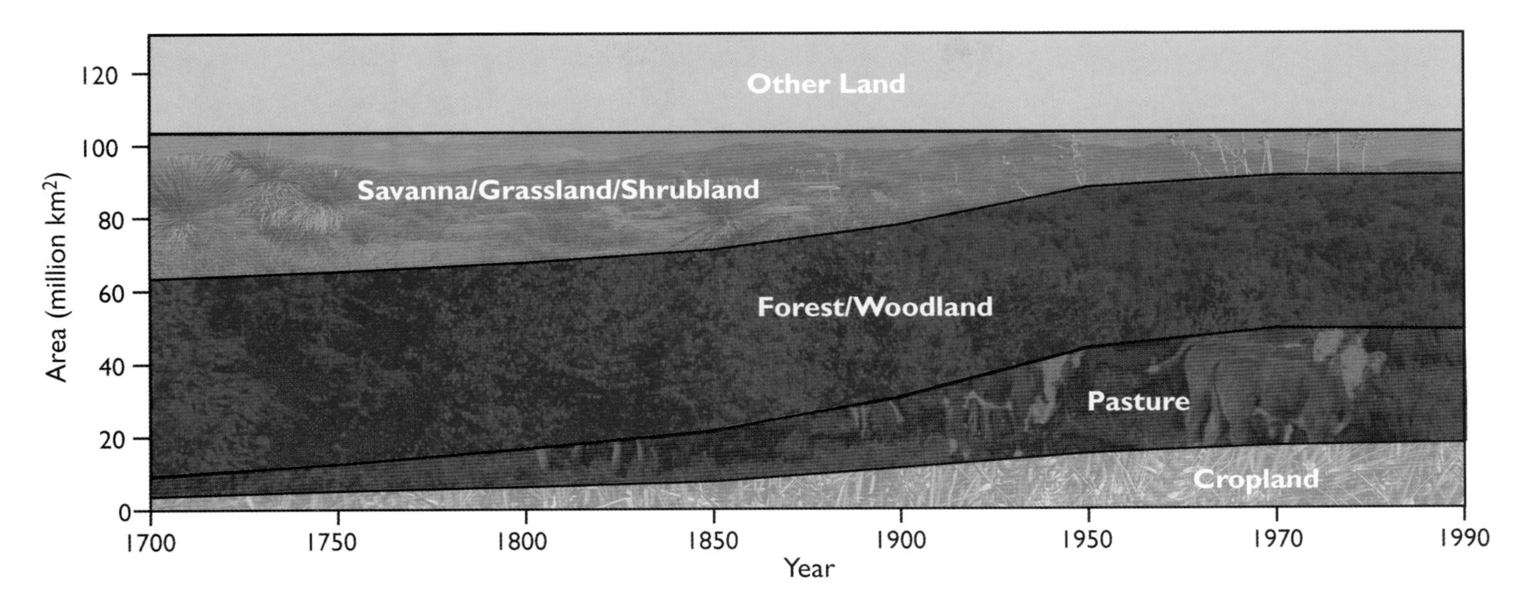


Changes in land cover during the last 300 years due to agricultural expansion. During the last 300 years, there has been a large increase in the amount of land devoted to agriculture (croplands and pastures)—coming at the expense of natural ecosystems. As human population and material consumption continue to increase, the pressure on our finite land base will also continue to increase. (Data from the Center for Sustainability and the Global Environment, University of Wisconsin. Background photographs courtesy USDA ARS Photo Unit.)

living trees (through decay or burning) into carbon dioxide—a greenhouse gas in the atmosphere that may be contributing to global warming. Changes in land cover also have profound impacts on climate through changes in how energy and water are released from the surface into the atmosphere, thereby changing the fueling of atmospheric circulations and the climate system.

Changes in land use and land cover can also affect the flow and quality of water traveling through a watershed. For example, deforestation may lead to increases in runoff and flooding, while high fertilizer use over croplands may lead to increased water pollution and the decline of healthy lakes, rivers, and coastal areas.

The future of the land is intimately tied to the future of the whole Earth system. On one hand, the way we use land affects the environment—across local, regional, and global scales. On the other hand, changes in the global environment may affect how our lands can be used and how we continue to live off them. On top of this, increases in global population and material consumption are increasing the pressure on this finite and immensely valuable resource.

One of the challenges facing our species is to learn to sustain our global land resources for ourselves, for our children, and for the rest of life on Earth.

David P. Roy
Christopher O. Justice

# The Changing Role of Fire on Earth

Fires lit to counteract bush encroachment in Madikwe Game Reserve, South Africa. Land managers and ecologists increasingly recognize the importance of fire in maintaining species biodiversity and the importance of small fires in reducing the risk of larger, more-damaging fires that can occur when small fires are suppressed and fuel accumulates. (Photograph provided by T. Landmann, CSIR, South Africa.)

Provided by Nature or stolen from the gods, fire has played a central role in human history. Humans have been using fire since at least 100,000 years ago; and archeological remains in Kenya suggest human exploitation of fire may be far more ancient, going back perhaps 1.6 million years. The use of fire was most likely an important impetus for the mental evolution of our ancestors and had a transforming effect on their culture and technology. Today, fire is a global phenomenon and an important agent of environmental change.

The causes of fire are many and varied. For a natural fire to occur, dead fuels such as grasses, twigs, and leaves must be present, and their moisture content must be low enough to sustain combustion. Lightning strikes were the primary source of ignition prior to

*Our Changing Planet*, ed. King, Parkinson, Partington and Williams
Published by Cambridge University Press 

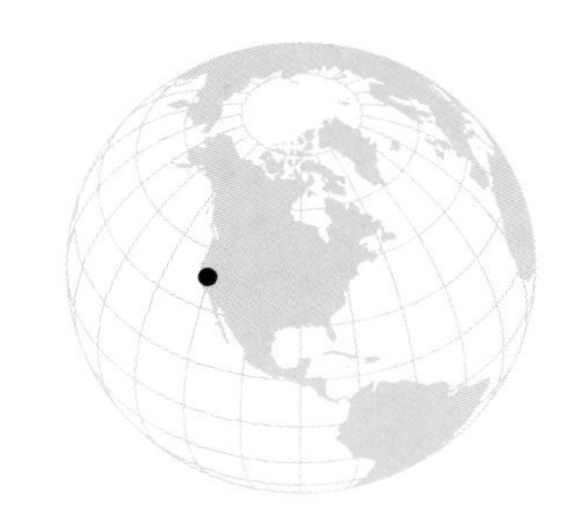

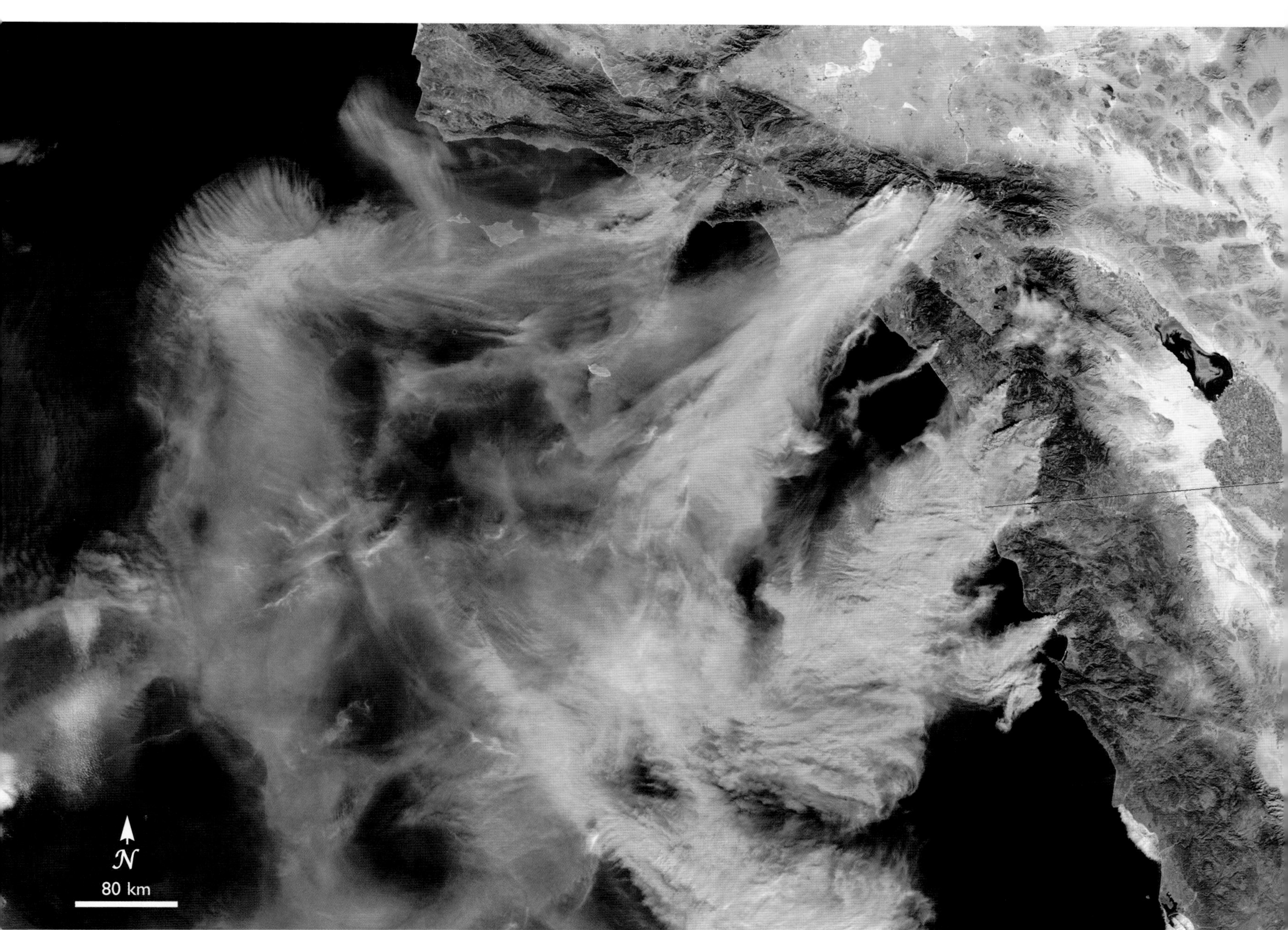


Fires and smoke in southern California and the northern Baja California Peninsula, October 26, 2003. Fires are burning in the vicinity of Los Angeles and San Diego. On this day, the Santa Ana winds helped fuel the fires, pushing their smoke plumes out over the Pacific. Images like this one provide a new tool that can help firefighters track the progress of fires and plan appropriate responses. The data are also useful once the fires are under control. Land managers can use information from satellite imagery to help plan the rehabilitation of burned areas and to protect water quality in the affected area. Scientists use fire observations to understand the effects of fire on local ecosystems and to estimate the amount of carbon dioxide and other greenhouse gases released by a fire, which can affect atmospheric composition and regional and global climate. (Data from the MODIS instrument on the Terra satellite.)

2005

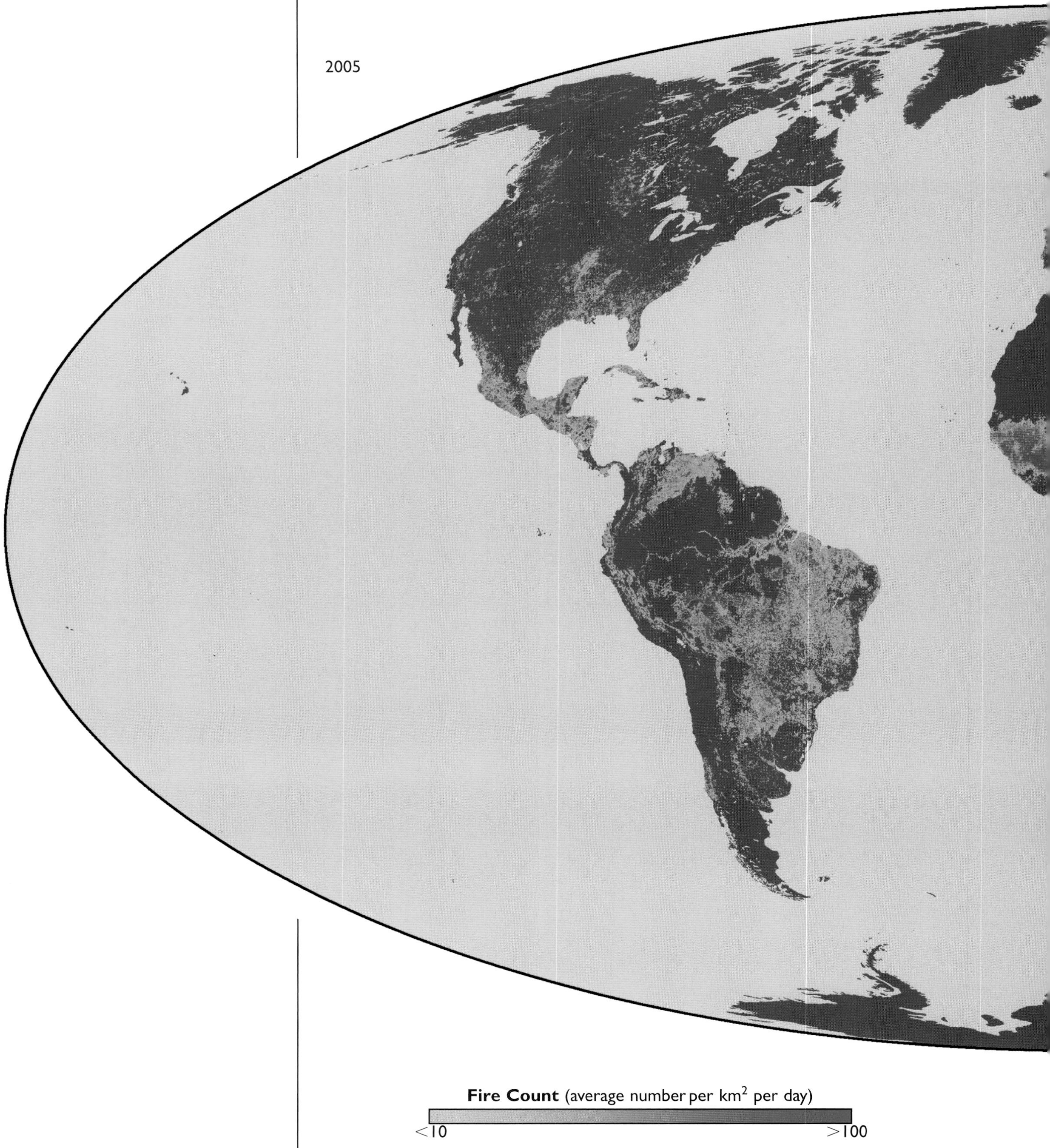

**Fire Count** (average number per $km^2$ per day)

<10 >100

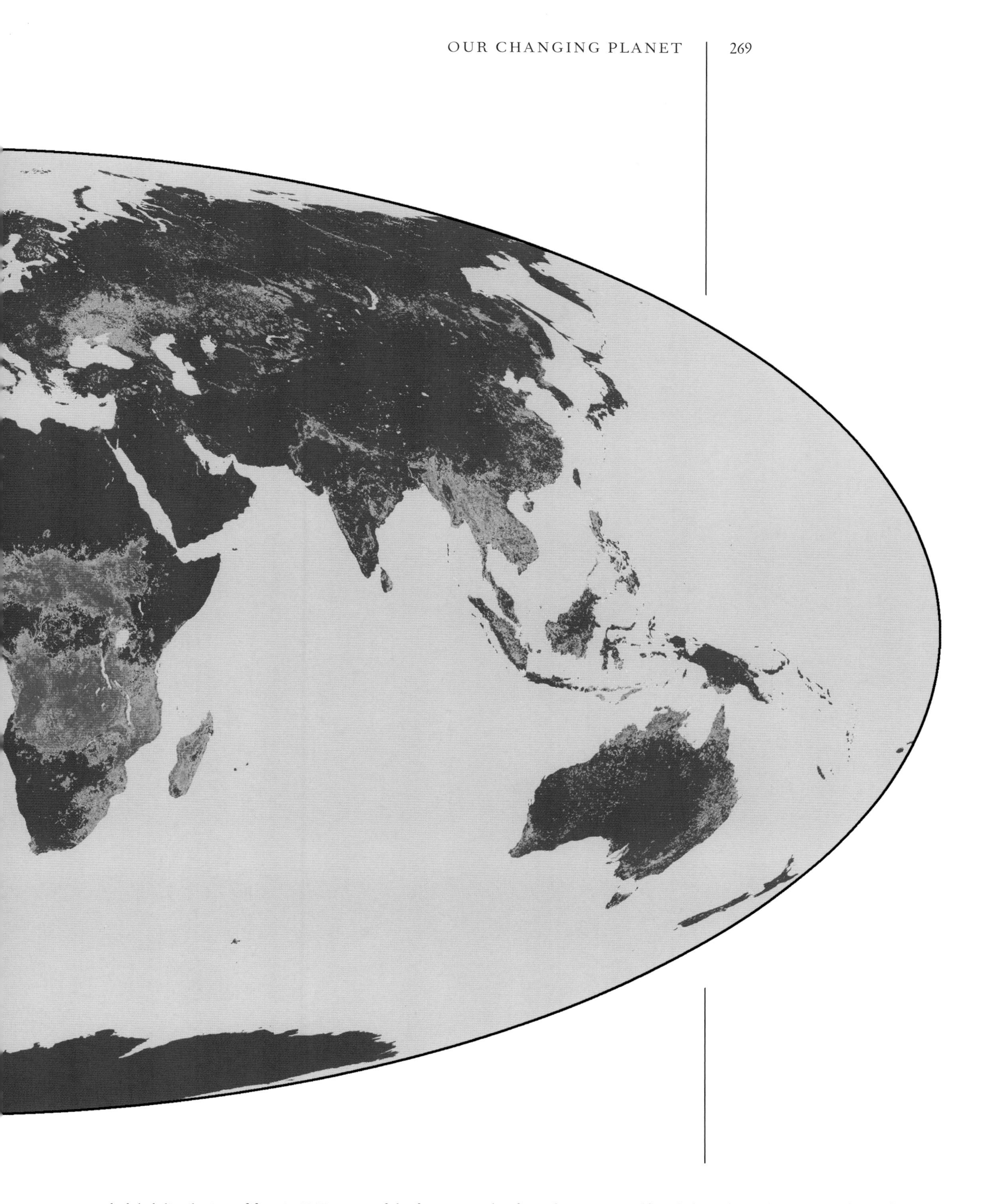

Annual global distribution of fires in 2005. Many of the fires in North Africa, the Persian Gulf, and along the Red Sea are stable gas flares associated with oil wells. (Data from the MODIS instruments on the Terra and Aqua satellites; figure provided by the MODIS Rapid Response Team.)

December 2004–February 2005

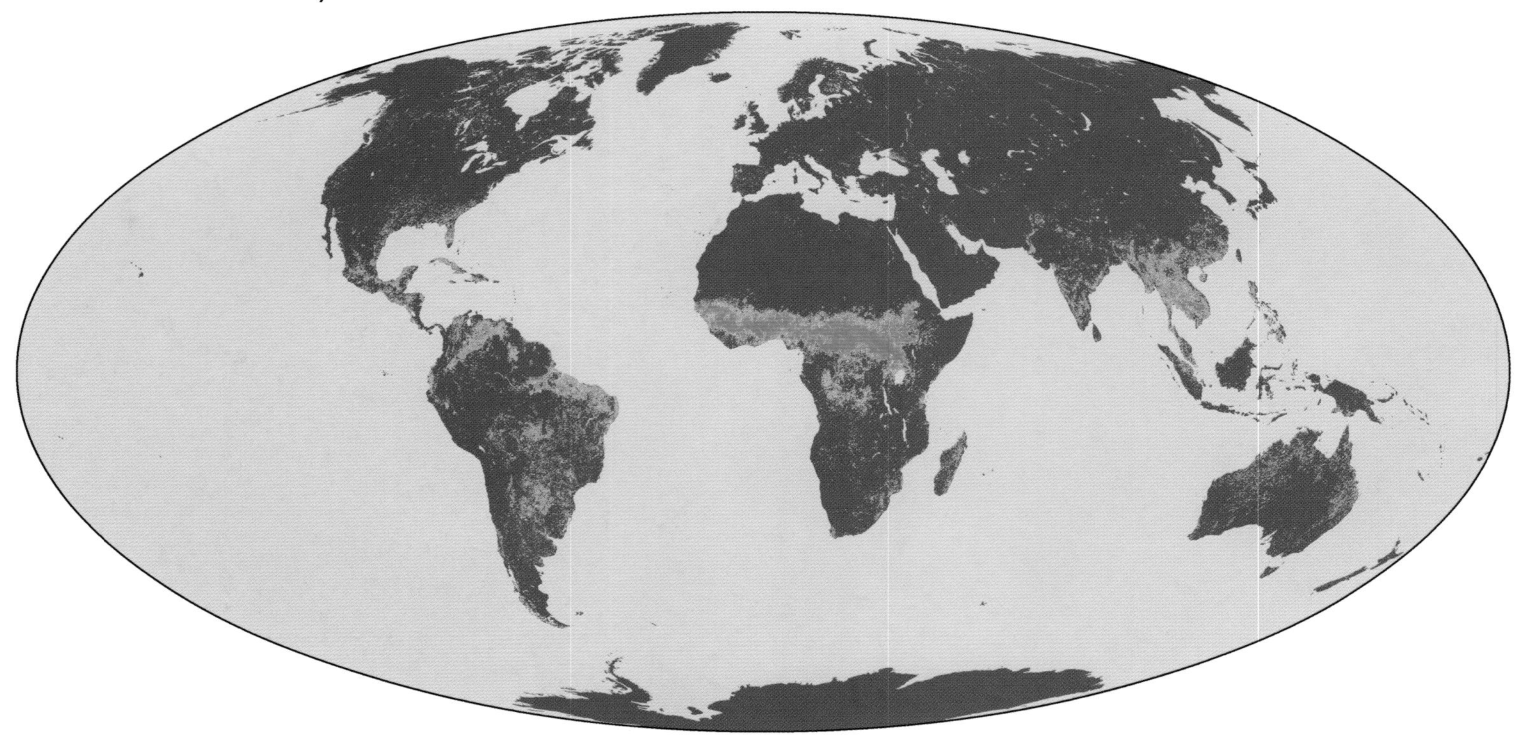

March–May 2005

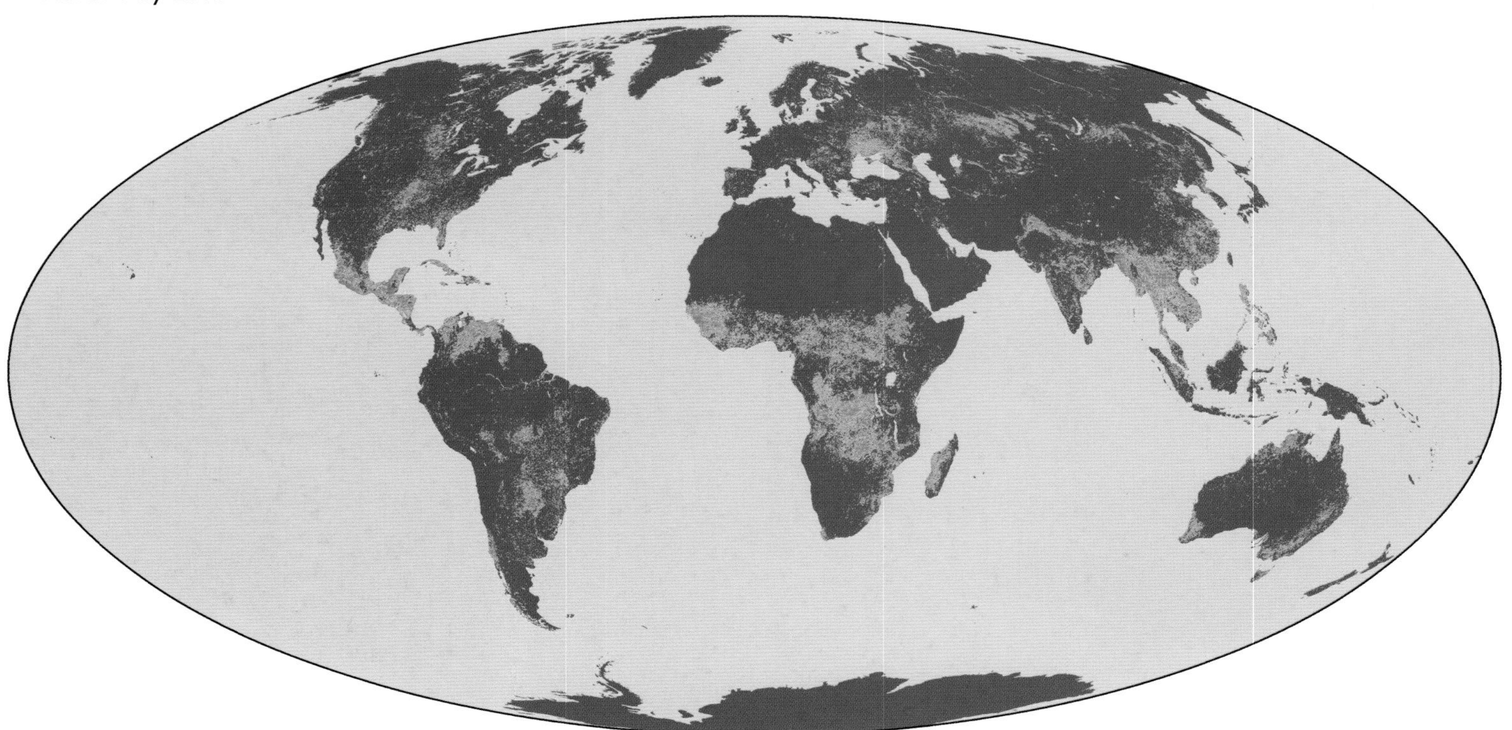

**Fire Count** (average number per km$^2$ per day)

<10 >100

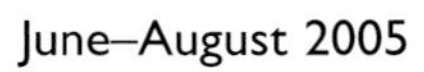


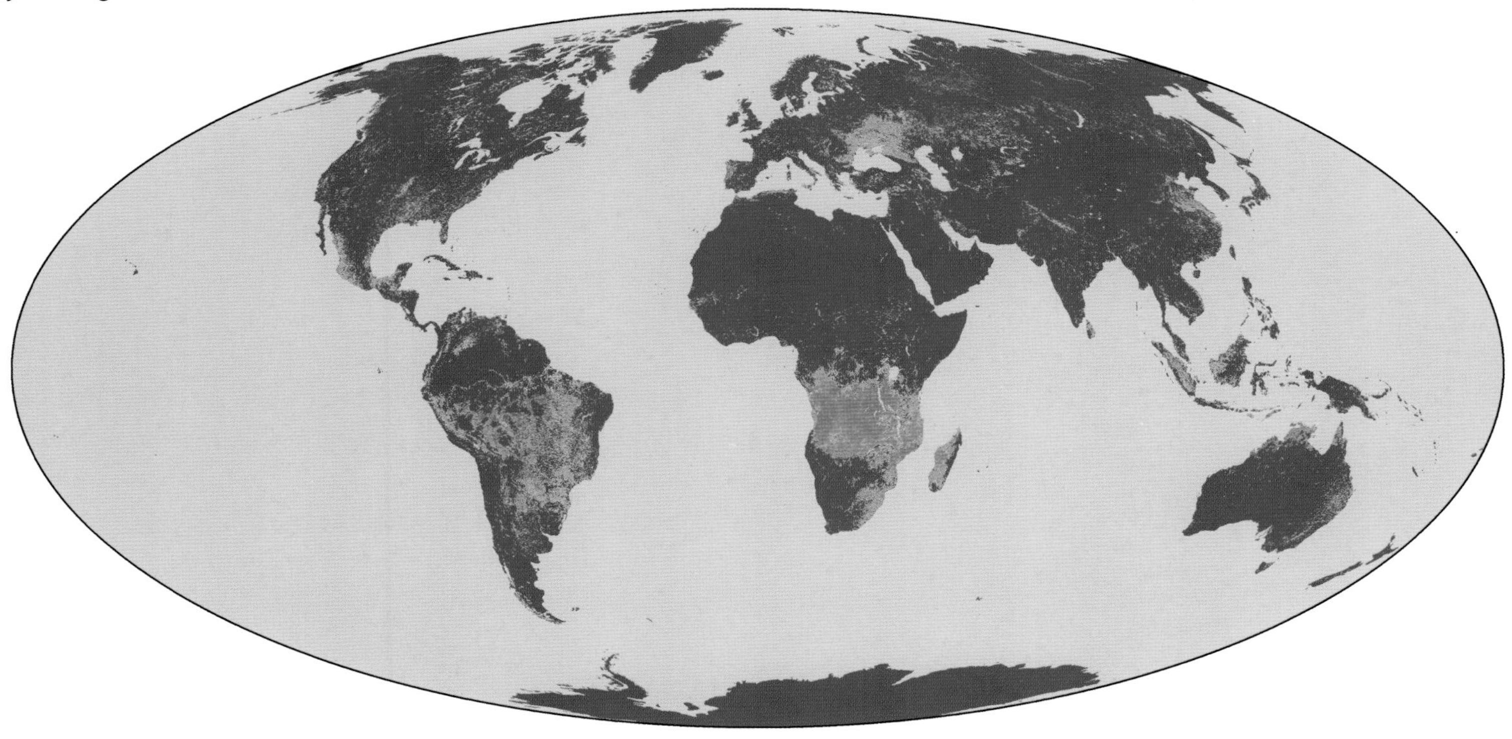

September–November 2005

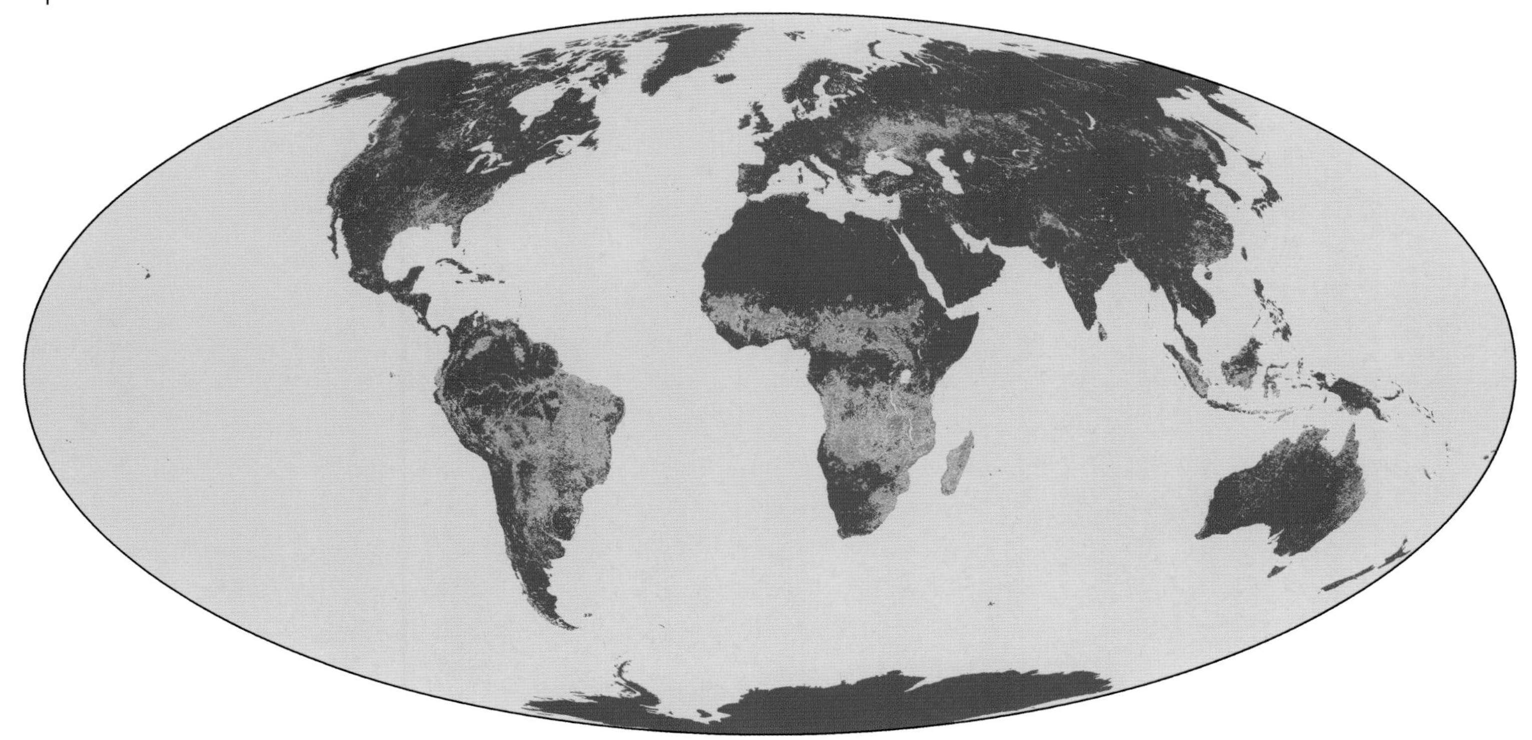

Seasonal global distribution of fires from December 2004 to November 2005. Many of the fires in North Africa, the Persian Gulf, and along the Red Sea are stable gas flares associated with oil wells. (Data from the MODIS instruments on the Terra and Aqua satellites; figure provided by the MODIS Rapid Response Team.)

humans and are still the major cause of fire in sparsely populated regions, such as the boreal forests of Eurasia and Canada. In most places, people are now the predominant cause of fire, whether intentional or accidental. Negative connotations often associated with fires, such as the loss of life and property, do not reflect the true role of fire in the balance of Nature. Fires occur every year in many parts of the world and extensively in Africa and Australia, where the annual long dry season produces ideal fire conditions. In these and all other fire prone environments, plants and certain animal species have evolved adaptations to cope with fire, such as seeds that require intense heat for germination, and many of these species become locally extinct when fire is excluded.

Fire is integral to many agricultural practices and is an important land management tool. In the humid tropics, subtropics and savannas, fires are lit for numerous reasons, including to clear land for cultivation, to provide nutrient-rich ash for crops, to maintain and improve pasture and grasslands for livestock grazing, to attract game, to drive animals during hunting, and to make fire breaks around fields and settlements. Because of increasing human population and demand for land, more fire-prone areas are becoming inhabited and more land is subject to fire management. Fire management policy is complex, and there are often conflicting environmental and economic considerations.

The frequency, intensity, seasonal timing, and type of fire that prevail in a region are referred to as the 'fire regime.' Scientists are anxious to understand whether the world's fire regimes are changing. How fire regimes will further change as human populations, their land use practices, and the climate change is unclear. These changes are likely to affect fire regimes directly, through changes in the ways fire is used, and indirectly, by modifying the environmental conditions and the amount of fuel available for fire. For example, in Indonesia extensive logging and traditional 'slash and burn' farming combined with extreme droughts are resulting in fires of an unprecedented extent in rain forest areas. The smoke from these fires has had severe impacts on regional air quality and has posed significant hazards to health and transportation.

It is thought that, on a global basis, at least 8 billion metric tons of dry biomass are burned by vegetation fires each year, releasing over 4 billion metric tons of carbon to the atmosphere. This is equivalent to about 70% of all human fossil fuel emissions from oil, gas, and coal powered vehicles, machines, homes, factories, and power stations. Intense, flaming fires release carbon dioxide, whereas smoldering fires release more carbon monoxide. Indonesian fires in 1997–1998 alone released the same amount of carbon into the atmosphere as Europe's annual carbon emissions from burning fossil fuels. Fires also release additional trace gases, which play key roles in atmospheric chemistry, and smoke and small particles, which affect the formation and physical properties of clouds. It is currently unclear the extent to which the gases and particles released by vegetation fires are contributing to global warming; there is the potential, perhaps, for a positive feedback loop in which more fires, arising from more frequent and intense drought, release more carbon to the atmosphere to cause yet more warming.

Spaceborne sensors provide a unique perspective with which to study and understand the global distribution and characteristics of fire. Dramatic new satellite maps showing fire activity across the entire Earth are providing a unique picture of seasonal and yearly fire activity. Using daily, global fire detections provided by the NASA Moderate Resolution Imaging Spectroradiometer (MODIS) on the Terra and Aqua satellites, scientists

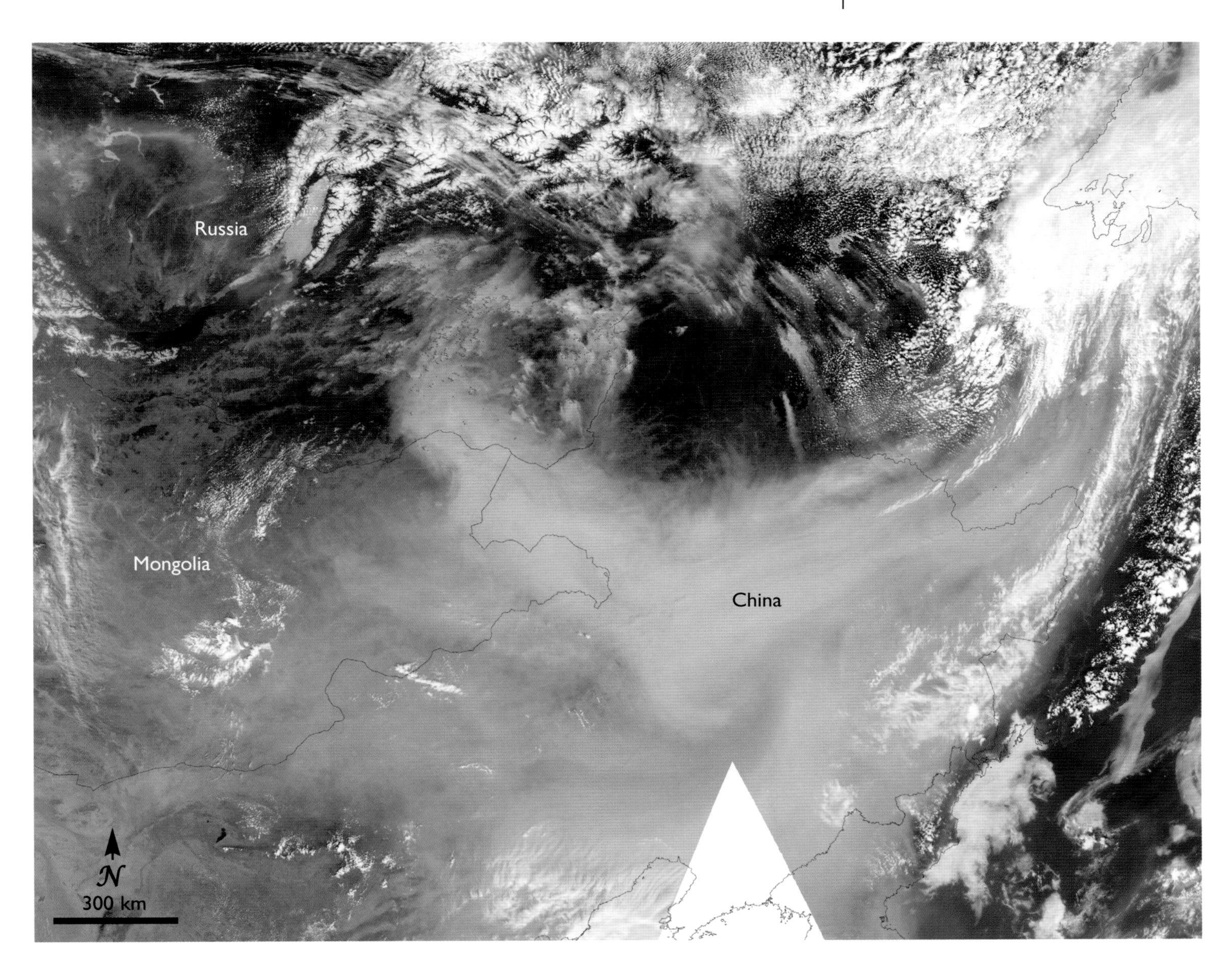


Fires and smoke in Russia and China, May 19, 2003. (Data from the MODIS instruments on the Terra and Aqua satellites; figure provided by the MODIS Rapid Response Team.)

have been mapping fire activity for the entire surface of the Earth every day since February 2000.

These global data sets are a milestone in the use of satellite data and the first step in creating a long-term fire record that is crucial for understanding the impact of fire on life and climate and how fire regimes are changing. MODIS data are also being used by fire managers in daily strategic planning of fire management. Such information will allow planners, managers, and policy makers the opportunity to understand fires in their environmental, economic, and social contexts, and to formulate their responses accordingly.

Never before have scientists had the opportunity to map the global occurrence of fire with such detail, accuracy, and frequency. The future challenge is to integrate these data with other information to understand the basic relationships between fire, population and land use, and climate, and to develop a better understanding of the underlying processes.

Kyle C. McDonald
Bruce Chapman
John S. Kimball
Reiner Zimmermann

# The Tropical Rain Forest: Threatened Powerhouse of the Biosphere

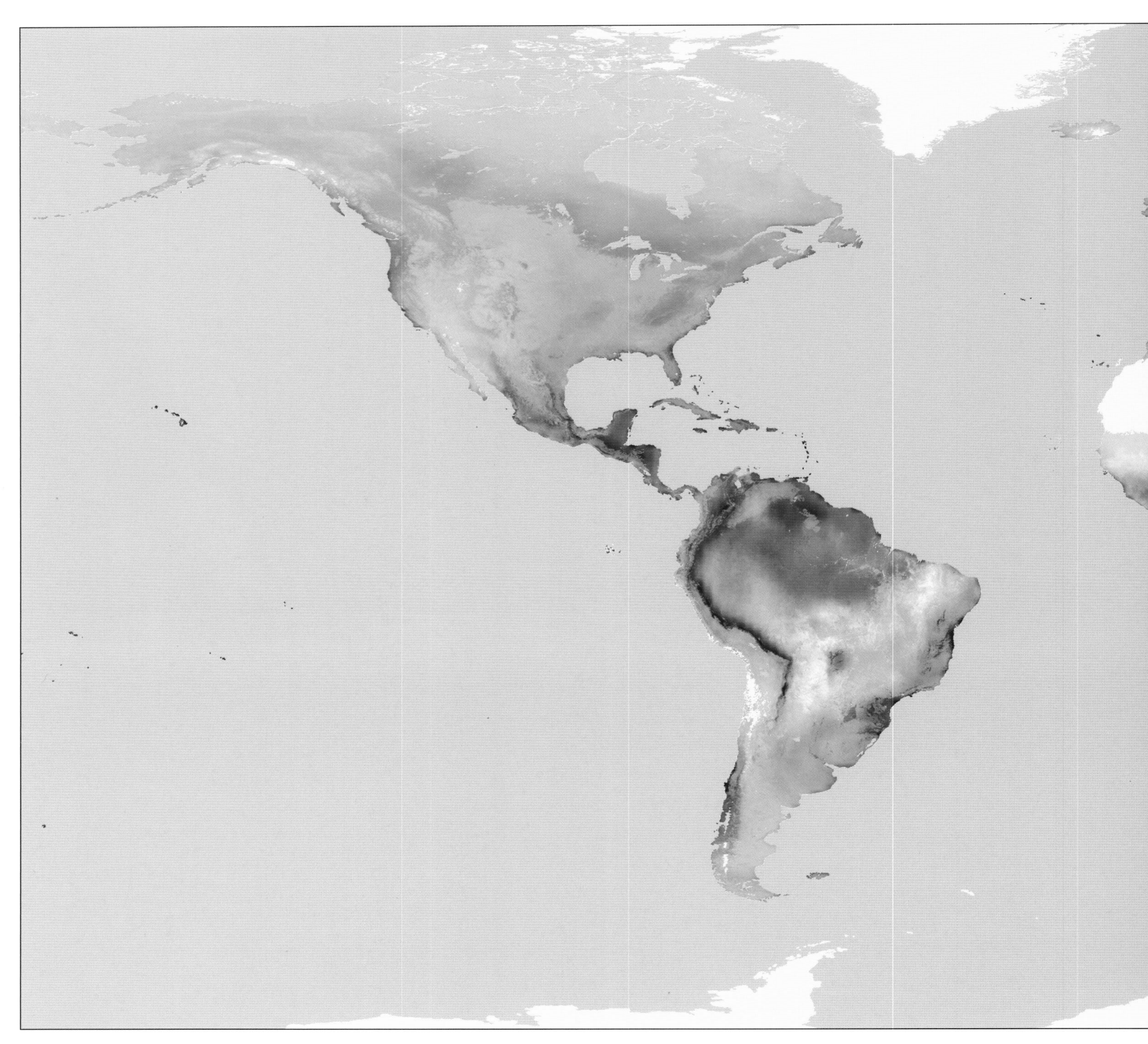

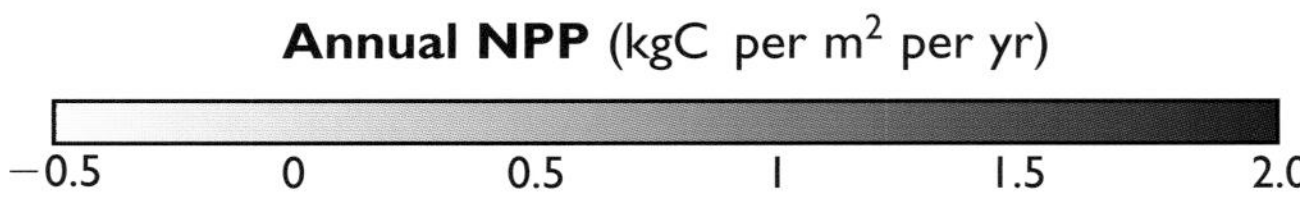


*Our Changing Planet*, ed. King, Parkinson, Partington and Williams
Published by Cambridge University Press 

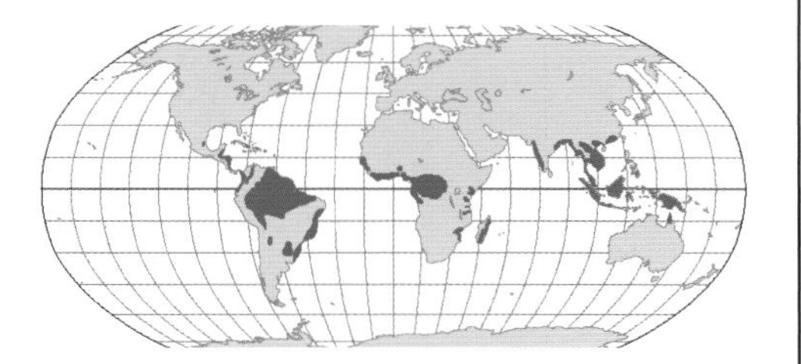

Location of the Earth's tropical rain forests, straddling the equatorial region between 23.5°S and 23.5°N. (Map courtesy Tropical Rain Forest Information Center.)

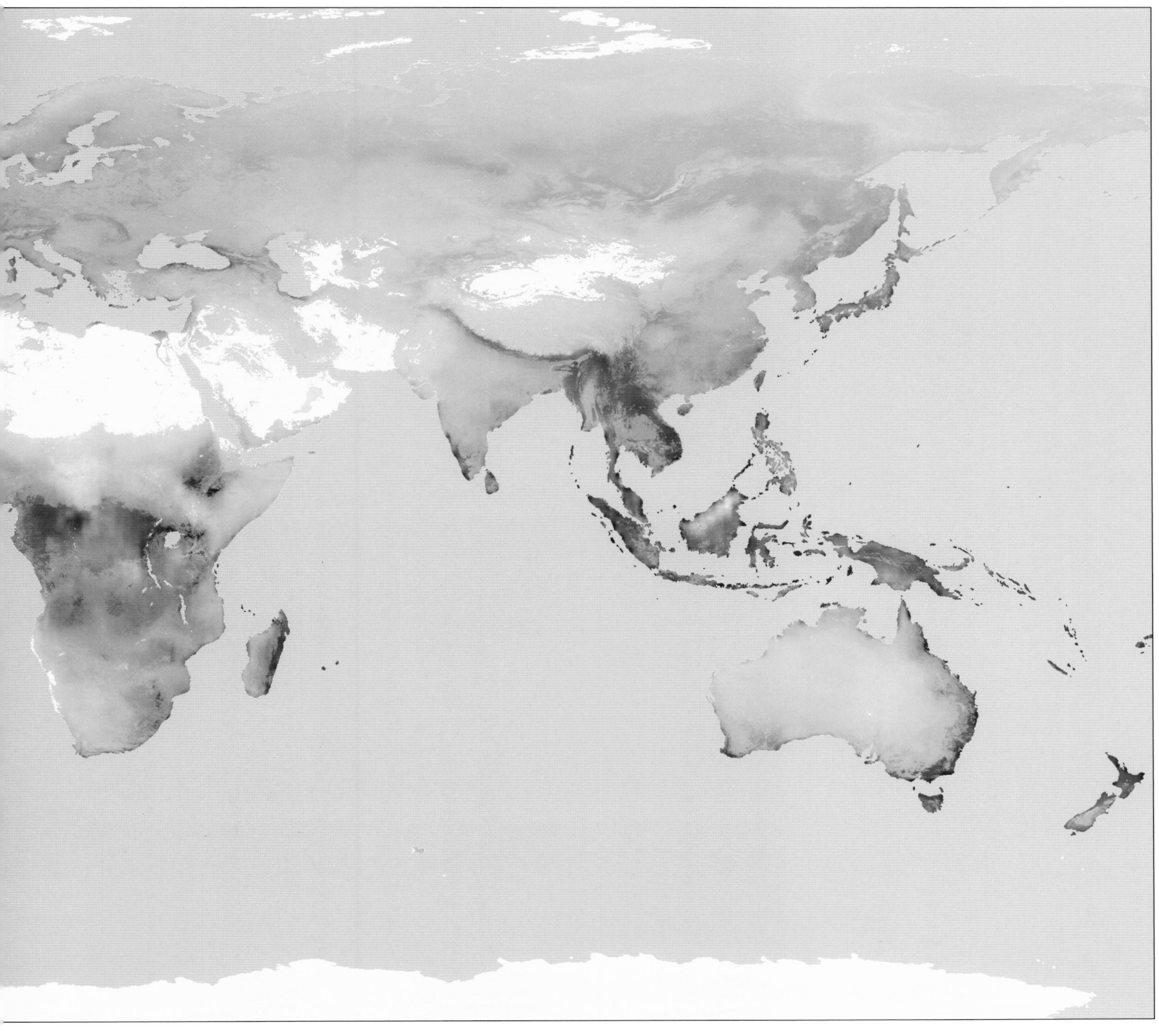

Global map of annual vegetation net primary productivity (NPP) for 2003. Landcover productivity is shown in kilograms of carbon taken up by the landscape per square meter per year by vegetation growth. The tropical rain forests reside in the band of highly productive land area straddling the equator. Uptake of carbon from the atmosphere and storage in above-ground biomass by these ecosystems dominates the global totals for land areas. (Data from the MODIS instrument on the Terra satellite.)

Burning tropical shrubland vegetation, called 'cerrado', near Brasilia, Brazil, June 25, 2002. This open grassland-woodland vegetation is common in south-central Brazil. Adapted to poor soils, it burns almost every dry season. Trees are fire resistant and resprout after the fire has passed through. (Photograph by Viviana Horna.)

Rain forests occur wherever forests grow and annual rainfall exceeds 2–3 m. Tropical rain forests straddle the Earth's equatorial regions, covering an area of roughly 10 million km$^2$. These regions represent only 7% of the global land area but contain more than half of all plant and animal species and provide the setting for major pumping mechanisms for the global cycling of water, energy, and carbon, mechanisms that are crucial to Earth's climate and life-sustaining processes. Consequently the health of the rain forests is of global concern, particularly since continued development and deforestation from logging and agriculture threaten these regions. The largest of the tropical rain forests is the Amazon rain

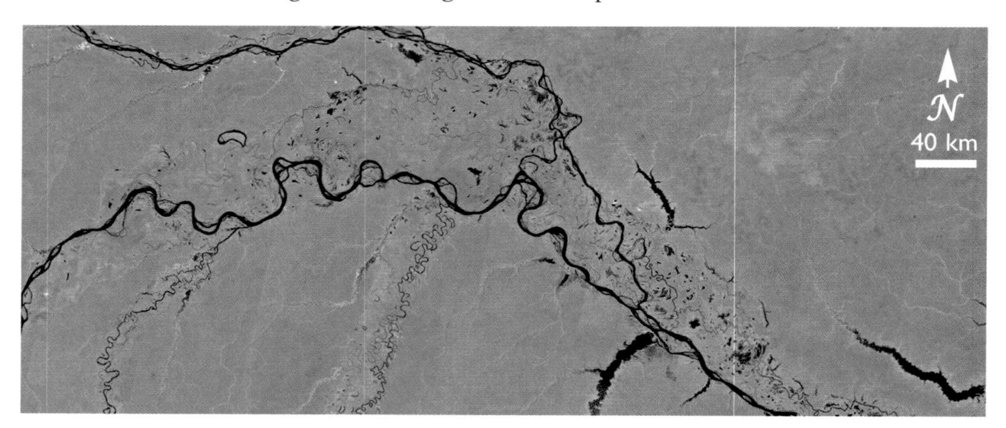


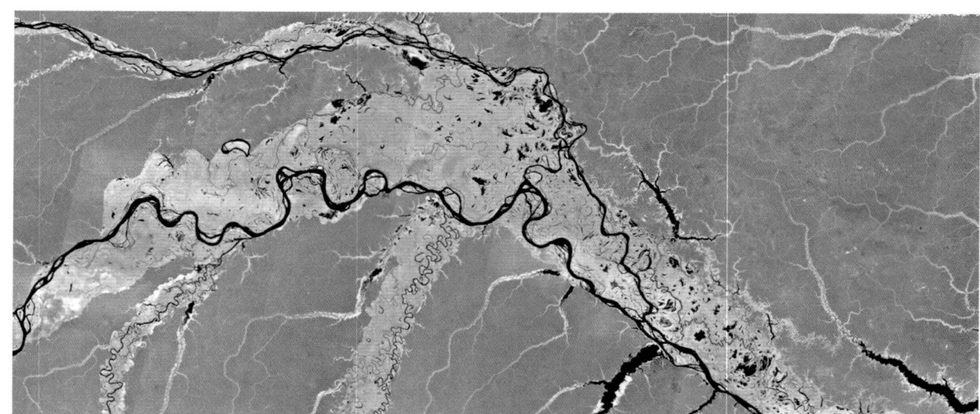

Radar images of rainy season inundation along the Amazon River. The top image is from the dry season, October 1995. The bottom image is from the rainy season, May 1996. Light areas are inundated forest, black is water, and medium-gray is forest. (Data from the SAR instrument on the Japanese Earth Resources Satellite (JERS-1); images © 1992–1998 NASDA/MITI, courtesy Global Rain Forest Mapping (GRFM) project.)

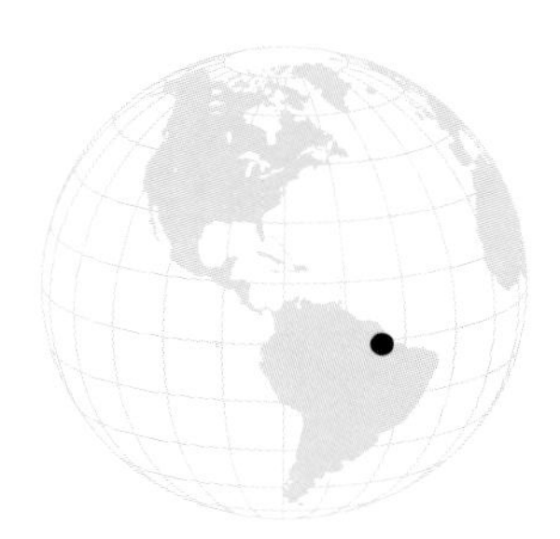

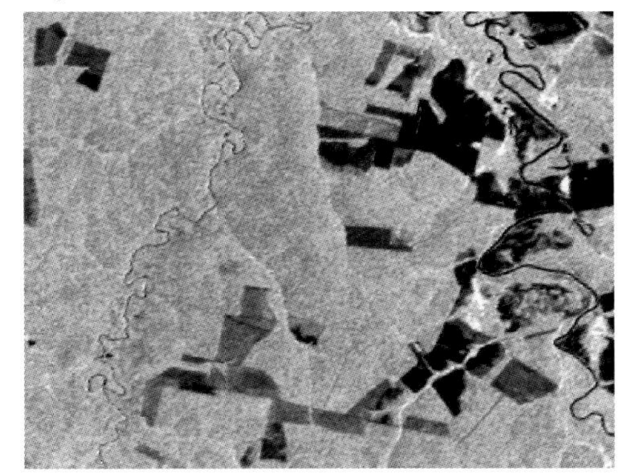


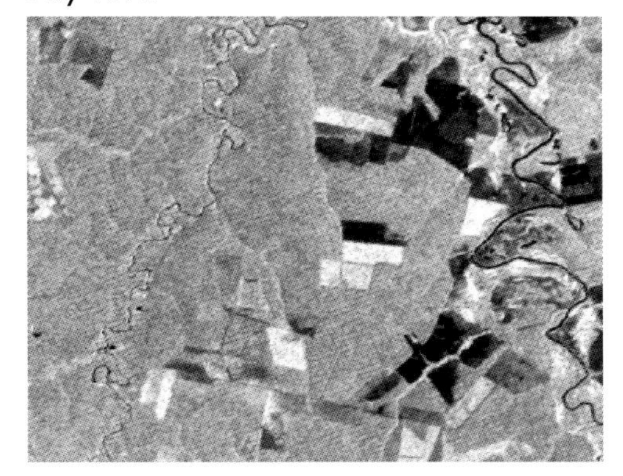


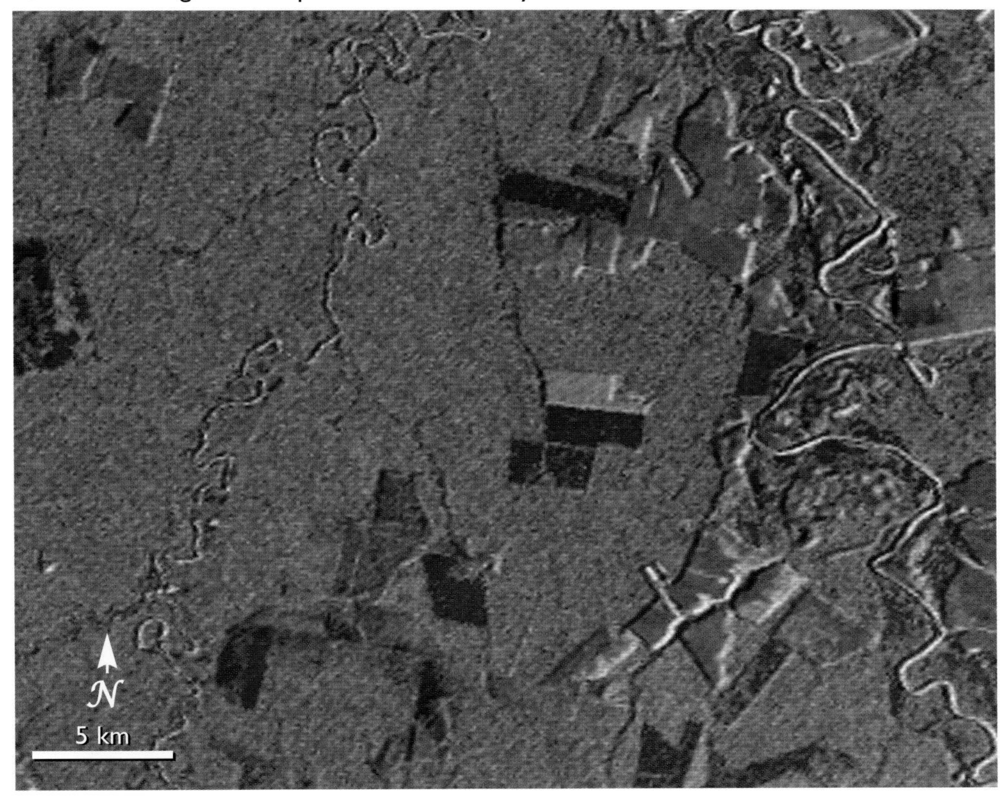


Deforestation activity in a 20 km × 30 km region of the Amazonian rain forest of South America between September 1995 and May 1996, demonstrating the powerful capability of satellite synthetic aperture radar for monitoring deforestation. Distributed through the area are regions deforested between the two observations shown in the top two images. The land cover change map derived from these two images illustrates the nature and progression of the deforestation activity. Felled trees which are left on the ground after clear-cutting can result in increased backscatter, indicated in red. Once the logs have been removed, the remaining low vegetation and open land surface results in decreased backscatter, indicated in gray, relative to the original forest. Green areas indicate regions that did not change between the observations. (Data from the SAR instrument on the Japanese Earth Resources Satellite (JERS-1); images © 1992–1998 NASDA/MITI, courtesy Global Rain Forest Mapping (GRFM) project; after Siqueria *et al.*, *Remote Sensing of Environment*, 2003.)

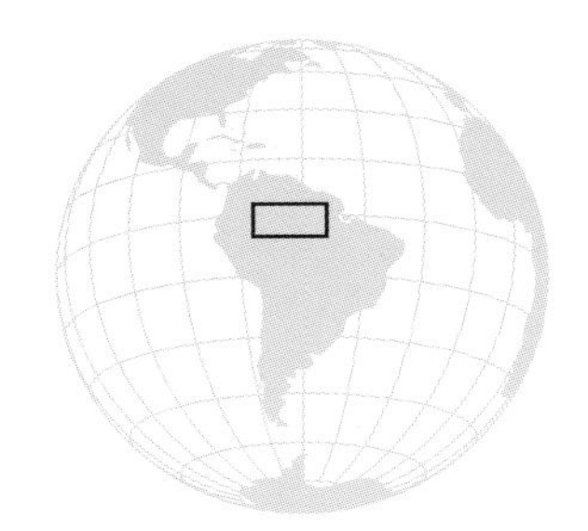

forest, located in the South American countries of Brazil, Bolivia, Colombia, Ecuador, and Peru.

One way to measure the health of rain forests is through the extent to which they have been disturbed from their natural state. Agricultural practices include shifting cultivation, whereby forests are cleared, the remaining area is burned, and new crops are established. One or more crops are then cultivated for 1–2 years until soils become leached of nutrients, after which the land is abandoned and additional forest is cleared. Tropical rain forests are also under threat from large-scale logging operations and associated development. The conversion of natural systems by human activity for various purposes may cause large transfers of carbon from forest ecosystems to the atmosphere. The United States Food and Agriculture Administration estimates that the global tropical deforestation rate was 180,000 $km^2/yr$ during the 1980s. This corresponds to an area roughly equivalent to the land area of Uruguay, or the state of North Dakota, being deforested every year.

Smoke rising from fires set to clear land for agriculture in the Brazilian Amazon on July 3, 2003. Dark green areas are rain forest remnants, brown areas are cleared fields, and dark gray regions are recently burned fields. (Photograph taken by astronauts aboard the International Space Station. Image courtesy Image Science and Analysis Laboratory, NASA Johnson Space Center, photo number ISS007-E-9313.)

Natural disturbances such as inundation can also be significant for the regional carbon balance. Seasonally inundated forests, woodlands, and grasslands are widespread in tropical rain forest basins. The duration and extent of wetland inundation are critical for the ecological, biogeochemical, and hydrological functioning of these ecosystems, as they impact the release of greenhouse gases to the atmosphere and the prospect of sustainable land use in these areas. Of particular interest is the role of wetlands in emitting methane, an important greenhouse gas. Seasonally inundated forests are also home to many unique plant and animal species, which have adapted to seasonal flood cycles.

'Varzea' forest on the Isla de Marchanteria in the Rio Solimoes at the onset of flooding, 10 km upstream from Manaus, Brazil, March 12, 1999. The water moves very slowly through the forest and carries driftwood and debris. The Solimoes River, the name of the Amazon River upstream of Manaus, carries sediment-rich waters from the Andes Mountains. (Photograph by Reiner Zimmermann.)

Scientists are employing a variety of satellite sensors for measuring and monitoring the dynamic state and health of the Earth's tropical forests. The information provided to date indicates that tropical forests play a critical role in regulating regional and global climate but are undergoing significant changes due to large-scale deforestation from logging, urban and agricultural expansion, and climate change. Monitoring these regions with satellite remote sensing will continue to provide comprehensive coverage of these changes and enable improved understanding of the critical role of tropical rain forests in the global biosphere.

Roraima Tepui, one of the best known of the tropical table mountains, February 13, 1989. At over 2,740 m (9,000 ft), its summit marks the border of Venezuela, Brazil, and Guyana. Remnants of tropical montane forest are seen in the foreground of the photograph, taken from the Venezuelan side of the mountain. Fires and climate change drive the advance of the savanna-dominated landscape. (Photograph by Reiner Zimmermann.)

Scene near Gran Sabana, Venezuela, February 12, 1989. A naturally occurring fire during the 1930s drove conversion of this landscape from tropical rain forest into grass savanna. Nutrient-poor soils and changes in seasonal rainfall patterns have prevented forest regrowth. (Photograph by Reiner Zimmermann.)

Liming Zhou
Robert K. Kaufmann
Compton J. Tucker
Ranga B. Myneni

# The Green Wave

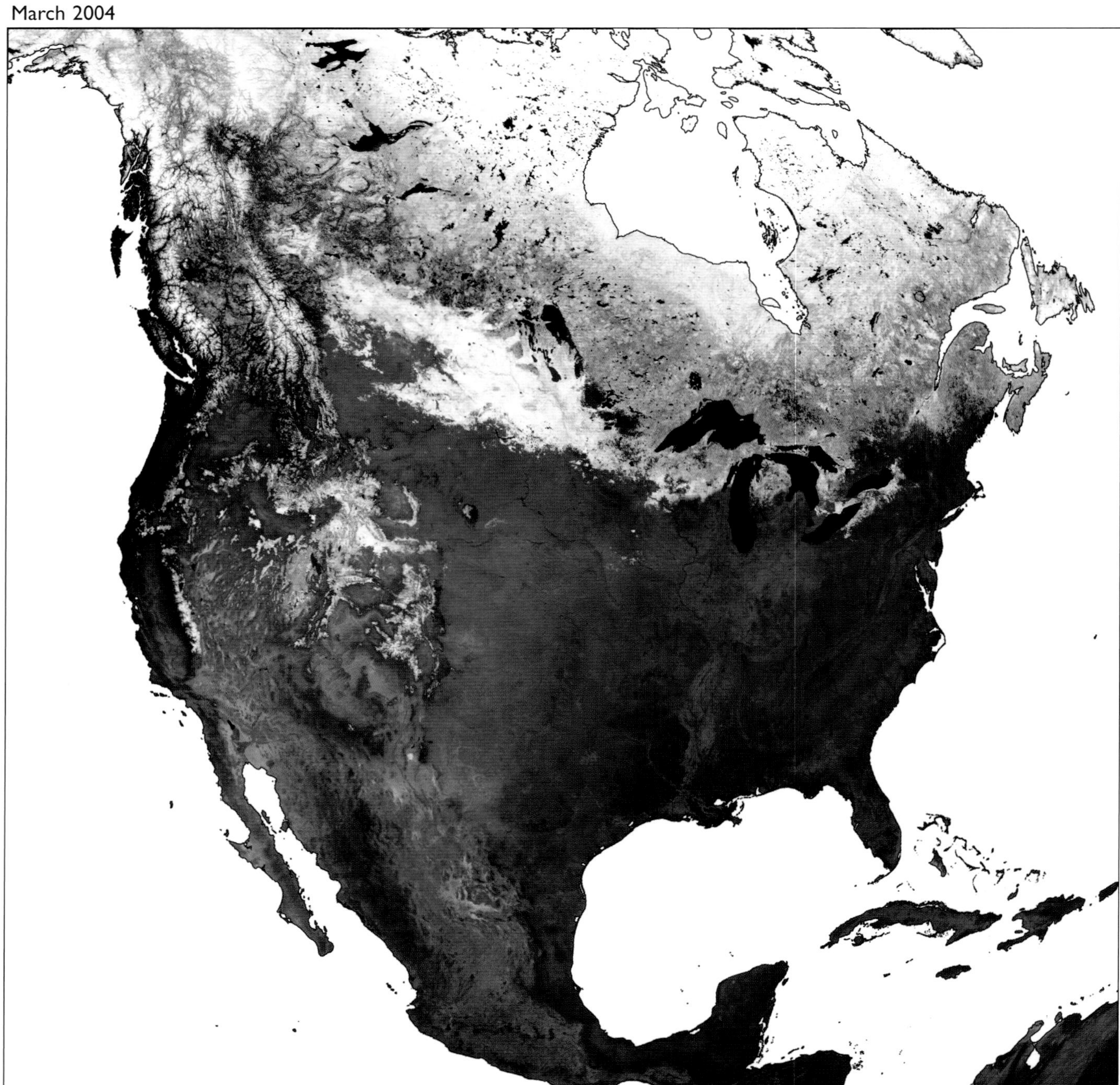


True-color images of North America comparing the appearance of the land in March and August 2004. (Reflectance data derived from the MODIS instrument on the Terra satellite; image courtesy NASA/SVS.)

*Our Changing Planet*, ed. King, Parkinson, Partington and Williams

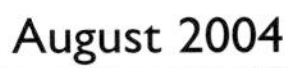


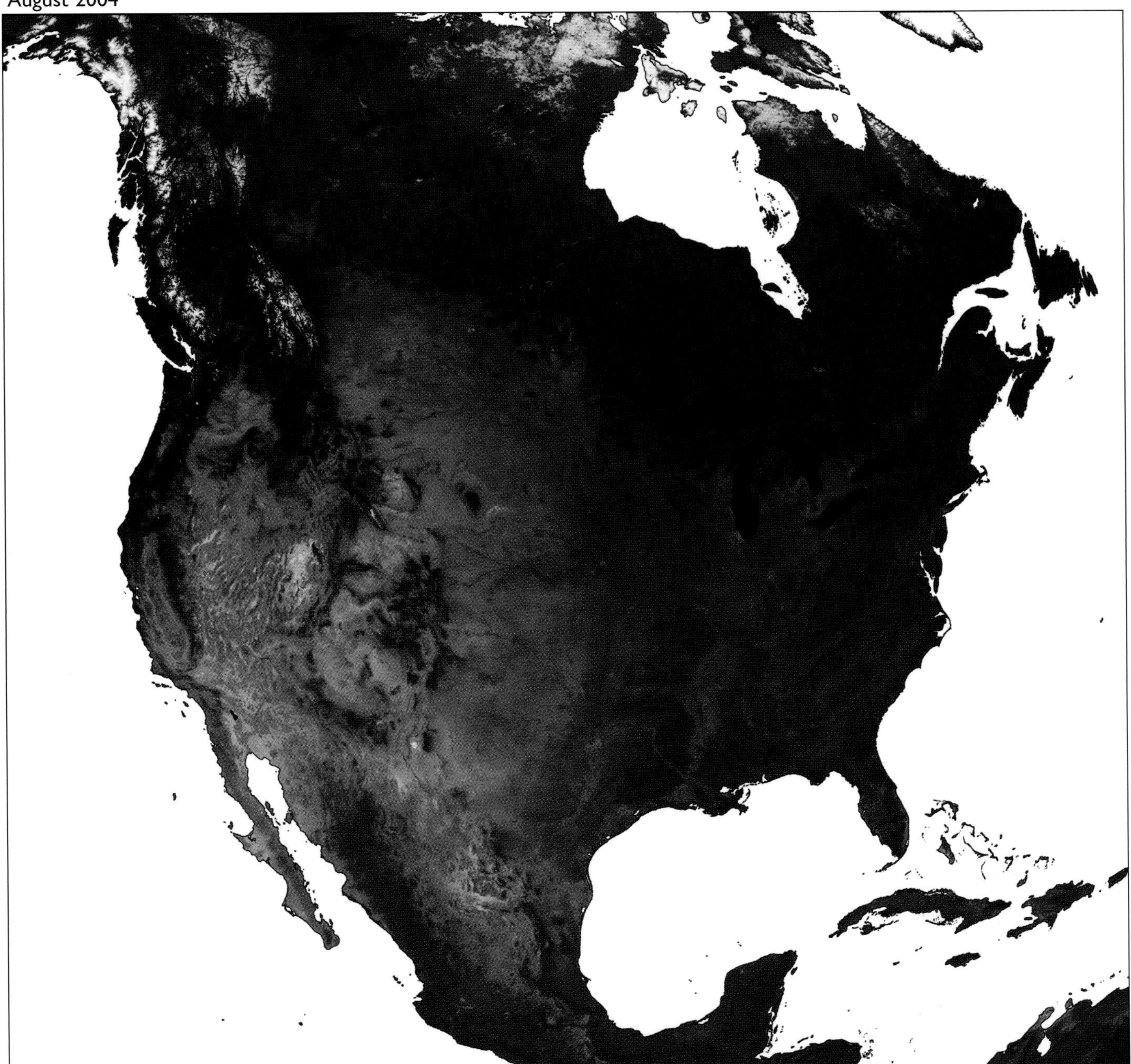

Plants play a significant role not only in providing animal habitats, food, fiber, fuel, and pharmaceuticals, but also in regulating the global environment and climate. They remove carbon dioxide from the atmosphere and store the carbon in wood through the process of photosynthesis. Carbon dioxide is a primary greenhouse gas and is suspected of being a major contributor to global warming as a result of its increased concentration in the atmosphere—a buildup primarily

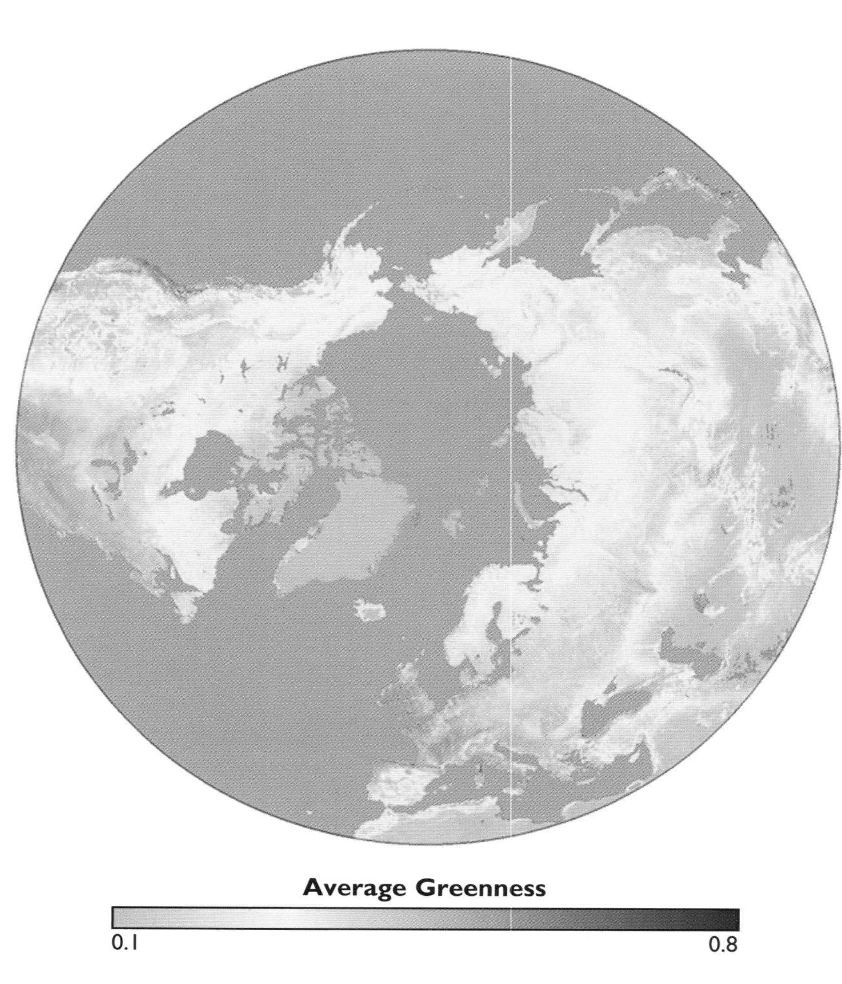


Average greenness over the period July 1981–December 1999. The greenness tones in northern Eurasia and North America depict greenness of the needle trees, in a unitless greenness index. (Data from the AVHRR instruments on the NOAA 7, 9, 11, and 14 satellites; data provided by the NASA GSFC Global Inventory Monitoring and Modeling Systems group.)

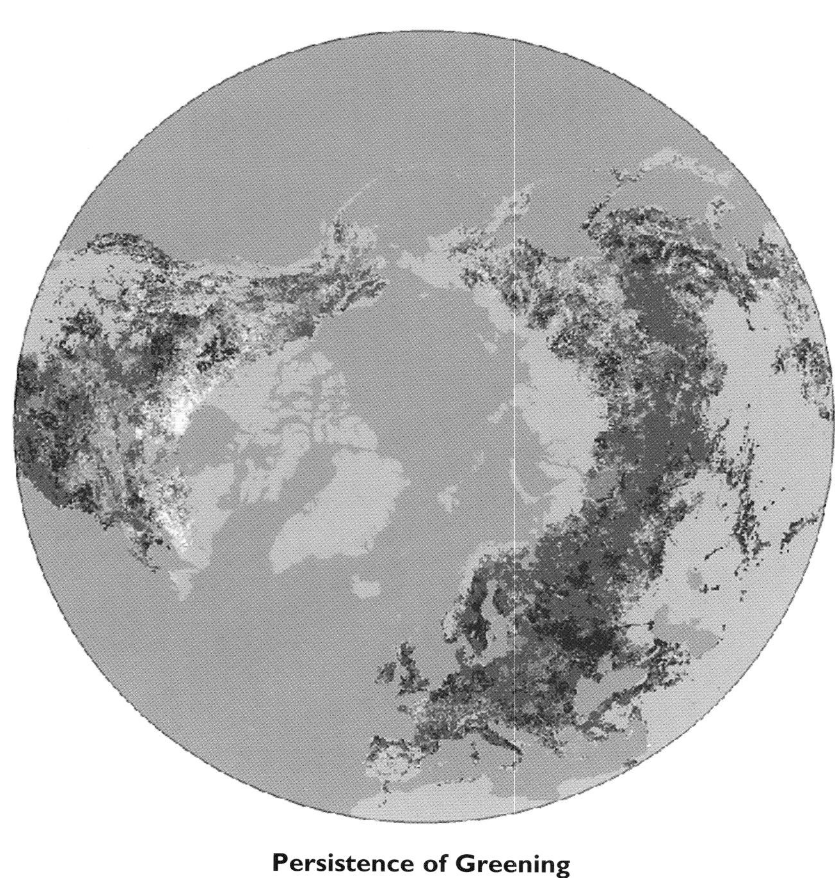


Persistence of greening poleward of 30°N over the period 1982–1999. The colors represent changes in vegetation lushness, ranging from white, denoting a low increase in the heartiness of vegetation, to green, denoting the highest increase in plant lushness. (Data from the AVHRR instruments on the NOAA 7, 9, 11, and 14 satellites; data provided by the NASA GSFC Global Inventory Monitoring and Modeling Systems group.)

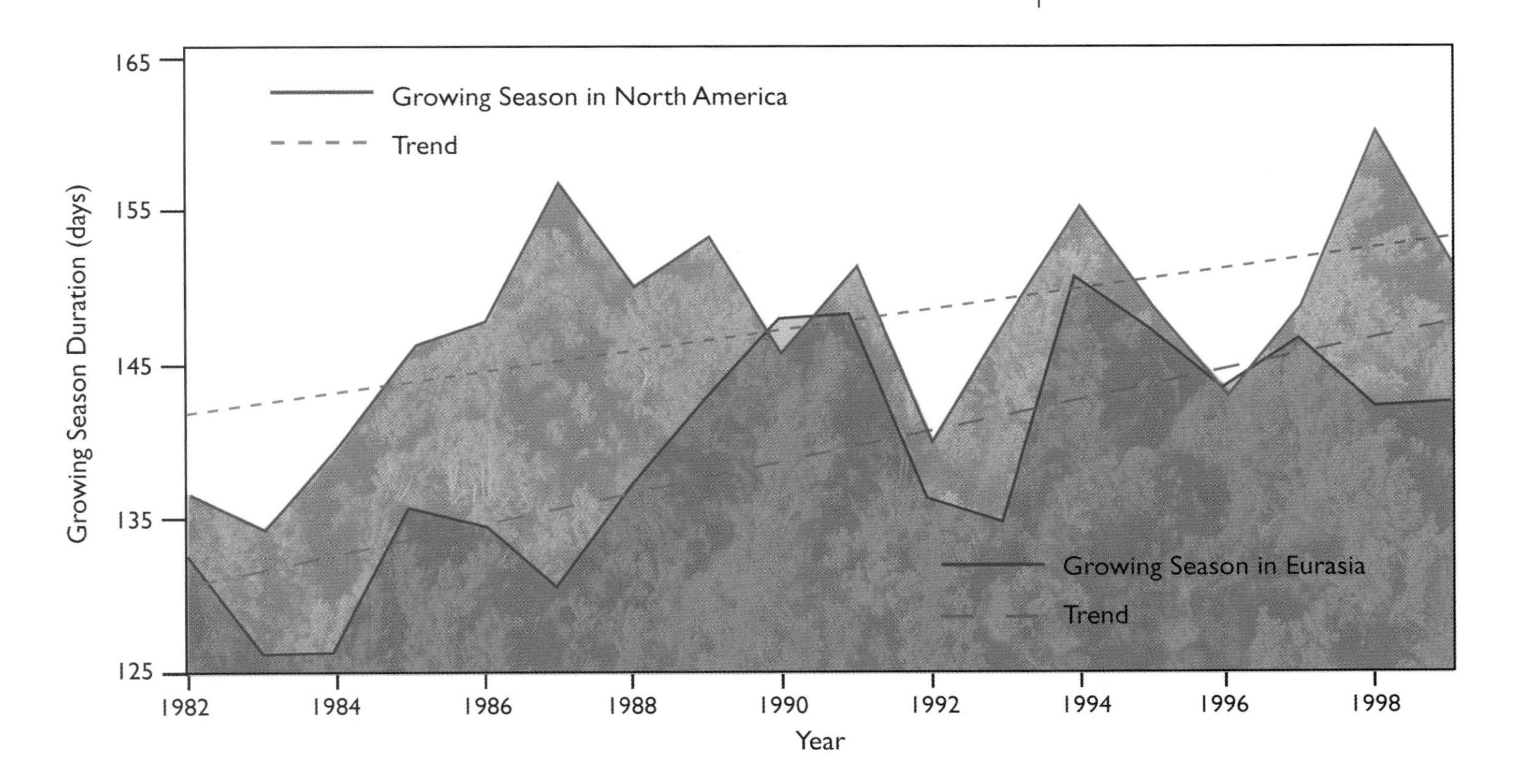


Length of the growing season in days, averaged over 40°N–70°N in North America (green plot) and Eurasia (blue plot). The plots show the time series of yearly growing season durations for the years 1982–1999, and the dashed red lines are the trends estimated by linear regression. The trend lines show, overall, a 12-day lengthening of the growing season in North America and an 18-day lengthening of the growing season in Eurasia. (Data from the AVHRR instruments on the NOAA 7, 9, 11, and 14 satellites.)

related to fossil fuel burning and deforestation. Scientists believe that global warming could result in serious consequences such as sea level rise, more extreme weather, and increased incidences of certain vector-borne diseases. Changes in plant activity may change the atmospheric carbon dioxide concentration. For this and other reasons, it is important to understand and monitor such changes.

The northern high latitudes warmed significantly during the 1980s and 1990s, especially during winter and spring. Associated with this warming are a reduction in annual snow cover and an earlier disappearance of snow in spring. So an immediate question is: how have these changes affected plant growth?

To answer this question, we need to analyze long-term observations of plant photosynthesis at the global scale. As field measurements are few, tedious and expensive to collect, we rely on satellites providing high quality global data at regular temporal intervals. Here we present evidence of a 'greener north' from analyses of 20 years of NASA satellite data.

Scientists use a greenness index derived from satellite data as a proxy for plant photosynthesis. This index uses red and near-infrared solar radiation reflected back to sensors aboard polar-orbiting satellites to gauge the abundance and energy absorption by leaf pigments such as chlorophyll. It is expressed on a scale from $-1$ to $+1$, with values generally between $-0.2$ and 0.1 for snow, inland water bodies, deserts, and exposed soils, and increasing from about 0.1 to 0.7 for increasing amounts of vegetation. The index shows greening and browning of plants as they relate to seasonal changes and conditions such as drought or abundant rainfall.

The satellite sensors observe every patch of land on the Earth at least once a day. The greenness data typically are of global extent, 8 km spatial resolution, and 10–15 day temporal frequency, with the record beginning in July 1981 and extending to the present. Processing of such massive amounts of data is a tedious task even on modern computers and requires special methods to correct for atmospheric blurring of the Earth's surface.

Amazon rainforest canopy in Napo, Equador taken on February 14, 2006. (Photograph by Michael King.)

Poleward of 30°N, temperature is a limiting factor for plant growth and there are fewer clouds than in tropical regions. Plant life north of 40°N, which represents a line stretching from New York to Madrid to Beijing, has been growing more vigorously since 1981. Eurasia seems to be greening more than North America, with more lush vegetation. On average, plant greenness level increased by about 8% in North America and 12% in Eurasia from 1982 to 1999.

Over the same period, the start of spring (when defined as the time of the start of vegetation green-up) advanced by about 8 days in North America and 6 days in Eurasia. The decline of plant growth in autumn was delayed by 4 days in North America and 12 days in Eurasia. Therefore, the growing season was almost 18 days longer, on average, in Eurasia in 1999 compared to 1982, while in North America, it became 12 days longer, as shown in the two line graphs.

What caused the greening in the north and the continental differences in plant greenness between North America and Eurasia? The primary suspects are the ever-rising temperatures possibly linked to the buildup of greenhouse gases in the atmosphere. The surface warming, recorded by several thousand meteorological stations around the world, is especially pronounced on land in the north during the past 25 years. It is not uniform, however; for instance, the warming rate in the United States is smaller than in most of the world. In fact, there is even a slight cooling trend in the eastern United States over the past 50 years.

Year-to-year changes in plant greenness between 40°N and 70°N recorded in satellite observations are tightly linked to year-to-year changes in temperature data. Evidently, these changes in greenness reflect changes in biological activity. Statistical analyses indicate a statistically meaningful relation between changes in greenness and land surface temperature. That is, the temporal changes and continental differences in greenness are consistent with ground-based measurements of temperature, an important determinant of biological activity.

These results suggest a warming-enhanced plant life in the north. They are in agreement with some ground-based evidence of changes of phenology in plants (trees bud and flowers bloom earlier) and animals (birds lay their eggs earlier, and birds and butterflies migrate farther north) in response to the warming.

These results are of great importance because of possible implications to the global carbon cycle. If the northern plants are greening, they may already be absorbing more carbon dioxide—a process that can influence global warming. However, more research is needed to determine how much carbon is being absorbed and how much longer it will continue to be absorbed.

ANNEMARIE SCHNEIDER

# Monitoring Urban Areas Globally and Locally

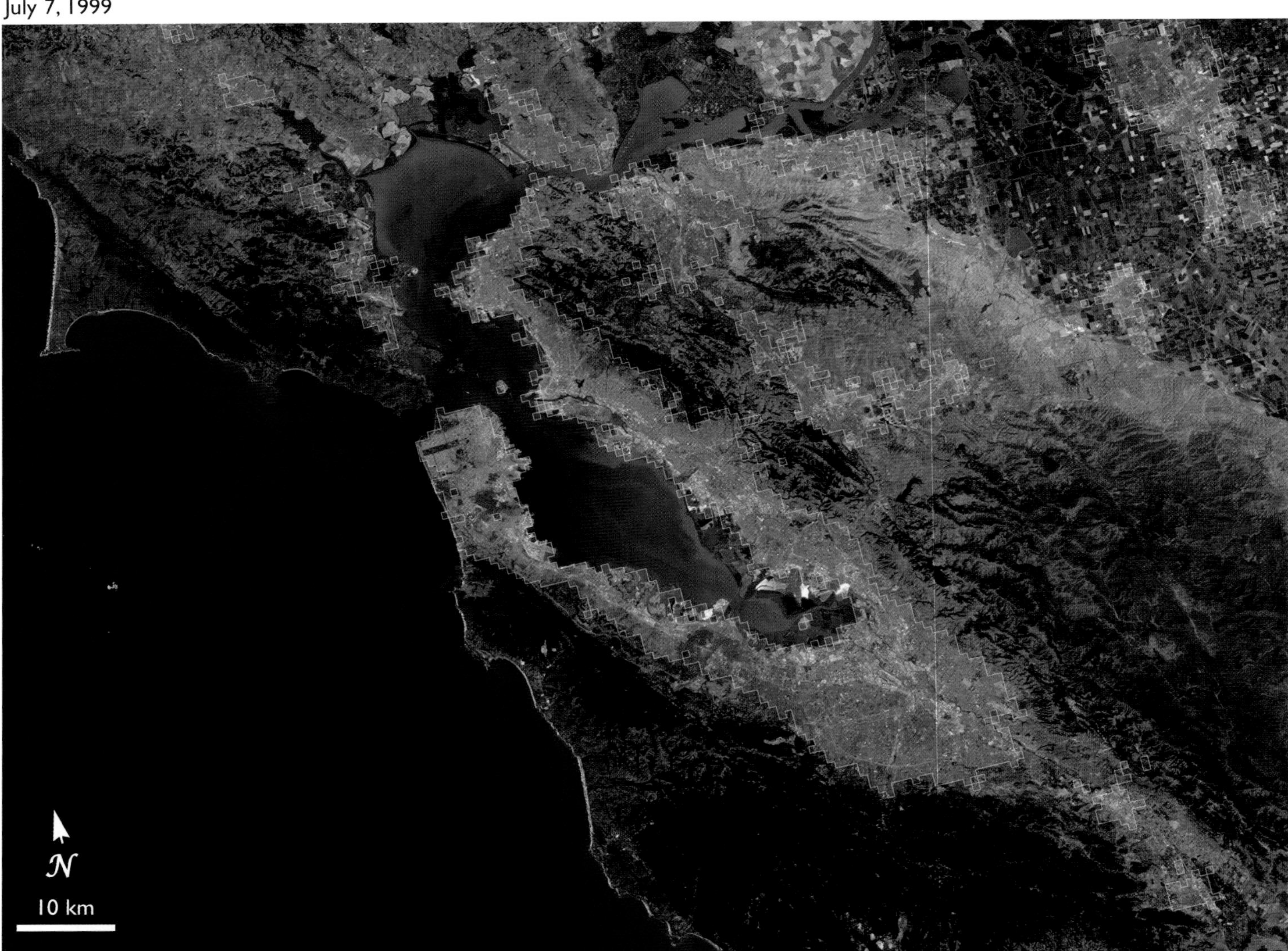


Satellite image of the San Francisco Bay area on July 7, 1999.Vegetation appears green, water blue, and bare land tan. (Data from the ETM+ instrument, bands 7 [infrared], 4 [near-infrared], and 2 [green], on the Landsat 7 satellite.)

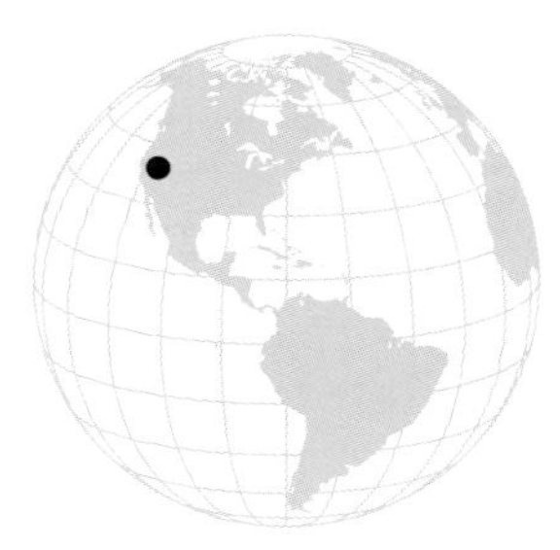

Urbanization—the transformation of land to urban uses—is one of the most significant and irreversible alterations of the Earth's surface. Throughout history, the reasons for urbanization have varied. Cities first began as small agglomerations of people that drew together for reasons of trade and defense, and could do so because of agricultural surplus. The Industrial Revolution brought the need for a greater urban population to fuel the industrial and commercial labor force and the need for a larger area of land to be converted to house those activities. As transportation and communication technologies advanced in the twentieth century, cities grew larger and larger, no longer limited by costs of moving goods or people. Today, a complex, dynamic, and interacting set of mechanisms cause urban growth, including not only economic reasons that push people into cities, but political, institutional, and cultural factors that make it easier and/or cheaper for land to be developed.

*Our Changing Planet*, ed. King, Parkinson, Partington and Williams
Published by Cambridge University Press 

Although it can be difficult to tease out which reasons are causing land conversion in any particular city, it is clear that an ever larger portion of the world's population is affected by urban changes. More than half of the Earth's inhabitants (3.3 billion people) now live in cities and metropolitan areas, and the percentage is expected to increase to over 60% in the next 20 years. By comparison, the Earth's human population was just 14% urban at the beginning of the twentieth century.

While urban land cover presently accounts for 1–2% of the Earth's total land area, this percentage also is growing rapidly. Cities have grown so significantly in the last decades that it is critical to have accurate and up-to-date maps to help monitor these changes at a global scale. Previously, very few sources of data on urban extent were available. Furthermore, although satellite imagery of nighttime lights now offers a striking depiction of human activity on the planet (see the 2-page sample global image in the next chapter), this imagery provides a representation of energy emissions, not a representation of the actual settlement patterns on the ground.

A combination of new satellite information and new data fusion methods has enhanced our ability to monitor urban areas. By linking population data and several types of satellite imagery, it is possible quickly and accurately to map the built environment across Earth. A small segment of this global urban map (the San Francisco Bay area, as illustrated) shows that the new map does a better job of characterizing the built-up areas than maps previously available.

Population and urban land area changes from 1990 to 2000, for a sample of 30 metropolitan areas around the globe. For each sample location, the line connects the two dots corresponding to the 1990 and 2000 data. (Land area data from the ETM+ instrument on the Landsat 7 satellite; population data from the following organizations: United States Census Bureau; Statistics Canada; Canadian Census at the University of Toronto; Instituto Nacional de Estadística, Geografía e Informática, Mexico; Instituto Brasileiro de Geografia e Estatística; Instituto Nacional de Estadística, Spain; Český Statistický Úřad, Czech Republic; Polska Główny Urząd Statystyczny, Poland; Ankara Devlet İstatistik Enstitüsü, Turkey; Central Agency for Public Mobilization and Statistics, Egypt; Kenya Central Bureau of Statistics; Registrar General and Census Commissioner of India; Chinese Bureau of Statistics; China Data Center at the University of Michigan.)

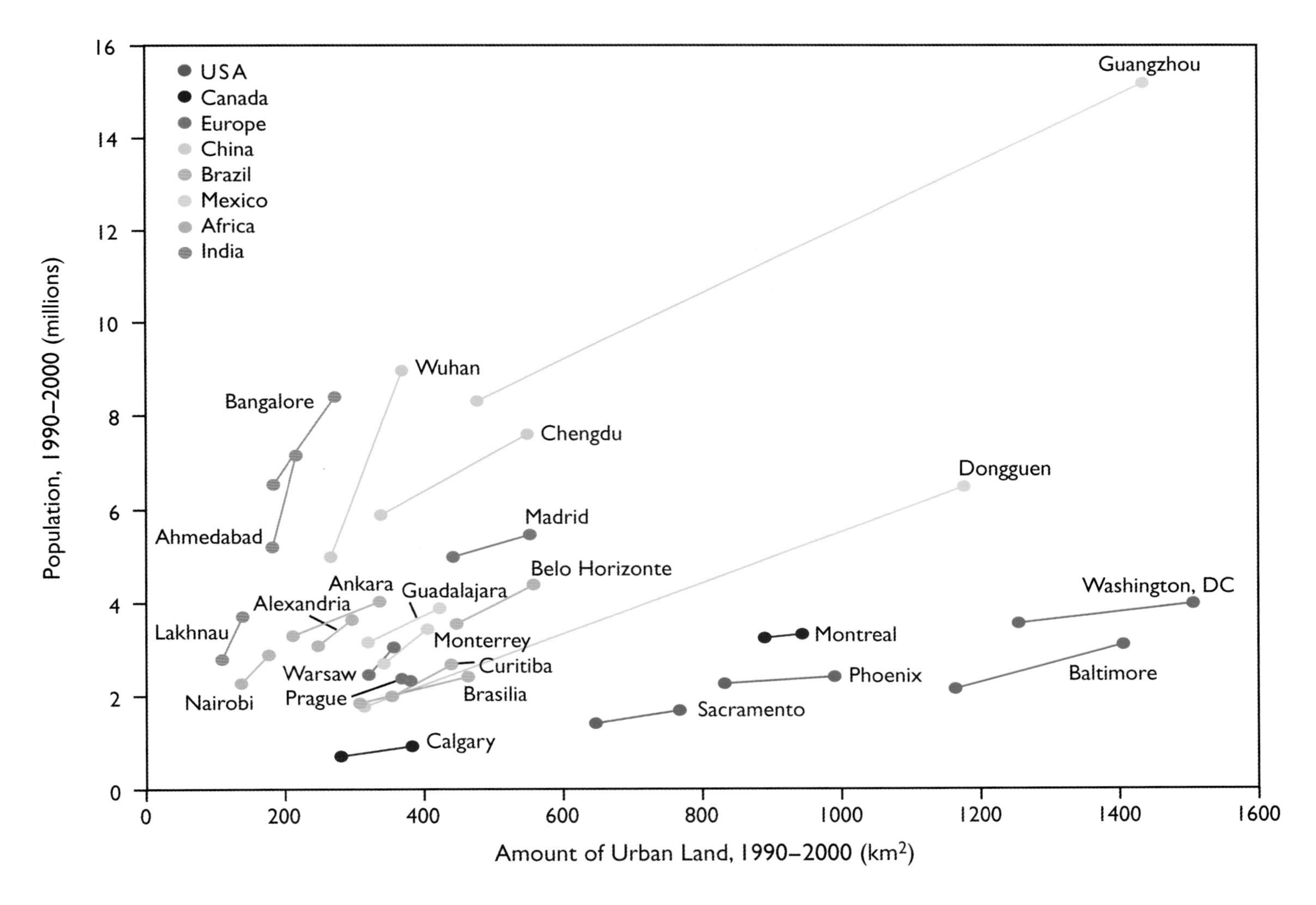

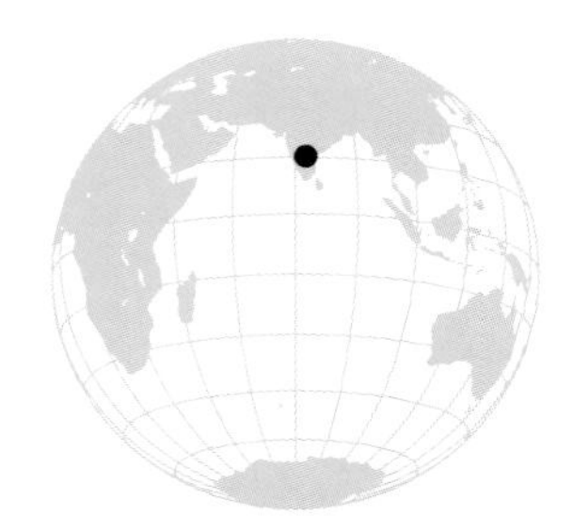

Bangalore, India, November 27, 2000

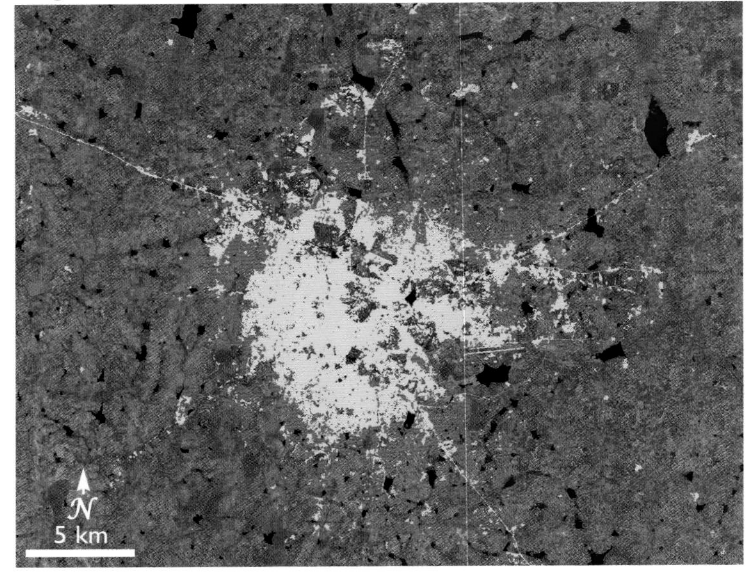


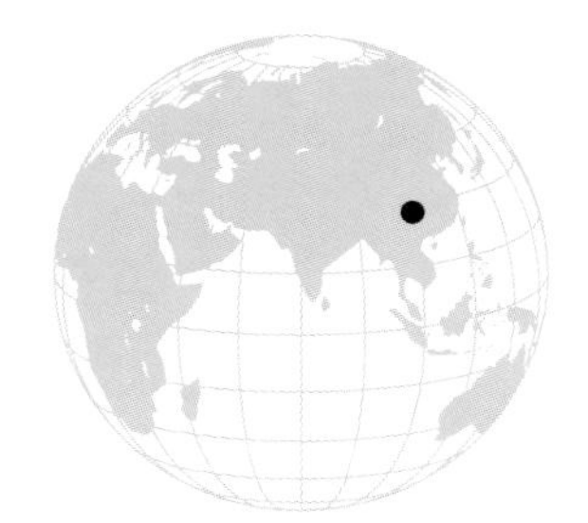

Chengdu, China, November 2, 2000

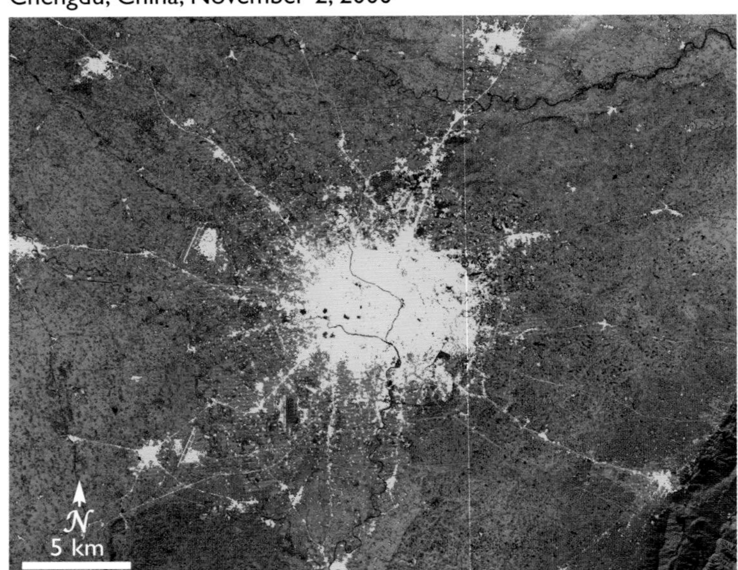


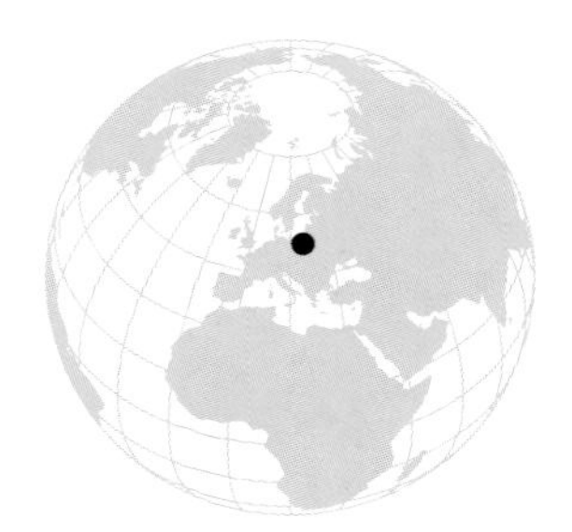

Warsaw, Poland, May 7, 2000

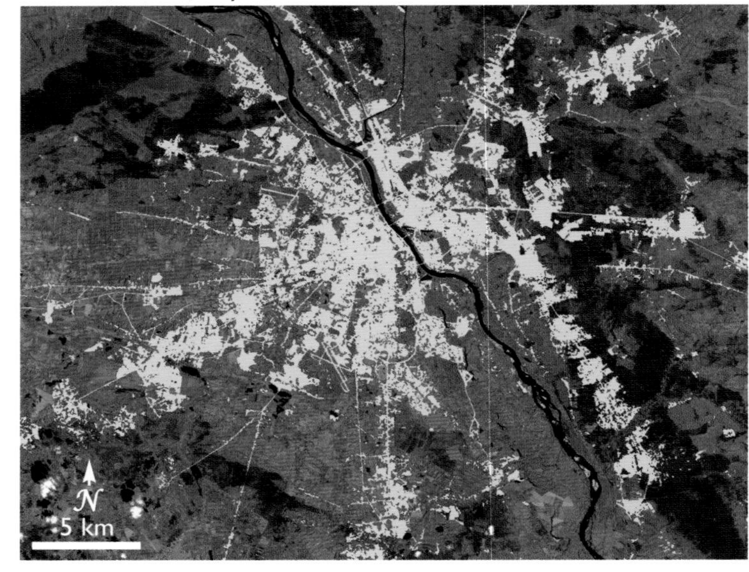

Sacramento, California, United States, July 9, 2000

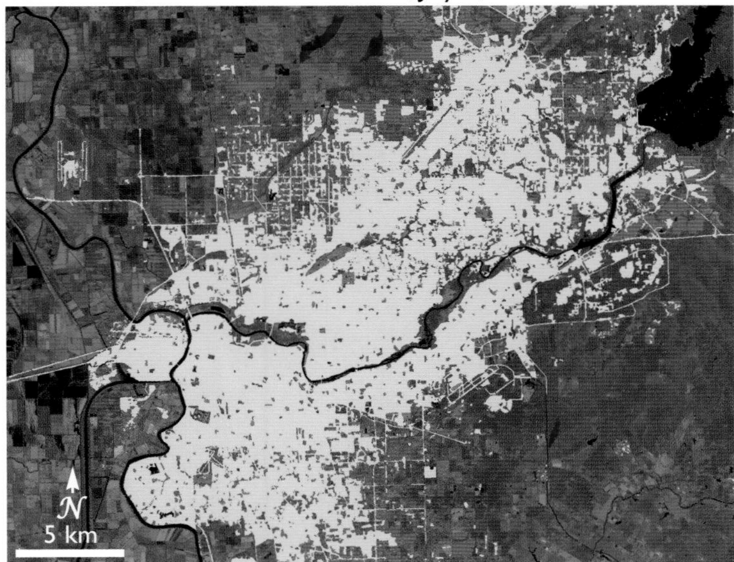


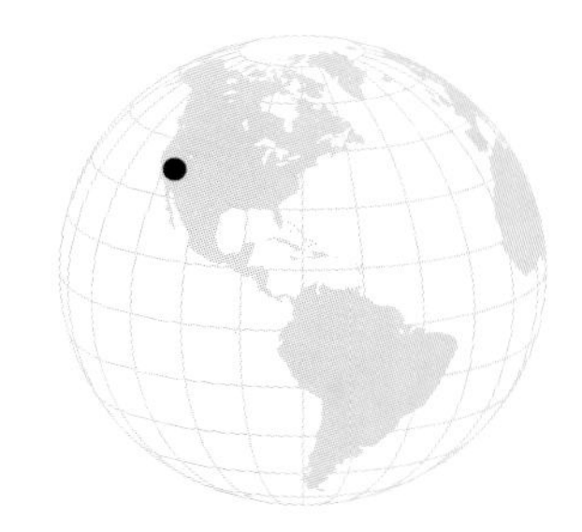

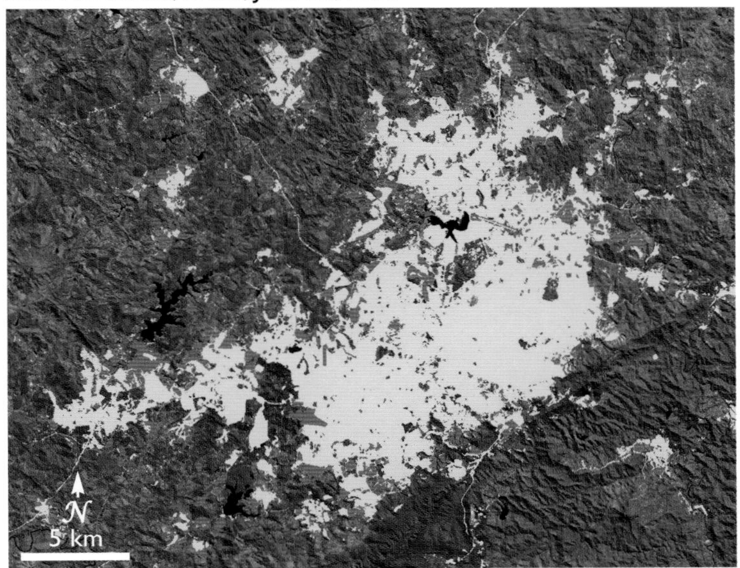


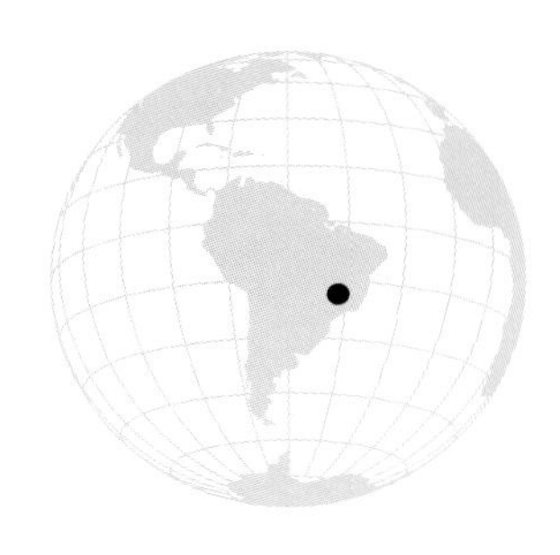

Images of land use change in 5 major cities. In each case, the base image is a satellite image from the indicated date, with vegetation represented in green, water in dark blue/black, and bare ground in purple. Maps of urban extent are overlaid on each image, with yellow depicting the amount of urban land cover circa 1990 and orange the amount of land developed between 1990 and 2000 (individual dates of the overlay information may vary by one year). (Data from the ETM+ instrument, bands 7, 4, and 2, on the Landsat 7 satellite.)

Global maps are useful for monitoring the extent of cities and suburban areas, but understanding land use change requires much more detailed local and regional data. By selecting a sample of 30 metropolitan areas and comparing how these cities have grown from 1990 to 2000, we can begin to understand how different the shapes and sizes of the world's cities really are. This spatially explicit information can then be combined with economic and population data to understand whether certain socioeconomic changes have created large amounts or specific patterns of development. One of the simplest questions regarding urban growth, for instance, is: what is the connection between population change and amounts of land developed in the city? Or: does a growing population indicate a corresponding amount of land conversion? In regard to this question, the plot illustrates that the relationship between population and land use change varies significantly depending on location. For instance, some cities in the United States (green lines) are much larger in area than their peers around the world, and these United States cities continue to experience vast amounts of land conversion without corresponding population increases. Second, cities in developing countries such

Townhouse complex, Greenbelt, Maryland, August 4, 2005, built in the 1970s on previously heavily forested land. (Photograph by Claire Parkinson.)

as China and India are making headway in matching the larger areas of more industrialized cities, but generally remain quite compact by comparison. Population growth in many cities in developing countries is still quite high because of high birthrates and high rates of migration into the cities. Finally, a large group of cities clusters in the middle of the plot, including cities in Europe, Brazil, Mexico, and Canada.

Looking at cities from this global perspective may provide several answers, but it also raises new questions. What is the path of urban development? Which, if any, additional metropolitan areas around the world will follow United States cities, expanding as far as 100 km from the downtown? Which, for geographic or other reasons, will follow a different trend, remaining more compact?

# Gray Wave of the Great Transformation

MARC L. IMHOFF
LAHOUARI BOUNOUA
J. MARSHALL SHEPHERD

Early morning fog engulfing the skyline of Houston, Texas, late fall 2001. (Photograph © 2005 Jim Olive.)

Land cover change driven by human activity has profoundly affected Earth's natural systems, with impacts ranging from a loss of biological diversity to changes in regional and global climate. The history of human land transformation is seen as a progression of civilization building, starting with fire management, herding practices, then the development of agriculture, and subsequently urbanization. This change has been so pervasive and progressed so rapidly, compared to natural processes, that scientists refer to it as 'the great transformation.'

*Our Changing Planet*, ed. King, Parkinson, Partington and Williams
Published by Cambridge University Press 

A composite image of the world at night, 2000. (Data from the OLS instrument on a DMSP satellite, provided by Christopher D. Elvidge, NOAA-NESDIS National Geophysical Data Center.)

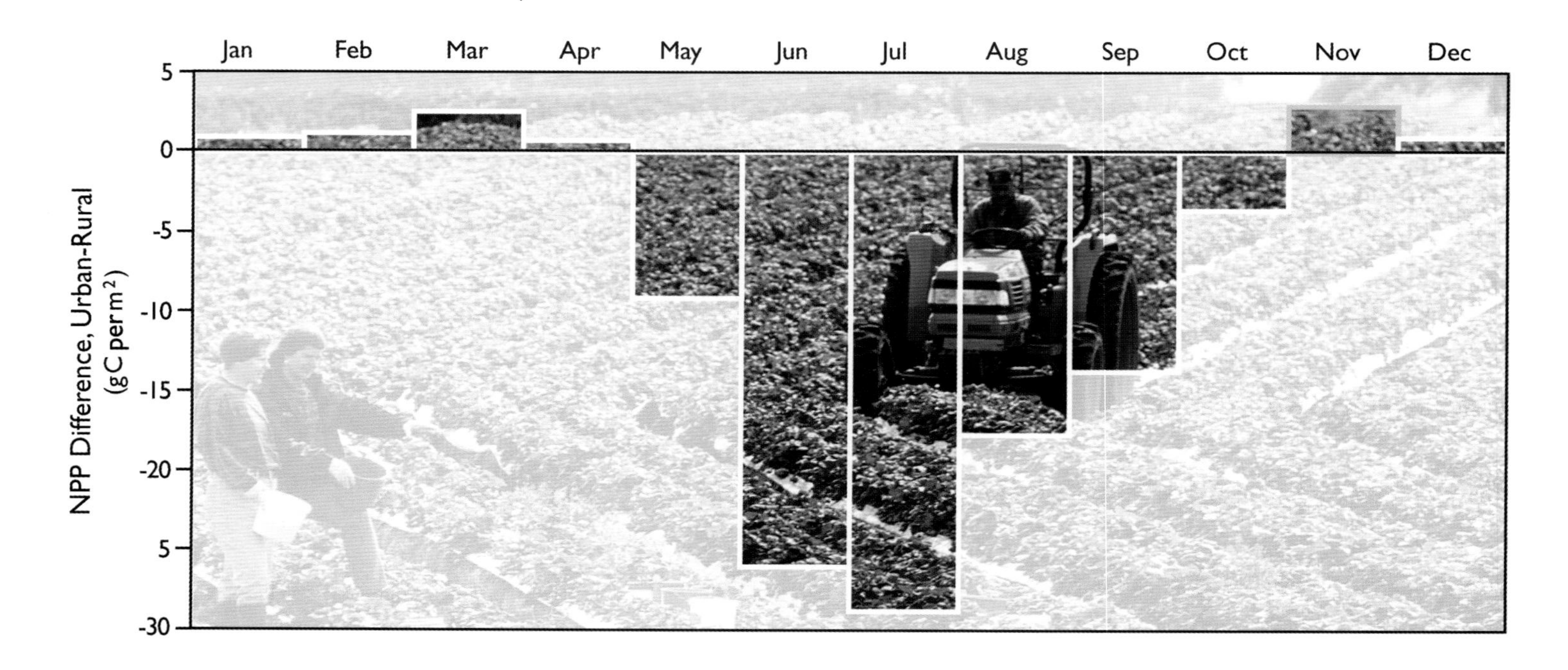


Difference in Net Primary Productivity (NPP) between urban and rural areas in the mid-Atlantic region of the United States, April 1992–March 1993. (Data from the AVHRR instrument on the NOAA 10, 11, or 12 satellite; background photograph by Scott Bauer, courtesy USDA ARS Photo Unit.)

Urban land transformation is being increasingly recognized as an important process in global change. Urbanization has accelerated with population growth and we are now in the mega-city era wherein cities with populations exceeding 10 million inhabitants exist throughout the globe. Cities of this size profoundly alter the land surface, affect local climate, and require vast amounts of energy and economic resources to sustain the infrastructure. Studying the interaction between urbanization and the environment is needed for maintaining important biological processes across the landscape as well as for developing healthy and efficient urban habitats.

Among the more important issues surrounding urbanization are how the size and density of cities affect temperature, rainfall, and plant growth. As cities grow, the vegetation cover is reduced as the amount of surface area covered by cement and asphalt increases. This change in surface composition is complex and creates conditions for urban heating that can cause heat related health problems within cities, increase the amount of particulates in the atmosphere, and potentially change weather patterns locally and regionally. Urbanization and the placement of urban developments on fertile soils also reduce the amount of total plant growth on the land surface, affecting both food production and the amount of carbon dioxide (a greenhouse gas implicated in global warming) in the atmosphere.

How do we study the complex suite of issues surrounding urbanization? The first requirement is to have a reasonably accurate global map showing the location and size of developed areas. However, defining exactly what is meant by 'urban' and then measuring it across the globe has been problematic; even in the United States, where modern census procedures are used, the definition and measurement are by no means straightforward. Nighttime images from space are providing a breakthrough in our ability to acquire and update global maps of urbanized land. Satellite sensors originally designed to map moonlit cloud cover for nighttime weather forecasting are sensitive to even a small amount of light. As illustrated in the large satellite nighttime image composite, the imagery collected by these satellites on cloud free nights shows a dramatic picture of global urbanization through the detection of city lights.

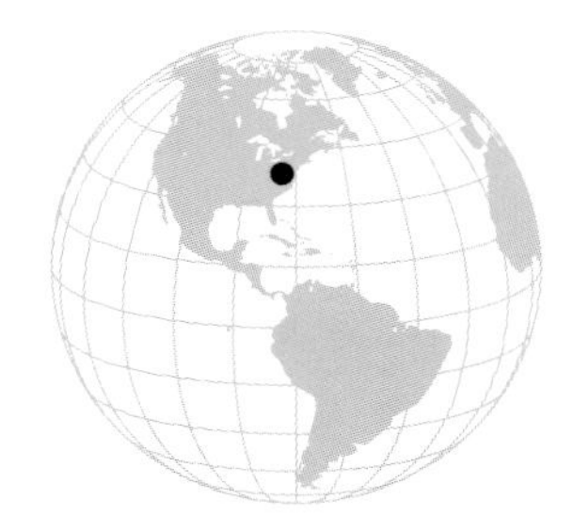

Combining the capabilities of different satellite systems allows us to answer questions about how urbanization affects the landscape. Daytime data collected from several satellites over the course of a single year or multiple years can be used to calculate the amount of plant growth. One of the ways scientists measure plant growth is by looking at the amount of carbon that was taken from the atmosphere and converted to plant tissue through the process of photosynthesis. This quantity is called net primary productivity (NPP) and can be presented in units of carbon. NPP is important not only because it is a part of the carbon cycle affecting global warming but also because it is the primary source of food energy for Earth's food web, including human beings. By merging the city lights imagery with satellite-based estimates of NPP, we can quantify the impact of urbanization on plant life and agricultural production for vast areas. For example, in the United States, urban land transformation has reduced NPP by 1.6%. While this may not seem like much, it is actually enough to offset the gains made by the conversion of land to agricultural use.  These results tell us that as populations grow and urban areas expand into agricultural areas we need to be aware of how this might affect future food security.

Combining nighttime images and daytime satellite data is also helping scientists understand urban heat islands and how they interact with the environment. For example, the graph shows the difference between the amount of plant growth (NPP) each month inside urban areas versus rural areas of the mid-Atlantic region of the United States, as determined from satellite data. Values above the 0 line indicate more NPP in urban areas. From November through April 'urban heat'

Satellite image of the urban heat islands of major cities in the northeast United States, September 27, 2005. A 3.9 μm channel was used to detect relatively cold and warm surfaces and thereby show the extents of the urban heat islands around cities such as New York, Washington, DC, and Richmond. (Data from the Imager instrument on the GOES 12 satellite.)

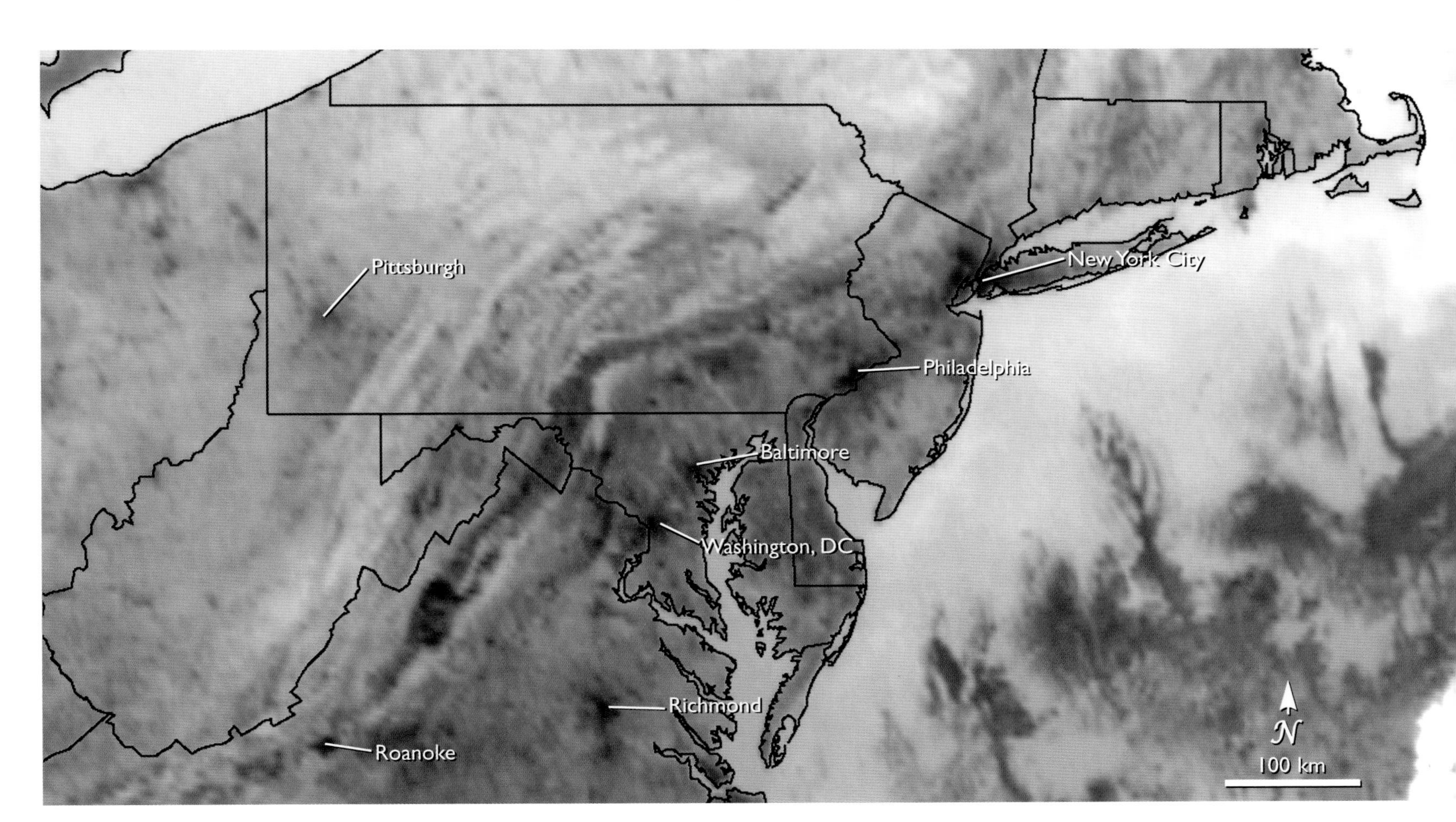


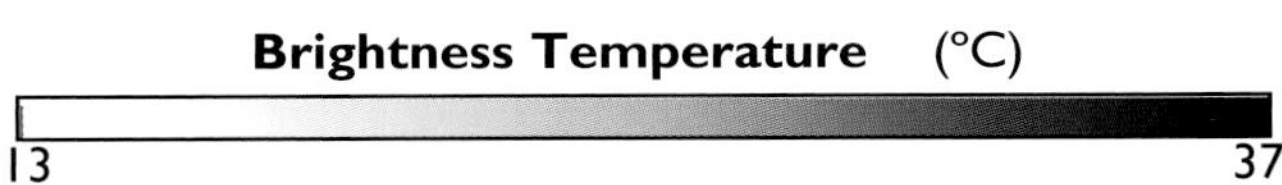

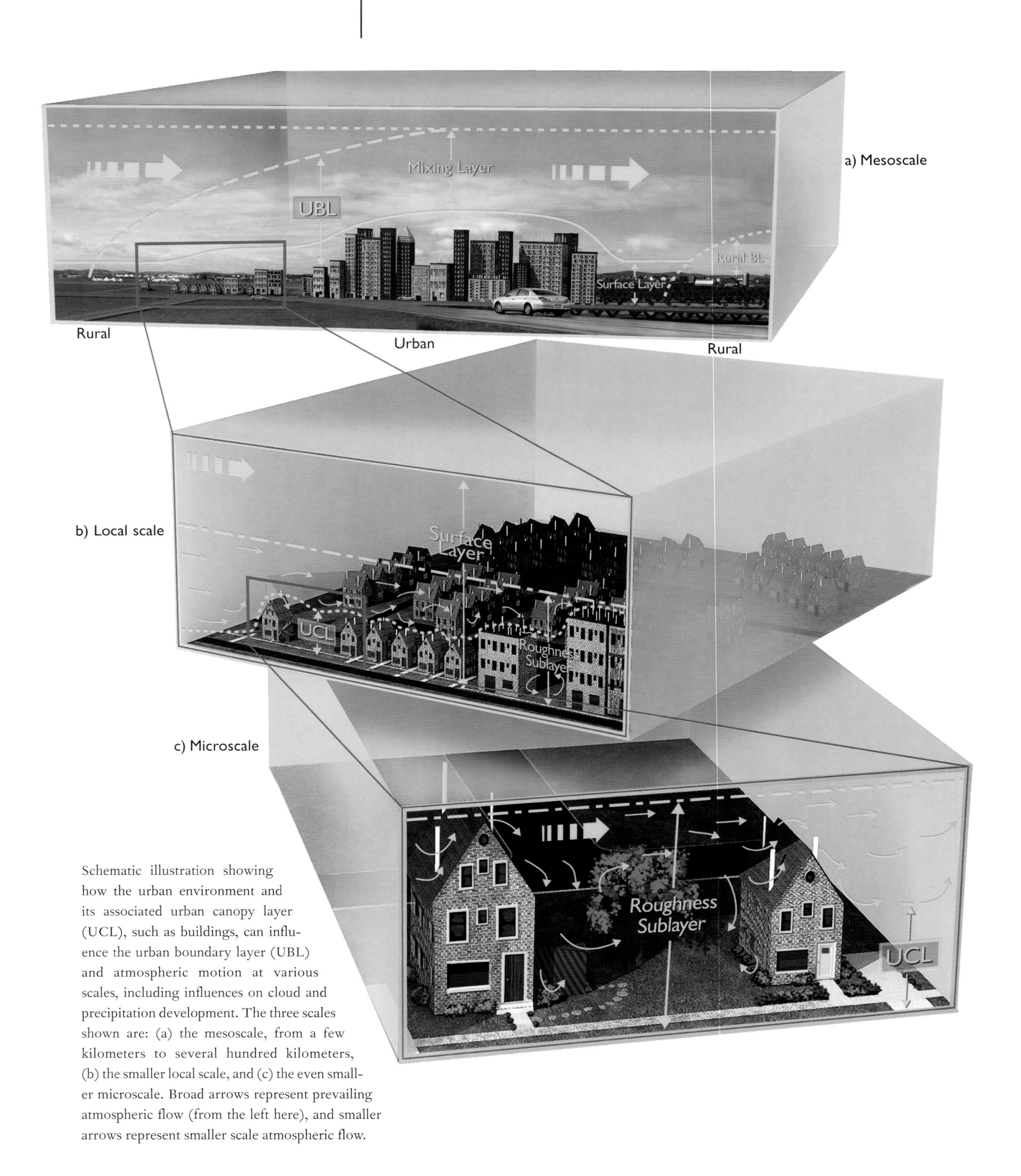


Schematic illustration showing how the urban environment and its associated urban canopy layer (UCL), such as buildings, can influence the urban boundary layer (UBL) and atmospheric motion at various scales, including influences on cloud and precipitation development. The three scales shown are: (a) the mesoscale, from a few kilometers to several hundred kilometers, (b) the smaller local scale, and (c) the even smaller microscale. Broad arrows represent prevailing atmospheric flow (from the left here), and smaller arrows represent smaller scale atmospheric flow.

extends the growing season for plants inside or very near the cities so that they are more productive than vegetation in the colder countryside. In the summer months, however, the loss of vegetation in the cities results in much lower NPP. This is more than enough to offset the extended growing season caused by urban heat, so that the annual total of NPP in a mid-Atlantic city is much lower than in a rural area of the same size.

Urban expansion and its associated urban heat islands also have measurable impacts on weather and climate processes. Scientists believe that the urban environment (with its associated urban heat islands, surface roughness from buildings, pollution, and other characteristics) affects local and regional temperature, wind patterns, and air quality, and may impact the global water cycle through affecting the development of clouds and precipitation in and around cities. But how do we prove and quantify these important interactions? Satellite observations, field observations, and numerical models are making it possible to address these issues. From these data scientists now know that the magnitude of urban heating (the difference between the average urban temperature and the average rural temperature in the same general vicinity) is typically proportional to the city size and most apparent after sunset.

Using data from the world's first satellite-based precipitation radar and ground-based rain gauges, scientists were able to link these urban environments with changes in rainfall. Results found for the metropolitan area around Houston (and other cities) showed elevated rainfall rates over and downwind of the city. The study also presented evidence that the increase in rainfall was linked to the urbanized land cover and not due exclusively to sea or bay breeze circulations. In order to develop, precipitating clouds require a few basic ingredients: moisture, a source of lift, and atmospheric instability (e.g., a blob of hot air will be locally buoyant relative to its surroundings). On moist warm-season days when large-scale atmospheric forcing is weak, the urban environment can contribute two of these ingredients, instability and a source of lift. In the Houston study, airborne thermal sensors clearly showed a distinct urban heat island created by the central business district. This part of the city is noticeably warmer than suburban and rural areas and is a phenomenon typical of most cities. The urban heat island destabilizes the boundary layer through vertical mixing and enhancement or creation of a direct thermodynamic circulation, the so-called 'urban heat island circulation'. Additionally, urban surfaces created by the large buildings in the central business areas tend to be aerodynamically rough, enhancing low-level convergence, which can provide the 'trigger' necessary to lift moist air to appropriate levels for condensation and formation of rain. If the air is adequately unstable, clouds and precipitation will form. In coastal environments like Houston, the convergence may be further enhanced by interactions with a sea breeze. Recent literature also suggests that urban aerosols (e.g., pollution) may modify cloud and precipitation processes, but there is still a degree of uncertainty in their net effect.

As the global population grows and urban areas expand, it becomes even more important that we understand the dynamic interactions between urban development and the environment. In the coming decade, scientists will continue to explore how this important land use affects plant growth, agricultural production, and weather and climate processes like precipitation.

# Urban Heat Islands

Dale A. Quattrochi
Maurice G. Estes, Jr.
Charles A. Laymon
William L. Crosson
Burgess F. Howell
Jeffrey C. Luvall
Douglas L. Rickman

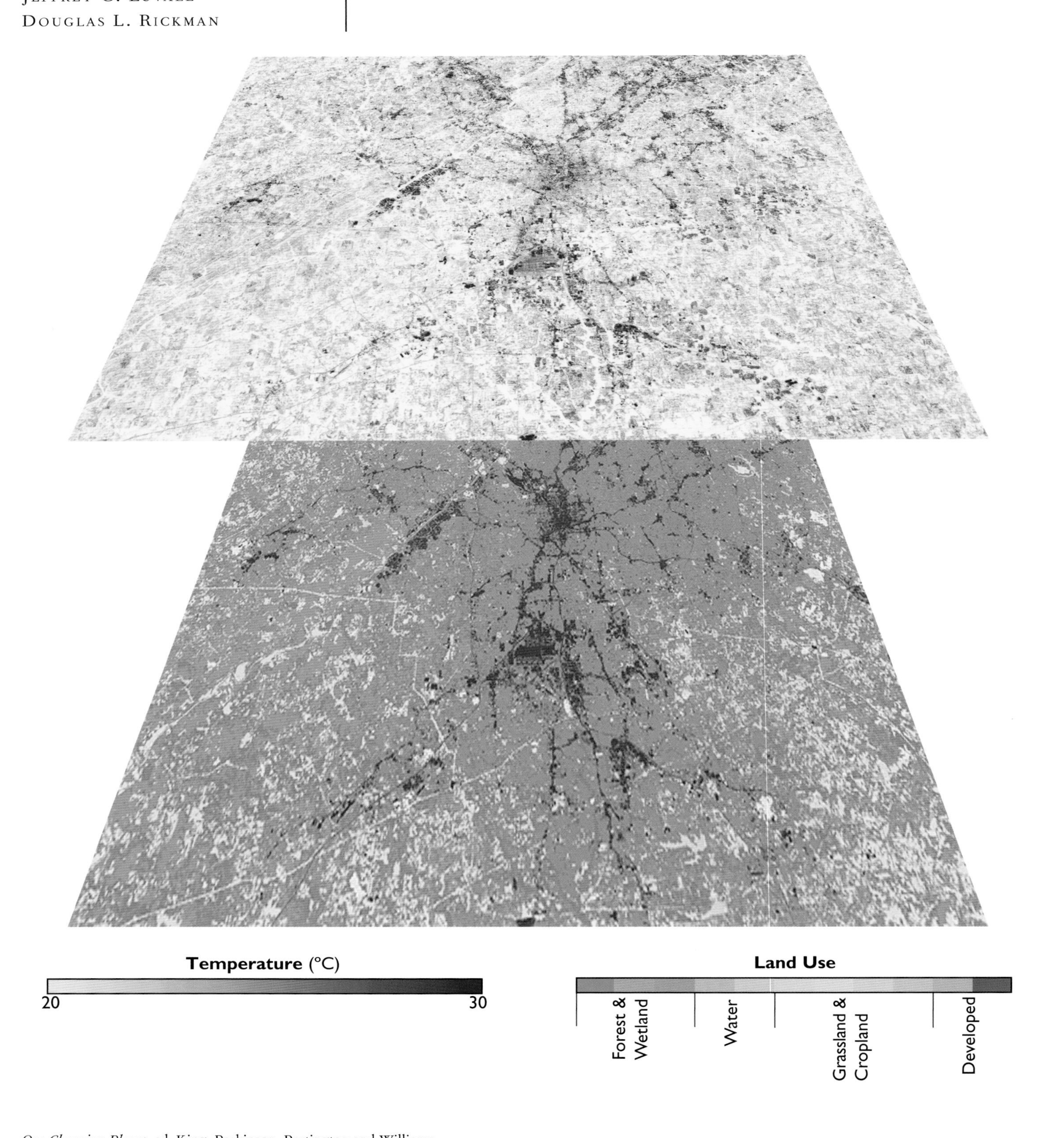


*Our Changing Planet*, ed. King, Parkinson, Partington and Williams
Published by Cambridge University Press 

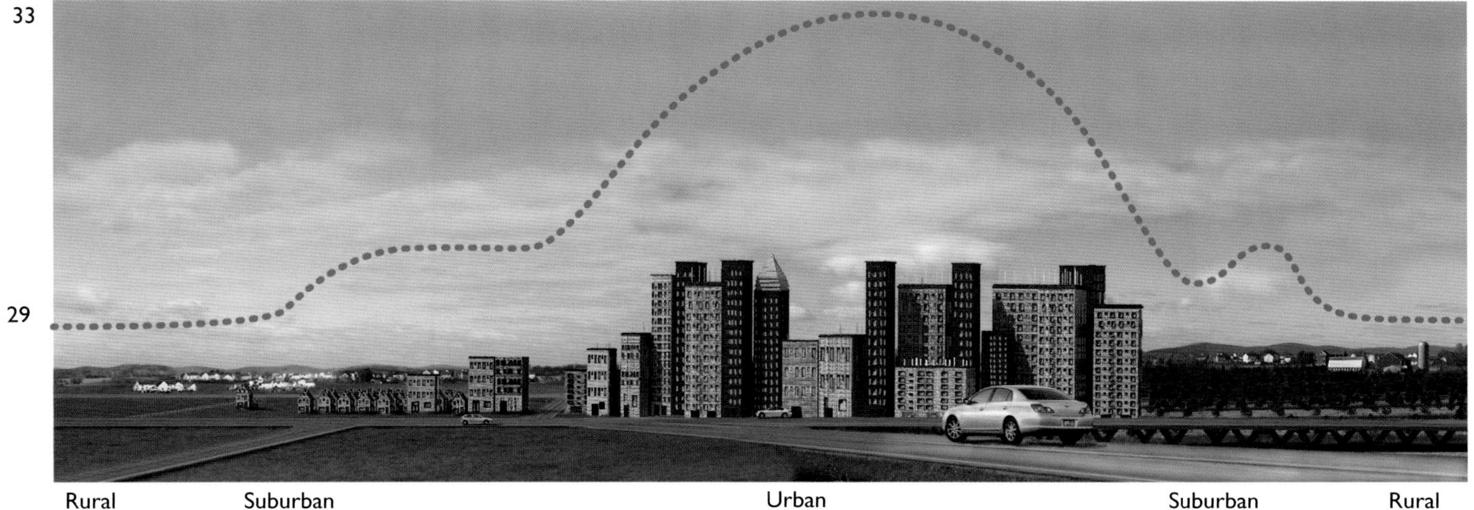


Diagram of an urban heat island profile for late afternoon temperatures. (Temperature profile taken from LLBL Heat Island Group.)

In the early 1800s, the eminent British scholar, Luke Howard, recognized that the urban center of London, England, was considerably warmer than the surrounding countryside. As urban areas across the world have grown into 'mega cities'—cities with populations of 10 million inhabitants or more—the temperature differences between urban areas and their rural counterparts have increased accordingly. This phenomenon has come to be known as the Urban Heat Island (UHI) effect, whereby surfaces typical of the city landscape, such as pavement, rooftops, or other 'non-natural' land covers, absorb solar radiation throughout the day and store it as heat energy. Typically, in the nighttime hours on clear nights, this stored energy is released into the lower atmosphere, forming a 'heat bubble' over the city. This dome of elevated air temperatures that presides over the city or the UHI is typically 1–3°C (2–5°F) above the temperatures over adjacent rural areas. Even more so, there can be a significant daytime heat island effect, particularly during spells of clear days during summertime.

Why is the study of the UHI important? After all, this phenomenon has been known since the early 1800s; why worry about it now? One of the main reasons why studying the UHI is important is because of the projections by the United Nations that by 2025, 60% or more of the world's population will live in cities. As urban areas become more populated, they have a major effect on the local, regional, and perhaps even the global environment. A particular consequence of urbanization is the deterioration of air quality over cities. The UHI has a significant impact on air quality by increasing the amount of ground-level ozone ($O_3$) over cities. However helpful ozone is high in the atmosphere, ozone at the ground level is a harmful air pollutant. We are unable to see it, smell it, or taste it; yet it has a serious impact on the human respiratory system. Ground-level ozone is formed by the interaction of two photochemical constituents in the presence of intense solar radiation: volatile organic compounds (VOCs) and nitrogen oxides ($NO_x$). VOCs are produced by anthropogenic sources such as solvents, cleaners, and paints and by natural vegetation sources (the latter resulting in biogenic VOCs). $NO_x$ are air pollutants produced from automobiles, smokestacks, and other point-source emissions. The catalyst that drives the production of $O_3$ out of this chemical 'soup' is sunlight, particularly during long, hot summer days. The UHI may be considered an additive background source of heat that enhances the production of $O_3$. This makes the UHI an important factor to study as cities around the world continue to grow and expand, both in population and in areal extent.

Opposite: Satellite-derived images of surface temperature (top) and land use (bottom) centered on Atlanta, Georgia, April 5, 2000 and 2001. Warmest temperatures are depicted by red and orange colors and cooler temperatures by yellow. Purple shades in the land cover image represent Atlanta urbanization, whereas green indicates vegetation and blue indicates water and wetlands. (Data from the TM instrument on the Landsat 5 satellite, courtesy NASA GSFC and USGS EROS; NLCD data from Seamless Data Distribution System, USGS EROS.)

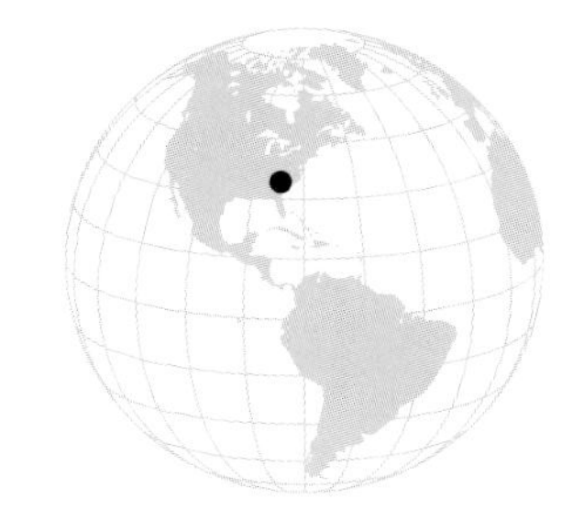

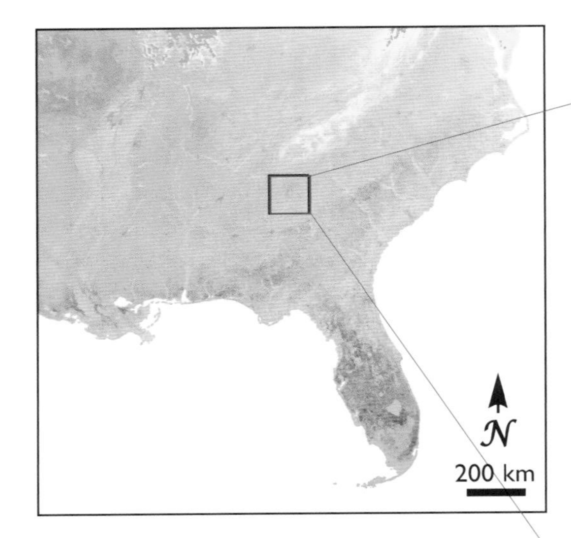


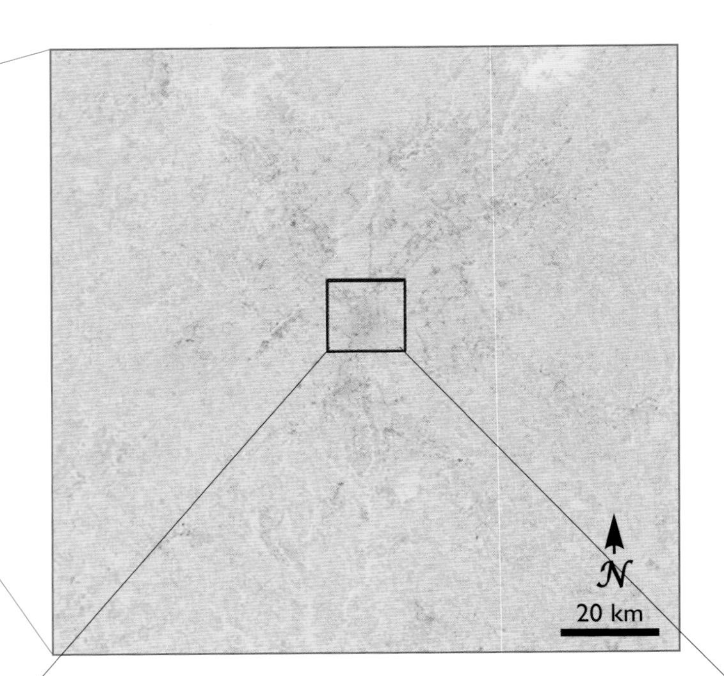

Color-coded images of land surface temperatures. Top left: the southeastern quadrant of the United States on April 5, 2000; the box highlights a region within the state of Georgia. Top right: the vicinity of Atlanta, within the state of Georgia on April 6, 2000. Bottom: the Central Business District of Atlanta on May 11, 1997, which overall is very warm and has a very low albedo, because of the low reflectivities of surfaces made of dark-colored materials such as tar and asphalt. (Top left: data from the MODIS instrument on the Terra satellite. Top right: data from the ETM+ instrument on the Landsat 7 satellite, from Laura Rocchio. Bottom: data from the ATLAS instrument on the NASA Learjet 23 aircraft.)

Although Luke Howard used a ground-based thermometer in measuring the air temperature differences between the city and the countryside, the UHI can be readily determined from satellite observations. We can also readily observe from satellites the relationship between the extent of urbanized areas and the elevated surface temperatures caused by the overall landscape characteristics of the city, as illustrated in the layered pair of Landsat images. Moreover, we can use thermal infrared (TIR) data from NASA aircraft to derive a very detailed depiction of 'what's hot and what's not' across the urban surface, as illustrated. These kinds of high resolution data are very useful for illustrating to urban planners, government officials, local decision-makers, and the general public, the extent to which warm non-natural surfaces are spread across the urban landscape and how the heat emanating from these surfaces drives the development of the UHI. Additionally, these data are illustrative for showing where 'Cool Community Strategies' can be useful for diminishing or mitigating the UHI. For example, trees can be planted in 'hot spots' in the downtown area to help cool the city surface.

Another factor that contributes to the development of the UHI is the reflectivity or 'albedo' of the surfaces in the city. Albedo is a measure of reflectivity that ranges from 0 to 1, with 0 for a surface that reflects no solar radiation (e.g., a black surface) and 1 for a surface that reflects all solar radiation. Surfaces that have low albedos become very hot in the presence of intense solar radiation and are direct contributors to the development of the UHI. We can use remote sensing data to observe and measure albedos and surface temperatures across the urban landscape for better understanding the causes of the UHI, as well as to illustrate ways potentially to mitigate its overall magnitude and impact on the urban environment.

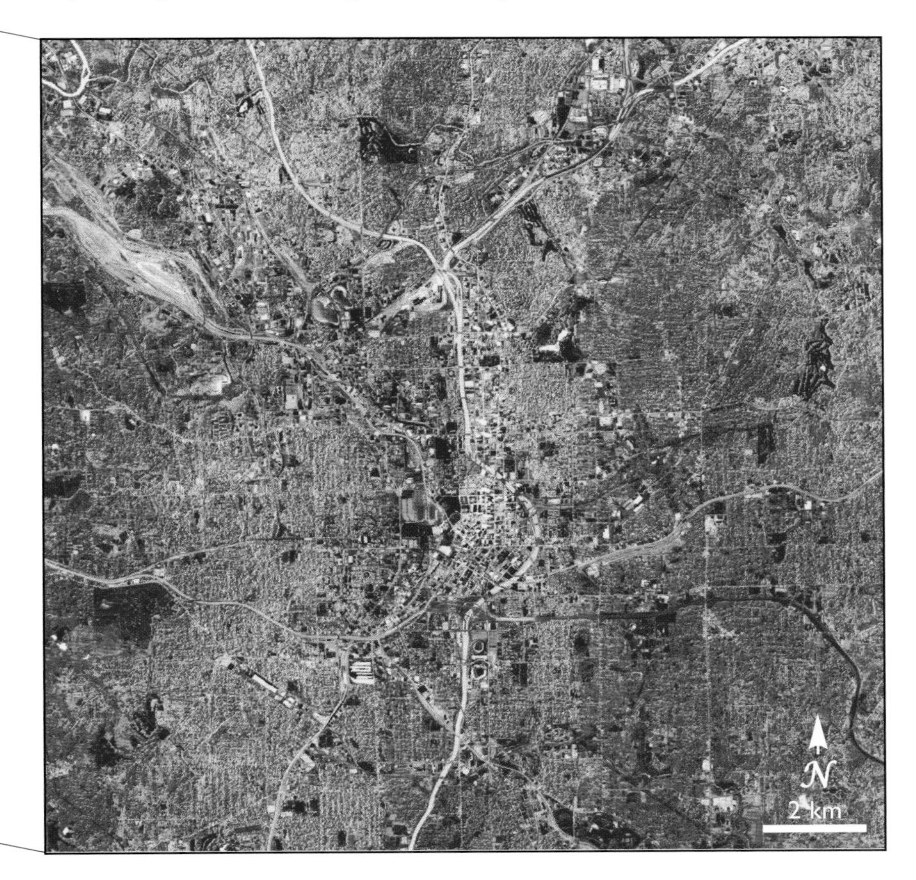


High-resolution color-coded image of albedos in and around Atlanta, May 11, 1997, from aircraft data. The Central Business District overall has a very low albedo, because of the low reflectivities of rooftops made of tar or other dark-colored materials and dark asphalt-covered roadways. (Data from the ATLAS instrument on the NASA Learjet 23 aircraft.)

Karen C. Seto

# Urbanization in China: The Pearl River Delta Example

The city of Shenzhen, China, on March 4, 2002. This metropolitan area was transformed from a small fishing village in 1973 to a modern metropolis by the end of the twentieth century. (Photograph by Karen Seto.)

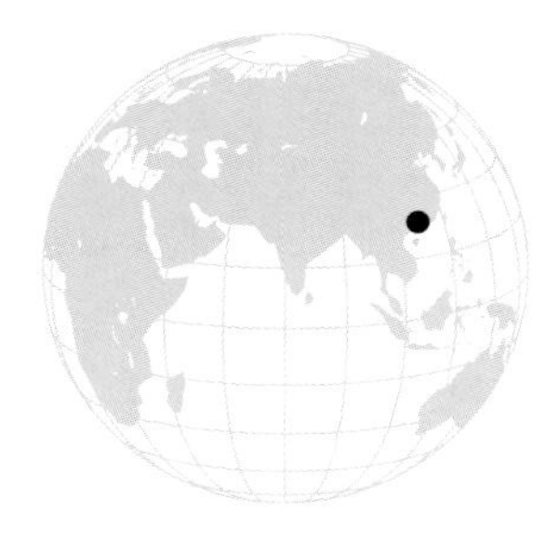

Bicycles are making way for cars as the departure of this iconic image signals China's entry into the twenty-first century. The Pearl River (Zhujiang) Delta in southern China exemplifies this ancient civilization's change from the pastoral to the urban. Over the last 2,000 years, this predominantly agricultural region has had a rich history of rice and fruit production, and is famous for its lychee (see photograph). During the Tang dynasty, Emperor Xuan Zong insisted that fresh lychees be transported by his fastest relay riders from the Pearl River Delta to his palace in Xi'an, the equivalent of the distance between Denver and Washington DC.

Although the lychee remains an important crop, the Delta's agricultural fields are giving way to modern supermarkets and highways. Policy reforms in the late 1970s and early 1980s spurred economic growth, which subsequently accelerated urbanization throughout the country. Satellite observations of the Delta show that urban areas have more than

*Our Changing Planet*, ed. King, Parkinson, Partington and Williams
Published by Cambridge University Press © Cambridge University Press 2007

Native to southern China, the lychee is a juicy and sweet fruit with a translucent inside and a firm but flexible red peel. (Photograph by Ian Maguire.)

quadrupled in size between the 1970s and 1990s, and that the combined metropolitan regions are now larger than the state of Rhode Island. The region has metamorphosed into a cosmopolitan cityscape complete with luxury golf courses, designer homes, and myriad skyscrapers.

Urbanization has affected not only the physical landscape but also the quality of people's lives, as evidenced by better housing, rises in income, increased motorization, more job opportunities, additional leisure time, and improvements in diet. The region's economy is evolving quickly to include increased services, high-tech industry, education, and tourism, all complementing its traditional manufacturing and agricultural bases.

The future holds multiple challenges for China, as the country strives to balance economic development with environmental conservation and the modern with the traditional. Despite physical and social changes, the Pearl River Delta remains the lychee capital of China, with farmers planting new orchards to continue its historical legacy.

Emperor Xuan Zong (AD 713–755). Xuan Zong was the most influential emperor during the Tang Dynasty (AD 618–907) and helped the Chinese empire reach unprecedented heights, with major advances in economics, trade, the arts, and the military. He was an accomplished artist and promoted painting, poetry, and sculpture. Emperor Xuan Zong's legacy prevails through countless poems, ballads, and novels inspired by his romance with Lady Yang Guifei.

Old farming practices juxtaposed against new skyscrapers, Shenzhen, China, March 10, 2002. (Photograph by Karen Seto.)

Comparison of land use around the Pearl River Delta in Southern China between December 10, 1988 (above) and March 3, 1996 (above opposite). (Data from the TM instrument on the Landsat 5 satellite.)

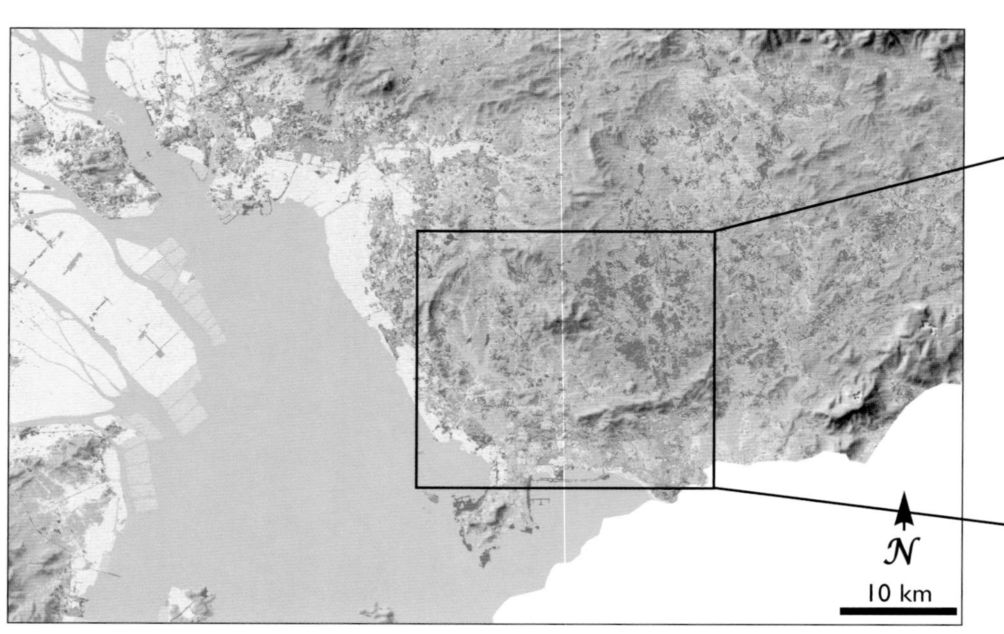


The areas of change in land use around the Pearl River Delta between 1988 and 1996. Lighter colors indicate those areas of land that did not change. The darker colors indicate changes. The legend, far right, shows the class of change. (Data from the TM instrument on the Landsat 5 satellite.)

March 3, 1996

N
5 km

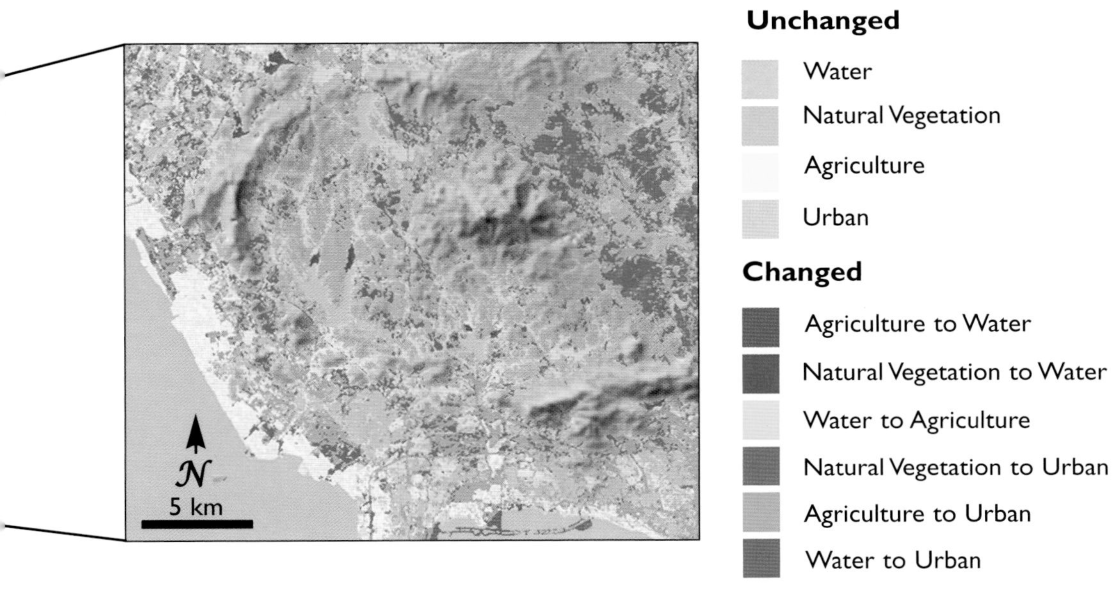
N
5 km
Unchanged
Water
Natural Vegetation
Agriculture
Urban
Changed
Agriculture to Water
Natural Vegetation to Water
Water to Agriculture
Natural Vegetation to Urban
Agriculture to Urban
Water to Urban

Mutlu Ozdogan
Curtis E. Woodcock
Guido D. Salvucci

# Water Issues in the Fertile Crescent: Irrigation in Southeastern Turkey

August 14, 1984

A portion of the Euphrates River in south-eastern Turkey, August 14, 1984, prior to the construction of the Atatürk Dam. (Data from the TM instrument on the Landsat 5 satellite.)

The story of controlling water for irrigated agriculture begins some 6,000 years ago on the plains between the Tigris and Euphrates Rivers, the heart of what would become the Fertile Crescent. Over the course of the next 4,000 years, irrigation contributed to the rise and fall of several major civilizations, from the Babylonians of Mesopotamia to the Egyptian kingdoms in the Nile Valley to the irrigation-based cultures of the American Southwest. Today, irrigation is practiced in virtually all parts of the globe (16% of all croplands) thanks to a comprehensive understanding of hydraulic principles and advances in engineering. Irrigated agriculture now accounts for more than 70% of freshwater withdrawn from lakes, rivers, and groundwater reservoirs, providing water for nearly half of the world's food supply. Yet questions remain as to how sustainable irrigated agriculture is in the face of rising human populations and pressures of economic development.

As our population increases and the push for economic development intensifies, it is highly likely that global agriculture will require expanded use of irrigation. A primary

*Our Changing Planet*, ed. King, Parkinson, Partington and Williams
Published by Cambridge University Press 

August 24, 2002

The same portion of the Euphrates River as in the August 14, 1984 image, but here for August 24, 2002, after the building of the Atatürk Dam and Reservoir. (Data from the ETM+ instrument on the Landsat 7 satellite.)

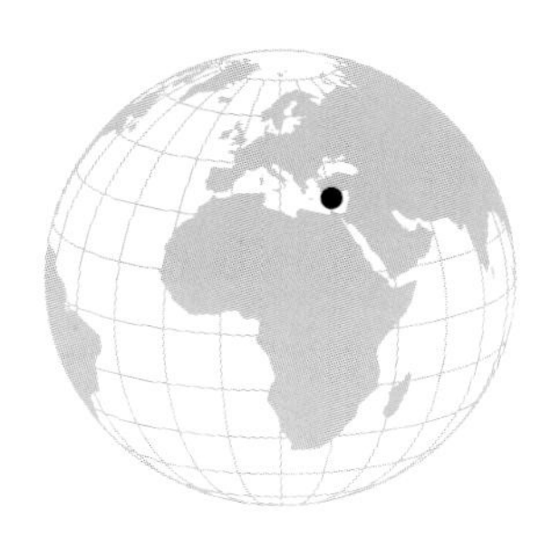

example of this expansion can be found in southeastern Turkey, where one of the biggest water development projects of recent times is underway—a series of dams referred to as the Southeastern Anatolia Project (or GAP in Turkish). The project consists of 22 dams harnessing the Tigris and Euphrates Rivers to provide power and irrigation water for an estimated 1.7 million hectares of land—an area comparable to 20 New York Cities—more than doubling Turkey's cotton production. The largest of these structures is the Atatürk Dam and reservoir. The dam was completed in 1990 and now supports extensive irrigation and a general increase in the quality of life in the region.

To improve our understanding of the sustainability of such irrigation-based development in this part of the Fertile Crescent, scientists today have a helping hand in space. Data from NASA's satellites show that the area of irrigated land grew by more than 3-fold in less than a decade in the Harran Plain, a prime agricultural site in southeastern Turkey.

August 23, 1993

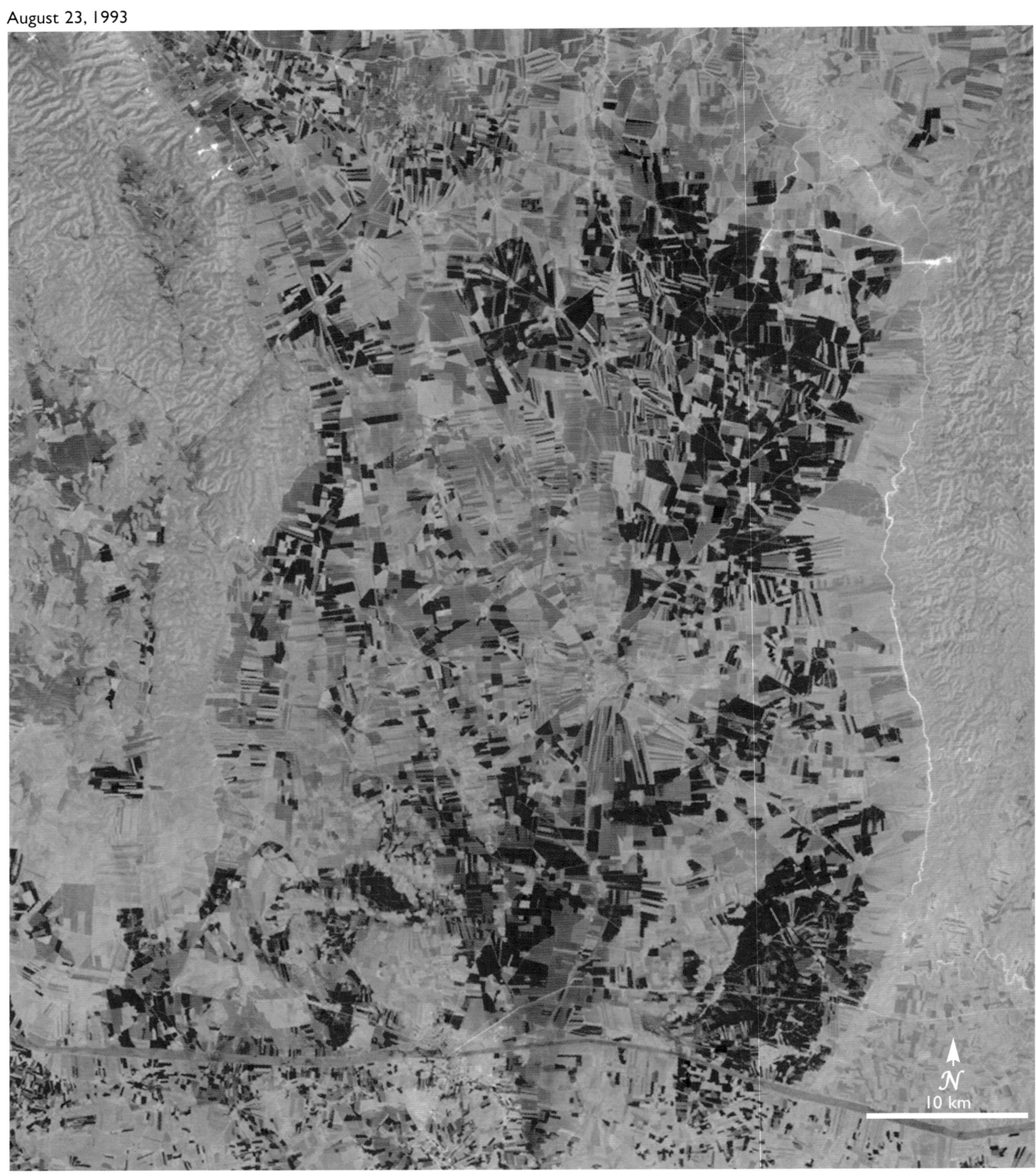

August 24, 2002

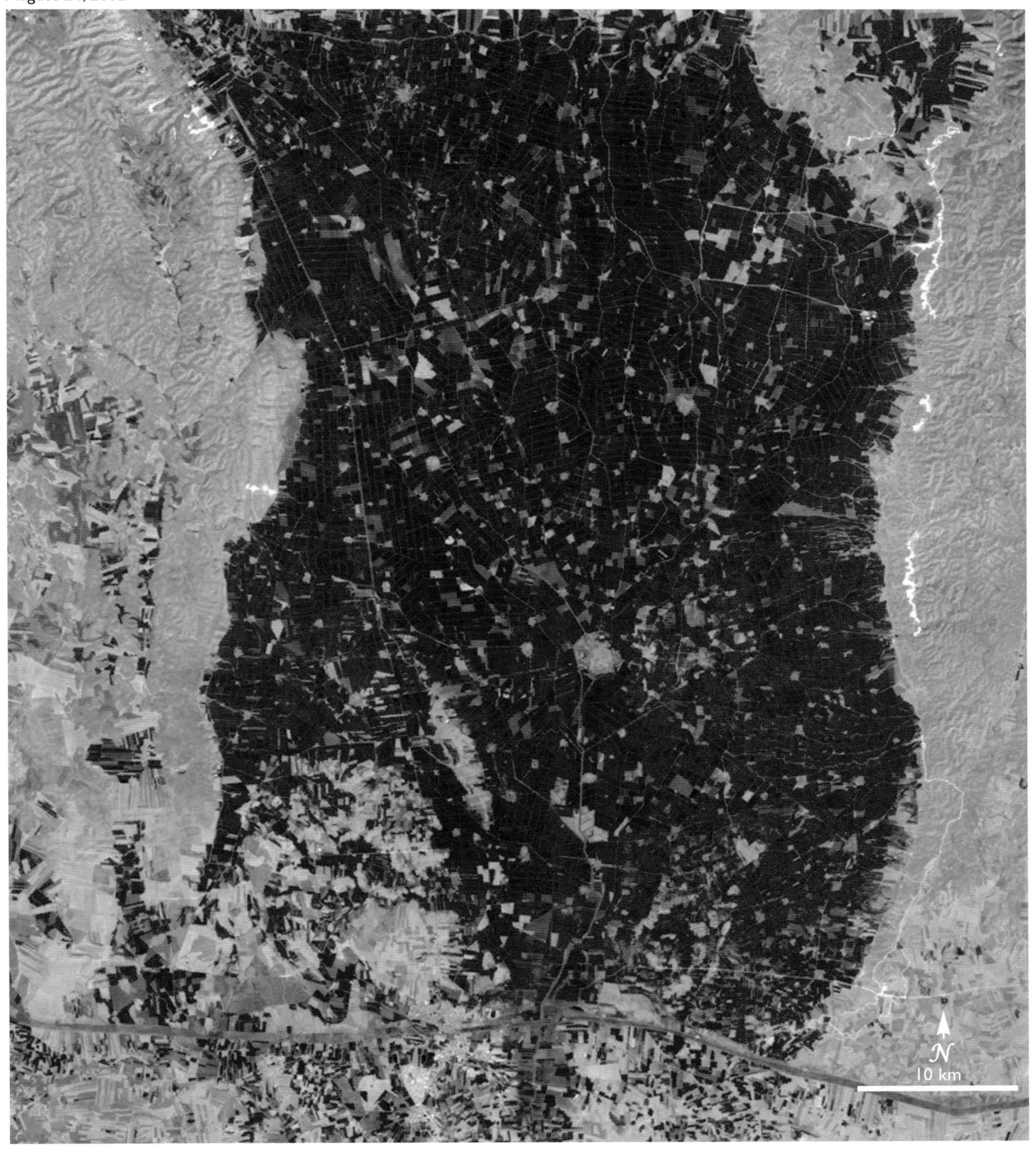


A comparative view of the Harran Plain, 1993 and 2002. The image opposite shows the effects on the land of traditional agricultural methods in 1993. The image above shows the effects of irrigation methods used since the construction of the Atatürk Reservoir. (Data from the TM instrument on the Landsat 5 satellite, opposite; data from the ETM+ instrument on the Landsat 7 satellite, above.)

Level furrow irrigation on a lettuce field in Yuma, Arizona, in 2002, illustrative of irrigation around the globe. (Photograph by Jeff Vanuga, courtesy Natural Resource Conservation Service.)

With the help of water from the Atatürk Reservoir, traditional agricultural methods that have existed for over 3,000 years have been replaced with methods involving irrigation. By artificially applying water to their fields, farmers can now grow popular 'cash crops' like cotton to improve their livelihood as well as meet the country's needs.

But the improvements brought about by irrigation are also disrupting the natural cycle of water in the region. The harvest of crops in the Harran Plain requires large quantities of water because agricultural plants require water in addition to carbon from the atmosphere to grow their plant tissues. When semi-arid agricultural lands such as the Harran Plain are artificially supplied with water, plants can evaporate so much water into the atmosphere that the climate is fundamentally altered. Among the more important issues surrounding the expansion of irrigated lands in southeastern Turkey is this alteration of the water cycle and a potential change in local and regional weather patterns. By combining data from NASA's satellites with local meteorological observations, it is clear that the atmosphere has

become more humid and wind speed has decreased, causing water evaporation to decrease as well. For now, this feedback loop is good news for the region, as less imported water is required to harvest the same amount of cotton. However, history reminds us that the same threats that undermined ancient irrigation civilizations—including build up of salts and degradation of productive soils—loom as current and future threats to sustainability.

Pie charts showing (left) the percent of global crop lands that are irrigated (blue) versus non-irrigated (yellow) and (right) the percent of the world's food supply coming from irrigated agriculture (blue) versus elsewhere (yellow). (Background photographs: left, by Jeff Vanuga, courtesy Natural Resource Conservation Service; right, by Peggy Greb, courtesy USDA ARS Photo Unit.)

Gregory T. Koeln
Yasmin Naficy
Claire L. Parkinson

# Chronicling the Destruction of Eden: The Draining of the Iraqi Marshes

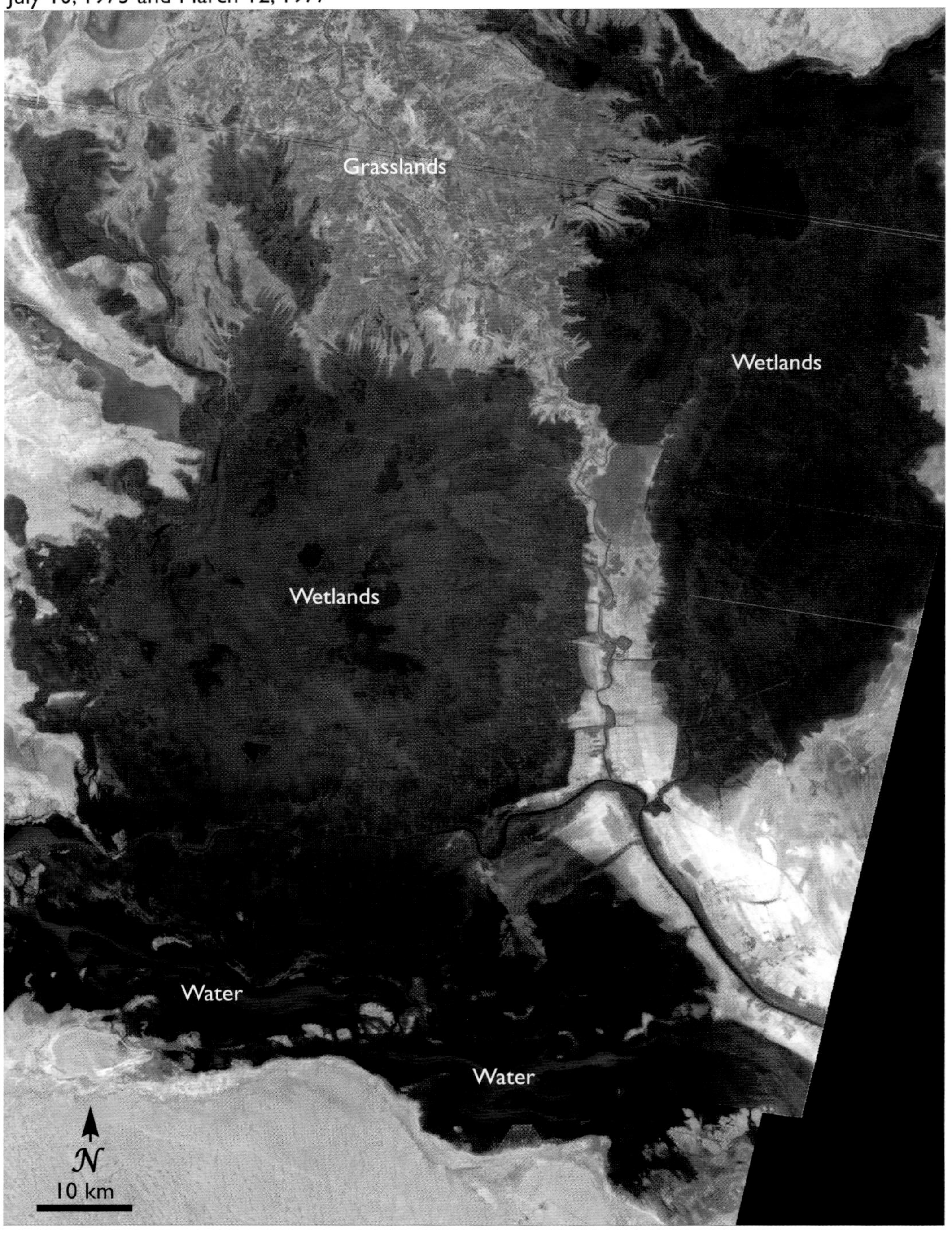


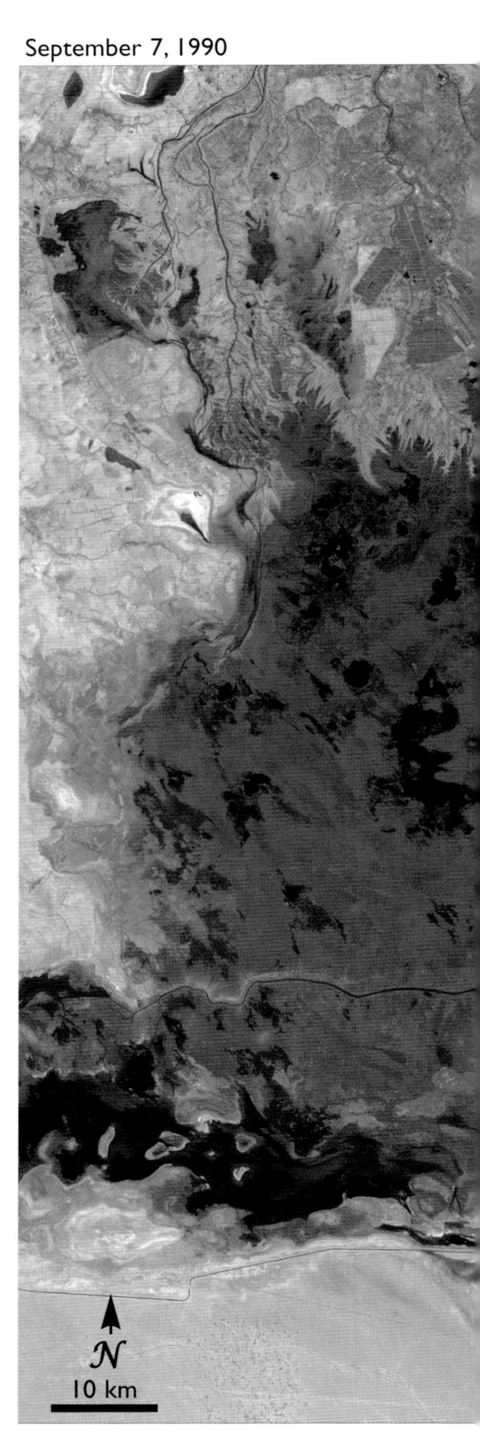


*Our Changing Planet*, ed. King, Parkinson, Partington and Williams
Published by Cambridge University Press 

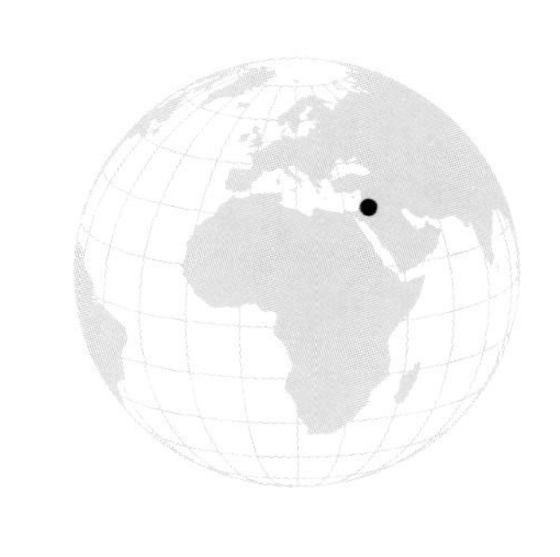

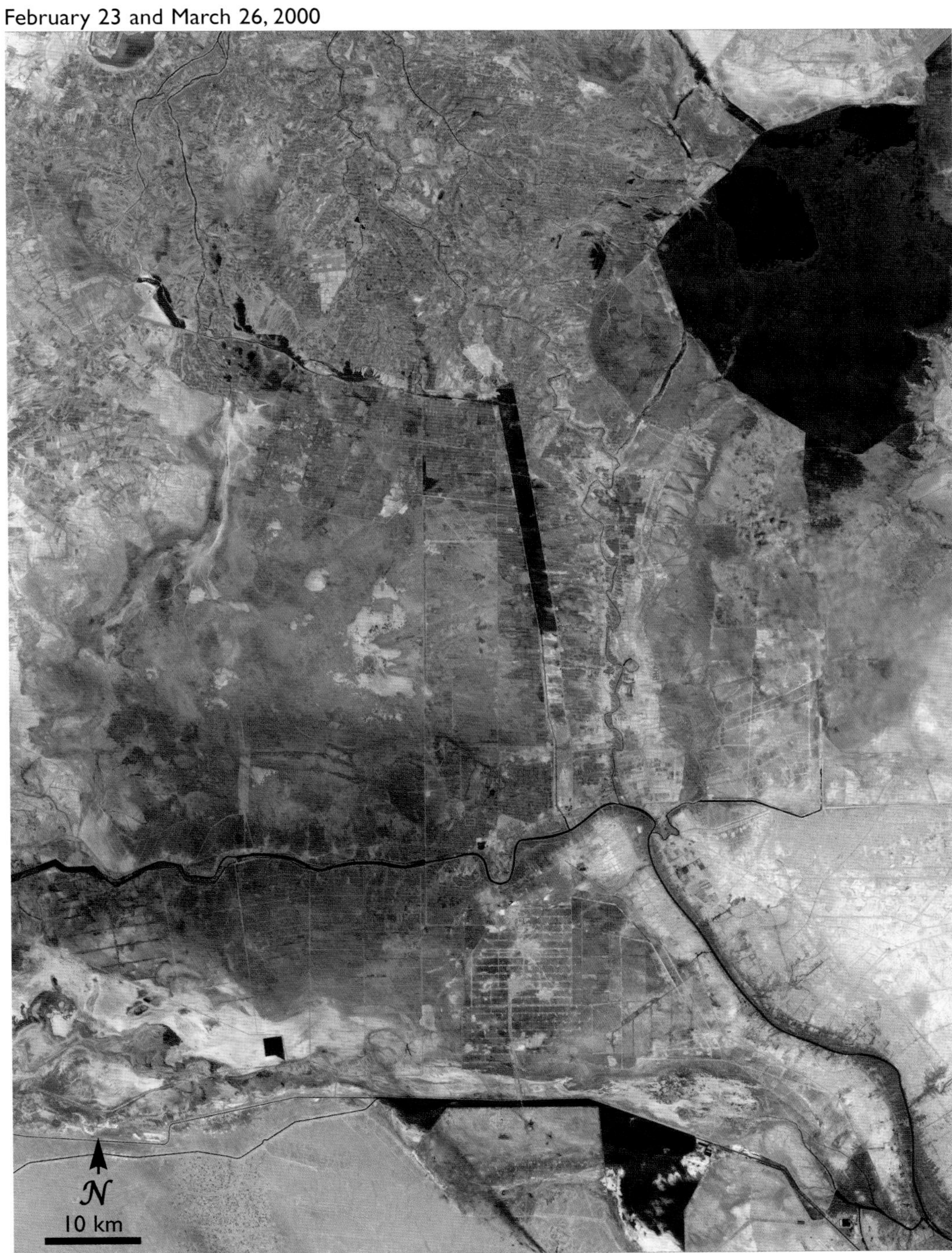


Satellite imagery of southern Iraq documenting the destruction of the Iraqi marshes. Data for the left and right images are merged mosaic pairs taken from different observation dates: the left image is merged from July 10, 1973 and March 12, 1977; the image on the right is from February 23 and March 26, 2000. The center image is taken from a single observation date, September 7, 1990. (Data for 1973 from the MSS instrument on the Landsat 1 satellite; data for 1977 from the MSS instrument on the Landsat 2 satellite; data for 1990 from the TM instrument on the Landsat 4 satellite; data for 2000 from the ETM+ instrument on the Landsat 7 satellite.)

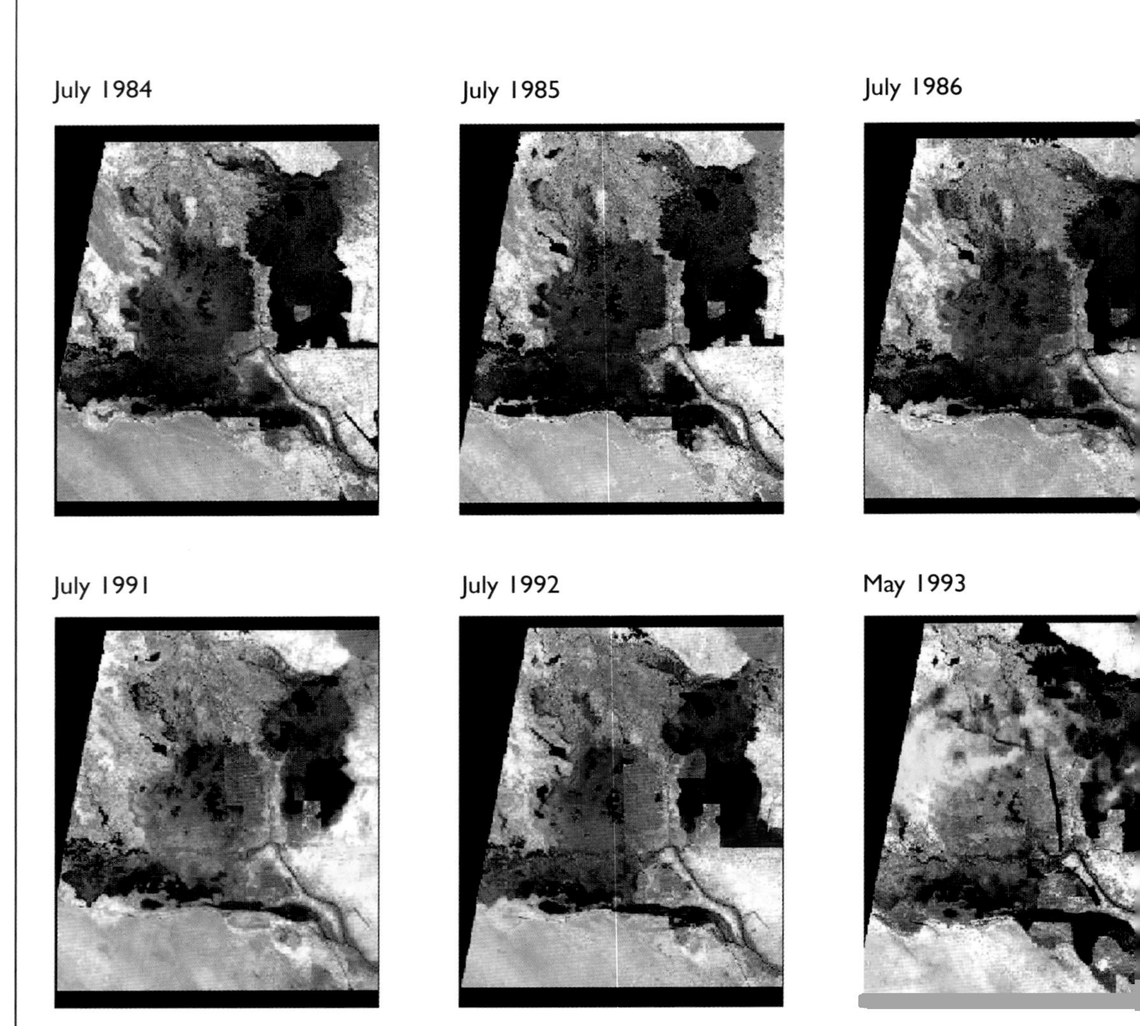


Changes in the Iraqi marshlands, 1984–2004, as documented through satellite imagery. Healthy, non-drained wetlands appear as shades of green, and water appears as shades of blue and black. Early evidence of the drainage of the marshes is shown in the reddish and brown shades near the center of the 1984 image, with striking evidence of the expansion of the drainage shown through the 1984–2000 sequence. By May 2000, the marshes are completely drained, with evidence of the return of water to the marshes shown in the May 2000 to February 2004 transition. Note that the color palette used for these images differs from the palette used for the images on the previous two pages. (Data for 1984–1987, 1994, and 1998 from the TM instrument on the Landsat 5 satellite; data for 1988–1993 from the TM instrument on the Landsat 4 satellite; and data for 2000 and 2004 from the ETM+ instrument on the Landsat 7 satellite.)

Cradled by the Tigris and Euphrates Rivers in southern Iraq, the Mesopotamian marshlands have—until modern times—brought prosperity and prolific life to an otherwise barren, arid region. The considerable ecological value of these marshes, once teeming with exotic and rare wildlife, is clear to scientists the world over. Serving as spawning and nursery grounds for fish populations in the Persian Gulf, the marshes have over the millennia replenished fish supplies upon which countless generations of local fishermen have depended. Moreover, this globally significant wetland once served as habitat and feeding grounds for 134 bird species as well as globally threatened mammals such as the gray wolf and the smooth-coated otter.

The marshland region—a virtual oasis within a vast desert—was a crucial stopping point on an inter continental route for migratory birds on their way from Russia to southern Africa. What remains of these wetlands (now approximately 10% of their original extent) supports almost the entire global population of two particular bird species, the Basrah reed warbler and Iraq babbler, and also most of the world population of gray hypocolius. According to a 2001 study by the United Nations Environment Programme (UNEP), the dominant wetland vegetation in the area is the common reed (*Pharagmites communis*), complemented by reed mace (*Typha augustata*). Salt-tolerant plants such as low sedges and bulrush (*Carex* and *Juncus* species, *Scripus brachyceras*) further populate the area. In the marshes' deeper waters, aquatic vegetation such as hornwort (*Ceratophyllum demersum*), eel grass (*Vallisneria* species) and pondweed (*Potamogeton* species), proliferate. Migratory

September 1987

October 1988

September 1990

1994

September 1998

May 2000

February 2004

birds were attracted to the shelter and feeding grounds offered by the once-generous marshland vegetation and aquatic features.

Not surprisingly, the social and economic value of the Iraqi marshlands was rooted heavily in the abundance and diversity of natural life, providing the resources for trade (for example, the marshland inhabitants farmed reeds and sold them to trading partners throughout Iraq and Kuwait). Wildlife such as fish, water buffalo, and waterfowl sustained many aspects of a rich indigenous society for thousands of years. Other rare wildlife, such as the soft-shelled turtle, the desert monitor, and the dragonfly, once inhabited the area in thriving numbers, but have now, like so many other original marshland species, become severely threatened.

The marshland was home in 4,000–2,000 BC to the Sumerians, renown for their early civilization and their cultural and scientific advances. However, according to UNEP, the more modern inhabitants of the Iraqi marshes lived in almost total isolation until World War I, when the central Iraqi government began to exert influence in the region. While many of these rural people migrated to urban areas from the marshlands in the 1930s, many others stayed and continued their subsistence lifestyle, passing on a valuable culture to the generations that came after them. These 'Marsh Arabs,' or Madaan, traditionally relied on numerous species of plants and animals once available in the marshes. Their peaceful existence, however, collided against damming of the inflowing rivers and the rule of a brutal political leader.

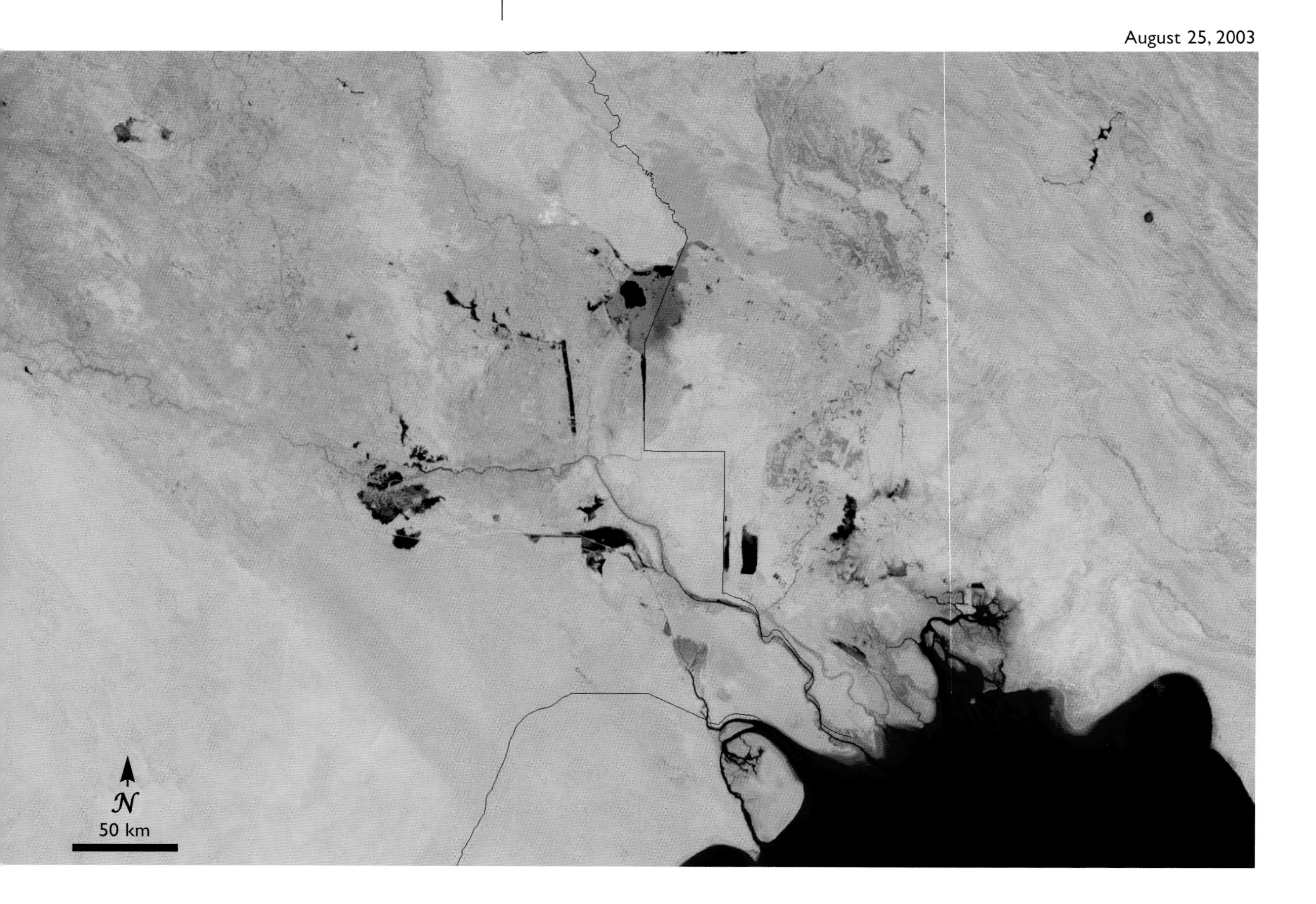


Rebounding of the Iraqi marshlands between August 25, 2003 (above) and August 25, 2005 (facing page). In these satellite images, water is blue or black, vegetation is green, and desert sand is pale tan or pink. Efforts to restore the marshlands have yielded visible results, with considerably more water and vegetation apparent in the 2005 image than in the 2003 image. (Data from the MODIS instrument on the Aqua satellite.)

The Iraqi marshlands are fed largely by the Tigris and Euphrates Rivers, which flow a considerable distance from their headwaters in Turkey before reaching the Iraqi marshes. Damming of the rivers upstream of the marshes beginning in the mid-twentieth century threatened their health. Then, in the early 1990s, the Marsh Arabs suffered a further major setback when then Iraqi dictator Saddam Hussein launched a massive wetlands drainage program to flush out Shiite rebels who sought a sanctuary in the marshes. Just as the legendary King of Babylon fled to the Mesopotamian marshes when threatened by the King of Assyria in ancient times, the Shiite rebels who had failed in their uprising against Saddam Hussein turned to these flourishing marshlands for relief. Saddam responded. "Saddam's engineers worked 24 hours per day for six months," writes civil engineer Sam Ali, "to build the 350-mile long Saddam River to divert water from the Euphrates." Saddam went on to order his engineers to implement the Iraqi Government Drainage Project, which centered around a series of canals with such names as 'Mother of All Battles Canal' and 'Loyalty to the Leader Canal.'

The Marsh inhabitants had been using the tranquil marshes to fulfill their needs for nourishment, trade, resources, and shelter for 5,000 years. Without the marshes, the very

August 25, 2005

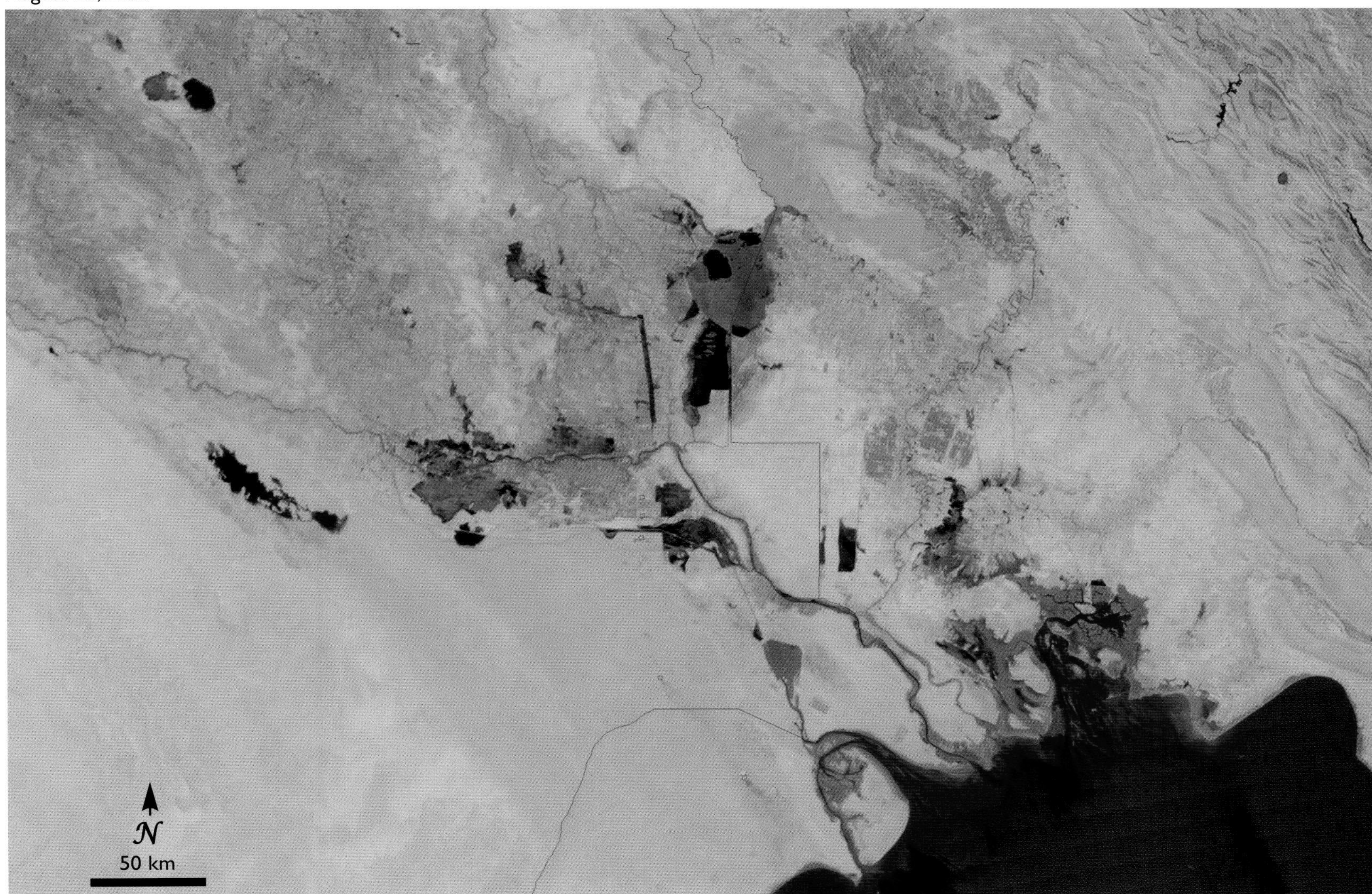


core of the indigenous Madaan population's existence was literally reduced to dust. As many as 70,000 Marsh Arab refugees fled to camps at the border with Iran. According to Ali, the original Madaan population was close to 450,000 as recently as the 1980s, but the wetland drainage project has since reduced their remaining population to less than 40,000. Upstream dams and drainage systems on the Tigris and Euphrates Rivers, as well as the politically motivated wetland drainage project, resulted in the desiccation of the once thriving Mesopotamian marshland—the very area many believe was the location of the Biblical 'Garden of Eden,' and the birthplace of the patriarch Abraham.

The socioeconomic and environmental impacts of Saddam's wetlands drainage project have been devastating. The livelihoods of hundreds of thousands of people were wiped out within a decade, and an internationally important ecosystem was so badly damaged that the impacts were visible from space.

While upstream dams in Turkey, Syria, and Iraq reduced the flow of the Tigris and Euphrates Rivers, and industrial activities polluted the water reaching the marshes, it was Saddam Hussein's decision to drain the marshlands entirely that led to potentially one

Typical settlement of Marsh Arabs, or the Madaan, 1979. (Photograph © 2006 Nik Wheeler.)

of the largest man-made environmental catastrophes in modern times. UNEP researcher Hassan Partow called "the destruction of an ecosystem of that scale ... phenomenal." Ali writes that the marshland drainage project "is having a global impact on climate and bird migration, endangering species of birds, fish and mammals."

The annihilation of the Iraqi marshlands is an illustrative symbol of humanity's destruction of 'Eden.' Fortunately, projects are now underway to rehabilitate the area and bring this valuable ecosystem and its people back to life. The MODIS image from 2005 reflects the return of some water to the southern portion of the Mesopotamian marshes. After the ousting of Saddam Hussein, ecologists and other technical specialists have entered Iraq to gather and record current environmental data and to launch on-the-ground projects to restore the Iraqi marshes. (Previous studies relied on satellite imagery without much ground-based information, or 'ground truth.')

Among the projects to restore the marshlands is the Eden Again Project, an effort led by the Iraq Foundation and contractor Psomas. The Eden Again Project has employed Geographical Information Systems (GIS) technology, as well as information gathered from former Iraqi citizens specializing in engineering, and historical data on the marshland's ecology.

Desiccation following the draining of the Iraqi marshes, in the central Iraqi marshlands, June 22, 2003. (Photograph by Curtis J. Richardson.)

In conclusion, with the assistance of satellite imagery, humans have been able to witness man-made destruction at a very large, ecosystem-level scale. With the establishment of a new Iraqi government, scientists are hopeful that the precious Iraqi marshlands can be restored, slowly but surely, and the satellite imagery reveals that some progress is already visible from space. In fact, the Saddam-led military inadvertently helped undo part of the wetland destruction when they opened dams north of the marshlands to confound Coalition troops in the Iraq War in 2003. In doing so, waters replenished the marshlands just in time for spring. Perhaps Nature will rebound, as it often does, despite the sometimes damaging activities of Humanity.

Philip P. Micklin

# Destruction of the Aral Sea

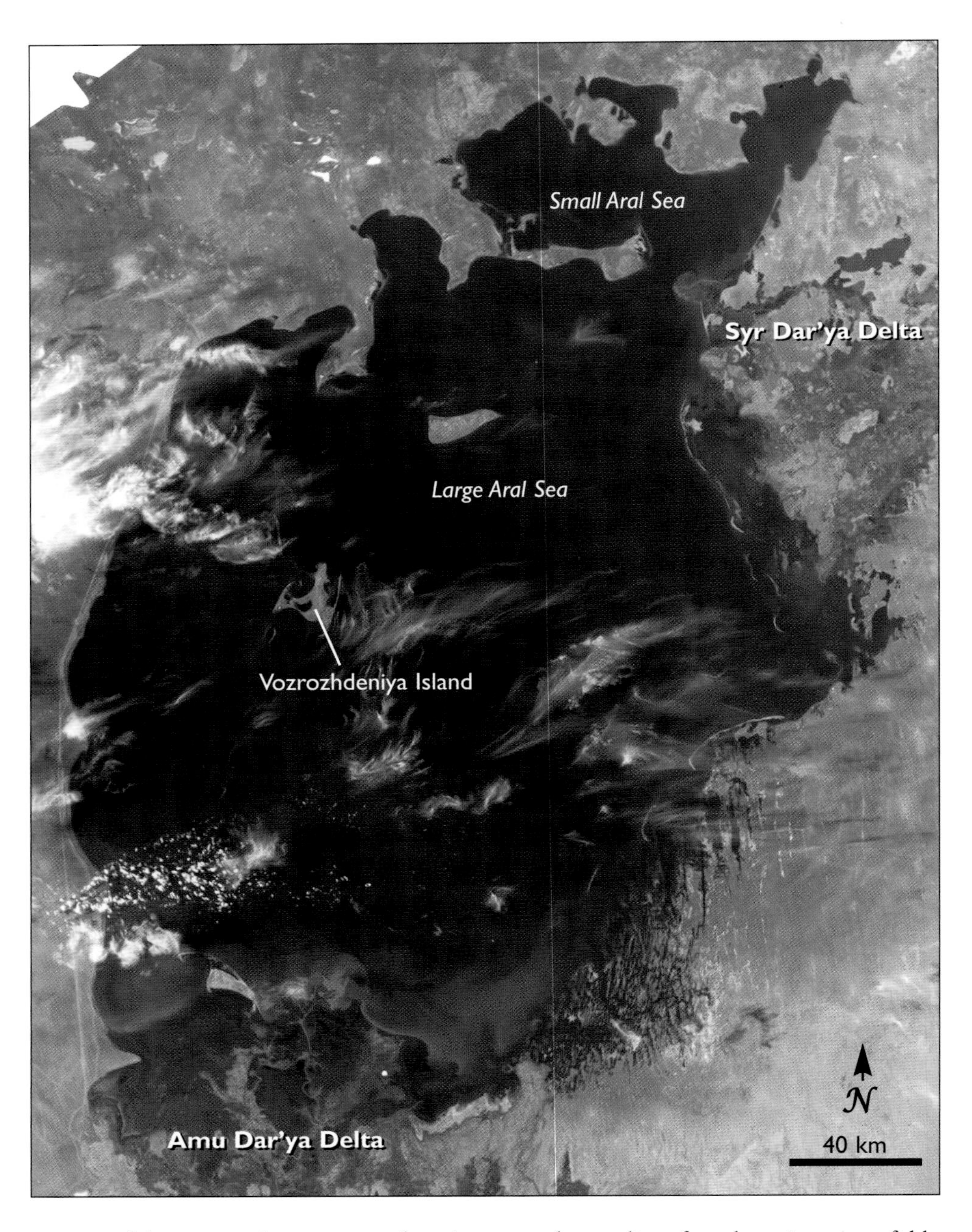


The Aral Sea, August 21, 1964. Although by this time navigation and the commercial fishery had suffered from the shrinkage of the sea, these activities were still viable. Vozrozhdeniya Island, with its biowarfare facility, was still small and well isolated from the mainland. (Photograph from the Argon satellite.)

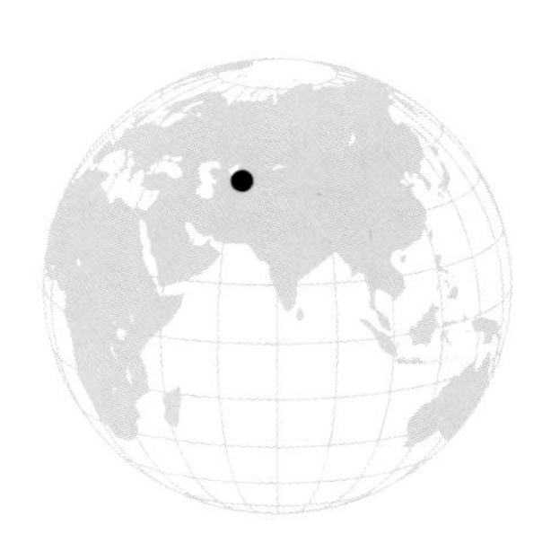

One of the greatest human-caused environmental tragedies of modern times is unfolding amidst the deserts of Central Asia. The Aral Sea—once the world's fourth largest lake in area—is drying at a rapid pace. Over the past 4 decades, this terminal lake (a water body with inflow but no outflow) lost more than 60% of its area and 80% of its volume while its salinity increased 6-fold. Excessive expansion of irrigation during the period when this region was part of the Soviet Union was the fundamental cause of the sea's desiccation. The main source of water for the sea, inflow from the Amu Dar'ya and Syr Dar'ya rivers, has declined markedly since 1960 as more and more water has been diverted to grow water-thirsty crops such as cotton and rice.

Numerous and severe environmental and human consequences have accompanied the sea's demise. The Aral's rich fishery, which in the 1950s directly and indirectly employed

*Our Changing Planet*, ed. King, Parkinson, Partington and Williams

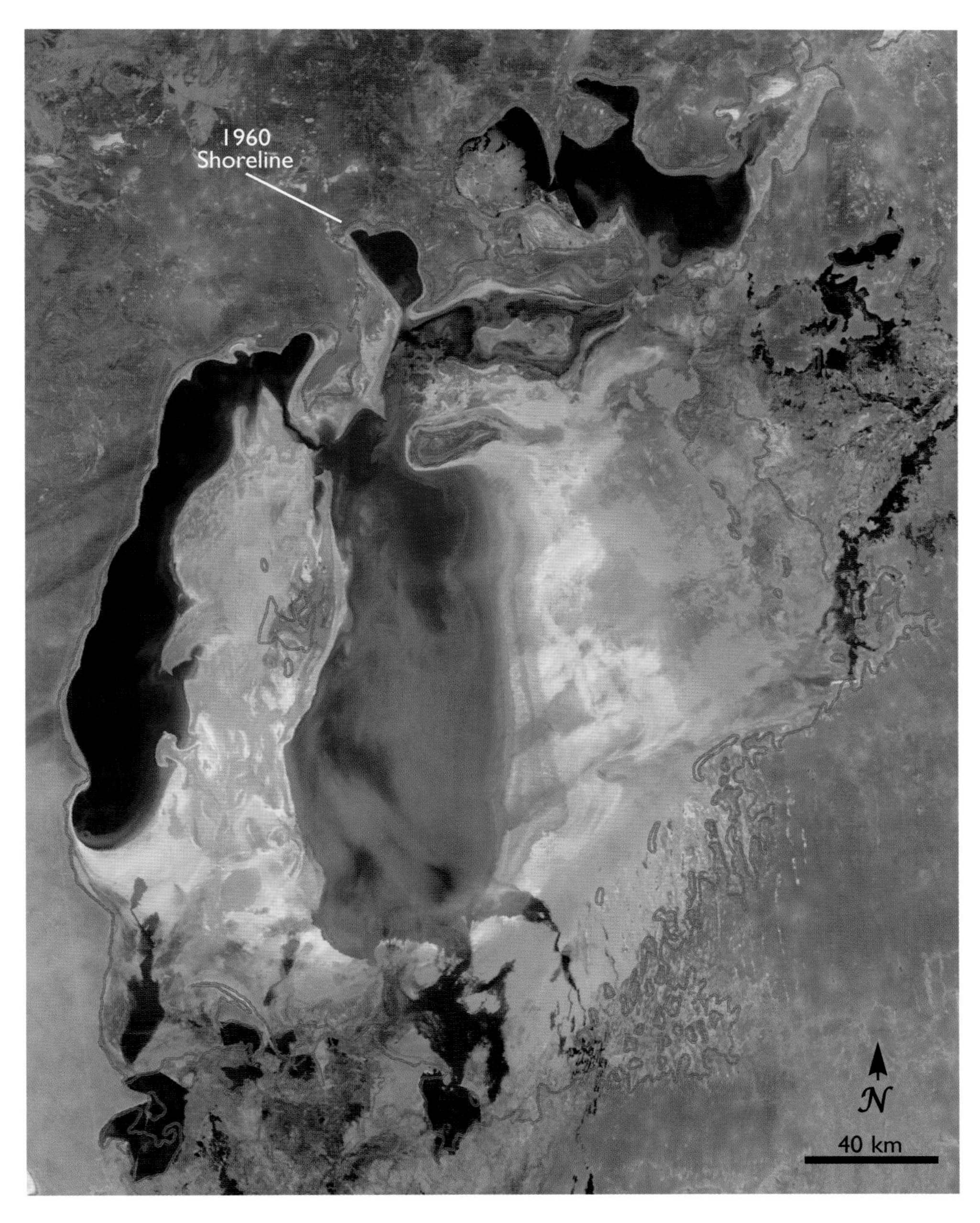


Natural color satellite image of the Aral Sea on April 9, 2006, with an overlay showing the shoreline of the sea in 1960. By 2006, lake level had fallen over 20 m since the early 1960s, and the Small and Large Seas had become separate lakes, although during the spring high flow season on the Syr Dar'ya, a channel still carried water from the Small to Large Sea. Vozrozhdeniya Island had by 2006 become a huge peninsula, which in the next few years will likely extend to divide the Large Sea into completely separate western and eastern portions. (Data from the MODIS instrument on the Terra satellite.)

60,000 people and provided a major source of protein for those living near the sea, entirely collapsed as salinity rose beyond the tolerance limit of commercially valuable indigenous species and as the shoreline retreated tens of kilometers from the major ports, preventing access to them by ships. The dried and heavily salinized bottom, now half the size of Maine, became a source region for frequent salt/dust storms depositing these materials hundreds (some claim thousands) of kilometers downwind. Salts, some of which are toxic to plants, animals, and people, falling on the biologically rich, agriculturally productive, and densely populated delta of the Amu Dar'ya to the sea's south have been particularly harmful. Reduced river flow and dropping groundwater levels also have promoted accelerating desertification in the Amu Dar'ya delta, with adverse impacts on native ecosystems and agriculture. The sea's shrinkage has shifted climate from more maritime to more continental in a

The Amu Dar'ya River 200 km south of its entrance into the Aral Sea, September 23, 2000. The once mighty Amu Dar'ya in its lower reaches is now only a trickle owing to heavy withdrawals for irrigation upstream. In 2000, a very dry year, little or no water reached the Aral Sea. (Photograph by Philip Micklin.)

several hundred kilometer wide zone around the sea such that summers are now hotter and dryer, winters colder, and the growing season shorter.

Perhaps the most ironic and dark consequence of the Aral's shrinkage is the story of Vozrozhdeniya (Resurrection) Island. The Soviet military in the late 1930s selected this, at the time, tiny, isolated island in the middle of the Aral Sea, as the primary testing ground for its super-secret biological weapons program. Here, in the decades after World War II until the late 1980s, they tested various genetically modified and 'weaponized' pathogens, including anthrax and plague. In 1988, according to former Soviet experts, hundreds of tons of anthrax powder were buried on the island. With the collapse of the USSR in 1991, the departing Soviet military took measures to decontaminate the island. As the sea shrank and became more shallow since the 1960s, Vozrozhdeniya grew in size and in 2001 united with the mainland to the south as a huge peninsula extending into the Aral Sea. The fear is

The Kiev, rusting on the dried bottom of what had been a portion of the Small Aral Sea, August 30, 2005. From the 1920s until the late 1970s, a fleet of transport and fishing vessels plied the waters of the Aral Sea. But as the sea dried, it became impossible to continue commercial fishing and shipping. By the early 1980s, these activities ceased. Ships, such as the one pictured, were grounded and allowed to rust in the desert sun or were (and still are) being cut up for scrap. (Photograph by Philip Micklin.)

Satellite image of the Aral Sea region, April 18, 2003, showing salt and dust from the dried bottom of the former sea being blown by strong northeast winds and carried across the sea and across the Amu Dar'ya delta. A loose layer of fine salt and dust particles that are very susceptible to long distance airborne transport covers extensive parts of the dried bottom. The dust and salt settle over a large area adjacent to the Aral Sea, harming natural ecosystems, crops, and human health. (Data from the MODIS instrument on the Aqua satellite.)

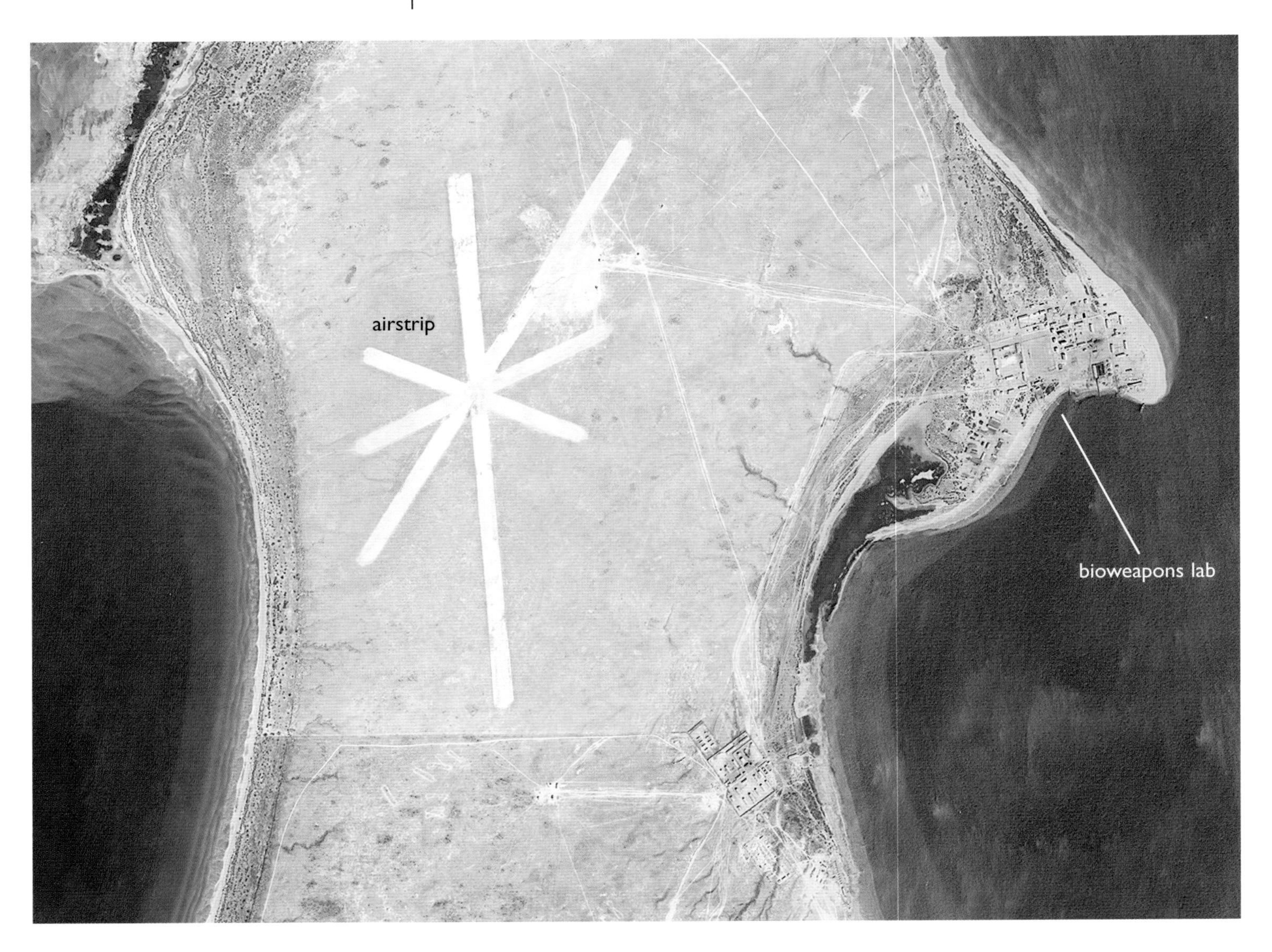


Satellite image of the former Soviet biowarfare test facility on Vozrozhdeniya Island in the Aral Sea on September 23, 1996. The star-shaped feature is the set of intersecting runways used to land transport planes arriving from biowarfare production facilities in other parts of the former Soviet Union. The facility was abandoned in the late 1980s, but there is fear that weaponized pathogenic organisms have survived and could escape to the mainland, as the Aral's shrinkage has transformed the island to a peninsula. (Image from the Corona satellite.)

that some weaponized organisms survived decontamination and could escape to the mainland via infected rodents or that terrorists might gain access to them. The United States has committed $6,000,000 to help the Government of Uzbekistan clean up the island and ensure that any surviving organisms are destroyed.

The future of the Aral Sea is clouded. The governments of the region, particularly those of Uzbekistan and Kazakhstan that border the sea, financially and technically aided by the international community, have cooperated since the early 1990s to implement health and medical measures to alleviate the worst problems for those residing in the vicinity, to slow the deterioration of ecological conditions in the river deltas, and to deliver more river flow to the deltas and the sea. However, these measures have had limited success. To stop the sea's continued desiccation and the deterioration of adjacent lands and substantially improve living conditions for the local inhabitants would require much greater efforts and investments than have been made or are contemplated. Although it may be possible with concentrated effort to partially rehabilitate parts of the sea (such as the separated northern portion), full restoration of the Aral is, sadly, only a dream for the distant future. The example of the Aral should serve as a warning to humankind of the folly of thoughtless, ill planned, large-scale interference in complicated natural and ecological systems, often leading to devastating and irreversible damage to both Nature and society.

April 14, 2007

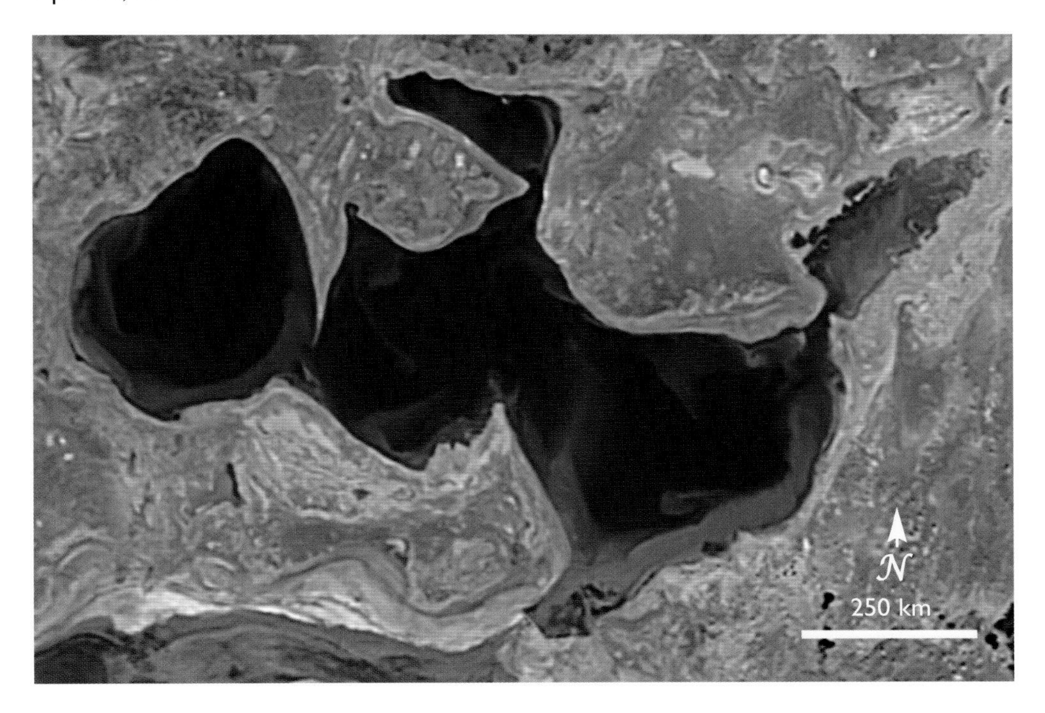


April 15, 2005

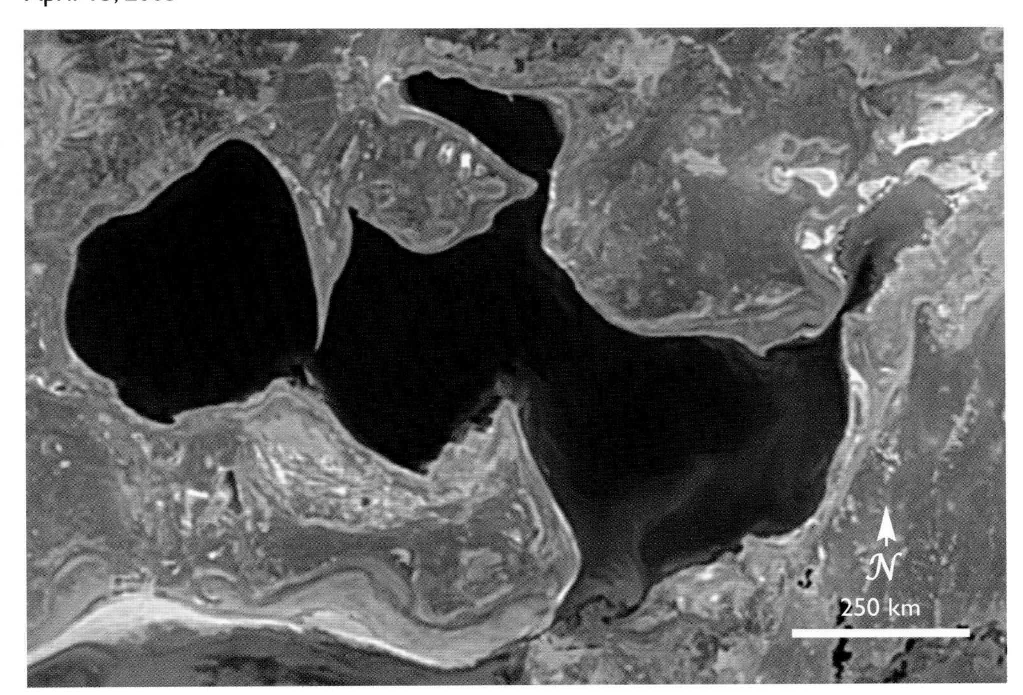


Partial recovery of the Small Aral Sea, between April 15, 2005 (bottom image) and April 14, 2007 (top image), following the construction of the Kok-Aral Dam in 2005. (Data from the MODIS instrument on the Terra satellite.)

Michael T. Coe
Charon Birkett

# The Desiccation of Lake Chad

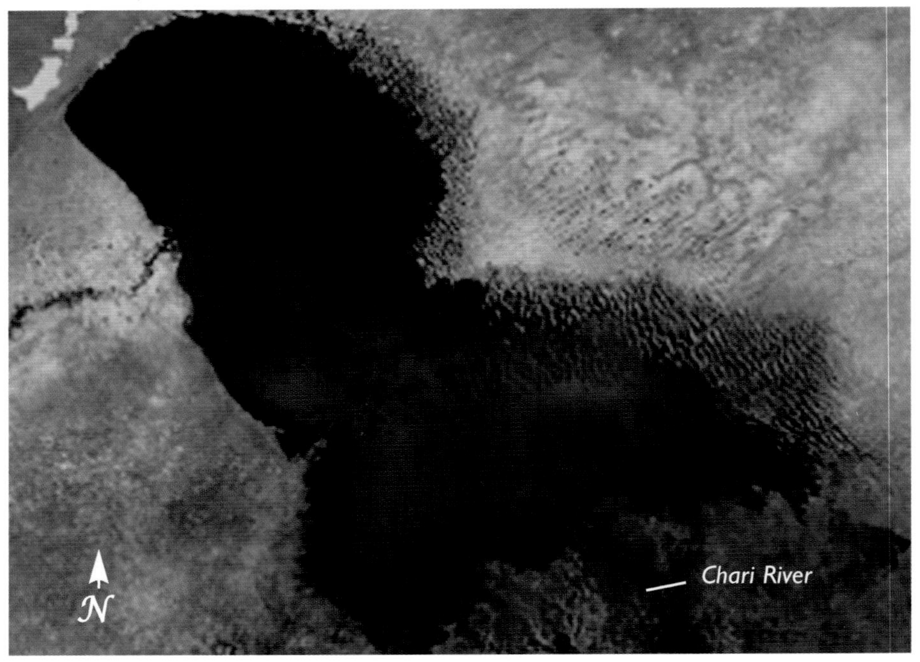


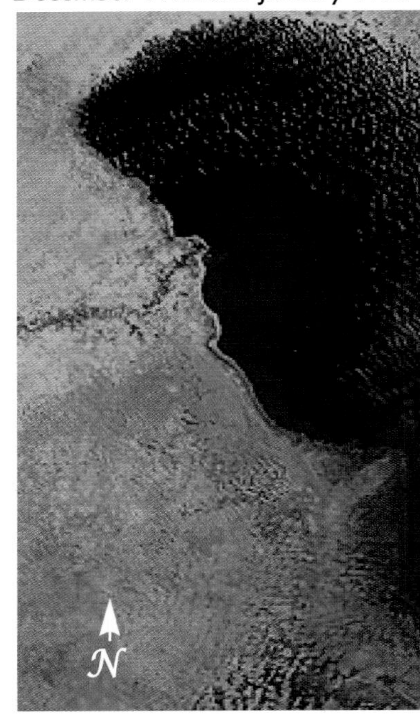


Lake Chad is the largest source of freshwater in northern Africa. It is situated on the edge of the Sahara desert in a semi-arid region known as the Sahel. Although the lake has a large catchment area of 2.5 million km$^2$, more than 90% of the lake's water comes from the Chari-Logone river system. This river system originates over 1,000 km away, in the more humid zones of the Adamawa Highlands in Cameroon and the Central African Republic. Because of the high temperatures and low humidity, all of the water that enters this shallow lake eventually evaporates. Despite this, the watershed of the lake is home to nearly 37 million people and its resources are shared by the peoples of Niger, Nigeria, Chad, and Cameroon.

Since 1968 the region has suffered from a persistent drought with below average rainfall. The years 1968–1974 and 1983–1987 were particularly severe. As a result, the lake shrank drastically, from an area of 25,000 km$^2$ in the 1960s to only 2,500 km$^2$ in the mid-1980s. Lake Chad was once the fourth largest lake in Africa, about the same size as Lake Ontario in North America. Today it is barely one-tenth that size and is the smallest in its recorded history.

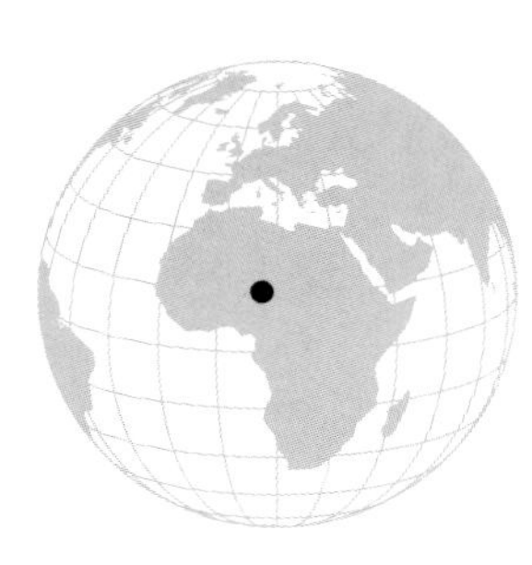

The desiccation of the lake has exposed ancient dunes, which now dominate the northern and eastern boundaries of the original coastline. The only remaining area of permanent open water exists in the southern region of the lake basin where it is surrounded by densely vegetated swamps and smaller pools. Each year, the waters from the Chari

*Our Changing Planet*, ed. King, Parkinson, Partington and Williams
Published by Cambridge University Press 

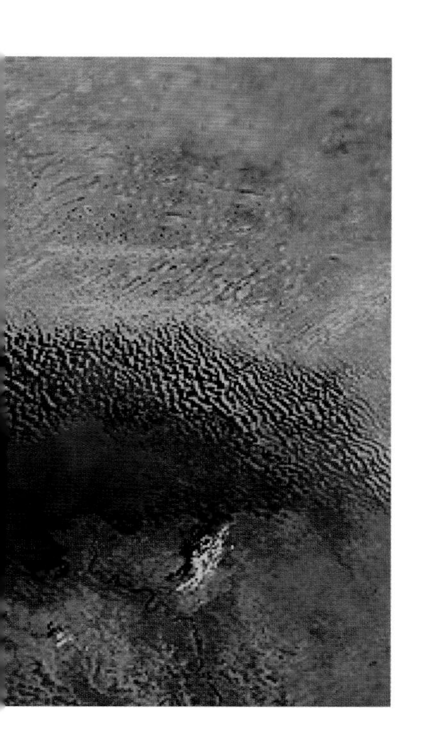

January 24, 31, and February 7, 1987

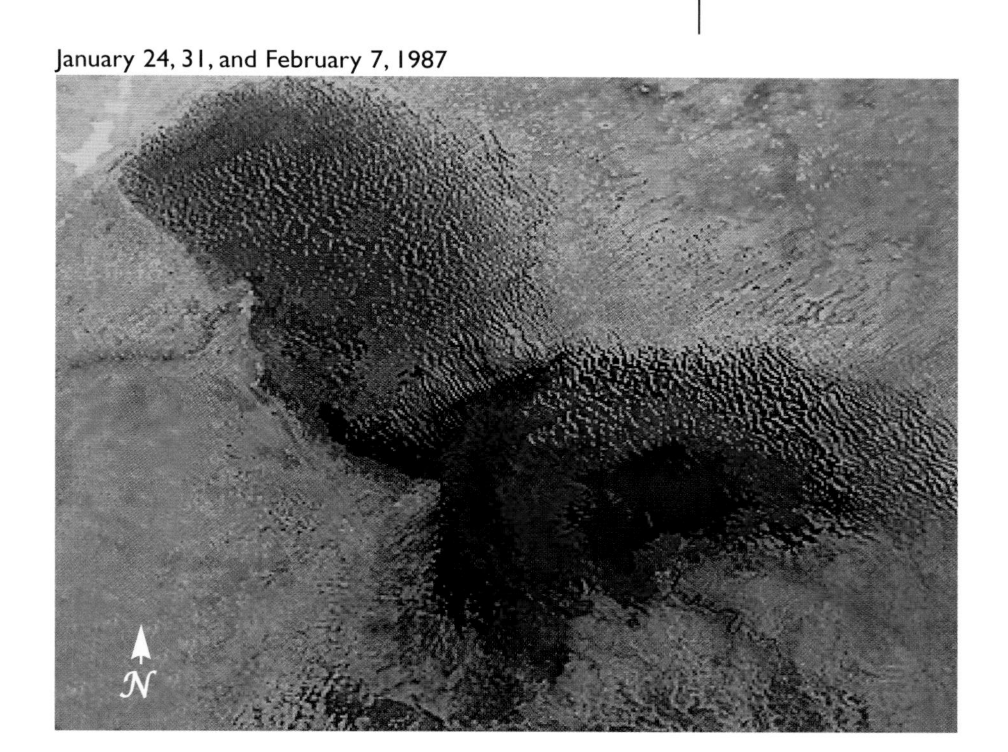


and Logone rivers flow into the lake basin from the south and gradually move northwards, supplying the permanent open water and seasonally inundating large marsh regions (as much as 5,000 km$^2$ in wetter years). At peak flood periods, the waters may cross a low grassy ridge known as the Grand Barrière, and enter the northern half of the lake basin.

A decrease in rainfall may not be the only factor in the desiccation of this lake. Poor farming practices, including the over-grazing of the delicate vegetation of the Sahel and the construction of large irrigation schemes, have also been blamed. However, it is very difficult to understand fully the impact of humans on the local climate and to quantify irrigation losses. How much of a role humans have played in this drought and the desiccation of the lake is still under discussion.

We do know, however, that the drought and resulting loss of water resources have had a tremendous impact on the sedentary and nomadic people who live in the region. A number of strategies have been developed to cope with the reduction in food and water. The introduction of flood retreat farming has been employed as one means of coping. For some residents this has brought modest prosperity as they have learned to exploit the seasonal variability of the lake level by fishing during the annual flood and farming during the flood recession. However, the continuing drought has led to a reduction of grazing lands, crop failures, and the collapse of irrigation projects. Even those who have prospered are at the mercy of the large fluctuations in the annual flood extent.

Satellite images of Lake Chad for October 31, 1963 (left), December 1972 and January 1973 (middle, a combined image from several dates in December and January), and January 24 and 31 and February 7, 1987 (right, a mosaic image). The large decrease in lake area can be seen clearly in these images, with the open water (shown in dark gray and black in the 1963 image) reduced in 1987 to the small blue area in the southern portion of the basin. The regions colored red in the middle and rightmost images are the seasonal vegetated wetlands that fluctuate greatly in area each year. The ancient sand dunes that have emerged from the lake can be seen along the northeast shore. (Data for 1963 from a single-frame camera on the United States photo-reconnaissance Argon satellite; data for 1972/1973 from the MSS instrument, bands 4, 2, and 1, on the Landsat 1 satellite; data for 1987 from the ETM instrument on the Landsat 5 satellite.)

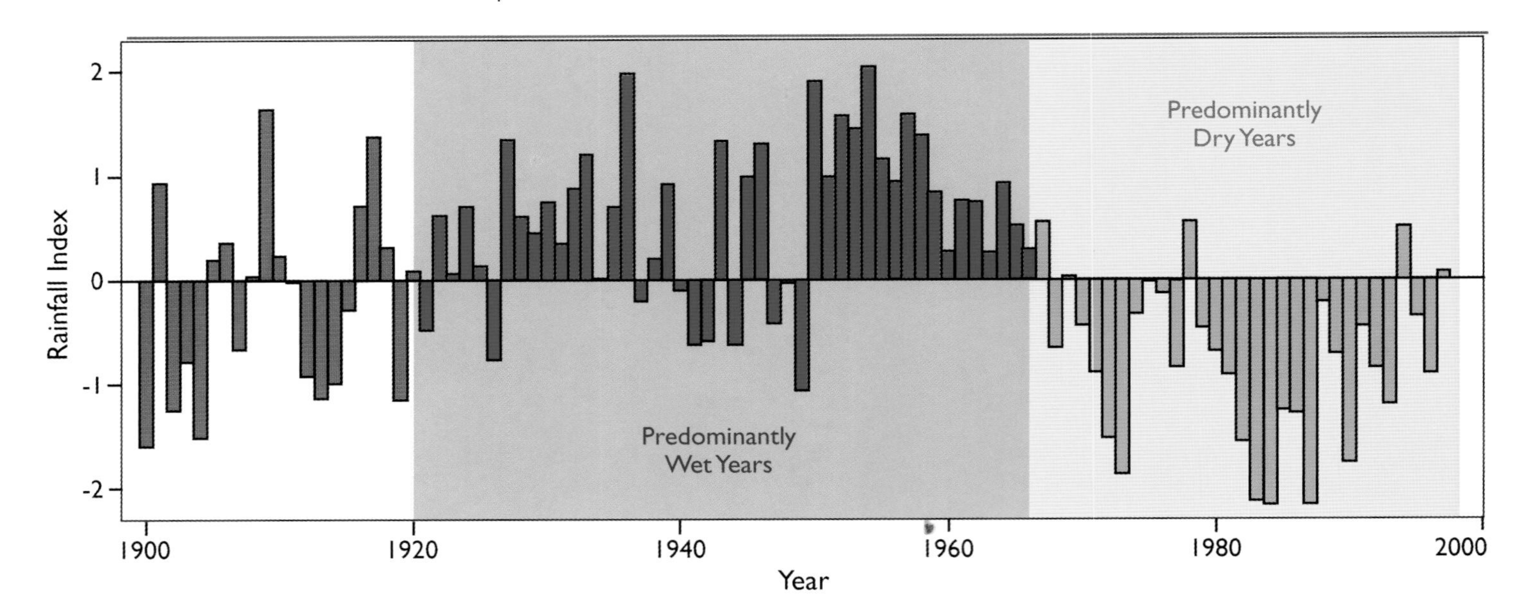


Annual rainfall variations in northern Africa from 1900 to 1998, in relative units. The persistent drought since 1968 is shown in the yellow box. (Modified from a figure by Guiling Wang, MIT PhD thesis, 2000.)

The observed (blue) and predicted (red) changing water levels of Lake Chad. By determining the relationship between the water levels of Lake Chad and the river levels of the Chari-Logone system 600 km upstream, both measured from satellite altimetry, scientists can predict the Lake Chad height variations 39 days in advance. Gaining knowledge of the expected amplitude of the floodwaters more than one month ahead is an example of the importance of satellite measurements as a tool for the local peoples. (Data from the NASA radar altimeter on the TOPEX/Poseidon satellite.)

Particularly important to the people of the region is the knowledge of when the floodwaters will arrive, how deep they will be, and over what extent the waters will travel across the lakebed. With a wealth of satellite information, the monitoring of such events has become a reality and the prediction of events a cause for current investigation. The changing extent of the lake can be observed with a variety of imaging data sets from sensors such as NASA's Landsat satellite, NOAA's AVHRR instrument, and NASA's MODIS instrument. The changing levels of the waters within the lake and its surrounding marsh can also be observed using current satellite radar altimeters on NASA/CNES's Jason-1, ESA's ENVISAT, and NOAA's GFO satellites. Scientists are learning how to combine the information from remote sensing systems with existing knowledge of the hydrology of the basin. They are also searching for ways to predict the availability of water resources across the changing seasons.

Regarding the future of Lake Chad: there are many plans to help provide water to the people of the region, including one to create a canal hundreds of miles long to divert water from the Congo River system in central Africa, where the rainfall is more regular, to the Lake Chad basin. However, these plans depend on large international loans, have unknown environmental impacts, and will take many years to complete. Therefore, unless the rains return soon, the future remains uncertain for the lake and those dependent on it.

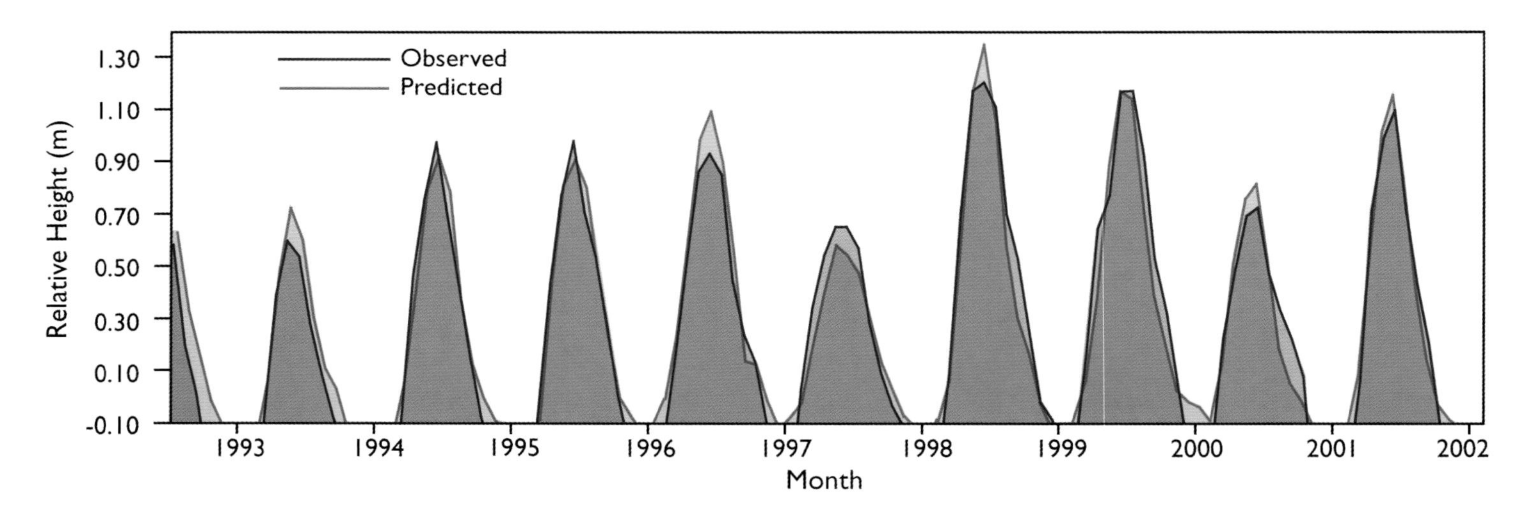

# Industrial Pollution in the Russian Arctic

W. Gareth Rees

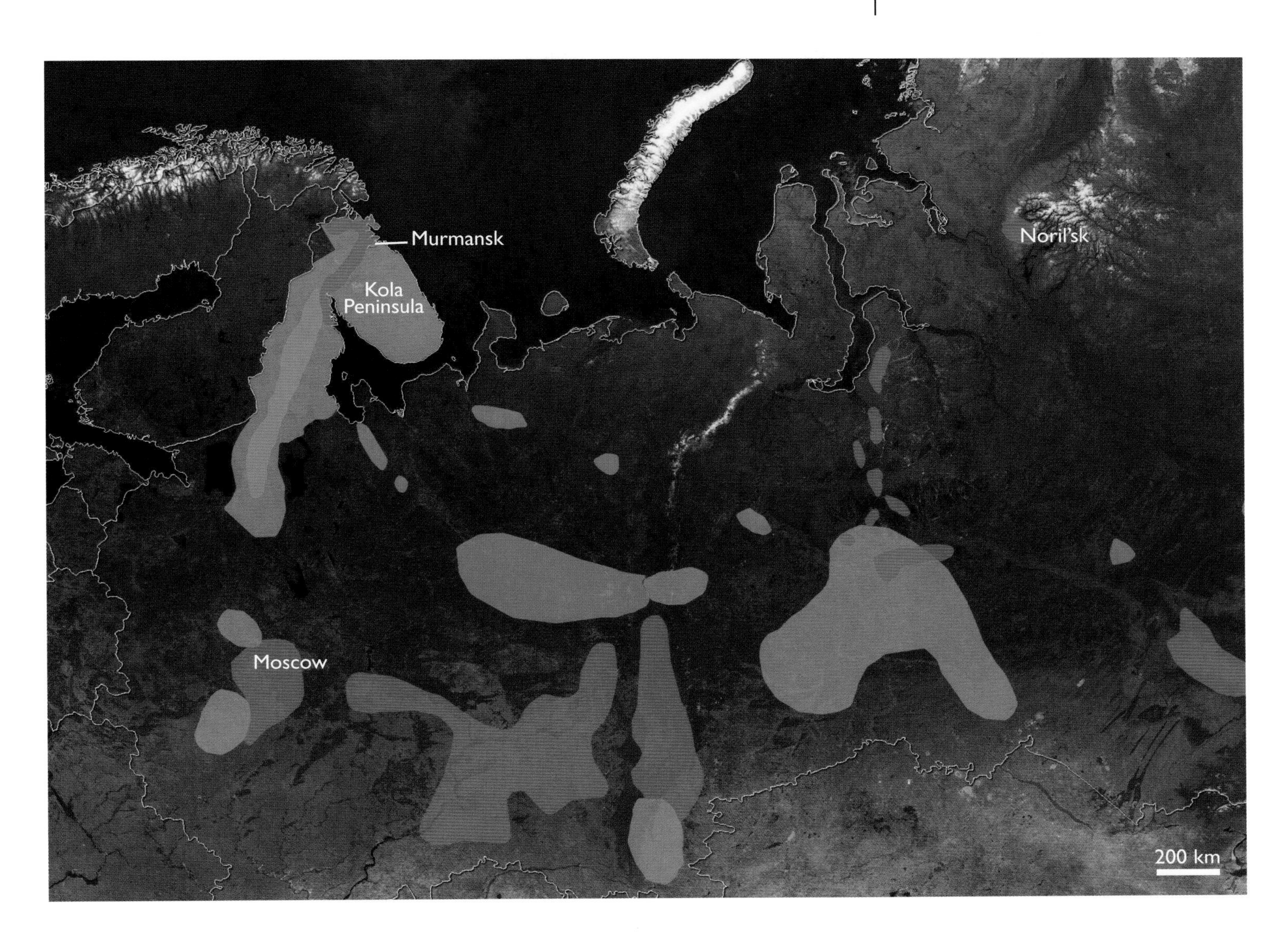


Environmental landscape impacts in the Russian North. Red: critically degraded areas; orange: areas affected by acid rain; bright green: areas of forest damage. (Based on data from the *Handbook of International Economic Statistics*, 1998.)

The Arctic has historically been viewed as a pristine wilderness, at least in the popular imagination, although more recently we have become familiar with the idea of industrial development in the North. What is perhaps less well known is the scale of industrialization in the Russian North, and the consequent scale of landscape change driven by industrial pollution.

The largest concentration of population and industrial activity in the Russian North is found in the Murmansk region on the Kola Peninsula. This region, with a population almost twice that of the state of Alaska, contains roughly a quarter of the people living in the Arctic. In terms of environmental impact, the most significant industrial activity is the smelting of non-ferrous metals such as copper, cobalt, and especially nickel. Smelting is carried out on a huge scale at Monchegorsk, Nikel', and Zapolyarnyy. The ores are rich in sulfur, and the smelters do not scrub the resultant sulfur dioxide from the emissions but release them to the atmosphere.

*Our Changing Planet*, ed. King, Parkinson, Partington and Williams
Published by Cambridge University Press 

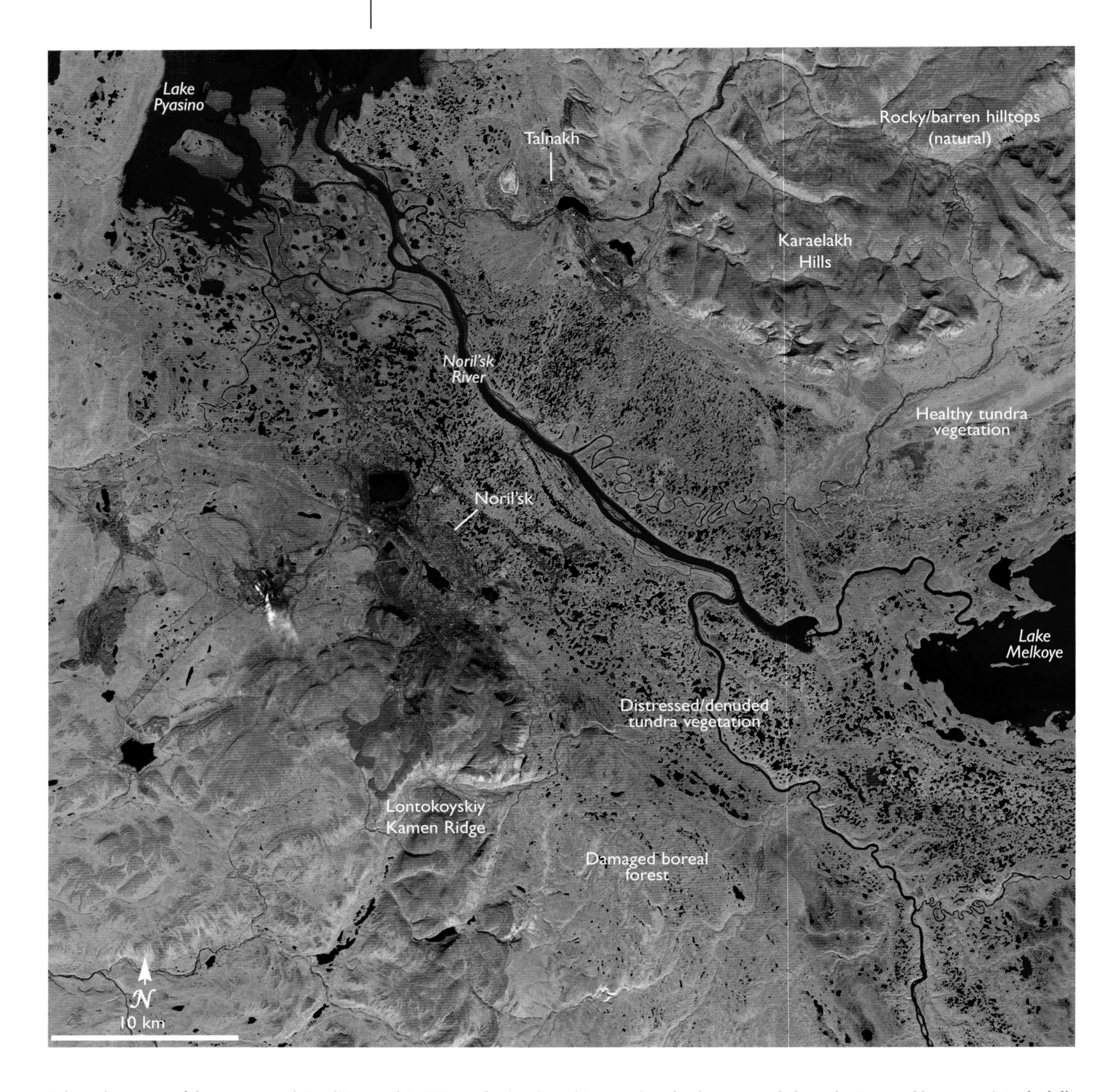


False-color image of the area around Noril'sk, April 9, 2006. In this band combination, bare land appears pale lavender (exposed barren rock in the hillside ridgeline running through the scene) to dark purple (landscape of Noril'sk and Talnakh). Brilliant greens show predominantly healthy Arctic tundra, riddled with characteristic deep blue small pothole lakes. The larger deep blues are rivers and lakes, including the large Lake Pyasino north of the city. Pale pink shows devegetated land, and the blue-white streams running from the city to the southeast (lower right) shows emissions from smoke stacks. The deep and pale pinks all downwind of the city, as well as the deep purple in the hillsides immediately outside Noril'sk, are moderately to severely damaged ecosystems. These include stressed tundra vegetation and denuded forests. Underground mines provide sulfide ores of copper and nickel which account for a sizeable fraction of Russia's entire national output of these metals. Sulfur dioxide emissions come from the city ore concentrators and smelters. (Data from the ETM+ instrument on the Landsat 7 satellite.)

Nickel smelter at Monchegorsk on the Kola Peninsula, July 1994. The smelter emits approximately 90,000 tons of sulfur into the atmosphere each year. (Photograph by Gareth Rees.)

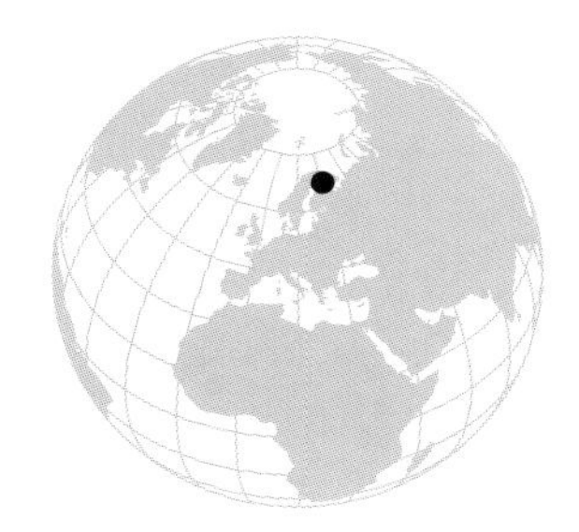

The three smelters on the Kola Peninsula emit a total of roughly 220,000 tons of sulfur into the atmosphere annually. The complex of smelters at Noril'sk, in north central Siberia, emits approximately 5 times this amount, over one million tons per annum. This is the largest single source of man-made sulfur dioxide in the world, responsible for over 1% of the global total emissions.

The impacts of these huge emissions of sulfur dioxide, and the associated emissions of heavy metals, are seen primarily in the effect they have on vegetation. Boreal forests exhibit a series of responses, including chlorosis, premature defoliation and desiccation (which renders them highly susceptible to fire). At extreme exposures, trees, shrubs, tundra vegetation, and even soil are destroyed, producing a 'technogenic barren.' Satellite remote sensing represents a powerful and efficient method of monitoring the nature and extent of the environmental damage caused by emissions from smelters. By comparing classified and interpreted images acquired over a period of time, it is possible to assess the dynamics of the pollution impact.

Although acidifying emissions from the smelting of non-ferrous metals are the major industrial source of landscape change in the Russian Arctic, they are not the only one. Mining operations, also predominantly concentrated on the Kola Peninsula, cause direct landscape damage through open-cast extraction and the dumping of mining waste, and the oil extraction industry, predominantly concentrated in the West Siberian basin, also leads to landscape damage and to oil spills. The notorious oil spill at Usinsk in 1994, which deposited 100,000 tons of oil, was probably not a rare event but typical of events that occur every 2–3 years. Satellite remote sensing is capable of monitoring all of these phenomena.

François Parthiot

# Oil Spills at Sea

Oil from the Prestige tanker washing up on the coast of Galicia, Spain, at Punta dos Remedios, December 2, 2002. (Photograph © 2002 Cedre.)

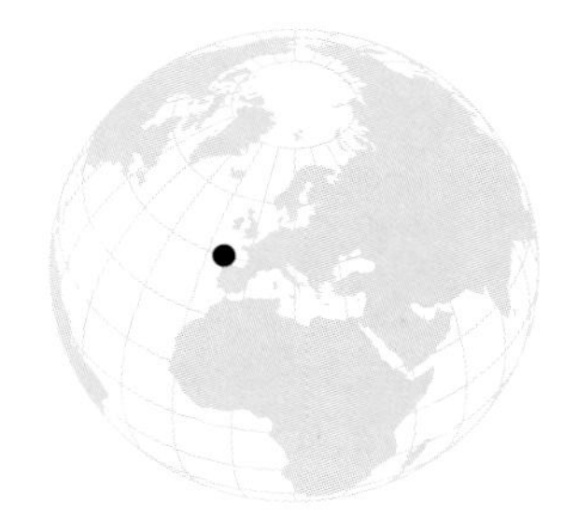

Oil releases due to all types of vessels at sea, whether operational or voluntary or accidental, are unfortunately quite frequent. They can have very large ecological and economic consequences, particularly in the case of a massive spill from a stricken tanker. The first such catastrophe occurred in 1967 when the Torrey Canyon oil tanker ran aground, releasing about 120,000 tons of crude oil and polluting pristine places along the coasts of Great Britain and France. This incident triggered international conventions that led to the present international oil pollution compensation system.

Shortly after the Torrey Canyon incident, war closed the Suez Canal for years, leading to a new and much longer oil route from the Persian Gulf to Europe, around South Africa. This induced a tremendous increase in the size of oil tankers, about 5 to 10 times in deadweight for the very large crude carriers (VLCCs), carrying up to 550,000 tons of oil

*Our Changing Planet*, ed. King, Parkinson, Partington and Williams
Published by Cambridge University Press 

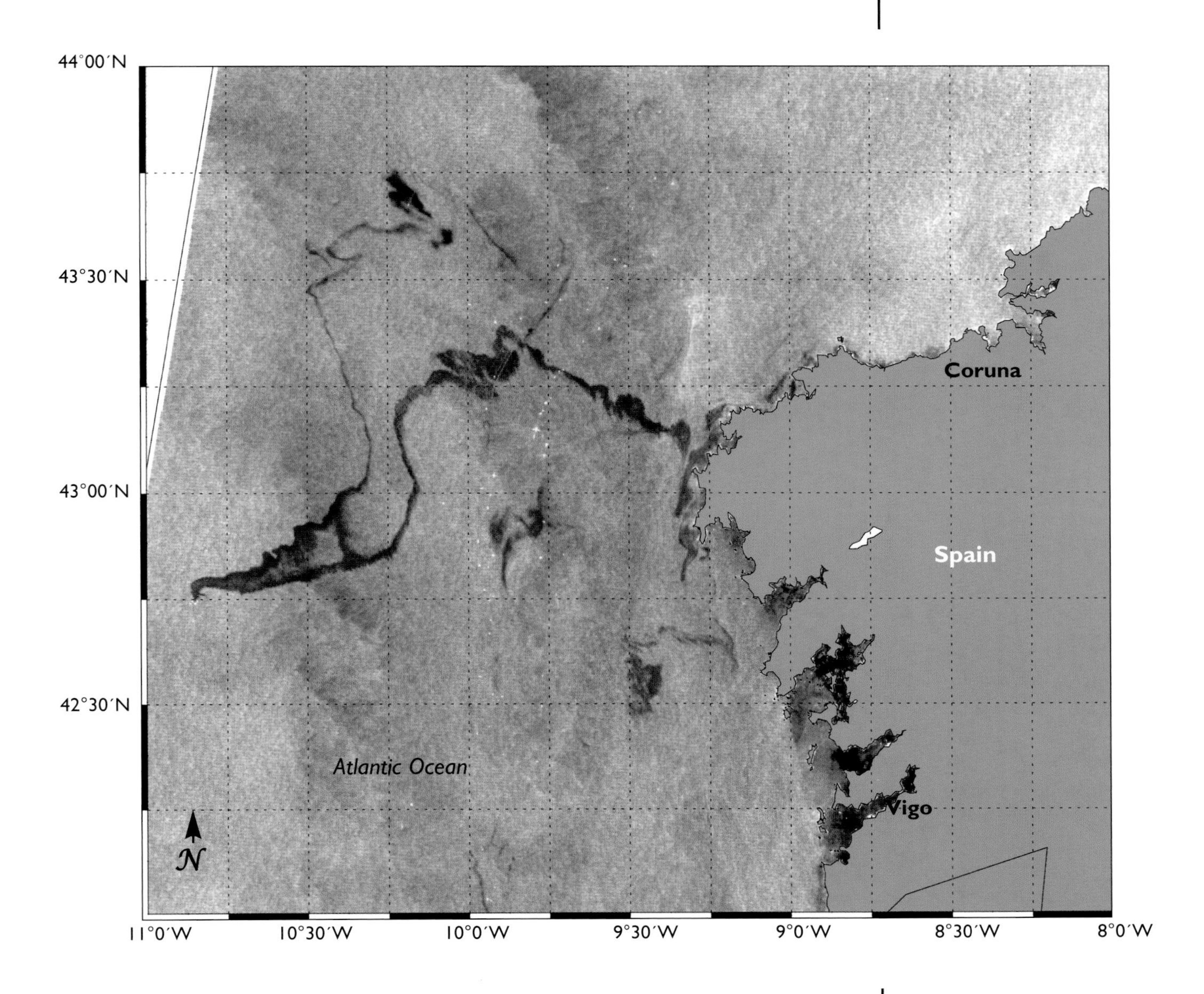


Pollution off the northwest coast of Spain, November 17, 2002, due to oil spilled from the Prestige tanker after suffering hull damage in rough seas. Two long divergent dark plumes appear in the wake of the crippled tanker (located at 42°45′N, 10°52′W), with some large polluted areas also prominently visible in the vicinity of the Galician coast. The vessel was being towed westwards away from the Spanish coasts, with continuous leakage of oil from the oil tanks. The Prestige sank in deep waters about 2 days later, spilling yet more oil during its descent. The term 'image intensity' here and on other images in this chapter is a dimensionless unit that corresponds to sea surface roughness, with high values signifying rougher seas. (Data from the SAR instrument on the ENVISAT satellite; copyright ESA 2002.)

onboard a single vessel! A long list of large oil spills followed, among them being those of the Amoco Cadiz (1978), Exxon Valdez (1989), Haven (1991), Aegean Sea (1992), Braer (1993), Sea Empress (1996), Erika (1999), and Prestige (2002).

A primary response of coastal countries to major spills has been oil slick monitoring carried out through aerial surveys. Although most of the surveys have been visual, radar imaging is also proving valuable, as the oil slicks smooth the sea surface, attenuating the radar echo and thus allowing the slicks to be highlighted as dark features on the radar images. Until late in the twentieth century, no satellite radar imaging was available, and it was difficult to have a global assessment of oil spills at sea. The situation has now changed, and coastal countries can increasingly benefit from a more efficient detection and monitoring of the oil slicks, with the help of satellite radar that can survey very large areas.

The value of satellite radar was demonstrated in 2002 when the oil spill disaster due to the Prestige tanker affected about 2,000 km of the European coast. Three days after the crippled

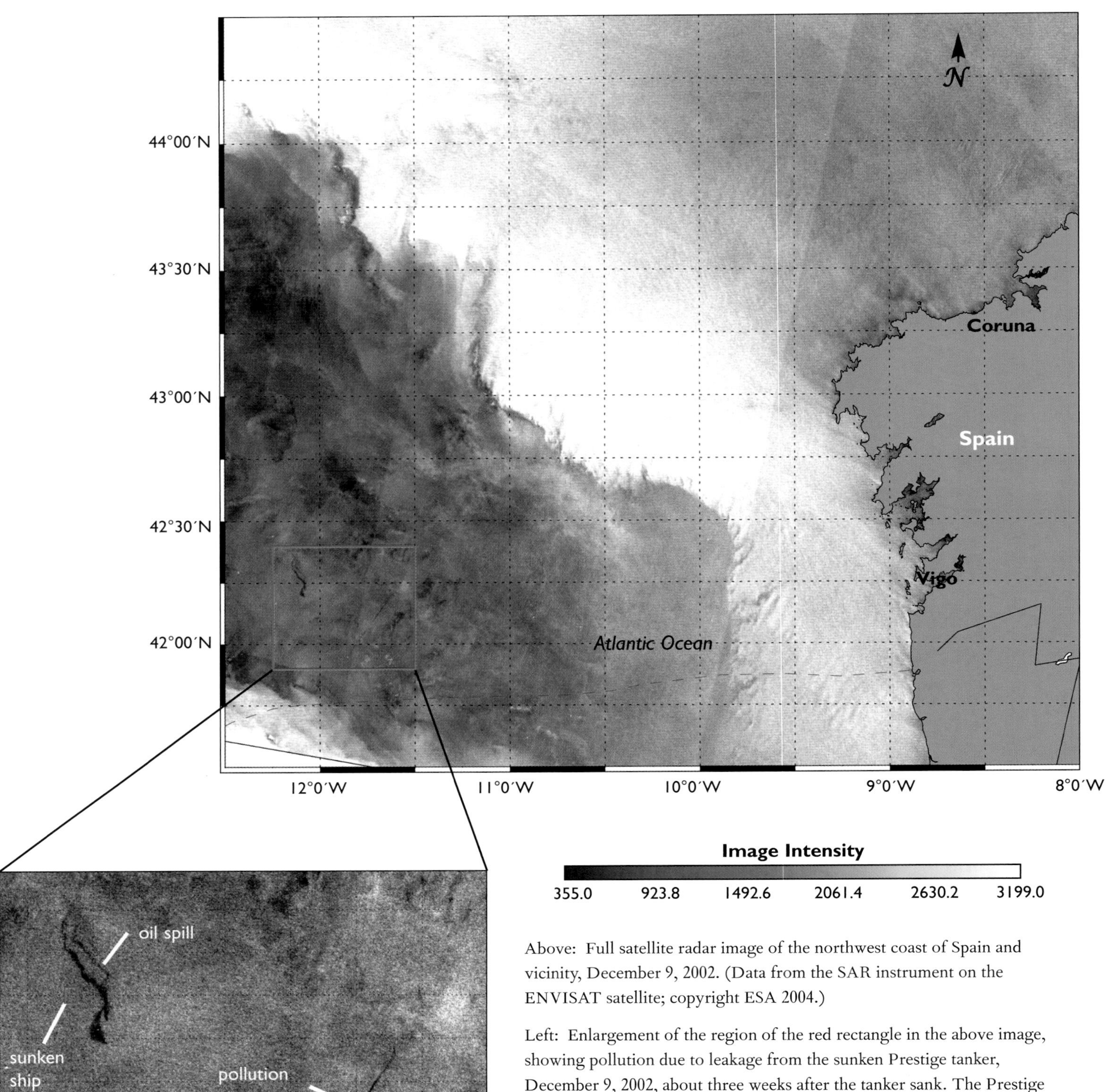


Above: Full satellite radar image of the northwest coast of Spain and vicinity, December 9, 2002. (Data from the SAR instrument on the ENVISAT satellite; copyright ESA 2004.)

Left: Enlargement of the region of the red rectangle in the above image, showing pollution due to leakage from the sunken Prestige tanker, December 9, 2002, about three weeks after the tanker sank. The Prestige wreck location is within the small red rectangle. Additional pollution due to a merchant vessel cruising nearby is also prominently visible, as a nearly linear diagonal strip across the right side of the image, starting at a white spot near the bottom of the image, marking the oil-discharging vessel itself. (Data from the SAR instrument on the ENVISAT satellite; copyright ESA 2002.)

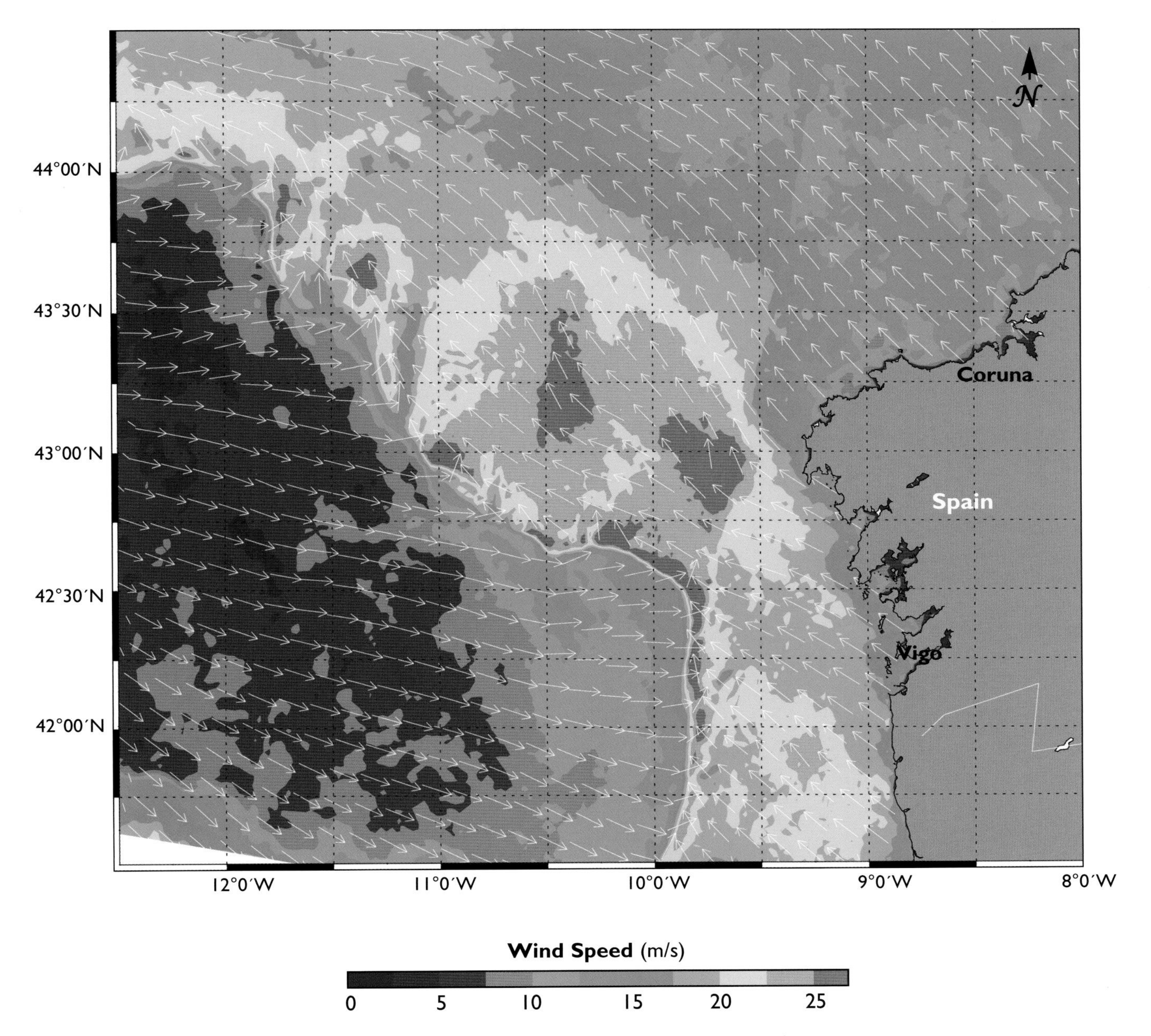


Wind pattern off the coast of northwest Spain, December 9, 2002, calculated from the radar data displayed in the corresponding radar image on the previous page. Wind knowledge is important in estimating the drift of the oil, and it allows a more efficient analysis of the radar images. Oil discharges are normally most visible if the wind speed is between 5 and 25 knots. (Data from the SAR instrument on the ENVISAT satellite, with data processing by Boost Technologies-France; copyright Boost Technologies 2004.)

vessel's call for assistance, a European satellite radar image, obtained through the International Charter "Space and Major Disaster," gave an excellent view of the large affected area of the ocean off the northwest coast of Spain (see the November 17, 2002 image). In fact, the satellite imagery was particularly important because dedicated surveillance aircraft were not able to give any radar view of the large oil polluted area during the first week of the incident.

Days after the Prestige sank, it appeared that the sunken tanker could still be leaking heavy fuel oil in spite of the very low temperature at the deep location of the wreck. This has also been confirmed by satellite radar imagery, which shows a small but noticeable oil slick in the vicinity of the wreck location (see the December 9, 2002 image). After a deep sea submersible investigated the wreck in detail two months later, it was estimated that the leaks were daily spilling about 100 tons of heavy fuel oil into the sea.

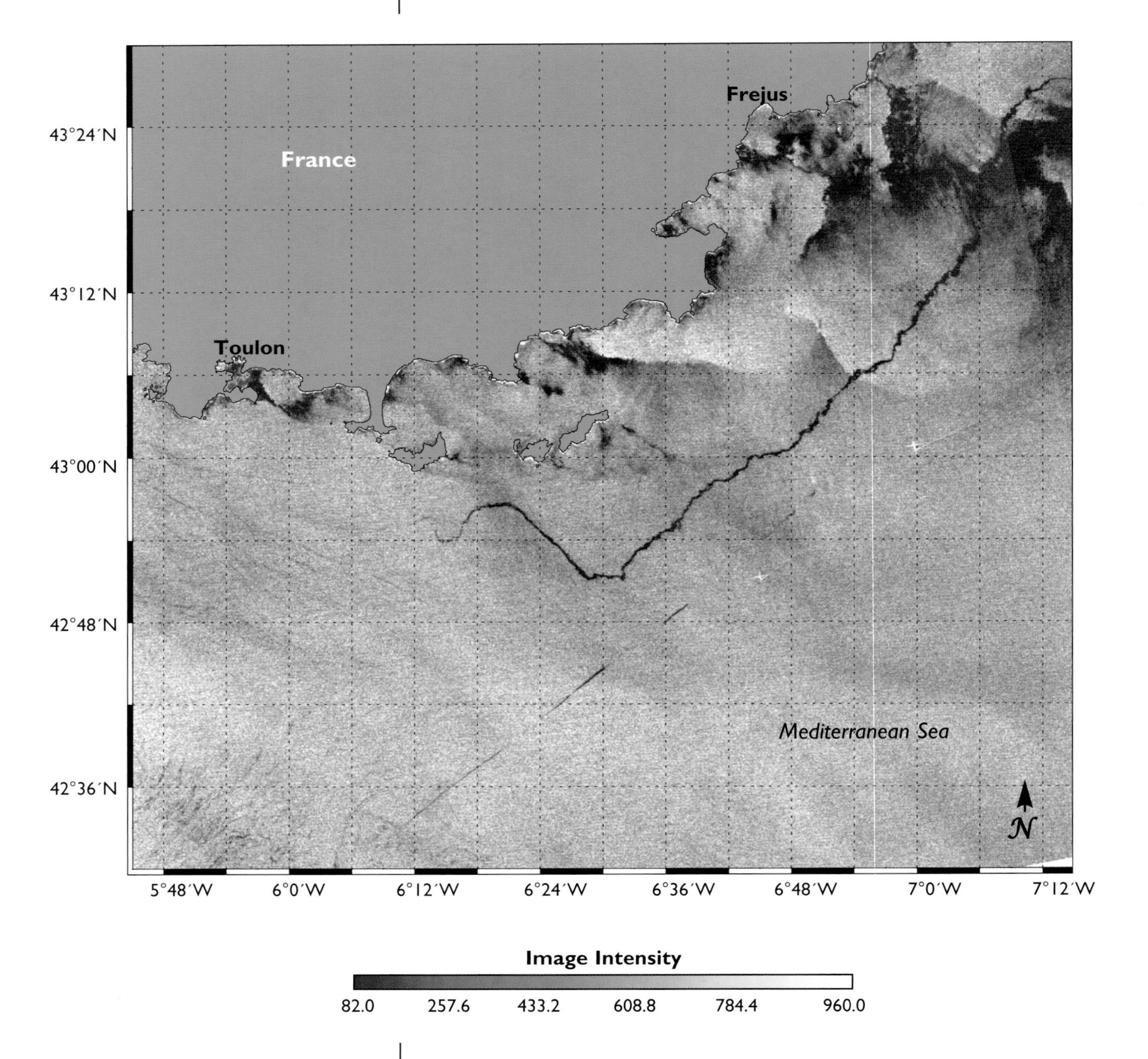


Oil pollution in the Mediterranean Sea, off the coast of southern France, August 3, 2004. On this radar image, close to the French Riviera, two different intentional and illegal spills are clearly visible. The older one is distorted by surface currents but is still quite dark; the second one is straighter and is not continuous, which is typical of several bilge water releases. (Data from the SAR instrument on the ENVISAT satellite; copyright ESA 2004.)

Satellite imagery can also be of value in monitoring voluntary, operational oil spills. The December 9, 2002 image, when analyzed in detail, notably shows a vessel that is intentionally releasing oil when cruising near the Prestige wreck location. Unfortunately, such discharges at sea are frequent in many places. It has been estimated that they could represent about half of all the oil discharged into the ocean due to human activities.

The Mediterranean Sea is subject to this kind of repeated pollution. The August 3, 2004 image illustrates that it can happen even in very sensitive areas, where it is strictly forbidden according to international conventions. Increased surveillance and deterrence, to which satellite radar imaging can efficiently contribute, is needed to better combat this form of pollution, which could easily be avoided.

# Epilogue

By now, you will have read or at least scanned this book and will be in the process of assessing what you have learned. For myself, I found out a lot about a number of completely new phenomena—new to me, that is—and filled in some otherwise vague notions of how things look or work. For example, the global patterns of CO and $NO_2$ are intriguing signs of industrial activity—someone down there (here) is working hard almost all the time.

But two things stand out. First, the planet is truly beautiful—small, precious, intricate, detailed and complex. In our ordinary lives, we seldom think of the world as a single, contained place, like a house or an island, but it is. It really is. It is finite, bounded and operates all by itself, like some perfect little toy. Why do we find it so beautiful? I honestly think that at some level we have evolved to see it that way. Humans, and every other living thing here, have evolved with the planet and are perfectly, exactly suited to its environment. So it's not surprising that we respond to views of the Earth with feelings of appreciation and a sense of the aesthetic—we belong here. This is home, and it is a perfect fit.

Second, and here we switch from art to science, the book is equivalent to a medical report on the health of the Earth. And here the news is mixed. Most doctors like to start with the good news: overall, the state of Life on Earth is good—humans have not caused catastrophic mass extinctions, and the atmosphere and oceans are not drenched in toxins. The world is currently very habitable. Most importantly, we have avoided blowing ourselves up, which would also mean taking most of the other creatures with us. And we can see the hand of man at work in positive ways: many areas of wilderness are now tended for agriculture to feed ourselves, without a significant negative impact on the global environment, and the great cities of Earth that house most of us are astounding in their own right. When you look at a city from space, you see a pattern of intersecting lines and polygons, straddling a river or wrapped around the curve of a bay—tiny patches of organization and effort grafted onto the sphere of the planet. Perhaps the cities represent mankind's greatest achievement so far; direct proof of our culture's energy, creativity and optimism. So far, so good.

But like a doctor gently warning his patient about early signs of problems due to overindulgence, the satellite data tell us clearly where we should be careful. The depletion of ozone is a global environmental problem that has been clearly recognized and, hopefully, contained. It will take decades for the damage to the ozone layer to be repaired, but it seems that we humans are on the right track with this issue at least. We noticed the problem, studied it carefully, and came up with a solution: the banning of certain harmful chlorofluorocarbon products used in air conditioners and aerosol cans. But the global warming caused by oil, coal, and gas burning is a far more dangerous and less tractable problem. More dangerous because large-scale climate changes would seriously dislocate our civilization and damage the environment that supports us, and less tractable because the miracle of economic development that has brought so many benefits to us is currently propelled by and addicted to fossil fuels. Some of the images in this book, and the graphs that summarize trends over

Astronaut Piers Sellers on a spacewalk at the International Space Station (ISS) during the STS-121 Shuttle mission, July 2006. Sellers is carrying a 75 lb grapple fixture—the object with a circular plate and gray support stand—to its new home on one of the ammonia tanks on the ISS. The Atlantic Ocean is 230 miles below; Sellers and the ISS are moving at 5 miles per second eastwards. (Image courtesy the Image Science & Analysis Laboratory, NASA Johnson Space Center.)

time, tell the story clearly—humans cannot continue to burn fossil fuels indefinitely without breakage. Something must be done, and a good first step is to raise public awareness of the problem. This book makes an excellent contribution in that respect.

A final thought: this place belongs to you, me, all of us. And we belong to it. Humans have changed from being just another species on Earth to being the primary homeowners and operators of the planet, and that implies certain obligations and responsibilities. On the whole, we are a sensible and considerate race and the more we understand and appreciate the workings of the planet, the more likely we are to take care of it.

Piers Sellers

# Appendix 1: Satellites and Satellite Orbits

MICHAEL D. KING

The Earth's atmosphere changes chemically and physically on widely varying timescales—ranging from minutes to decades—and is therefore a challenge to observe precisely over the entire globe. Land and ocean surfaces, including the semi-permanent ice caps and seasonally varying sea ice in Polar regions, also change dynamically, with sometimes strikingly large short-term changes as well as seasonal and interannual variations. Earth-observing satellites enable the sampling of the vast expanse of the Earth's surface and ever-changing atmosphere every day with highly sophisticated remote sensing instruments. 'Remote sensing' refers to the use of devices to observe or measure things from a distance without disturbing the intervening medium. Depending upon the measurement objectives, Earth-observing satellites primarily fly in one of three kinds of orbits: (i) a near-polar, sun-synchronous orbit to allow their sensors to observe the entire globe at the same solar time each day, (ii) a mid-inclination orbit to focus their sensors on the equatorial region and lower to middle latitudes where the observations are made at different times of day to better sample time-varying phenomena such as clouds, and (iii) a geosynchronous orbit located at one longitude over the Equator to observe the same 'sector' of the Earth from the same vantage point but every 15–30 minutes throughout the day and night.

## The Orbits

As the Earth rotates throughout the day, sun-synchronous and mid-inclination orbits continue to circle the globe producing anywhere from 12.8 to 15.7 orbits a day from their

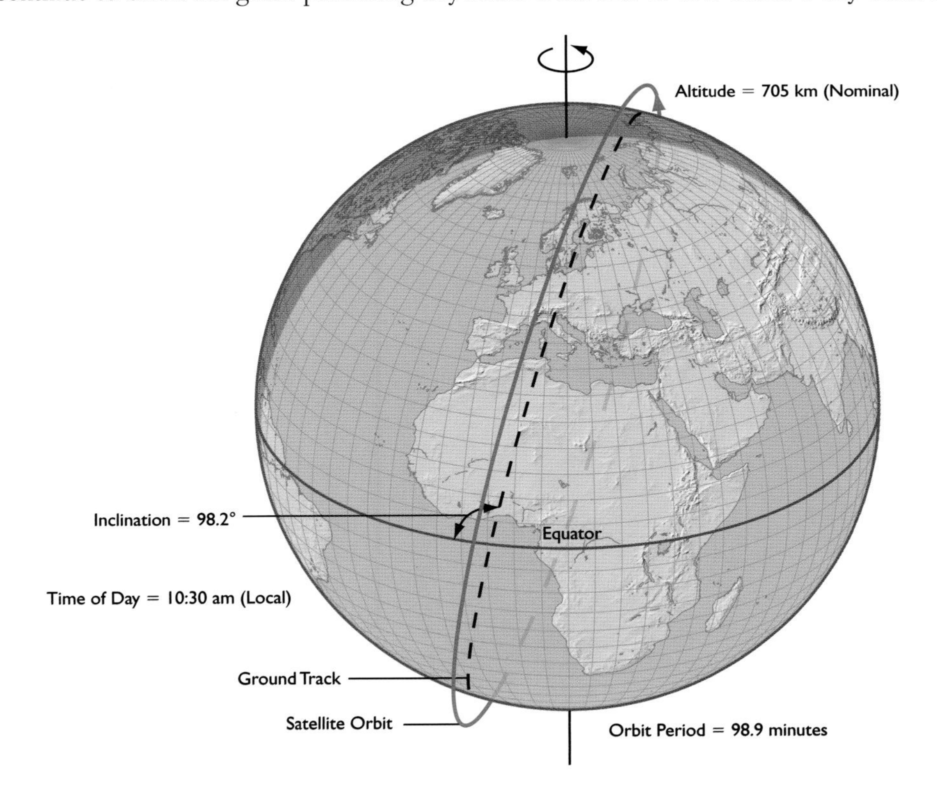


Example of a sun-synchronous polar orbit with an orbital altitude of 705 km above the Earth's surface, an orbital period of 98.9 minutes, resulting in 14.3 orbits per day, and an orbital inclination of 98.2° at the Equator. Sun-synchronous orbits are specified in terms of the time of day they cross the equator on the sunlit side of the orbit (10:30 am local time in this Terra example).

*Our Changing Planet*, ed. King, Parkinson, Partington and Williams
Published by Cambridge University Press 

One-day (July 11, 2005; left) and two-day (July 11 and 12, 2005; right) coverage of the MODIS sensor on the sun-synchronous Aqua satellite, whose orbital inclination of 98.2° provides near-global coverage each day due to the wide 2,330 km swath of the MODIS sensor. In two days there is complete global coverage with multiple observations of polar latitudes. The daily composite consists of daytime observations only.

One-day (March 15, 2007; left) and two-day (March 15 and 16, 2007; right) coverage of the TMI sensor on the mid-inclination TRMM satellite, whose orbital inclination of 35° restricts the orbital coverage to tropical latitudes from about 35°N to 35°S. The daily composite consists of both daytime and nighttime observations.

orbital altitudes of 1,336 to 350 km, respectively. Depending on the swath width of the satellite sensor, observations of the entire Earth's surface can be acquired in anywhere from 2 to 16 days. The widely used MODIS sensor aboard the Terra and Aqua satellites has a swath width of 2,330 km from an orbital altitude of 705 km, whereas the Landsat 7 ETM+ sensor, in contrast, images a region approximately 185 km in width and only when observations are acquired (~16% of the time).

Satellite sensors with high spatial resolution (15 m per pixel, or 'picture element'), can discern objects as small as, say, a large house. Other satellite sensors designed to make continental or global-scale observations have spatial resolutions that can discern objects down to 500 m per pixel to perhaps 20 km per pixel.

## The Satellites

Although many satellites use instruments that provide images in narrow wavelength bands in the ultraviolet, visible, and thermal infrared, others use microwave and radio

| Satellite | Operator | Altitude (km) | Inclination (°) | Purpose |
|---|---|---|---|---|
| *Mid-inclination* | | | | |
| TRMM | USA/NASA Japan/JAXA | 350–402 | 35 | Precipitation, lightning, Earth radiation budget |
| UARS | USA/NASA | 585 | 57 | Stratospheric chemistry |
| OrbView-1 | USA/GeoEye | 740 | 70 | Lightning |
| Geosat Follow-on | USA/Navy | 788 | 108.1 | Ocean surface topography |
| TOPEX/Poseidon | USA/NASA France/CNES | 1,336 | 66 | Ocean surface topography |
| Jason-1 | USA/NASA France/CNES | 1,336 | 66 | Ocean surface topography |
| *Sun-synchronous* | | | | |
| JERS-1 | Japan/JAXA | 580 | 98.0 | Flooding, forest mapping, sea ice |
| Landsat 7 | USA/USGS | 705 | 98.2 | Land-surface imaging, volcanoes, glaciers |
| Terra | USA/NASA | 705 | 98.2 | Land-surface imaging, Earth radiation budget, ocean color and sea surface temperature, carbon monoxide, aerosol and cloud properties |
| Aqua | USA/NASA | 705 | 98.2 | Atmospheric temperature and moisture profiles, sea surface temperature, fire, aerosol and cloud properties, ocean color, land properties, sea ice |
| ERS-2 | Europe/ESA | 785 | 98.5 | Atmospheric chemistry, oil spills |
| Envisat | Europe/ESA | 800 | 98.6 | Ocean surface topography, flooding, ocean color and sea surface temperature, atmospheric chemistry, sea ice, ocean wind speed and direction |
| QuikScat | USA/NASA | 803 | 98.6 | Ocean wind speed and direction |
| NOAA-14 | USA/NOAA | 844 | 98.7 | Sea surface temperature, atmospheric temperature, clouds |
| F15 | USA/DoD | 850 | 98.8 | Night lights, sea ice, sea surface temperature |
| *Geosynchronous* | | | | |
| GOES-12 | USA/NOAA | 35,800 | 0 | Hurricanes, water vapor, severe weather |

Satellites, operators, and purposes for a selection of the 48 different international Earth-observing satellites used to provide illustrations in this book.

frequencies. These are all digital or electronic measurements of energy reflected or emitted by the Earth and its atmosphere. They are not, in general, digital or analog photographs, per se, but can be reconstructed to produce 'images' using a combination of red, green, and blue bands analogous to the workings of a television. Simple image reconstruction as described here is often used to illustrate natural hazards such as fires, dust and smoke, floods, severe storms such as hurricanes, droughts, and snow, volcanic eruptions, clouds and fog, and phytoplankton blooms over the ocean.

## Principles of Remote Sensing

The power of satellite remote sensing, however, arises not from the beauty of its global Earth-imaging capability, but from the quantitative analysis of the reflected and emitted signal from the Earth-atmosphere system using well-understood principles of physics. By carefully selecting the wavelength region most sensitive to: (i) scattering by dust or aerosol particles, (ii) scattering and absorption by cloud drops and ice crystals, (iii) reflection from blue or green

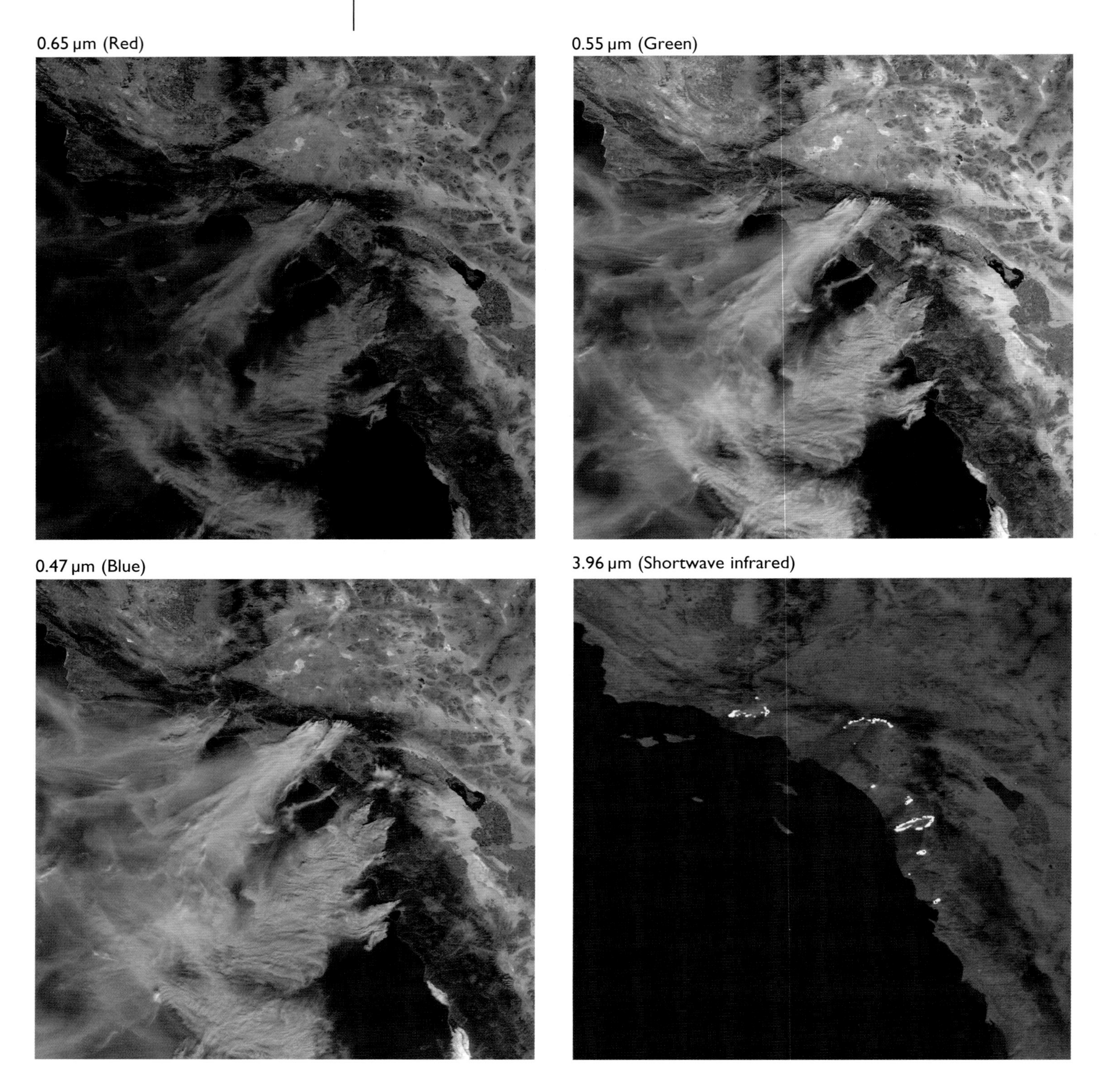


Satellite imagery from individual MODIS bands during a fire and smoke outbreak in southern California and the northern Baja California Peninsula on October 26, 2003. Images are atmospherically corrected at red (top left), green (top right), blue (lower left) and shortwave infrared (lower right) bands. (Data from the MODIS instrument on the Terra satellite.)

ocean water, (iv) emission of land and ocean surfaces as a function of temperature, (v) emission by sea ice in contrast to salt water in the world's oceans, (vi) reflection by green vegetation, (vii) absorption by atmospheric gases such as ozone, water vapor, and nitrogen oxides, (viii) emission by the Earth's atmosphere as a function of wavelength in the thermal infrared, and (ix) microwave emission by rain drops, it is possible to ascribe the satellite-measured signal to a wide variety of properties of the Earth-atmosphere system. Monitoring these properties over time also allows extreme events and change detection to be observed in a uniform manner without regard to national boundaries and natural barriers, like rivers.

True color image of a fire and smoke outbreak in southern California and the northern Baja California Peninsula on October 26, 2003 constructed by combining individual MODIS bands (see images on facing page) and overlaying the hot fires from the 3.96 μm band. (Data from the MODIS instrument on the Terra satellite.)

## Remote Sensing Methods

Satellite sensors are often classified into two kinds of devices: (i) passive instruments that use natural sources of energy, such as reflected sunlight or emission of thermal radiation in the infrared or microwave frequencies, and (ii) active instruments that provide their own source of illumination, such as lidar (laser radar) and radar.

### *Passive Remote Sensing*

There are a wide variety of passive remote sensing methods available to scientists today. Perhaps the most well-known would be the 'pushbroom' radiometer, which is analogous to a

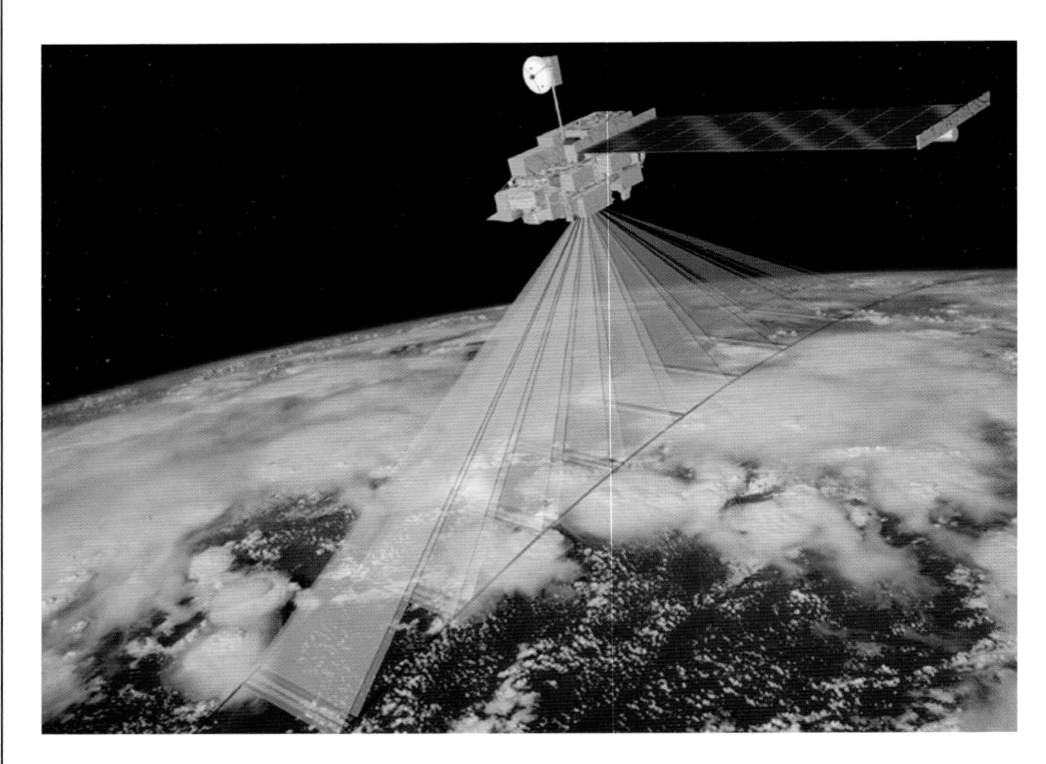

Artist's rendition of the MISR instrument on the Terra satellite, which produces images of the Earth at nine discrete look angles using multispectral pushbroom cameras during the daytime portion of each Earth orbit. The swath width of MISR is 400 km and thus requires nine days to obtain complete global coverage.

digital camera whereby an image is produced through the motion of the satellite along its orbit. In this radiometer design, individual pixels are recorded in a line across the swath width of the imager, and can be recorded at multiple narrowband wavelength intervals, if desired. MISR is an example of such a sensor, and it flies on the Terra spacecraft, where an image is produced by this sensor through a combination of a linear charge coupled device (CCD) detector and camera plus the motion of the satellite. MISR data are used to study aerosol particles in the Earth's atmosphere and the non-uniform reflection of the Earth's surface and clouds, since MISR carries nine cameras oriented in different viewing directions. This 'multi-angular' remote sensing approach complements multi-spectral observations by providing additional information about the physical structure of the surface and atmosphere. This technique also allows global stereo mapping of cloud heights and volcanic plumes from the apparent motion of discernible features as viewed from different telescope directions.

Many of the remote sensing instruments in wide use today are 'whiskbroom' radiometers whereby the image is produced through the use of a scan mirror and telescope with wavelengths selected through the use of narrowband interference filters, prisms, or diffraction gratings. Thus an image is produced through the sweeping of the scan mirror across the scene, similar to the sweeping of a whiskbroom. Among the commonly used whiskbroom scanners are the ETM+ on Landsat 7, MODIS on Terra and Aqua, and the thermal infrared subsystem of the ASTER instrument on Terra.

### *Active Remote Sensing*

The TRMM satellite carries a precipitation radar that is the first such radar to be flown in space. It was developed by the Japanese Space Agency (JAXA) and uses an active microwave radar to enable precipitation to be measured down to a minimum detectable level of about 0.5 mm/hr without the use of a rotating radar dish. By transmitting its own microwave signal to the surface and measuring the return echo and the time it takes for the radar signal to reach the rain drops and return to the satellite, it is possible to measure the vertical distribution of rain rate at a vertical resolution of 250 m from the surface to about 15 km. By flying the NASA-developed TRMM satellite at an altitude as low as 350 km (early in the mission; later raised to 402 km), the spatial resolution of the radar signal at the surface is 4.3 km. If the satellite were to be raised even higher in orbital altitude, the spatial resolution of the return from the surface would result in a larger footprint size and less spatial detail.

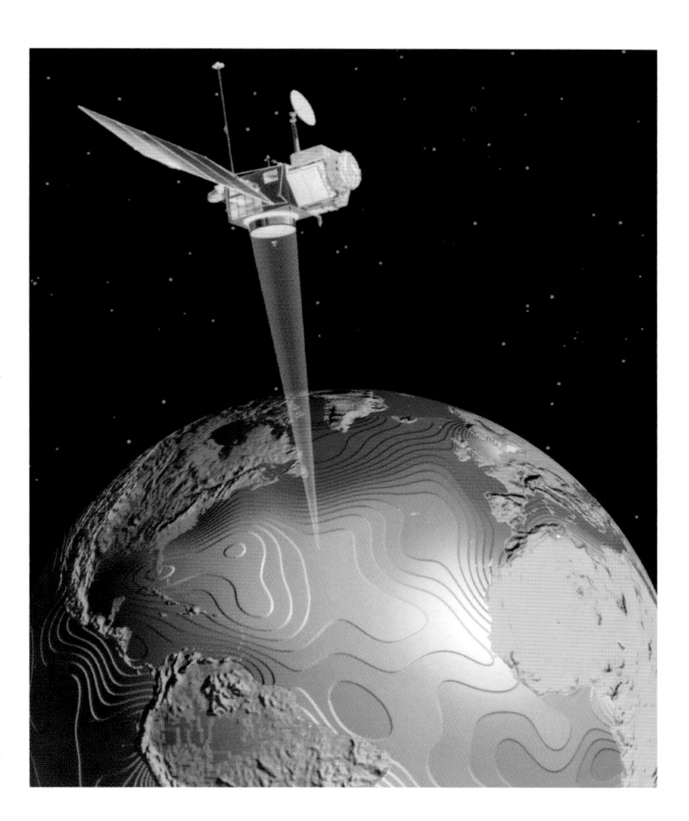

Artist's rendition of TOPEX/Poseidon making measurements of ocean surface topography using the method of radar altimetry.

Another active sensing method in wide use today is radar altimetry, whereby a nadir-looking microwave radar is transmitted from a high-flying satellite to the surface and the time delay to its return is measured. Ocean altimetry typically uses a microwave signal at 13.575 GHz (~2.2 cm wavelength), enabling the ocean surface height relative to an Earth-fixed coordinate to be measured to an accuracy of 3–4 cm. This accuracy requires corrections for time delays in propagating through the ionosphere as well as atmospheric water vapor, but this level of accuracy permits studies of the shape of the ocean surface globally, including ocean currents, sea level height, and tides. Radar altimeters have been developed by NASA and CNES, the French space agency, and flown on the joint TOPEX/Poseidon and Jason-1 satellites. They have also been developed by ESA and flown on their ERS-1, ERS-2, and Envisat satellites. Together, radar altimeters have received wide use for monitoring the spatial and temporal change in sea level globally, especially for assessing fluctuations in the ocean surface as a result of large-scale El Niño and La Niña events, and have also been used to observe changes in the elevation of the polar ice caps.

Another unique active remote sensing method that has received wide use in recent years is scatterometry, yet another application of Ku-band (13.4 GHz, 2.24 cm) microwave radiation. In this case, however, the radar beam is not pointed straight down, as in radar altimetry, but rather uses two rotating pencil-beam antennas that transmit microwave radiation to the ocean surface and measure reflection by the small capillary waves that ride on top of larger ocean waves. By looking at the differences in return signal from both beams that illuminate the ocean surface at different look angles, it is possible to acquire high-resolution, continuous, all-weather measurements of near-surface wind speed and direction over the ice-free global oceans, as the size and orientation of the capillary waves can be related to wind speed and direction, respectively. Examples of scatterometry are the SeaWinds instruments that have flown on the Japanese ADEOS-II satellite and NASA's QuikScat satellite, illustrations from which can be found in the chapter entitled "Winds Over Ocean" on page 168.

Satellite remote sensing has grown in complexity and capability since the first Earth-observing satellite was launched to take pictures of cloud cover from high above the Earth's surface in April 1960. Over the decades, the art and science of space-based observations of the Earth expanded, and methods were developed to take advantage of sensitive regions of the electromagnetic spectrum other than visible light, thereby allowing us to observe aspects of our planet that we would not ordinarily see. Many new methods have been employed since the 1960s with some of the active sensor applications described above having been developing much later as a direct consequence of advances in technology. The vast majority of quantitative remote sensing applications date from the mid-1970s and have been developed by many different nations of the world. Many of these applications and analysis methods have further benefited from the revolution of highly sophisticated computer technology that enables more rapid analysis and visual interpretation of the vast amounts of digital Earth observations available on a daily basis worldwide. Satellites will continue to be used to monitor and help understand the planet.

# Appendix 2: Glossary

## A

**Abyssal:** the deeper regions near or at the bottom of oceans, typically at depths of 2,000 to 6,000 m.

**Active microwave:** an instrument that transmits ***microwave radiation*** and then detects the reflected signal.

**Adsorbing:** a gas or a liquid that accumulates on the surface of a solid or a liquid, forming a molecular or atomic film. In contrast, absorption (the more familiar process) refers to a substance diffusing into a liquid or solid to form a solution.

**Aerosol:** extremely small solid particles, or very small liquid droplets, suspended in a gas.

**Aerosol index:** a measure of how much the wavelength dependence of backscattered ***ultraviolet radiation*** from an atmosphere containing ***aerosols*** differs from that of a pure molecular atmosphere. This uses the property that absorbing aerosols (dust and smoke) enhance ***reflectance*** of a molecular atmosphere in the ***ultraviolet*** wavelength region, which enables dust and smoke to be distinguished over bright desert surfaces.

**Albedo:** the fraction of the total solar ***radiation*** incident on a target that is reflected by it.

**Algorithm:** a step-by-step mathematical procedure for solving a problem.

**All-sky conditions:** conditions associated with almost, or all, the entire hemisphere of sky above the observation point on the Earth's surface.

**Altimeter:** an instrument to measure the distance between the instrument (e.g., on an aircraft or satellite) and a specified surface. That measurement, coupled with the knowledge of the position of the platform and sensor, enables determination of the surface topography.

**Altocumulus clouds:** middle level clouds occurring as roll-like patches or puffs forming waves or parallel bands. These clouds are found between 2,400 and 6,100 m above the Earth's surface.

**Anabatic wind:** a wind that blows up a slope.

**Anomaly:** the difference between the average measurement and an individual measurement.

**Antarctic circumpolar belt:** the portions of the Atlantic, Indian, and Pacific oceans that extend from about 50°S to 60°S and form an oceanic belt around the Antarctic continent.

**Antarctic circumpolar wave:** a wave-like pattern of related atmosphere, ocean, and ice conditions, which travels eastward around Antarctica over a period of several years (about 8 years in recent times). There are postulated links to the ***El Niño-Southern Oscillation*** phenomenon.

**Anthropogenic:** effects or processes that are due to human activity.

**Anti-cyclonic:** atmospheric motions around areas of ***high pressure***, being clockwise in the Northern Hemisphere and counter-clockwise in the Southern Hemisphere.

**Arthropod:** the animal grouping which includes insects, arachnids, crustaceans, and similar animals. Arthropods have a segmented body with appendages and a hard external skeleton.

**Atmospheric correction:** the process of analyzing ***remote-sensing*** data to remove the effects of ***electromagnetic*** scattering in the atmosphere on measurements from the Earth's surface.

**Austral:** relating to the Southern Hemisphere (as in 'austral summer' being the months of December to February).

*Our Changing Planet*, ed. King, Parkinson, Partington and Williams
Published by Cambridge University Press © Cambridge University Press 2007

# B

**Backscatter:** the reflection of waves, particles, or signals in the direction from which they originated.

**Bajos:** seasonally flooded swamps in Guatemala.

**Bathymetry:** underwater topography, which is normally expressed as water depth.

**Bergy bit:** a very small ***iceberg***.

**Biochemical:** chemical processes and transformations in living organisms.

**Biochemical cycle:** pathways by which the chemicals in the ***biosphere*** are transported and transformed.

**Biodiversity:** the diversity of living organisms, typically measured by the number of species in the region of interest.

**Biogeochemical cycle:** the cyclical path along which solid, liquid, and gaseous materials flow through the ***biosphere***, ***lithosphere***, ***hydrosphere***, and atmosphere.

**Biological productivity:** the amount of organic matter, carbon, or energy content that is accumulated during a given time period; this includes ***primary productivity*** and secondary productivity.

**Biomass:** the total dry organic matter or stored energy content of living organisms that is present at a specific time in a defined unit (ecosystem, crop, etc.) at or near the Earth's surface.

**Biosphere:** the global ecological system, encompassing all living beings.

**Blue ice:** glacial ice visible at the surface of a ***glacier***, ***ice cap***, or ***ice sheet***. Blue ice is present only in areas where there is no accumulation of snow on the surface and its blue color is caused by absorption of red and yellow light.

**Bodele Depression:** part of the low-lying former bed of Lake Chad in Africa. Thought to be the largest source of airborne dust on Earth.

**Bolometer:** a detector of ***infrared radiation***.

**Boreal Forest:** the belt of mainly coniferous trees that grow across North America and Eurasia.

**Bottom water:** the lowest layers of ocean water. This water is distinguished from shallower, overlying waters by its high density and by characteristic temperatures, salinities, and oxygen content.

**Boundary layer:** the layer of fluid adjacent to a bounding surface, such as the boundary layer in the atmosphere, which is the bottom layer of the troposphere (extending to the ground) and is affected by ***diurnal*** heat, moisture or momentum transfer to, or from, the surface of the Earth.

**Brackish water:** water that has salinity values between that of fresh water and seawater. It usually results from the mixing of seawater and fresh water, as occurs in estuaries.

**Brightness temperature:** a measure of the intensity of ***radiation*** thermally emitted by an object.

**Brine:** water that is saturated, or nearly saturated, with salt.

# C

**Calcium carbonate:** a chemical compound with the formula $CaCO_3$. It is the main component of seashells and is found in many rocks, including limestone, chalk, and marble.

**Caldera:** a volcanic feature formed by the collapse of a volcano into itself, making it a large, specific form of volcanic crater.

**Calibration:** the process of setting or correcting a measuring device so that the measurements conform to an accurate standard.

**Capillary waves:** small surface waves caused by wind interaction with the water surface.

**Carbon cycle:** the movement of carbon among all reservoirs (storages, especially the atmosphere, terrestrial ***biosphere***, oceans, and sediments, the latter including ***fossil fuels***).

**Carbon dioxide:** a chemical compound with formula $CO_2$. $CO_2$ is present in the atmosphere in low concentrations and acts as a ***greenhouse gas***. Plants utilize $CO_2$ during ***photosynthesis***, and animals exhale $CO_2$ during respiration.

**Carbon emission:** the emission of carbon, often limited to ***carbon dioxide***, and often expressed in tons of carbon dioxide equivalent; used as the basis for carbon emissions trading and the control of ***greenhouse gas*** emissions.

**Carbon monoxide:** a chemical compound with the formula CO. It is a colorless, odorless, and tasteless gas, which results from the incomplete combustion of carbon-containing compounds, such as occurs in internal combustion engines.

**Catalytic cycle:** chemical reactions where a substance called a 'catalyst' enables the reaction to proceed at a faster rate or under different conditions than otherwise possible, but which itself remains unchanged at the end of the cycle (although it might be temporarily altered during the cycle). Catalytic cycles frequently occur in biochemistry and organometallic chemistry.

**Circumpolar belt:** see ***Antarctic circumpolar belt***.

**Chlorophyll:** the chemical compound present in plants that absorbs the energy from sunlight and uses it to create carbohydrates from ***carbon dioxide*** and water. This process, known as ***photosynthesis***, sustains life in all plants.

**Chlorophyll-*a* pigment:** the pigment that captures the light energy necessary for ***photosynthesis*** to occur.

**Chlorine monoxide:** a chemical compound with the formula ClO. It is the main chemical for destroying ozone and is highly reactive.

**Chlorofluorocarbons (CFCs):** a family of inert, non-toxic, and easily liquefied chemicals used in refrigeration, air conditioning, packaging, and insulation, or as solvents or ***aerosol*** propellants. Because they are not destroyed in the lower atmosphere, they drift into the upper atmosphere, where, given suitable conditions, their chlorine compounds destroy ***ozone***.

**Cirrocumulus clouds:** high clouds that are composed of ice crystals and exhibit a wavelike pattern of small, white, rounded patches rarely covering the entire sky.

**Cirrostratus clouds:** high, thin, layered clouds composed of ice crystals that form a thin white veil over the sky, at altitudes from 6,000 to 12,000 m. Cirrostratus clouds are nearly transparent, so that light from the Sun or reflected from the Moon readily shines through them.

**Cirrus clouds:** high, thin, layered clouds composed of ice crystals and occurring as delicate, 'silky' strands that are streaks of falling ice crystals blown laterally by strong winds. Cirrus clouds usually occur at altitudes above 8,000 m.

**Cirrus uncinus clouds:** a variety of ***cirrus*** cloud found at very high altitudes, often referred to as mare's tails.

**Climatology:** the study of climate.

**Cloud-condensation nuclei:** tiny solid and liquid particles that promote the condensation of water vapor at temperatures both above and below the freezing point of water. Cloud-condensation nuclei, generally 2–10 μm in radius, provide relatively large surface areas on which condensation or deposition can initially take place.

**Coccolith:** individual plates of ***calcium carbonate*** formed by ***coccolithophores*** that are arranged around them.

**Coccolithophore:** single-celled algae, or ***phytoplankton***, consisting of calcium carbonate plates (or scales) known as ***coccoliths***. They are marine in origin and found in large numbers throughout the surface ***euphotic zone*** of the ocean. When they die and sink through the water column they form an important part of deep-sea sediment.

**Coherent:** ***electromagnetic radiation*** having a ***phase*** that is not random, as in ***laser*** and ***synthetic aperture radar*** sensors.

**Coherence:** a measure of the degree of non-randomness in two or more sets of waves, including ***electromagnetic radiation***.

**Convection:** the internal movement of currents within fluids (i.e., liquids and gases, such as the ocean and the atmosphere respectively), driven by gravity and differential heating or cooling.

**Crevasse:** a crack in fresh water ice (***glaciers***, ***ice caps***, and ***ice sheets***) typically formed as a result of stress created from different rates of ice movement.

**Cryosphere:** the portion of the Earth that is in a frozen state, including snow, ice, and permafrost.

**Cumulonimbus clouds:** thunderstorm clouds that form as a consequence of deep ***convection*** in the atmosphere; they have tops that sometimes reach altitudes of 20,000 m or more. The upper portions of these clouds are anvil-shaped.

**Cumulus clouds:** vertically developed clouds that form as a consequence of the updraft in ***convection*** currents. Cumulus clouds consist of mostly water droplets and resemble balls of cotton floating in the sky.

**Cumulus congestus clouds:** a towering variety of ***cumulus*** cloud associated with strong ***convection***.

**Cumulus humilis clouds:** fair-weather ***cumulus*** clouds.

**Current meter:** an instrument used to measure currents in the ocean or other bodies of water. Examples are the Savonius rotor current meter or the acoustic Doppler current profiler.

**Cyanobacteria:** bacteria that are blue-green in color because they obtain their energy through ***photosynthesis***. They are often referred to as blue-green algae.

**Cyclone:** a weather system characterized by relatively low surface-air pressure compared with the surrounding air: same as a ***low-pressure system***. Viewed from above, surface winds blow counterclockwise and inward in the Northern Hemisphere but clockwise and inward in the Southern Hemisphere.

**Cyclonic:** atmospheric motions around ***low-pressure systems***, being counterclockwise in the Northern Hemisphere and clockwise in the Southern Hemisphere.

# D

**Decadal:** related to a 10-year period.

**Deep ocean trench:** long and narrow topographic sea-bed features found between lithospheric plates that are related to ***plate tectonics***.

**Desiccation:** the state of extreme dryness, or the process of extreme drying.

**Detraining:** the transfer of air from an organized air current to the surrounding atmosphere; the opposite of ***entraining***.

**Dew point:** the temperature to which air must be cooled (at constant barometric pressure and water vapor content) to achieve saturation of water vapor. At this temperature the water vapor condenses and forms dew.

**Diazotrophs:** bacteria that fix nitrogen gas and convert it into a more stable nitrogen compound, such as ammonia, nitrate, or nitrogen dioxide.

**Dinoflagellate:** any of a class of single-cell marine ***plankton*** organisms having characteristics of both plants and animals.

**Dipole:** paired equal and opposite phenomena; e.g., an electric dipole is a separation of negative and positive charge. In climate, the term is used to describe a pair of often widely separated locations that are closely linked, but have opposite climate characteristics, an example being the ***North Atlantic Oscillation***, with low sea level pressure over Iceland and high sea level pressure over the Azores.

**Diurnal:** related to the period of a day, e.g., recurring on a daily basis.

**Dobson unit:** a measurement unit often used in indicating the amount of ***ozone*** in the atmosphere, named after G. M. B. Dobson for his pioneering work on the ***stratospheric ozone*** layer in the 1920s. A Dobson measurement indicates how thick the layer of ozone would be if all the ***ozone*** in the column of the atmosphere at the point of the measurement were brought down to sea level. One Dobson unit corresponds to a thickness of 0.001 cm.

**Doldrums:** see ***Intertropical Convergence Zone***.

**Doppler effect:** a shift in frequency (or phase) of an ***electromagnetic*** or sound wave due to the relative movement of the source and the observer.

**Doppler frequency:** the observed apparent ***electromagnetic*** or sound wave frequency of a moving object that is shifted from the actual frequency due to the ***Doppler effect***.

**Ductile:** characterized by yielding under stress (as opposed to brittle fracture). Gold, copper, and aluminum are highly ductile metals.

# E

**Easterly winds:** winds that blow from the east.

**Ebola hemorrhagic fever:** an often deadly disease caused by a group of viruses belonging to genus Ebola virus, family Filoviridae. The disease encompasses a range of symptoms including vomiting, diarrhea, changes in skin color, general body pain, internal and external bleeding, and fever. The virus is named after the Ebola River in the African country of the Democratic Republic of the Congo (formerly Zaïre), near the site of the first outbreaks.

**Echo sounder:** see ***sounder***.

**Eddy:** a portion of a fluid within a larger fluid mass that has its own structure and life history; ocean eddies frequently form on the side of, or within, a main current as a result of a disturbance or encountering an obstacle. Ocean eddies often exhibit a circular flow.

**Elastic waves:** a disturbance that propagates through, or on the surface of, a medium without causing permanent deformation of any point in the medium.

**Electromagnetic band:** a range of wavelengths forming part of the ***electromagnetic spectrum***.

**Electromagnetic pulse:** a short duration burst of ***electromagnetic radiation***.

**Electromagnetic radiation:** a self-propagating wave characterized by electric and magnetic elements.

**Electromagnetic spectrum:** the range of wavelengths or frequencies associated with ***electromagnetic radiation***. Most Earth-observing satellite sensors operate in the wavelength range from visible radiation through ***infrared*** radiation to ***microwave radiation***.

**Electromagnetic waves:** waves of ***electromagnetic radiation*** formed by the coupling of an electric field with a magnetic field.

**Electron microscope:** an instrument that uses a beam of highly energetic electrons to examine objects on a very fine scale.

**El Niño:** an anomalous warming of ocean surface waters in the eastern tropical Pacific, generally brought on by the suppression of upwelling off the coasts of Ecuador and northern Peru, and along the Equator east of the International Dateline. El Niño conditions typically last for 12 to 18 months and occur every 3 to 7 years. El Niño is accompanied by weather extremes in various parts of the world.

**El Niño-Southern Oscillation (ENSO):** the term for the coupled ocean-atmosphere interactions in the tropical Pacific characterized by episodes of anomalously high sea surface temperatures in the equatorial and tropical eastern Pacific. It is associated with large swings in surface air pressure between the eastern and western tropical Pacific and is the most prominent source of interannual variability in weather and climate around the world.

**Entraining:** the mixing of surrounding or ambient air into a pre-existing organized air current so that the surrounding air becomes part of the current; the opposite of ***detraining***.

**Euphotic zone:** the upper part of the ocean that receives incoming sunlight.

**Exosphere:** the outermost layer of the atmosphere. The air density in the exosphere is extremely low, and in fact the layer's lower boundary, generally at an altitude of 500–1,000 km, is defined based on the probability that an upward moving particle would collide with another particle before escaping out of the atmosphere altogether.

# F

**Fault:** a crack or fracture in the Earth's surface along which earthquakes can occur. ***Magma*** from the Earth's interior can rise up through faults.

**Fault zone:** a region on the Earth's surface that experiences the direct effects of ***seismic events*** related to ***tectonic*** activity.

**Field of view:** the area that a sensor views at a single moment in time.

**Fire regime:** the characterization of fires by a typical frequency, intensity, seasonal timing, and type.

**First-year sea ice:** sea ice that has not (yet) survived a summer melt season.

**Flux:** the flow of a quantity, such as energy, mass, or momentum, per unit area per unit time.

**Fossil fuel:** any ***hydrocarbon*** deposit that can be burned for heat or power, such as petroleum, coal, and natural gas.

**Frazil ice:** a collection of loose, randomly oriented needle-shaped ice crystals in water. Typically an early stage in the development of ***sea ice***.

# G

**Geodesy**: the study of the Earth's gravitational field and related phenomena, such as Earth tides.

**Geoid:** a surface that on average coincides with the mean ocean surface, and over which the force of gravity is the same.

**Geographical information system:** a system for creating and managing spatial data and associated attributes. In the strictest sense, it is a computer-based system capable of integrating, storing, editing, analyzing, and displaying geographically referenced information.

**Geosphere:** the solid Earth, including continental and oceanic crust.

**Geo-stationary satellite:** a satellite in an orbit around the Earth that keeps it continuously above the same location on the Earth's surface; this location is necessarily somewhere along the equator.

**Germination:** the emergence of growth from a resting stage, such as the sprouting of a seedling from a flowering plant's seed.

**Glacier:** a large mass of ice that flows under the influence of gravity. Typically a glacier refers to a body of ice that is smaller than an ***ice cap***.

**Global conveyor belt:** the global oceanic set of currents and ***thermohaline circulation*** that connects the ocean's surface and deep waters, transporting heat and salt on a planetary scale.

**Global Positioning System:** a satellite navigation system used for determining precise location and providing a highly accurate time reference almost anywhere on Earth or in Earth orbit.

**Greenhouse effect:** the heating of the Earth's surface and lower atmosphere as a consequence of differences in atmospheric transparency to ***electromagnetic radiation***. The atmosphere is nearly transparent to incoming ***solar radiation***, but much less transparent to outgoing ***infrared radiation***. Terrestrial ***infrared radiation*** is absorbed and radiated principally by ***water vapor*** and, to a lesser extent, by ***carbon dioxide*** and other ***trace gases***, thereby slowing the loss of heat to space by the Earth-atmosphere system and significantly elevating the average temperature of the Earth's surface over what it would be without the atmosphere's trace gases.

**Greenhouse gas:** those gases, such as ***water vapor***, ***carbon dioxide***, tropospheric ***ozone***, ***nitrous oxide***, ***methane***, and ***chlorofluorocarbons***, that are largely transparent to ***solar radiation***, but opaque to outgoing ***longwave radiation***. Their action is similar to that of glass in a greenhouse. Most of the incoming solar radiation is allowed through to the Earth's surface, but some of the outgoing ***longwave (infrared) radiation*** is absorbed and re-emitted by the greenhouse gases. The effect of this is to warm the surface and lower atmosphere of the Earth.

**Ground track:** the trajectory of a satellite projected onto the Earth directly beneath the orbit.

**Groundwater:** water found beneath the surface of the Earth, in contrast to ***runoff***, which is found at the Earth's surface.

**Gyre:** see ***vortex***.

# H

**High-pressure system:** an atmospheric mass with relatively high pressure compared to its surroundings. There is downward movement of the air in the center of a high-pressure system. The motion around a high-pressure system is ***anti-cyclonic***.

**Horse latitudes:** a nautical term describing the subtropical latitudes between 30° and 35° both north and south of the Equator, where the winds are light and the weather is hot and dry, caused by descending air.

**Hydrocarbon:** a chemical compound consisting only of carbon and hydrogen. Hydrocarbons are combustible and are the main components of ***fossil fuels***.

**Hydrologic cycle:** the circulation of water through the Earth's ***hydrosphere***, driven by the energy from ***solar radiation***.

**Hydrology:** the study of the movement, distribution, and quality of water; used most often in the context of water on land surfaces.

**Hydrosphere:** the component of the Earth system that includes water in all three phases (ice, liquid, and vapor).

**Hydrothermal:** heat-driven transport of water, such as that caused by the opening of vents in the ocean floor during ***tectonic*** activity.

# I

**Iceberg:** a mass of fresh-water ice originally derived from a ***glacier***, ***ice cap***, or ***ice sheet*** and found in a body of liquid water, such as an ocean or lake.

**Ice cap:** a dome-shaped grounded ice mass that covers less than 50,000 $km^2$ of land area (i.e., smaller than an ***ice sheet***), but is in general sufficiently large to cover mountains rather than to be constrained by them.

**Ice concentration:** the percent areal coverage of ***sea ice*** over an ocean area; generally expressed as a percentage or in tenths or eighths.

**Ice edge:** the location where ice meets open water.

**Ice floe:** a body of sea ice that has a degree of rigidity and is identifiably distinct from neighboring ice.

**Ice pack:** ***sea ice*** that is closely packed and forms a near-continuous cover of the ocean or lake surface (i.e., having high ***ice concentration***).

**Ice sheet:** a dome-shaped ice mass that covers more than 50,000 $km^2$ of land area. The two remaining ice sheets on Earth today are the Antarctic and Greenland ice sheets. The Antarctic ice sheet is sometimes divided into the West Antarctic and East Antarctic ice sheets, the former being a ***marine ice sheet***, grounded largely below sea level.

**Ice shelf:** a part of an ***ice cap*** or ***ice sheet*** that is floating on liquid water (generally the ocean).

**Ice stream:** a fast moving 'river' of ice within an ***ice sheet***; an ice stream is sometimes separated from neighboring slower-moving ice by ***crevasses***.

**Ice tongue:** part of a ***glacier*** that extends onto the ocean and hence is floating.

**Igneous rock:** rock formed when ***magma*** cools and solidifies either below the surface as intrusive rocks or on the surface as ***lava***.

**Ignimbrite rock:** rock formed by the deposition and consolidation of ash flows and ***pyroclastic flows***.

**Infrared radiation:** the portion of the ***electromagnetic spectrum*** with wavelengths between approximately 0.75 μm and 1 mm.

**Insulator:** a material that reduces the ***flux*** of heat, sound and/or ***electromagnetic radiation***.

**Interferometry:** see ***synthetic aperture radar interferometry***.

**Interseismic period:** the relatively quiescent period between ***seismic events***.

**Intertropical Convergence Zone:** a narrow, discontinuous belt of ***convection*** clouds and thunderstorms running approximately parallel to the equator and marking the convergence of the trade winds of the two hemispheres. This zone moves north and south with the seasons.

**Ion:** an atom or group of atoms that has gained or lost electrons, thereby becoming negatively or positively charged, respectively.

**Ionization:** the process of converting an atom or molecule into an ***ion*** by removing the balance between protons and electrons.

**Ionosphere:** the rarefied region of the Earth's atmosphere at altitudes of approximately 60–400 km, where the Sun's energy ensures that particles can remain charged (in the form of ***ions***).

**Irradiance:** the total incident power of ***electromagnetic radiation*** over the complete ***electromagnetic spectrum*** expressed per unit area.

## K

**Katabatic wind:** a wind that blows down a topographic incline such as a hill, mountain, or ***glacier***. Katabatic winds are created by the impact of gravity on relatively cold, dense air.

## L

**Lahar:** a torrential gravity-driven flow of water-saturated debris, or mudflow, down the slope of a volcano.

**La Niña:** an episode of strong trade winds and resulting strong upwelling and unusually low sea-surface temperatures in the central and eastern areas of the tropical Pacific. These episodes are linked to particular sets of climate conditions elsewhere around the Earth. La Niña is essentially the opposite of ***El Niño***.

**Laser:** an acronym for 'light amplification by stimulated emission of radiation'; it refers to optical ***electromagnetic radiation*** that is transmitted ***coherently***.

**Latent heat:** the energy that is used to change the phase of a physical state of a material (e.g., liquid, gaseous), but not the temperature. Hence the term 'latent', meaning 'hidden'. Latent heat contrasts with ***sensible heat***.

**Lava:** molten rock (***magma***) on the Earth's surface.

**Leaching:** the process of extracting a substance from a solid material, such as soil, by dissolving it in a liquid, such as water.

**Leads:** waterways between floes in a ***sea ice*** cover; leads allow significant heat and moisture exchanges between the ocean and atmosphere, with especially large heat fluxes from the ocean to the atmosphere in winter.

**Lithosphere:** the solid, outermost shell of the Earth, including the uppermost ***mantle***. Its thickness varies from about 50 km (under deep ocean) to 150 km (under the continents).

**Little Ice Age:** a relatively cool interval from about AD 1400 to AD 1890, when average temperatures were lower over many areas of the Earth, especially in Europe, and when many mountain glaciers advanced. The Little Ice Age followed the Medieval Warm Period.

**Longwave radiation:** ***radiation*** with wavelengths greater than 4 μm; often used to define the ***radiation*** emitted from Earth into space.

**Low-pressure system:** an atmospheric mass with relatively low pressure compared to its surroundings. There is upward movement of the air in the center of a low-pressure system. The motion around a low-pressure system is ***cyclonic***.

## M

**Magma:** molten rock originating from beneath the surface of the Earth.

**Magma chamber:** a large underground pool of ***magma***.

**Magnetic pole:** each of two points where the magnetic field lines around the Earth descend vertically into the surface of the Earth, at high latitudes in the Northern and in the Southern Hemispheres, respectively. The poles vary in position over time due to drift in the Earth's magnetic field.

**Mantle:** the layer of the Earth's interior between the core and the crust, ranging from depths of approximately 40–2,900 km.

**Marine ice sheet:** an ***ice sheet*** that is largely grounded below sea level, with floating margins around at least a portion of its edge. The West Antarctic ***ice sheet*** is a marine ice sheet.

**Maya civilization**: a Mesoamerican civilization that reached its peak of development between AD 250 and AD 900. It was noted for the sophistication of its culture, which included written language, art, monumental architecture, and mathematical and astronomical systems.

**Mega-city:** a city with a population in excess of 10 million.

**Megaton:** a unit of mass equivalent to one million metric tons, or $10^9$ kg.

**Melt pond:** a mass of liquid water on top of ice, formed from melting.

**Meningitis:** an illness involving inflammation of membranes that surround the central nervous system, caused by bacteria or viral infections elsewhere in the body that have spread into the blood. Symptoms include fever, vomiting, stiff neck, and headaches.

**Mesosphere:** a region of the atmosphere which extends from immediately above the ***stratosphere***, at about 50 km altitude, to the thermosphere, at about 85 km altitude.

**Metamorphic rock:** one of three major rock categories, resulting from the transformation of either of the other two rock types (***sedimentary*** or ***igneous***) when these are subjected to sufficient heat and/or pressure to cause the rock to change its physical and/or chemical form. The process is called 'metamorphism', which means 'change in form'.

**Methane:** a chemical compound with formula $CH_4$. It is the principal component of natural gas and is an important ***greenhouse gas***.

**Microphysical:** relating to molecular, atomic, and smaller processes.

**Microwave radiation:** ***electromagnetic radiation*** with wavelengths between about 1 mm and 1 m.

**Mixed layer:** a layer within the ocean or atmosphere which is fairly uniform in terms of density and other properties. In the context of the ocean, this is the uppermost layer maintained by mixing due mainly to the penetration of energy from wind stress at the surface. The depth of the ocean mixed layer varies widely, due to different wind and other conditions, but is generally between 25 and 200 m. In the context of the lower atmosphere, the mixed layer is a result of ***convection***.

**Montreal Protocol:** a major international agreement signed in September 1987, to limit further production of chemicals that contribute to the depletion of the stratospheric ***ozone*** layer.

**Moraine:** rock debris left after erosion, transportation, and deposition by ***glacier*** ice.

**Multi-spectral:** observations at multiple wavelengths of the ***electromagnetic spectrum***, such as ***ultraviolet***, visible, and ***thermal infrared*** bands.

**Multi-year sea ice:** sea ice that has survived at least one summer of melt. Second-year sea ice, which has survived only one summer, is sometimes distinguished from older multi-year sea ice.

# N

**Nadir:** satellite sub-point located directly below a satellite on the Earth's surface.

**Near-infrared:** the part of the ***electromagnetic spectrum*** ranging from 0.75 to 1.4 $\mu$m in wavelength.

**Negative feedback:** a response of a system or object to an event that serves to dampen the event itself, or dampen its impact. Thus, negative feedback can be considered to be a 'stabilizing' process. It contrasts with ***positive feedback***.

**Net primary production:** the mass of organic, carbon-based material produced from ***carbon dioxide*** by organisms, minus that which is lost during respiration by the organisms; a measure of biological productivity and net carbon fixation. Measured per unit area and per unit time, e.g., kg of carbon per $m^2$ per year.

**Net radiation:** the difference between the incoming and outgoing ***radiation***. Between 40°N and 40°S the Earth experiences a net gain of energy. Poleward of 40°N and 40°S there is a net loss of ***radiation***. There is therefore a transfer of energy from the tropical and sub-tropical regions to the cooler regions to compensate for this, via ocean currents and the atmosphere.

**Nilas:** a thin elastic sheet of ***sea ice***, thin enough to bend with ocean waves and less than about 10 cm in thickness. It is one of many types of young ***sea ice***.

**Nitrogen oxide:** oxygen compounds of nitrogen that form a component of air pollutants produced by automobile manufacture, smoke stacks, and other industrial emissions.

**Nitrous oxide:** a chemical compound with the formula $N_2O$. In the atmosphere, it is a powerful ***greenhouse gas***.

**Normalized difference vegetation index:** a measure of the productivity and health of vegetation calculated from the visible and ***near-infrared*** light reflected by vegetation. This uses the property that green leaves normally have greater ***reflectance*** in the ***near-infrared*** than in the visible range of the ***electromagnetic spectrum***.

**Northerly winds:** winds that derive from the north.

**Northwest Passage:** a sea route through the Arctic Archipelago of Canada. From the end of the fifteenth century, explorers hoped that such a passage would be revealed as a commercial sea route to the Pacific from the Atlantic.

**North Atlantic Oscillation:** a 'see-saw' variation in air pressure between Iceland (the Icelandic ***low-pressure system***) and the Azores (the Bermuda-Azores subtropical ***high-pressure system***). The North Atlantic Oscillation is linked to the climate of eastern North America and much of Europe and North Africa.

**North Atlantic Oscillation index:** a measure of the state of the ***North Atlantic Oscillation***. During some years, the pressure difference is relatively large, leading to a positive index value, and during other years the difference is relatively small, leading to a negative index value. During the latter years of the twentieth century, the index was more often positive than negative.

**Nutrient:** any substance assimilated by living organisms that can be used to provide energy or to promote growth.

# O

**Optical:** relating to visible, ***infrared*** and ***ultraviolet radiation*** (or light).

**Optical thickness:** a dimensionless quantity that measures the cumulative depletion of a beam of ***radiation*** that would be experienced in passing through a layer directed straight downward; the optical thickness thus depends on the physical constitution (***aerosol*** particles, cloud crystals, drops, and/or droplets), the form, the concentration, and the vertical extent of the scattering and absorbing medium, and the wavelength. In quantitative terms, an optical thickness of 1.0 corresponds to the reduction of a beam of light to about 37% of the incident light at the top of the layer.

**Organic compounds:** any member of a large class of chemical compounds whose molecules contain carbon and hydrogen.

**Outlet glacier:** a valley ***glacier*** that drains an ***ice sheet*** or ***ice cap***.

**Overturning:** the vertical mixing of water in the ocean, which is an important characteristic of ***thermohaline circulation***.

**Oxidant:** a substance that causes ***oxidation***.

**Oxidation:** the loss of one or more electrons by an atom, ion, or molecule during a chemical reaction. At a simple level, it may be regarded as a chemical reaction involving the addition of oxygen to, or loss of hydrogen from, a compound.

**Ozone:** a ***trace gas*** made up of three atoms of oxygen, with formula $O_3$. In the ***stratosphere***, it occurs naturally and provides a protective layer shielding the Earth from ***ultraviolet radiation*** and its otherwise harmful health effects on humans and the environment. In the ***troposphere***, it is a chemical ***oxidant*** and major component of ***photochemical smog***. Ozone is an effective ***greenhouse gas***.

**Ozone hole:** a reduction in ***stratospheric ozone*** that, in recent decades, has occurred during spring over the Antarctic. Chlorine-containing molecules are formed during the dark winter on the surface of small ***polar stratospheric cloud*** particles that can only form in the intense cold of the polar winter. When the sunlight returns in spring, these chlorine-containing molecules release reactive chlorine that destroys the ozone. The chlorine is derived from ***chlorofluorocarbons*** that have previously been broken up in sunlit regions.

# P

**Paleo-spreading direction:** the direction of spreading of the sea floor, often inferred from the age of fossils that appear in the sea bed.

**Panchromatic:** a sensor that records data across a single, wide range of the ***electromagnetic spectrum***, normally the visible part, in contrast to ***multispectral***.

**Passive microwave radiation:** ***microwave radiation*** emitted naturally by various surfaces and targets.

**Passive microwave instrument:** an instrument that measures ***microwave radiation*** using passive means, meaning that, rather than sending a signal out, it receives ***radiation*** coming to it naturally.

**Pathogenic organism:** organisms, including bacteria and viruses, capable of causing disease in their hosts (e.g., humans).

**Perennial sea ice:** see ***multi-year sea ice***.

**Permafrost:** ground at, or close to, the Earth's surface that has been frozen for two or more years.

**Phase:** the phase of ***electromagnetic radiation*** is the position of an ***electromagnetic wave*** within its (repeating) cycle, measured in angular terms. The phase associated with a complete cycle, where the wave repeats itself, is 360°. ***Electromagnetic radiation*** is ***coherent*** when the phase is not random, so that constructive and destructive interference of the waves occurs. Alternatively, phase refers to the ***thermodynamic*** state of water, such as liquid water, ***water vapor***, or ice.

**Phenology:** the study of relationships between recurring natural phenomena related to living organisms, such as the appearance of leaves, animal migrating patterns, etc., and those factors that influence the timing, such as climate.

**Photo micrography:** the practice of photographing very small, e.g., microscopic, objects.

**Photochemical smog:** air pollution associated with industrial activities and motor vehicles that can contain, among other constituents, ***volatile organic compounds*** and ***tropospheric ozone***.

**Photosynthesis:** the synthesis of sugar, which plants use for energy, from light, ***carbon dioxide*** and water, with oxygen as a waste product. In plants, and some algae, the conversion of sunlight into chemical energy is achieved by ***chlorophyll***.

**Phytoplankton:** that portion of the ***plankton*** community made up of microscopic plants that live in the ocean (e.g., algae and diatoms) and provide their energy from ***photosynthesis***. See also ***zooplankton***.

**Pixel:** a 'picture element,' corresponding to the basic building block of the image. In a digital image, the pixel is the smallest sample of an image and is characterized by an area and position.

**Planetary albedo:** the fraction of incident ***solar radiation*** that is reflected by a planet and returned to space. The planetary ***albedo*** of the Earth varies with space and time, and depends on clouds in the atmosphere and the underlying ground surface. The planetary albedo of the Earth as a whole is approximately 31%, most of which is due to backscatter from clouds.

**Planetary waves:** long atmospheric and ocean waves that girdle the planet. Atmospheric planetary waves are found in the ***stratosphere***.

**Plankton:** organisms that drift, mainly passively, in the ocean. Plankton can be sub-divided into ***phytoplankton*** and ***zooplankton***. Many, but not all, are microscopic in size.

**Plant transpiration:** see ***transpiration***.

**Plate:** see ***tectonic plate***.

**Plate boundary:** narrow zones along the margins of ***tectonic plates*** where the results of plate-tectonic forces are most evident, as in active fault movements.

**Plate tectonics:** the movement of ***tectonic plates*** of the ***lithosphere*** that float on the underlying, less rigid part of the Earth (called the asthenosphere). There are seven major tectonic plates and many minor plates. These plates move in relation to one another and ***tectonic*** activity is concentrated around the ***plate boundaries***.

**Polar convergence:** the region where sub-polar and polar water masses meet in the ocean.

**Polar-orbiting satellite:** satellites in relatively low-altitude orbits that pass near the north and south geographical poles. The Earth rotates through the plane of the satellite's orbit, which is typically at an altitude of about 800 to 1,000 km.

**Polar stratospheric clouds:** clouds that are found in the winter polar ***stratosphere*** at altitudes of 15,000 to 25,000 m. Stratospheric chlorine, which originates largely from industrial processes, is converted on the surface of these clouds to forms that are highly reactive with ***ozone***. Thus, polar stratospheric clouds are implicated in the formation of ***ozone holes***.

**Polar vortex:** persistent, large-scale ***cyclonic*** atmospheric circulations located near the Earth's poles, in the middle and upper ***troposphere***, and in the ***stratosphere***. They are strongest in intensity during winter.

**Polynya:** an area of open water within the ***ice pack***.

**Positive feedback:** a response of a system or object to an event that serves to enhance the event itself, or enhance its impact. Thus, positive feedback can be considered to be an 'amplifying' process. It contrasts with ***negative feedback***.

**Precipitation:** any form of water in liquid or solid form that is delivered to the surface of the Earth, including rain, sleet, hail, or snow.

**Primary productivity:** the mass of organic, carbon-based material produced by organisms from ***carbon dioxide***, measured as the rate of carbon fixation. On land, almost all primary production is carried out by plants. In the ocean, almost all primary production is carried out by algae. It is measured per unit area and per unit time, e.g., kg of carbon per $m^2$ per year.

**Proxy indicator:** a measurement that is well correlated to, and hence serves as an approximation for, another measurement that is itself difficult to obtain directly. An example would be the width of tree rings serving as an indicator of past temperatures.

**Pyroclastic:** pertaining to fragmented rock material, such as volcanic fragments, crystals, ash, pumice, and glass shards, emanating from a volcanic explosion or a volcanic vent.

**Pyroclastic flow:** a turbulent mixture of hot gases and unsorted ***pyroclastic*** material that can move in a hot, fluid form at high speed (80 to 160 km/hr). The term also can refer to the resulting deposit.

# R

**Radar:** a device or system consisting usually of a synchronized transmitter and receiver that emits radio waves (***microwaves***) and processes their reflections for display or analysis. The term is derived from the expression 'Radio Detection and Ranging.'

**Radar interferometry:** see ***Synthetic Aperture Radar interferometry***.

**Radar interferogram:** an image of interference patterns generated from two ***Synthetic Aperture Radar*** images, used in the process of ***Synthetic Aperture Radar interferometry***.

**Radiance:** the total amount of ***electromagnetic*** energy radiating in a particular direction. The units are W/m per steradian.

**Radiant energy:** the energy associated with, or transported by, ***electromagnetic waves***. See also ***radiance***.

**Radiation:** see ***electromagnetic radiation***.

**Radiogenic heat:** the heat generated by the decay of radioactive isotopes within the Earth.

**Radiometer:** an instrument that is designed to accurately measure ***radiant energy***.

**Rainforest:** forests growing under the influence of annual ***precipitation*** that exceeds 2 m.

**Reflectance:** the ratio of reflected power to incident power.

**Reflectivity:** the intrinsic ***reflectance*** of a material, for example from a material sufficiently thick that the ***reflectance*** does not change with increasing thickness.

**Relative humidity:** a measure of the mass of moisture in the air compared to the maximum amount that the air could hold at the given temperature, expressed as a percentage.

**Remote sensing:** the use of a sensor to detect and characterize a target at a distance. Earth remote sensing typically involves the use of satellites or aircraft, but ground-based remote sensing is also used to infer properties of the Earth's atmosphere by looking upward from the ground.

**Resolution:** a measurement of the level of detail that can be observed using a sensor, defined as the minimum distance of separation between two targets at which they can be distinguished.

**Rift:** in the context of geology, this is a narrow region where the Earth's crust and ***lithosphere*** are moving apart.

**Rise:** in the context of oceanography, this is a peak in topography beneath the ocean (as in 'Maud Rise').

**Roaring forties:** the name given to the latitudes between 40°S and 50°S, so called because of the strong ***westerly winds***.

**Running mean:** a form of smoothing data, where the smoothing is carried out by averaging. If there are 100 points in sequence of time, and the running mean involves averaging 10 points, then the average is calculated for the points 1 to 10 inclusive, then for points 2 to 11, then points 3 to 12, etc.

**Runoff:** ***precipitation*** that does not enter the ground but flows over the land surface.

## S

**Sahel:** the semi-arid transition zone in Africa between the Sahara Desert to the north and the more fertile savanna region to the south.

**Santa Ana winds:** warm, dry winds in southern California that blow out of the desert towards the coast in autumn and early winter. They are a result of the development of high pressure over Nevada and adjacent areas.

**Scan:** as the satellite moves along its orbit, an imaging sensor achieves coverage of the surface in the direction perpendicular to the ***ground track*** by the use of a scan. The scan may, for example, be achieved by the use of mirrors that rotate to obtain light from different positions along the scan, or may be achieved by the use of a set of reception devices each directed towards different positions along the scan. The distance associated with the scan defines the image ***swath***.

**Sea ice:** any ice formed from the freezing of seawater.

**Seasonal sea ice:** see ***first-year sea ice***.

**Sedimentary rock:** one of the three main rock groups (along with ***igneous*** and ***metamorphic*** rocks). Sedimentary rock is mainly formed by compaction of the weathered remains of other rocks along with the results of biological activity, resulting in fossils.

**Seismic:** relating to an earthquake or Earth vibration.

**Seismic event:** a movement in the Earth's crust resulting in ***fault*** disturbance.

**Seismic wave:** a wave that travels through the Earth, or along its surface, in response to an earthquake or other Earth movement.

**Sensible heat:** heat energy that is absorbed or transported as a result of variations in temperature, via conduction, convection, or both, and not associated with a change of state (e.g., gas, liquid, solid). Sensible heat contrasts with ***latent heat***.

**Sensible heat polynya:** a ***polynya*** that is maintained by the transport of heat from depth in the ocean to the surface.

**Shear margin:** in the context of ***ice sheets*** or ***ice caps***, a shear margin is found along the edge of fast-moving ***glaciers*** and ***ice streams***. Shear is the result of stress that builds up between the fast-moving ice and the neighboring slow-moving ice. Often associated with areas of major ***crevasses***.

**Shortwave radiation:** ***radiation*** in the visible, ***near-infrared*** and near-***ultraviolet*** parts of the ***electromagnetic spectrum***, less than about 4 $\mu$m in wavelength.

**Slash and burn:** an agricultural practice used in forested areas, involving the clearance of an area of forest (with the possible exception of large trees that may be killed standing). The timber is used for firewood, charcoal or construction. After a few weeks the, by now, dry vegetation is burned. Plots are then cultivated. The number of seasons that the land may support cultivation varies depending on location, but normally after a few seasons, fertility declines and the land is abandoned. Slash and burn may then continue elsewhere.

**Snow megadunes:** large dune-like structures of snow found on the Antarctic ***ice sheet***. They typically have heights of a few meters and are separated by a few kilometers.

**Solar constant:** the amount of ***solar radiation*** reaching the Earth from the Sun at a vertical angle, which is approximately 1,370 W/m. It is not, in fact, truly constant and variations are detectable.

**Solar radiation:** ***electromagnetic radiation*** emitted by the Sun, about half of which is in the visible part of the ***electromagnetic spectrum***.

**Sounder:** a device that measures the vertical properties of a medium, such as the atmosphere or ice, by sending sound or ***electromagnetic radiation*** into the medium and measuring the return signal, or by detecting the thermal emission of a medium at various infrared frequencies.

**Southerly winds:** winds that derive from the south.

**Spectral reflectance:** the ***reflectance*** measured at different positions along the ***electromagnetic spectrum***.

**Speed of light:** the speed of light in a vacuum is 299,793,458 m/sec.

**Stick-slip:** intermittent and jerky sliding between two surfaces in contact and subject to a sliding force.

**Storm surge:** a greater than normal movement of water onto the coast associated with forcing from strong winds around an intense ***low-pressure*** weather system. A storm surge may result in coastal flooding.

**Stratification:** the building up of layers in a material or fluid, for example relating to bodies of water in the ocean, air masses in the atmosphere, or sediment deposition. Stratification implies layering in which the layers have distinct characteristics.

**Stratiform:** a layer form. In the context of the atmosphere, it is used to indicate clouds with extensive horizontal development as opposed to vertical development.

**Stratocumulus:** large dark, rounded cloud masses, usually in groups, lines, or waves, and usually located below 2,400 m in altitude.

**Stratosphere:** the layer of the atmosphere immediately above the ***troposphere***, extending from a height of between 10 and 17 km, to a height of about 50 km. The stratosphere is characterized by having temperature profiles in which the temperatures increase or stay constant with increasing height. The ***ozone*** layer is found here.

**Subduction:** the process in which one ***tectonic plate*** is pushed downward beneath another ***tectonic plate***.

**Succession:** in the context of ecosystems, succession is the generally predictable series of changes in an ecosystem over time after a disturbance, such as a fire or hurricane.

**Sulfur dioxide:** a chemical compound with the formula $SO_2$, produced by volcanos and various industrial processes.

**Sun glint:** the direct reflection of incoming sunlight from the ocean surface back to an observer or measuring device.

**Sun-synchronous:** a satellite orbit that observes any given location on the Earth at the same local solar time each time it is observed, so that there is similar solar illumination geometry associated with each pass.

**Surface mixed layer:** the layer between the ocean surface and a depth usually ranging between 25 and 200 m, throughout which the density is about the same as at the surface.

**Swath:** the area on the Earth's surface that is imaged by a satellite sensor as it moves along its orbit. The swath follows the direction of the satellite projected onto the Earth's surface, either below or to one side of the satellite. For sensors that ***scan***, the width of the swath is determined by a single ***scan***.

**Symbiotic:** an interaction between two organisms that is beneficial to one or both, and harmful to neither.

**Synoptic:** a comprehensive and wide coverage view of conditions associated with a single point in time. This term is often used in the context of describing atmospheric conditions.

**Synthetic Aperture Radar:** a sensor that transmits ***microwave radiation*** to the Earth's surface and processes the reflected ***radiation*** to provide a detailed ***radar*** image of the surface. The term 'synthetic aperture' refers to the use of the motion of the satellite (or aircraft) to create the effect of a long antenna that in turn provides the focus needed to image the surface in detail in that direction. The synthetic aperture radar is a ***coherent*** measurement system. As well as recording the amplitude of the ***radar backscatter***, it records the ***phase***.

**Synthetic Aperture Radar interferometry:** a technique that uses sets of ***synthetic aperture radar*** images to measure with great precision surface topography and displacement, for example related to earthquakes. The technique works by using interference between the images, which is possible because ***synthetic aperture radar*** is a ***coherent*** imaging system. The ***phase*** information retrieved from interference of the ***radiation*** from pairs of images, together with precise knowledge of the positions of the satellite at the times associated with the imaging, enables the position of each image pixel to be resolved in three dimensions.

# T

**Tectonics:** see ***Plate tectonics***.

**Tectonic plate:** a region of the Earth's crust or ***lithosphere***, that floats on fluid-like material underneath. There are 7 major plates and many more minor plates, with thicknesses of approximately 100 km. Each plate forms the surface stage of a process of vertical ***convection*** in which material from beneath the ***lithosphere*** is brought to the surface at divergent boundaries between the plates and returned to beneath the surface at convergent boundaries. ***Seismic events*** are most common in regions close to the boundaries between the tectonic plates.

**Tectonic stress:** stress in the Earth's surface related to the movement of ***tectonic plates***.

**Tephra:** fragments of volcanic rock or ***lava*** that are blasted into the air by volcanic explosions or eruptions, ranging in size from less than 2 mm (ash) to more than 1 m in diameter.

**Tendril:** a specialized component of a plant (e.g., a stem, leaf or stalk) with a threadlike shape that is used for support or attachment to other objects.

**Terminal lake:** a lake with inflow of water, but no outflow.

**Thermal infrared:** the portion of the ***electromagnetic spectrum*** that includes wavelengths between 3.5 and 20 $\mu$m.

**Thermal radiation:** the radiation emitted from a body as a result of its temperature.

**Thermodynamic:** changes to a material or system that relate to the movement of energy, or heat. The Laws of Thermodynamics describe how energy can be exchanged between physical systems as heat or work.

**Thermohaline circulation:** the vertical movements of currents within the oceans driven by density variations. The term is derived from *thermo-* for heat and *-haline* for salt, which together determine the density of sea water. Key elements in this process include the formation and melting of ***sea ice*** that results in the exchange of salt with the ocean surface waters, and the cooling of surface water transported from the equatorial regions via currents such as the Gulf Stream. Thermohaline circulation is global in its impact and hence is linked to climate.

**Thermosphere:** the outermost layer of the atmosphere, above the ***mesosphere***. The thermosphere extends upward from the top of the mesosphere, at approximately 85 km, and has temperatures generally increasing with height. The thermosphere includes most of the ***ionosphere*** and all of the ***exosphere***.

**Total ozone:** the total amount of ***ozone*** in a vertical column from the surface to the top of the atmosphere.

**Toxin:** a substance poisonous to humans or other animals and produced by living cells or organisms.

**Trace gas:** a minor constituent of the atmosphere, normally considered to be less than 1% in terms of total mass. The most important trace gases and compounds contributing to the ***greenhouse effect*** are ***water vapor***, ***methane***, ***ozone***, ***carbon dioxide***, ***nitrous oxide***, and ***chlorofluorocarbons***.

**Trade winds:** winds that flow from east to west, and towards the Equator, in tropical and sub-tropical regions.

**Transform fault:** ***faults*** that are associated with two ***tectonic plates*** sliding against each other. An example is the San Andreas fault.

**Transpiration:** the evaporation of water directly from plants to the atmosphere.

**Trichodesmium:** a genus of ***cyanobacteria***, often called sea sawdust and first described by Captain James Cook in Australia, that fixes nitrogen gas and converts it to ammonium.

**Tropical cyclone:** tropical ***low-pressure systems*** that are given their high energy and intensity by the heat released when moist air rises and the water vapor therein condenses. Their energy is such that they can generate particularly strong winds.

**Tropopause:** the boundary between the upper ***troposphere*** and the lower ***stratosphere*** that varies in altitude between approximately 7 km at the poles and 17 km at the Equator.

**Troposphere:** the lowest atmospheric layer, between the surface and the ***tropopause***.

**Tsunami:** a great sea wave produced by abrupt submarine events such as earthquakes, volcanic eruptions, or large landslides. Over the deep ocean, a tsunami can travel at 700 km / hr, or more.

**Turbulent:** flow of a fluid (e.g., air or water) that has chaotic characteristics.

**Typhoon:** a ***tropical cyclone*** with winds that are sustained at 33 m/sec or more.

## U

**Ultraviolet radiation:** ***electromagnetic radiation*** in the wavelength band 0.01 to 0.38 μm. In the ***electromagnetic spectrum***, ultraviolet radiation falls between X-ray radiation and visible radiation.

**Upwelling:** the vertical motion of water in the ocean by which deeper water of lower temperature and greater density moves toward the surface of the ocean.

**Urban boundary layer:** the layer of air impacted directly by an urban environment and marked at its upper boundary by a sharp discontinuity in atmospheric properties and conditions.

**Urban canopy layer:** the layer of air in the urban area beneath the mean height of the buildings and other significant structures, including trees.

**Urban heat island:** a built up area such as a city that is significantly warmer than its surroundings.

## V

**Vector-borne disease:** a disease in which the ***pathogenic organism*** is transmitted from an infected individual to another individual via insects and other ***arthropods***, sometimes with other animals involved as part of the chain of transmission.

**Ventilation:** the replacement of existing fluid or gas by fresh fluid or gas, respectively; in the context of the ocean, ventilation is the process by which water is transferred from the surface mixed layer to the interior ocean.

**Vertical mixing:** the process of mixing two fluids in the vertical direction, for example through ***convection***.

**Volatile organic compound:** chemicals containing carbon that can easily vaporize and enter the atmosphere. Volatile ***organic compounds*** include a very wide range of individual substances, such as ***hydrocarbons***. Volatile organic compounds are pollutants and some are ***greenhouse gases***.

**Vortex:** a rotational movement in a fluid or gas, which can be present in the atmosphere or ocean.

## W

**Water-leaving radiance:** light that radiates upward from the sea surface, due to the scattering of sunlight incident on the ocean.

**Water vapor:** water present in the atmosphere in gaseous form.

**Watershed:** the region of land from which water drains into a specified body of water, often, but not always, the ocean.

**Westerly winds:** winds that derive from the west.

**Wind vector:** data that indicate both wind direction and speed, for example indicated by an arrow on a map.

## Z

**Zooplankton:** invertebrate ***plankton*** that feed on other plankton. See also ***phytoplankton***.

# Appendix 3: Acronyms

| | |
|---|---|
| **ADEOS** | Advanced Earth Observing Satellite (a Japanese satellite series) |
| **AMSR-E** | Advanced Microwave Scanning Radiometer for EOS (a Japanese instrument on NASA's Aqua satellite) |
| **ARS** | Agricultural Research Service (part of USDA) |
| **ASTER** | Advanced Spaceborne Thermal Emission and Reflection Radiometer (a Japanese instrument on NASA's Terra satellite) |
| **ATLAS** | Advanced Thermal and Land Applications Sensor (an instrument on the NASA Learjet aircraft) |
| **AVHRR** | Advanced Very High Resolution Radiometer (an instrument on the NOAA series of satellites) |
| **CCD** | Charge coupled device |
| **CCN** | Cloud condensation nuclei |
| **CERES** | Clouds and the Earth's Radiant Energy System (an instrument on the TRMM, Terra, and Aqua satellites) |
| **CFC** | Chlorofluorocarbon |
| **CIRES** | Cooperative Institute for Research in Environmental Sciences (at the University of Colorado) |
| **CNES** | Centre National d'Etudes Spatiales (France's National Center for Space Studies) |
| **CPI** | Cloud Particle Imager (an instrument on the SPEC Learjet aircraft) |
| **CPR** | Cardiopulmonary resuscitation |
| **CSIR** | Council for Scientific and Industrial Research (in South Africa) |
| **CZCS** | Coastal Zone Color Scanner (an instrument on the Nimbus 7 satellite) |
| **DC** | District of Columbia |
| **DMSP** | Defense Meteorological Satellite Program (a United States satellite series) |
| **DoD** | Department of Defense (United States) |
| **DRC** | Democratic Republic of the Congo |
| **DU** | Dobson Unit |
| **EBO** | Ebola (common abbreviation for identification of Ebola strains) |
| **ECMWF** | European Centre for Medium Range Weather Forecasts |
| **EDC** | EROS Data Center |
| **ENSO** | El Niño Southern Oscillation |
| **ENVISAT** | Environmental Satellite (an ESA satellite) |
| **EOS** | Earth Observing System (a NASA-centered international program) |
| **ER2** | A NASA high-altitude research aircraft (modified Lockheed U-2) |
| **ERBE** | Earth Radiation Budget Experiment |
| **ERBS** | Earth Radiation Budget Satellite |
| **EROS** | Earth Resources Observation System (in USGS) |
| **ERS** | European Remote Sensing Satellite (an ESA satellite series) |
| **ERSDAC** | Earth Remote Sensing Data Analysis Center (Japan) |
| **ESA** | European Space Agency |
| **ESMR** | Electrically Scanning Microwave Radiometer (an instrument on the Nimbus 5 satellite) |
| **ESTAR** | Electronically Scanned Thinned Array Radiometer |
| **ETM+** | Enhanced Thematic Mapper Plus (an instrument on the Landsat 7 satellite) |
| **FEMA** | Federal Emergency Management Agency (United States) |
| **GAP** | Southeastern Anatolia Project (Turkey; Turkish acronym) |

| | |
|---|---|
| **GEOSAT** | Geodetic Satellite (a United States Navy satellite) |
| **GFO** | GEOSAT Follow-On (a United States Navy satellite) |
| **GIS** | Geographical Information Systems |
| **GMT** | Greenwich Mean Time |
| **GOES** | Geostationary Operational Environmental Satellite |
| **GOME** | Global Ozone Monitoring Experiment (an instrument on the ERS-2 satellite) |
| **GPS** | Global Positioning System |
| **GRACE** | Gravity Recovery and Climate Experiment |
| **GRFM** | Global Rain Forest Mapping project |
| **GSFC** | Goddard Space Flight Center |
| **HIRS** | High Resolution Infrared Radiation Sounder (an instrument on the Nimbus 6 satellite and the TIROS-N and NOAA series of satellites) |
| **ISCCP** | International Satellite Cloud Climatology Project |
| **ITCZ** | Intertropical Convergence Zone |
| **JAROS** | Japan Resources Observation System |
| **JAXA** | Japan Aerospace Exploration Agency |
| **JERS** | Japanese Earth Resources Satellite |
| **LAGEOS** | Laser Geodynamics Satellite |
| **LH** | Latent heat |
| **LIS** | Lightning Imaging Sensor (an instrument on the TRMM satellite) |
| **LLBL** | Low Latitude Boundary Layer |
| **LW** | Longwave |
| **METI** | Ministry of Economy, Trade and Industry (Japan) |
| **MISR** | Multi-angle Imaging SpectroRadiometer (an instrument on the Terra satellite) |
| **MIT** | Massachusetts Institute of Technology |
| **MITI** | Ministry of International Trade and Industry (Japan) |
| **MLS** | Microwave Limb Sounder (an instrument on the UARS and Aura satellites) |
| **MODIS** | Moderate Resolution Imaging Spectroradiometer (an instrument on the Terra and Aqua satellites) |
| **MOPITT** | Measurements of Pollution in the Troposphere (an instrument on the Terra satellite) |
| **MOST Model** | Method of Splitting Tsunami Model |
| **MSS** | Multispectral Scanner (an instrument on the Landsat 1–5 satellites) |
| **MSU** | Microwave Sounding Unit (an instrument on the TIROS and NOAA series of satellites) |
| **NAO** | North Atlantic Oscillation |
| **NASA** | National Aeronautics and Space Administration (United States) |
| **NASDA** | National Space Development Agency (Japan; merged into JAXA on October 1, 2003) |
| **NDVI** | Normalized Difference Vegetation Index |
| **NLCD** | National Land Cover Data |
| **NOAA** | National Oceanic and Atmospheric Administration (United States) |
| **NPP** | Net primary productivity |
| **OLS** | Operational Linescan System (an instrument on DMSP satellites) |
| **OMI** | Ozone Monitoring Instrument (a Dutch/Finnish instrument on NASA's Aura satellite) |
| **OTD** | Optical Transient Detector (an instrument on the OrbView-1 satellite) |
| **PhD** | Doctor of Philosophy |
| **PR** | Precipitation Radar (an instrument on the TRMM satellite) |

**PSC** Polar Stratospheric Cloud

**QuikScat** Quick Scatterometer
**RADARSAT** Radar Satellite (a Canadian satellite series)

**SAR** Synthetic Aperture Radar
**SCIAMACHY** Scanning Imaging Absorption Spectrometer for Atmospheric Chartography (an instrument on the ENVISAT satellite)
**SeaWiFS** Sea-viewing Wide Field-of-view Sensor (an instrument on the OrbView-2 satellite)
**SH** Sensible heat
**SMMR** Scanning Multichannel Microwave Radiometer (an instrument on the Nimbus 7 satellite)
**SPEC** Stratton Park Engineering Company (manufacturer of the SPEC Learjet aircraft)
**SPOT** Systéme Pour d'Observation de la Terre (a French Space Agency satellite series)
**SSH** Sea surface height
**SSMI** Special Sensor Microwave Imager (an instrument on the DMSP series of satellites)
**SST** Sea surface temperature
**STS** Space Transportation System
**SVS** Scientific Visualization Studio (at NASA GSFC)
**SW** Shortwave

**TIR** Thermal infrared
**TIROS** Television Infrared Observation Satellite
**TM** Thematic Mapper (an instrument on the Landsat 4 and 5 satellites)
**TMI** TRMM Microwave Imager (an instrument on the TRMM satellite)
**TOMS** Total Ozone Mapping Spectrometer (an instrument on the Nimbus 7, Meteor 3, and Earth Probe satellites)
**TOPEX** Ocean Topography Experiment
**TRMM** Tropical Rainfall Measuring Mission (a joint NASA/JAXA satellite mission)

**UARS** Upper Atmosphere Research Satellite
**UBL** Urban boundary layer
**UCL** Urban canopy layer
**UHI** Urban heat island
**UK** United Kingdom
**UNEP** United Nations Environment Programme
**USDA** United States Department of Agriculture
**USGS** United States Geological Survey
**UTC** Universal Time Coordinate

**VIRS** Visible Infrared Scanner (an instrument on the TRMM satellite)
**VLCC** Very large crude (oil) carrier
**VOC** Volatile organic compound

**WOF** William O. Field

## Chemical Symbols

| | |
|---|---|
| **C** | carbon |
| **CFC** | chlorofluorocarbon |
| **Cl** | chlorine |
| **ClO** | chlorine monoxide |
| **CO** | carbon monoxide |
| **$CO_2$** | carbon dioxide |
| **NO** | nitrogen monoxide |
| **$NO_2$** | nitrogen dioxide |
| **$NO_X$** | nitrogen oxides |
| **O, $O_2$** | oxygen (atomic and molecular) |
| **$O_3$** | ozone |
| **$SO_2$** | sulfur dioxide |

## Units

| | |
|---|---|
| **°C** | degrees Celsius (or centigrade) |
| **cm** | centimeter |
| **E** | east |
| **°F** | degrees Fahrenheit |
| **ft** | feet |
| **hPa** | hectoPascal (100 Pascals) |
| **hr** | hour |
| **kg** | kilogram |
| **km** | kilometer |
| **m** | meter |
| **mg** | milligram |
| **mi** | mile |
| **mm** | millimeter |
| **μm** | micrometer |
| **N** | north |
| **ppbv** | parts per billion by volume |
| **ppm** | parts per million |
| **ppmv** | parts per million by volume |
| **s** | second |
| **S** | south |
| **W** | west or Watt, depending on context |
| **yr** | year |
| **%** | percent (parts per hundred) |

# Appendix 4: Contributors

James G. Acker
Goddard Earth Sciences Data and Information Services Center, Code 610.2
NASA Goddard Space Flight Center
Greenbelt, MD 20771, USA

Robert F. Adler
Mesoscale Atmospheric Processes Branch, Code 613.1
NASA Goddard Space Flight Center
Greenbelt, MD 20771, USA

William M. Balch
Bigelow Laboratory for Ocean Sciences
180 McKown Point Road, P.O. Box 475
W. Boothbay Harbor, ME 04575, USA

Steffen Beirle
Max Planck Institut für Chemie
Joh.-Joachim-Becher-Weg 27
D-55128 Mainz, Germany

Robert A. Bindschadler
Hydrospheric and Biospheric Sciences Laboratory, Code 614
NASA Goddard Space Flight Center
Greenbelt, MD 20771, USA

Charon Birkett
Earth System Science Interdisciplinary Center
2207 Computer and Space Sciences Building #224
University of Maryland
College Park, MD 20742-2465, USA

Lahouari Bounoua
Biospheric Sciences Branch, Code 614.4
NASA Goddard Space Flight Center
Greenbelt, MD 20771, USA

G. Robert Brakenridge
Dartmouth Flood Observatory
Department of Geography
Dartmouth College
Hanover, NH 03755, USA

Dennis E. Buechler
University of Alabama in Huntsville
National Space Science and Technology Center
320 Sparkman Drive
Huntsville, AL 35805, USA

Jennifer P. Cannizzaro
College of Marine Science
University of South Florida
140 Seventh Avenue South
St. Petersburg, FL 33701, USA

Sébastien Caquard
Geomatics and Cartographic Research Centre (GCRC)
Department of Geography
1125 Colonel By Drive
Ottawa (ON) K1S 5B6, Canada

Kendall L. Carder
College of Marine Science
University of South Florida
140 Seventh Avenue South
St. Petersburg, FL 33701, USA

Bruce Chapman
Mail Stop 300-319, Jet Propulsion Laboratory
California Institute of Technology
4800 Oak Grove Drive
Pasadena, CA 91109, USA

Michael T. Coe
Woods Hole Research Center
149 Woods Hole Road
Falmouth, MA 02540, USA

Josefino C. Comiso
Cryospheric Sciences Branch, Code 614.1
NASA Goddard Space Flight Center
Greenbelt, MD 20771, USA

Brian A. Cosgrove
Hydrological Sciences Branch, Code 614.3
NASA Goddard Space Flight Center
Greenbelt, MD 20771, USA

William L. Crosson
Universities Space Research Association
National Space Science and Technology Center
320 Sparkman Drive
Huntsville, AL 35805, USA

Scott Curtis
Atmospheric Science Laboratory
Department of Geography
East Carolina University
Brewster A-232
Greenville, NC 27858, USA

David J. Diner
Mail Stop 169-237, Jet Propulsion Laboratory
California Institute of Technology
4800 Oak Grove Drive
Pasadena, CA 91109, USA

Mark R. Drinkwater
Earth Observation Programmes
European Space Agency, ESTEC
Postbus 299
2200 AG Noordwijk, The Netherlands

David P. Edwards
National Center for Atmospheric Research
P. O. Box 3000
Boulder, CO 80307, USA

Maurice G. Estes, Jr.
Universities Space Research Association
National Space Science and Technology Center
320 Sparkman Drive
Huntsville, AL 35805, USA

Gene C. Feldman
Ocean Biology Processing Group, Code 614.8
NASA Goddard Space Flight Center
Greenbelt, MD 20771, USA

Jonathan A. Foley
Center for Sustainability and the Global
Environment (SAGE)
Nelson Institute for Environmental Studies
University of Wisconsin, Madison
1710 University Avenue
Madison, WI 53726, USA

James L. Foster
Hydrological Sciences Branch, Code 614.3
NASA Goddard Space Flight Center
Greenbelt, MD 20771, USA

Lee-Lueng Fu
Mail Stop 300-323, Jet Propulsion Laboratory
California Institute of Technology
4800 Oak Grove Drive
Pasadena, CA 91109, USA

Qiang Fu
Department of Atmospheric Sciences
Box 351640, University of Washington
Seattle, WA 98195, USA

Holly K. Gibbs
Center for Sustainability and the Global
Environment (SAGE)
Nelson Institute for Environmental Studies
University of Wisconsin, Madison
1710 University Avenue, Madison, WI 53726, USA

James F. Gleason
Atmospheric Chemistry and Dynamics
Branch, Code 613.3
NASA Goddard Space Flight Center
Greenbelt, MD 20771, USA

Steven J. Goodman
Earth Science Office
National Space Science and Technology Center
NASA George C. Marshall Space Flight Center
320 Sparkman Drive
Huntsville, AL 35805, USA

Dorothy K. Hall
Cryospheric Sciences Branch, Code 614.1
NASA Goddard Space Flight Center
Greenbelt, MD 20771, USA

Jeffrey B. Halverson
Department of Geography and
Environmental Systems
University of Maryland Baltimore County
211-K Sondheim Hall
Baltimore, MD 21228-4664, USA

Burgess F. Howell
Universities Space Research Association
National Space Science and Technology Center
320 Sparkman Drive
Huntsville, AL 35805, USA

N. Christina Hsu
Climate and Radiation Branch, Code 613.2
NASA Goddard Space Flight Center
Greenbelt, MD 20771, USA

George J. Huffman
Science Systems and Applications, Inc.
Mesoscale Atmospheric Processes Branch
Code 613.1
NASA Goddard Space Flight Center
Greenbelt, MD 20771, USA

Marc L. Imhoff
Biospheric Sciences Branch, Code 614.4
NASA Goddard Space Flight Center
Greenbelt, MD 20771, USA

Daniel E. Irwin
Earth Science Office
National Space Science and Technology Center
NASA George C. Marshall Space Flight Center
320 Sparkman Drive
Huntsville, AL 35805, USA

Thomas J. Jackson
USDA ARS Hydrology and Remote
Sensing Lab
104 Building 007 BARC-West
Beltsville, MD 20705, USA

Michael F. Jasinski
Hydrological Sciences Branch, Code 614.3
NASA Goddard Space Flight Center
Greenbelt, MD 20771, USA

Kenneth C. Jezek
Byrd Polar Research Center
The Ohio State University
1090 Carmack Road
Columbus, OH 43210, USA

Celeste M. Johanson
Department of Atmospheric Sciences
Box 351640, University of Washington
Seattle, WA 98195, USA

Ian Joughin
Polar Science Center
Applied Physics Laboratory
University of Washington
1013 NE 40th Street
Seattle, WA 98105-6698, USA

Christopher O. Justice
Department of Geography, Le Frak Hall
University of Maryland
College Park, MD 20742, USA

Robert K. Kaufmann
Department of Geography and Center for
Energy and Environmental Studies
Boston University
675 Commonwealth Avenue
Boston, MA 02215, USA

John S. Kimball
Flathead Lake Biological Station and NTSG
The University of Montana
311 Biostation Lane
Polson, MT 59860-9659, USA

Michael D. King
Earth Sciences Division, Code 610
NASA Goddard Space Flight Center
Greenbelt, MD 20771, USA

Gregory T. Koeln
Environmental and GIS Services
MDA Federal Inc.
(formerly Earth Satellite Corporation)
6011 Executive Blvd., Suite 400
Rockville, MD 20852-3804, USA

Charles A. Laymon
Universities Space Research Association
National Space Science and Technology Center
320 Sparkman Drive
Huntsville, AL 35805, USA

Billiana Leff
National Geographic Society
1145 17th Street, NW
Washington, DC 20036-4688, USA

Gang Liu
NOAA/NESDIS, SSMC1, #5310
1335 East-West Highway
Silver Spring, MD 20910, USA

W. Timothy Liu
Mail Stop 300-323, Jet Propulsion Laboratory
California Institute of Technology
4800 Oak Grove Drive
Pasadena, CA 91109, USA

Norman G. Loeb
Climate Science Branch, Mail Stop 420
NASA Langley Research Center
Hampton, VA 23681-2199, USA

Jeffrey C. Luvall
Earth Science Office
National Space Science and Technology Center
NASA George C. Marshall Space Flight Center
320 Sparkman Drive
Huntsville, AL 35805, USA

Eugene W. McCaul, Jr.
Universities Space Research Association
National Space Science and Technology Center
320 Sparkman Drive
Huntsville, AL 35805, USA

Charles R. McClain
Ocean Biology Processing Group, Code 614.8
NASA Goddard Space Flight Center
Greenbelt, MD 20771, USA

Kyle C. McDonald
Mail Stop 300-233, Jet Propulsion Laboratory
California Institute of Technology
4800 Oak Grove Drive
Pasadena, CA 91109-8099, USA

Richard D. McPeters
Atmospheric Chemistry and Dynamics Branch
Code 613.3
NASA Goddard Space Flight Center
Greenbelt, MD 20771, USA

W. Paul Menzel
Department of Atmospheric and Oceanic Sciences
1225 West Dayton Street
University of Wisconsin
Madison, WI 53706-1695, USA

Philip P. Micklin
Geography Department
Western Michigan University
1903 West Michigan Avenue
Kalamazoo, MI 49008, USA

Peter J. Minnett
Division of Meteorology and Physical
Oceanography, Rosenstiel School of Marine and
Atmospheric Science
University of Miami
4600 Rickenbacker Causeway
Miami, FL 33149-1098, USA

Patrick Minnis
Climate Science Branch, Mail Stop 420
NASA Langley Research Center
Hampton, VA 23681-2199, USA

Gary T. Mitchum
College of Marine Science
University of South Florida
140 Seventh Avenue South
St. Petersburg, FL 33701, USA

Peter J. Mouginis-Mark
Hawaii Institute Geophysics and Planetology
University of Hawaii
1680 East-West Road, POST 504
Honolulu, HI 96822, USA

Ranga B. Myneni
Department of Geography
Boston University
675 Commonwealth Avenue
Boston, MA 02215, USA

Yasmin Naficy
5505 Connecticut Ave., NW, #210
Washington, DC 20015, USA

R. Steven Nerem
Colorado Center for Astrodynamics Research
Department of Aerospace Engineering Sciences
University of Colorado, 429 UCB
Boulder, CO 80309-0429, USA

Manfred Owe
Hydrological Sciences Branch, Code 614.3
NASA Goddard Space Flight Center
Greenbelt, MD 20771, USA

Mutlu Ozdogan
Center for Remote Sensing
Boston University
725 Commonwealth Avenue
Boston, MA 02215, USA
(now at the Hydrological Sciences Branch,
Code 614.3
NASA Goddard Space Flight Center
Greenbelt, MD 20771, USA)

Claire L. Parkinson
Cryospheric Sciences Branch, Code 614.1
NASA Goddard Space Flight Center
Greenbelt, MD 20771, USA

François Parthiot
Cedre, c/o Centre IFREMER de Toulon
Zone portuaire de Bregaillon
83507 La Seyne sur Mer
Cedex, France

Kim C. Partington
Polar Imaging Ltd
Stoke Road, Hurstbourne Tarrant
Andover, Hampshire, SP11 0BA
United Kingdom

Jorge E. Pinzon
Science Systems & Applications, Inc., Code 614.4
NASA Goddard Space Flight Center
Greenbelt, MD 20771, USA

Steven Platnick
Climate and Radiation Branch, Code 613.2
NASA Goddard Space Flight Center
Greenbelt, MD 20771, USA

Ulrich Platt
Institut für Umweltphysik
Universität Heidelberg
Im Neuenheimer Feld 229
D-69120 Heidelberg, Germany

Dale A. Quattrochi
Earth Science Office
National Space Science and Technology Center
NASA George C. Marshall Space Flight Center
320 Sparkman Drive
Huntsville, AL 35805, USA

Navin Ramankutty
Department of Geography
McGill University
805 Sherbrooke Street W.
Montréal, QC H3A 2K6, Canada

Richard D. Ray
Space Geodesy Branch, Code 698
NASA Goddard Space Flight Center
Greenbelt, MD 20771, USA

W. Gareth Rees
Scott Polar Research Institute
Lensfield Road
Cambridge CB2 1ER, United Kingdom

Douglas L. Rickman
Earth Science Office
National Space Science and Technology Center
NASA George C. Marshall Space Flight Center
320 Sparkman Drive
Huntsville, AL 35805, USA

Eric Rignot
Mail Stop 300-227, Jet Propulsion Laboratory
California Institute of Technology
4800 Oak Grove Drive
Pasadena, CA 91109-8099, USA

Ian S. Robinson
National Oceanography Centre Southampton
Waterfront Campus, European Way
Southampton, SO14 3ZH, United Kingdom

David P. Roy
Geographic Information Science Center of Excellence
South Dakota State University
Brookings, SD 57007, USA

Guido D. Salvucci
Department of Geography, Boston University
685 Commonwealth Avenue
Boston, MA 02215, USA

David T. Sandwell
Scripps Institution of Oceanography
La Jolla, CA 92093-0225, USA

Michelle L. Santee
Mail Stop 183-701, Jet Propulsion Laboratory
California Institute of Technology
4800 Oak Grove Drive
Pasadena, CA 91109, USA

Crystal B. Schaaf
Department of Geography
Boston University
675 Commonwealth Avenue
Boston, MA 02215, USA

Remko Scharroo
Altimetrics LLC
330a Parsonage Road
Cornish, NH 03745, USA

Annemarie Schneider
Department of Geography and Center
for Remote Sensing, Boston University
675 Commonwealth Avenue
Boston, MA 02215, USA
(currently at Department of Geography University
of California, Santa Barbara, Ellison Hall 5703
Santa Barbara, CA 93106-4060, USA)

Piers J. Sellers
Astronaut Office, Code CB
NASA Lyndon B. Johnson Space Center
Houston, TX 77058, USA

Karen C. Seto
Building 320 Stanford University
Stanford, CA 94305-2115, USA

Thomas L. Sever
Earth Science Office
National Space Science and Technology Center
NASA George C. Marshall Space Flight Center
320 Sparkman Drive
Huntsville, AL 35805, USA

J. Marshall Shepherd
Department of Geography
University of Georgia
Athens, GA 30602-2502, USA

Bridget R. Smith
Institute of Geophysics and Planetary Physics
Scripps Institution of Oceanography
9500 Gilman Drive MC 0225
La Jolla, CA 92093-0225, USA

Brian J. Soden
Division of Meteorology and Physical
Oceanography, Rosenstiel School of
Marine and Atmospheric Science
University of Miami
4600 Rickenbacker Causeway
Miami, FL 33149-1098, USA

Alan E. Strong
NOAA/NESDIS, SSMC1, #5304
1335 East-West Highway
Silver Spring, MD 20910-3226, USA

Robert H. Thomas
EG&G Services, NASA Wallops Flight Facility
Wallops Island, VA 23337, USA

Si-Chee Tsay
Climate and Radiation Branch, Code 613.2
NASA Goddard Space Flight Center
Greenbelt, MD 20771, USA

Compton J. Tucker
Hydrospheric and Biospheric Sciences
Laboratory, Code 614
NASA Goddard Space Flight Center
Greenbelt, MD 20771, USA

Thomas Wagner
Max Planck Institut für Chemie
Joh.-Joachim-Becher-Weg 27
D-55128 Mainz, Germany

John J. Walsh
College of Marine Science
University of South Florida
140 Seventh Avenue South
St. Petersburg, FL 33701, USA

Zhengming Wan
Institute for Computational Earth System Science
University of California, Santa Barbara
Ellison Hall 6717
Santa Barbara, CA 93106-3060, USA

Mark O. Wenig
University of Maryland Baltimore County
Atmospheric Chemistry and Dynamics Branch
Code 613.3
NASA Goddard Space Flight Center
Greenbelt, MD 20771, USA

Richard S. Williams, Jr.
US Geological Survey
Woods Hole Science Center
384 Woods Hole Road
Woods Hole, MA 02543-1598, USA

Robin G. Williams
532 Gravilla Street
La Jolla, CA 92037, USA

Takmeng Wong
Climate Science Branch, Mail Stop 420
NASA Langley Research Center
Hampton, VA 23681-2199, USA

Curtis E. Woodcock
Department of Geography and Environment
Boston University
675 Commonwealth Avenue
Boston, MA 02215, USA

David K. Woolf
Environmental Research Institute
North Highland College
UHI Millennium Institute, Castle Street
Thurso, KW14 7JD, United Kingdom

Donald P. Wylie
Space Science and Engineering Center
1225 West Dayton Street
University of Wisconsin
Madison, WI 53706-1695, USA

Xiaosu Xie
Mail Stop 300-323, Jet Propulsion Laboratory
California Institute of Technology
4800 Oak Grove Drive
Pasadena, CA 91109, USA

Liming Zhou
School of Earth and Atmospheric Sciences
Georgia Institute of Technology
311 Ferst Drive
Atlanta, GA 30332-0340, USA

Reiner Zimmermann
Forest Ecology and Remote Sensing Group
Ecological-Botanical Gardens
University of Bayreuth, Universitätsstr. 30
D-95440 Bayreuth, Germany

# Index

# About the Editors

Michael D. King is Senior Project Scientist of NASA's Earth Observing System, based at NASA's Goddard Space Flight Center, and a Principal Investigator of the Moderate Resolution Imaging Spectroradiometer (MODIS) on the Terra and Aqua satellites. His research has centered on the remote sensing of aerosol and cloud optical properties and their effects on climate using ground-based, aircraft, and satellite sensors. Dr. King's research experience includes conceiving, developing, and operating multispectral scanning radiometers from a number of aircraft platforms in field experiments ranging from Arctic stratus clouds to smoke from the Kuwait oil fires and biomass burning in Brazil and southern Africa.

Claire L. Parkinson is a climatologist at NASA's Goddard Space Flight Center and the Project Scientist for the Aqua satellite mission, focused on the water cycle and other aspects of global climate. Her research emphasis has been on polar sea ice and its connections to the rest of the climate system and to climate change, with a particular emphasis on satellite remote sensing. She has also developed a computer model of sea ice and has done field work in both the Arctic and the Antarctic. She is the lead author of an atlas of Arctic sea ice and a co-author of two other sea ice atlases, and she has written books on satellite Earth observations and the history of science.

Kim C. Partington is Managing Director of Polar Imaging Limited in the United Kingdom. He has been an Associate Professor at the University of Alaska Fairbanks, the Chief Scientist at the U.S. National Ice Center, and the Program Manager of the Polar Research Program at NASA Headquarters, as well as Managing Director and Co-Founder of Vexcel UK, a company focusing on polar and marine applications of remote sensing. His interests extend from development of sensor concepts associated with future satellite missions to the operational and scientific exploitation of data from current satellite missions.

Robin G. Williams is an oceanographer based in La Jolla, California and the volcanic island of Montserrat, where he owns and operates an environmental studies field center and hostel. He has an academic background in oceanography and meteorology and his research interests include ocean waves, sea ice, and climate studies of the Caribbean. He has participated in several field programs and led one to the Arctic and one to South Georgia in the Southern Ocean. He has been involved with climate-related research for 20 years.

FREEPORT MEMORIAL LIBRARY
3 1489 00574 5144